Sustainable Energy Storage in the Scope of Circular Economy

Sustainable Energy Storage in the Scope of Circular Economy

Advanced Materials and Device Design

Edited by

Carlos Miguel Costa
University of Minho,
Centre of Physics,
Braga, Portugal

Renato Gonçalves
University of Minho,
Centre of Chemistry,
Braga, Portugal

Senentxu Lanceros-Méndez
BCMaterials,
Basque Center for Materials,
Applications and Nanostructures,
UPV/EHU Science Park,
Leioa, Spain

and

IKERBASQUE,
Basque Foundation for Science,
Bilbao, Spain

WILEY

This edition first published 2023

Registered Offices
John Wiley & Sons, Inc., 111 River Street, Hoboken, NJ 07030, USA
John Wiley & Sons Ltd, The Atrium, Southern Gate, Chichester, West Sussex, PO19 8SQ, UK

For details of our global editorial offices, customer services, and more information about Wiley products visit us at www.wiley.com.

Wiley also publishes its books in a variety of electronic formats and by print-on-demand. Some content that appears in standard print versions of this book may not be available in other formats.

Library of Congress Cataloging-in-Publication Data
Names: Costa, Carlos Miguel, 1991– editor. | Renato Gonçalves, editor. | Senentxu Lanceros-Méndez, editor.
Title: Sustainable energy storage in the scope of circular economy : advanced materials and device design / edited by Carlos Miguel Costa, Renato Gonçalves, and Senentxu Lanceros-Méndez.
Description: Hoboken, NJ, USA : Wiley, 2023. | Includes bibliographical references and index.
Identifiers: LCCN 2022049605 (print) | LCCN 2022049606 (ebook) | ISBN 9781119817680 (hardback) | ISBN 9781119817697 (adobe pdf) | ISBN 9781119817703 (epub)
Subjects: LCSH: Electric batteries–Materials. | Electric batteries–Recycling. | Green products.
Classification: LCC TK2910 .S87 2023 (print) | LCC TK2910 (ebook) | DDC 621.31/2420286–dc23/eng/20221221
LC record available at https://lccn.loc.gov/2022049605
LC ebook record available at https://lccn.loc.gov/2022049606

Cover Design: Wiley
Cover Image: © petovarga/Shutterstock

Set in 9.5/12.5pt STIXTwoText by Straive, Pondicherry, India

Contents

List of Contributors

Reeya Agarwal
Department of Electronics and
Communication Engineering
National Institute of Technology
Patna, India

Paul Albertus
Department of Chemical and
Biomolecular Engineering
Maryland Energy Innovation Institute
University of Maryland
UMD College Park, USA

Linda Barelli
Dipartmento di Iingegberia
Università degli Studi di Perugia (UNIPG)
Perugia, Italy

Manuel Baumann
Institute for Technology Assessment and
System Analysis (ITAS)
Karlsruhe Institute of Technology (KIT)
Karlsruhe, Germany

Eric Carmona
Department of Chemical and
Biomolecular Engineering
Maryland Energy Innovation Institute
University of Maryland
UMD College Park, USA

Hélder Castro
Instituto de Soldadura e Qualidade
Grijó, Portugal

Gavin Collis
CSIRO, Manufacturing
Clayton
VIC, Australia

Vitor Correia
Algoritmi Research Center
University of Minho
Campus de Azurém
Guimarães, Portugal

Carlos Miguel Costa
Physics Centre of Minho and Porto
Universities (CF-UM-UP)

and

Laboratory of Physics for Materials and
Emergent Technologies, LapMET

and

Institute of Science and Innovation
for Bio-Sustainability (IB-S)
University of Minho
Braga, Portugal

Yunfa Dong
National Key Laboratory of Science and
Technology on Advanced Composites in
Special Environments, and Center for
Composite Materials and Structures
Harbin Institute of Technology
Harbin, China

Hüseyin Ersoy
Institute for Technology Assessment and System Analysis (ITAS)
Karlsruhe Institute of Technology (KIT)
Karlsruhe, Germany

Juan P. Esquivel
Instituto de Microelectrónica de Barcelona, IMB-CNM (CSIC)
Bellaterra
Barcelona, Spain

BCMaterials, Basque Centre for Materials, Applications and Nanostructures
UPV/EHU Science Park
Leioa, Spain

and

IKERBASQUE, Basque Foundation for Science
Bilbao, Spain

Arkaitz Fidalgo-Marijuan
BCMaterials, Basque Center for Materials, Applications and Nanostructures
UPV/EHU Science Park
Leioa, Spain

Eider Goikolea
Organic and Inorganic Chemistry Department
Faculty of Science and Technology
University of the Basque Country UPV/EHU
Leioa, Spain

Diana Golodnitsky
School of Chemistry
Tel Aviv University
Tel Aviv, Israel

Renato Gonçalves
Center of Chemistry
University of Minho
Braga, Portugal

Ram K. Gupta
Department of Chemistry, Kansas Polymer Research Center
Pittsburg State University
Pittsburg
KS, USA

Kailot C. Harris
Department of Chemical and Biomolecular Engineering
Maryland Energy Innovation Institute
University of Maryland
UMD College Park, USA

Weidong He
National Key Laboratory of Science and Technology on Advanced Composites in Special Environments, and Center for Composite Materials and Structures
Harbin Institute of Technology
Harbin, China

Manab Kundu
Department of Chemistry
University Building
SRM Institute of Science and Technology
Kattankulathur
Tamil Nadu, India

Senentxu Lanceros-Méndez
BCMaterials, Basque Center for Materials, Applications and Nanostructures
UPV/EHU Science Park
Leioa, Spain

and

IKERBASQUE, Basque Foundation for Science
Bilbao, Spain

Karl Larson
Department of Chemical and Biomolecular Engineering
Maryland Energy Innovation Institute
University of Maryland
UMD College Park, USA

Qun Li
National Key Laboratory of Science and Technology on Advanced Composites in Special Environments, and Center for Composite Materials and Structures
Harbin Institute of Technology
Harbin, China

Peter Mahon
Swinburne University of Technology
Faculty of Science, Engineering and Technology
Hawthorn
VIC, Australia

Malachi Noked
Department of Chemistry
Bar-Ilan University
Ramat Gan, Israel

Marina Navarro-Segarra
Instituto de Microelectrónica de Barcelona, IMB-CNM (CSIC)
Bellaterra
Barcelona, Spain

Stefano Passerini
Helmholtz Institute Ulm for Electrochemical Energy Storage (HIU)
Karlsruhe Institute of Technology (KIT)
Ulm, Germany

Oliver Pohl
CSIRO, Energy
Clayton
VIC, Australia

and

Swinburne University of Technology
Faculty of Science, Engineering and Technology
Hawthorn
VIC, Australia

Heftsi Ragones
Faculty of Engineering
Holon Institute of Technology
Holon, Israel

Thomas Rüther
CSIRO, Energy
Clayton
VIC, Australia

Manuel Salado
BCMaterials, Basque Center for Materials, Applications and Nanostructures
UPV/EHU Science Park
Leioa, Spain

Ahmed E. Shalan
BCMaterials, Basque Center for Materials, Applications and Nanostructures
UPV/EHU Science Park, Barrio
Leioa, Spain

and

Central Metallurgical Research and Development Institute (CMRDI)
Helwan
Cairo, Egypt

Rosy Sharma
Department of Chemistry
Indian Institute of Technology (Banaras Hindu University)
Varanasi
Uttar Pradesh, India

Sangeeta Singh
Department of Electronics and Communication Engineering
National Institute of Technology
Patna, India

Ela Strauss
Israel Science Foundation
Jerusalem, Israel

Shilpi Sengupta
Department of Chemistry
SRM Institute of Science and Technology
Kattankulathur
Tamil Nadu, India

Muhammad R. Sulaiman
Department of Plastic Engineering and
Department of Chemistry
Kansas Technology Center
Pittsburg State University
Pittsburg
KS, USA

Ana J. Sobrido
School of Engineering and
Materials Science
Queen Mary University of London
London, UK

Michael Thielke
School of Engineering and
Materials Science
Queen Mary University of London
London, UK

Yifeng Wang
BCMaterials, Basque Center for Materials,
Applications and Nanostructures
UPV/EHU Science Park
Leioa, Spain

and

Organic and Inorganic Chemistry
Department
Faculty of Science and Technology
University of the Basque
Country UPV/EHU
Leioa, Spain

Marcel Weil
Institute for Technology Assessment
and System Analysis (ITAS)
Karlsruhe Institute of
Technology (KIT)
Karlsruhe, Germany

and

Helmholtz Institute Ulm for
Electrochemical Energy Storage (HIU)
Karlsruhe Institute of Technology (KIT)
Ulm, Germany

Mary York
Department of Chemical and Biomolecular
Engineering
Maryland Energy Innovation Institute
University of Maryland
UMD College Park, USA

Qi Zhang
BCMaterials, Basque Center for Materials,
Applications and Nanostructures
UPV/EHU Science Park
Leioa, Spain

and

IKERBASQUE, Basque Foundation
for Science
Bilbao, Spain

Shijie Zhong
National Key Laboratory of Science and
Technology on Advanced Composites in
Special Environments, and Center for
Composite Materials and Structures
Harbin Institute of Technology
Harbin, China

Preface

"Energy and persistence conquer all things."

Benjamin Franklin (1706–1790)

Considering the United Nations Agenda 2030 for Sustainable Development and the urgent need to reduce the environmental impact of current and emerging technologies, the development of sustainable smart and multifunctional materials and processes is essential for an increasing number of application areas.

One of the most relevant areas is energy, based both on the increasing energy demand that must be supported by a growing share of renewable energy sources and energy storage systems and in the electrification and digitalization of the economy in the scope of the Industry 4.0 and Internet of Things (IoT) paradigms, demanding ubiquitous sensing, data communication systems, and mobile electronic devices, also relying in energy storage systems.

In this book, the current state of the art on energy storage devices based on sustainable materials is addressed within the framework of the circular economy.

Further, the main applications being developed for these energy storage devices as well as the main advantages and remaining challenges in this research field are presented.

The first chapter provides an overview on circular economy, including the general concepts, definition, benefits, and origins. The principles of the circular economy are presented and their relevance in the context of energy is discussed. Also, the approach based on the sustainability for energy field, with special focus on battery devices, is described.

Chapter two presents the energy carrier properties of reactive metals and their application in energy storage applications. The economic and environmental implications of reactive metals are also discussed.

Chapters three, four and five focuses on sustainable battery materials, covering materials for lithium-ion batteries, beyond lithium, and biodegradable batteries, respectively.

Chapter six is dedicated to sustainable materials for supercapacitor devices, with special attention to electrodes and electrolyte materials. Recent advances on nanomaterials for electrodes are presented and discussed.

Chapters seven and eight focuses on sustainable approaches for fuel cells, describing the materials for fuel cell devices and the recent advances in microbial fuel cells, respectively.

Chapter nine presents multifunctional sustainable materials for energy storage based on biomass-derived electrodes and traditional carbon electrodes for redox flow batteries.

Chapters ten and eleven focuses on sustainable devices and their corresponding design, covering sensors/actuators and devices in the scope of Internet of Things (IoT), respectively.

Chapter twelve reports on waste prevention for energy storage devices based on second-life use of lithium-ion batteries.

Chapter thirteen describes the recycling procedures for energy storage devices with a particular focus on the implementation of the electric vehicle.

Finally, chapter fourteen summarizes some of the open questions, needs, and future trends in the scope of the principles of sustainability of materials and processes applied to energy storage systems.

This book was only possible by the excellent contributions of the authors who accepted the invitation to write different chapters in this relevant, needed, and timely research and technological field that encompasses sustainability and energy storage systems. The editors are grateful for the time and effort to write these magnificent chapters that will serve as a reference for the state of the art and as a guide for further developments in this area. For us, it was real pleasure and an honor to have contacted, to have worked together, and to provide all together the important milestone that this book represents!

Also, this book would not have been possible without the support, understanding, and dedication of our research group and also of the research centers at the University of Minho, Braga, Portugal and BCMateriais, Basque Center for Materials, Applications and Nanostructures, Leioa, Spain, to which the different editors belong. Pushing science and technology forward through the idealization, writing and materialization of a book is one of the most beautiful and relevant efforts to share with you all!

Last but not least, we truly thank the excellent support from the team from Wiley: from the first contacts with Sarah Higginbotham and Sakeena Quraishi to the latter support from Stefanie Volk and Jenny Cossham. The continuous support, technical expertise, patience, and kindness were essential to make this book come true. It has been a pleasure working with you all!

Finally, being a book that covers sustainable materials for batteries, supercapacitors, and fuel cells; energy storage devices for sensors, biomedical, and wearable applications, as well as recycling and utilization, the book can be used on a variety of different scopes: from researcher and professionals aiming to enter any of the different fields covered by the book to courses at universities focusing on sustainable energy. It is also suitable for political organizations, companies, and NGOs focusing on this subject and/or organizing specific funding schemes and/or development strategies in the area. The editors truly believe that this book will become a milestone to further foster increasing scientific and engineering efforts based on sustainability and circular economy paradigms applied to energy storage devices, urgently needed to successfully tackle some of the main challenges society is facing nowadays: energy and environment.

Carlos Miguel Costa, Renato Gonçalves, and
Senentxu Lanceros-Méndez
Braga, Portugal
10 November 2022

Part I

Introduction

1

The Central Role of Energy in the Scope of Circular Economy and Sustainable Approaches in Energy Generation and Storage

Renato Gonçalves[1], Arkaitz Fidalgo-Marijuan[2], Carlos Miguel Costa[3,4,5], and Senentxu Lanceros-Méndez[2,6]

[1] *Centre of Chemistry, University of Minho, 4710-057, Braga, Portugal*
[2] *BCMaterials, Basque Center for Materials, Applications and Nanostructures, UPV/EHU Science Park, 48940, Leioa, Spain*
[3] *Physics Centre of Minho and Porto Universities (CF-UM-UP), University of Minho, 4710-057, Braga, Portugal*
[4] *Laboratory of Physics for Materials and Emergent Technologies, LapMET, University of Minho, 4710-057, Portugal*
[5] *Institute of Science and Innovation for Bio-Sustainability (IB-S), University of Minho, 4710-057, Portugal*
[6] *IKERBASQUE, Basque Foundation for Science, 48009, Bilbao, Spain*

1.1 Introduction

Since the last century, it is observed that fast economic growth worldwide has also led to a strong environmental impact. Thus, efforts from different world countries are being directed toward the development of more sustainable technologies. In this context, the attention to the circular economy concept increased, where the waste of resources and environmental pollution matters have been intensively addressed in a variety of ways. Different countries have been developing approaches to increase the implementation of the circular economy paradigm in their own country to substitute the current linear economy model, which leads to a strong environmental impact [1]. Different countries and geographical areas such as China, the United States, the European Union, and India have implemented targets and conditions to improve the circular economy not only for their individual country but also to interrelate which other countries create a world circular economy based on exchange and product valorization [2].

The general concept of circular economy is based on the fact that economic activity contributes to the overall sustainability of the system. The circular economy concept recognizes how important the functioning of the economy is at any level – large and small businesses, organizations and individuals, and globally and locally. Thinking about the subject, it is realized that nature has no room for waste or trash. That is, all the elements in nature are related in a delicate balance. Thus, the circular economy must be a system of use of resources where the reduction, reuse, and recycling of elements predominate. Considering this, the circular economy model is based on consumption and production in relation to concepts such as sharing, leasing, reusing, repairing, refurbishing, and

Sustainable Energy Storage in the Scope of Circular Economy: Advanced Materials and Device Design, First Edition. Edited by Carlos Miguel Costa, Renato Gonçalves, and Senentxu Lanceros-Méndez.

recycling, preserving the materials and products as long as possible and extending their life cycle. Through the reduction of waste and extending the material's life cycle, the circular economy can also provide further value.

Unlike the traditional model of linear economy, where natural resources and energy are turned into products that are ultimately destined to become waste based on a take-make-consume-throw-away pattern [3], a circular economy employs reuse, sharing, repair, refurbishment, remanufacturing, and recycling to create a closed-loop system, minimizing the use of resource inputs and the creation of waste, pollution, and carbon emissions [4]. In this sense, the circular economy model relies on large quantities of cheaper, easily accessible materials and energy, and aims to keep products, materials, equipment, and infrastructure [5] in use for longer time, thus improving the productivity of the resources. Waste materials and energy should become input for other processes through waste valorization: either as a component for another industrial process or as regenerative resources for nature (Figure 1.1).

A circular economy approach aims to preserve the value of products, materials, and resources for as long as possible. Following the procedures of fewer products discarded and fewer materials extracted, least energy is lost and the environment is preserved. Thus, the circular economy, in order to tackle global challenges like climate change, biodiversity loss, waste, and pollution, is a framework of three principles, driven by design: eliminate waste and pollution, keep products and materials in use, and regenerate natural systems [6].

The circular economy proposes an opportunity to restructure our economy. This comprehensive strategy that aims to reduce both the input of raw materials as well as the production of waste closes the cycles or economic and ecological flows of resources, providing the benefits summarized in Table 1.1 for the economy, the environment, and the community.

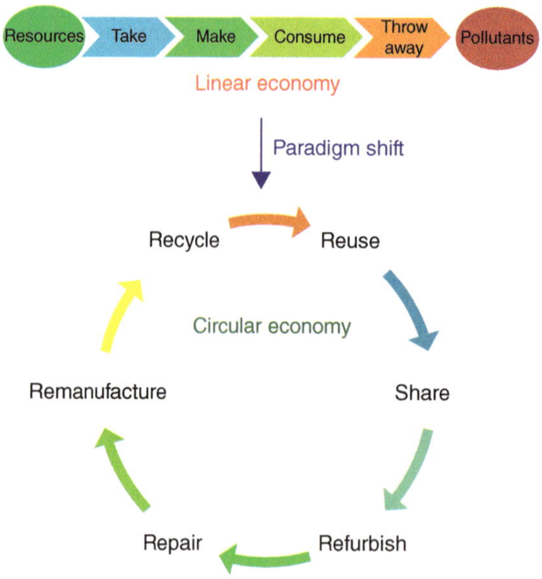

Figure 1.1 An illustration showing the paradigm shift from the take-make-consume-throw-away pattern of the linear economy to the circular economy approach.

Table 1.1 Summary of the benefits provided by the circular economy in the three main areas of action.

Benefit area	
Economy	Creates balanced wealth
	Opportunities for local jobs and social integration
	Reduces expenses and investments. More innovative and efficient ways of producing and consuming
	Reorients the production of the countries, protecting businesses against scarcity of resources and volatile prices
Environment	Reduces the use of resources
	Optimization of waste management which boosts recycling and reduces landfill
	Limits energy consumption, fewer production processes require less energy
	Benefits for the environment in terms of climate and biodiversity, air, soil, and water preservation
Community	Allows the change of consumption habits.
	Creates awareness
	Balances society with the economy and the environment

1.2 Circular Economy and the Central Role of Energy

The notion of circularity has its origin in the historical as well as in the philosophical field. The idea of feedback and cycles in world systems is old and arises in various philosophical schools. This concept regained interest at the end of World War II, when computer studies of nonlinear systems revealed the complex, connected, and unpredictable nature of our world. Those studies led to the conclusion that the circular economy concept is more like a metabolism than a machine. In this scope, recent technological advances have the power to support the transition to a circular economy through modeling and computer simulations, allied to the new materials and production technologies. Finally, the term *circular economy* was first used in Western literature in 1980 [7] to describe a closed system of interactions between the economy and the environment. Therefore, the circular economy is part of the study of feedback from nonlinear systems, including living systems.

More recently, the idea and concept of circular economy have been introduced in the guidelines and action plans of institutions worldwide. In 2015, the European Commission adopted an ambitious "Circular Economy Package." An EU Action Plan for the circular economy establishes a specific program of actions outlining measures that cover the entire product life cycle: from production and consumption to waste management and the market for secondary raw materials. On 4 March 2019, the European Commission adopted a comprehensive report on the implementation of the Circular Economy Action Plan.

On 11 March 2020, the European Commission adopted a new Circular Economy Action Plan – one of the main building blocks of the European Green Deal, Europe's new agenda for sustainable growth. The new Action Plan announces initiatives along the entire life cycle of products, targeting for example their design, promoting circular economy processes, fostering sustainable consumption, and aiming to ensure that the resources used are kept in the EU economy for as long as possible. It introduces legislative and non-legislative measures targeting areas where action at the EU level brings real added value.

In addition, the circular economy has strong synergies with the EU's objectives on climate and energy and with the Commission's package on "Clean Energy for all Europeans." The circular economy is also instrumental in supporting the EU's commitments on sustainability, as outlined in the communication "Next steps for a sustainable European future" and in particular to reach Sustainable Development Goal 12 "Responsible consumption and production" [8]. China has established guidelines and regulations at a multi-level scale from micro to macro in a top-down structure. The *Circular Economy Promotion Law,* published in 2008, is one of the many efforts that promote the 3R policy (reduce, reuse, and recycle). With the "National Development and Reform Commission (NDRC)" in 2016, digital solutions have into account the circular economy to design products and new business models. In 2018, the "*Circular Economy Promotion Law*" was revised highlighting the recycling processes until 2020, by the domestic–international circulation [9]. Latin American countries are organizing primary work on circular economy roadmap since 2020 based on policies related to landfill bans, extended producer responsibility, and renewable energy end-of-use [10]. The US *National Recycling Strategy* implements a national recycling rate goal of 50% by 2030. This goal is based on five strategies: support recycling, standardize measurements, reduce contamination from recycling, improve markets for recycling commodities, and increase materials recycling collection and infrastructures [11].

Although some circular economy practices are already well established in some places, most definitions and actions relate to perspective possibilities and strategies, rather than to what actually exists. Multiple interrelated concepts such as loop closing, ecodesign, industrial ecology, industrial symbiosis, life cycle analysis, and performance economy have contributed to the concept of a circular economy. Nonetheless, a strong policy direction is likely needed for effective implementation [12].

The general objective of the circular economy is to obtain products and manufacturing processes that consume little energy and do not generate waste or garbage that negatively affect society and the environment. In order to meet this important challenge, there are different components or principles that define how the circular economy should be implemented:

1) **The waste becomes a resource:** it is the main characteristic. All biodegradable materials return to nature and the non-biodegradable materials are reused.
2) **The second use:** to reintroduce into the economic circuit those products that no longer correspond to the initial needs of the consumers.
3) **Reuse:** reuse certain residues or certain parts of them, which can still work for the production of new products.
4) **Repair:** find a second life for damaged products.
5) **Recycling:** use the materials found in waste.
6) **Valorization:** take advantage of waste that cannot be recycled.
7) **The economy of functionality:** the circular economy proposes to eliminate the sale of products in many cases to implement a system of renting goods. When the product completes its main function, it returns to the company, which will disassemble it to reuse its valid parts.

8) **Energy from renewable sources:** elimination of fossil fuels to produce the product, reuse, and recycle.
9) **Eco-conception:** considers the environmental impacts throughout the life cycle of a product and integrates them from its conception.
10) **Industrial and territorial ecology:** establishment of a mode of industrial organization in the same territory characterized by optimized management of stocks and flows of materials, energy, and services.

These ten principles are interrelated in the concept of circular economy as represented in Figure 1.2.

As a consequence of the 10 principles, oil and gas energy resources are incompatible with the idea of a circular economy, since they represent "development that meets the needs of the present while compromising the ability of future generations to meet their own needs" [13]. A sustainable circular economy can only be powered by renewable energies, such as wind, solar, hydropower, and geothermal, among others [14].

What gives entities the ability to achieve "net zero" carbon emissions, is that they can offset their fossil fuel consumption by removing carbon from the atmosphere. While this is a necessary first step, it has been also stated that in order to create a truly circular economy we should adopt the concept of "true zero" as opposed to "net zero," which is eliminating fossil fuel consumption entirely so that all energy is produced from renewable sources [15, 16].

Current growth projections in the renewable energy industry expect a significant amount of energy and raw materials to manufacture and maintain these renewable systems. Nevertheless, "due to the emissions attributed to fossil-fuel electricity generation, the overall carbon footprint of renewable energy technologies is significantly lower than for fossil-fuel generation over the respective systems lifespan" [17]. However, there are still

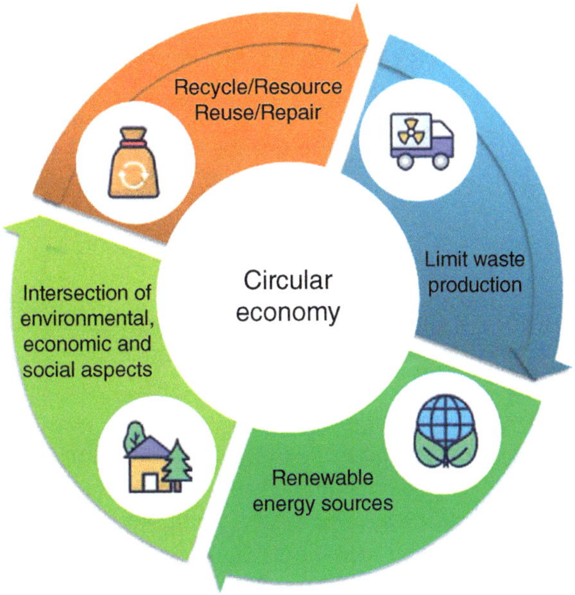

Figure 1.2 Illustration of the relationship between the basic principles of the circular economy.

linear trajectories when establishing renewable energy systems that should be assessed in order to fully transition to a circular economy.

In addition to the perspective of renewable energies, different regulations across the world have been applied, such as the European Commission, in the field of batteries and accumulators, which aims to ensure that batteries and accumulators placed in the EU market are sustainable and safe throughout their entire life cycle. This regulation focuses on the importance of sustainable energy in the area of batteries and accumulators, as they will play an essential role to ensure that many daily-used products, appliances, and services work properly anywhere and anytime, constituting an indispensable energy source in our society. In the last couple of years, approximately 800 000 tons of automotive batteries, 190 000 tons of industrial batteries, and 160 000 tons of consumer batteries enter the European Union every year, so that, in order to achieve the Green Deal, sustainable batteries for a circular and climate neutral economy are necessary.

1.3 The Central Role of Energy in the Scope of Sustainability

Companies are playing a key role in energy transition by contributing to energy security and decarbonisation. In the face of growing political pressure and increasing levels of regulation, investors on the one hand and citizens on the other are demanding energy supply organizations to address climate change and sustainability. These trends continue to compel energy companies to systematically pursue their transition to affordable, reliable, and more sustainable energy and to set ambitious targets in terms of reducing CO_2 emissions. In this sense, circular approaches decouple economic activity from the consumption of materials and energy by building closed cycles in which waste is minimized or even eliminated, and in which resources are reused, including carbon. They do this by using resources efficiently, prioritizing renewable inputs, maximizing the effective life of a product, and capturing and reusing what was previously considered waste. Obviously, not all actors in the energy sector have the opportunity to aim for full circularity. But sticking to circular thinking as the core approach will lead to powerful new insights and levers for efficiency and costing. The increasing use of renewable energy and energy storage can bring new perspectives for a sustainable approach (Figure 1.3) [18].

1.3.1 Energy Generation

Industries with operational models centered on the extraction and use of fossil fuels are inherently linear and hard to be compatible with the idea of a sustainable circular economy. However, companies in these industries are able to incorporate cyclic elements into their operations and, in some cases, have adopted technological advances to reduce the impact of fossil fuel use. For example, several initiatives are underway to explore the potential for reusing oil fields and offshore assets after drilling operations cease. One of the most common initiatives is the reuse of platforms to support offshore wind projects [19]. In the Mediterranean, a system of turbines is being developed that could be used to turn decommissioned oil platforms into islands of renewable energy [20]. Further, there are different

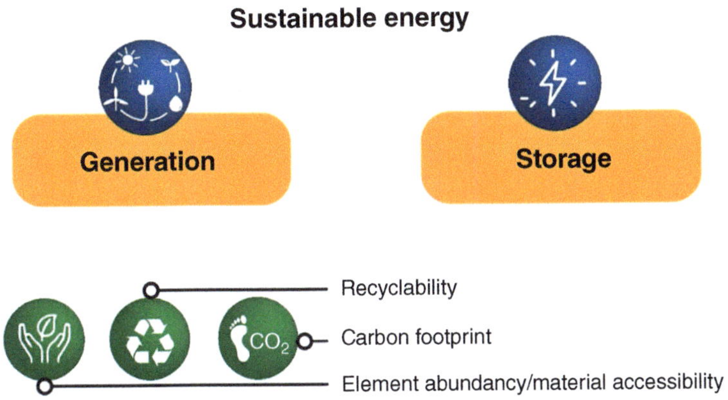

Figure 1.3 Sustainable energy insights for energy generation and storage.

projects intending to use discontinued offshore oil and gas fields to store CO_2 permanently. It is estimated to store approximately 3.5 million tons of CO_2 per year by 2030 [21]. In this regard, other initiatives are experimenting with ways to capture greenhouse gases that would otherwise be released into the atmosphere from factories and power plants. These gases are being used for enhanced oil recovery, injecting these gases into aging oil fields in an attempt to pump out the remaining supply [21, 22].

On the other hand, different organizations are focusing on technologies to generate energy based on gas with low carbon emissions. To do this, natural gas is burned with pure oxygen instead of air to generate electricity without emitting CO_2. This process also avoids generating nitrogen oxides, the main air and health pollutant emitted by gas plants [23]. Early developments point to a 20% increase in power output compared to a conventional plant using the same amount of fuel [24].

Circularity provides a strategic and effective way to identify opportunities for both cost savings and value forming. In some cases, switching to renewable energy as an input is the first step. As technologies advance, the adoption of zero-emission energy is becoming more widespread. This shift to renewable energy is a key driver of the circular economy, and a growing number of organizations are seeking to minimize prices, increase sustainability, and create long-term strategic costs through the shift. Several aspects are driving renewable energy adoption, and the economic dimension is also becoming less of a barrier not only in the power generation sector but also for end users [25]. Examples of organizations continuing to move toward renewable energy in order to achieve a sustainable approach span continents and sectors: in the mining industry, which has generally relied on natural gas or diesel to power its operations, significant investments in solar energy and battery technology have been announced for ore extraction operations. The integration of photovoltaic solar energy combined with large-scale battery storage will eventually meet 25–30% of stationary energy needs from solar generation [26, 27]. More ambitious plans aim to achieve 100% energy from renewable sources by mid-2020s [28]. Similarly, many refineries and chemical plants are moving to solar power to satisfy the energy demand [29, 30].

Although the change to solar energy and batteries is one of the key steps to achieve sustainable energy production in all dimensions, without a commitment to better recycling, the adoption of renewable energy will come with ever-increasing material footprints. Solar panel and battery technology present challenges and opportunities. In 2016, the International Renewable Energy Agency (IRENA) estimated that there was around 250 000 tons of solar panel waste on the planet. The IRENA study shows that many technical obstacles need to be overcome before closed circularity for solar panels is feasible. But the raw materials already have the opportunity to be recycled at a rate of 65–70% in weight [31].

According to the International Energy Agency, the energy obtained from wind renewable sources in 2020 is responsible for around 1500 TWh of the total world energy production and it is estimated that in 2030 this production reaches 8000 TWh. This technology increased significantly over the last few years contributing to the reduction of carbon emissions. The lifetime of a wind turbine is around 25 years reaching in some cases 35 years. Nevertheless, this increase leads to higher consumption of wind turbines that require a huge number of different components. To prevent these components from going to landfills, it is urgent to develop new components capable of being incorporated into the circular economy. The circular economy for this system is focused on recycling, upcycling, extending life cycles, and a combination of new and fewer materials. To overcome these obstacles, some initiatives have been started as increasing funding for resources and development (R&D) and incentives to use recycled composite materials [25].

1.3.2 Energy Storage

In the case of battery technologies, the increasing use of battery technology is at the forefront of renewable energy storage and transportation decarbonisation strategies. But battery production itself has a significant carbon footprint. The CO_2 footprint from electric vehicle fabrication, for example, is higher than for an internal combustion engine vehicle fabrication. Nevertheless, the lower direct/indirect emissions during the usage of the electric vehicle result in a lower overall CO_2 footprint [32]. Battery circular economic chain should be considered in the context of the end market and assessed against industry benefits for power generation and transportation.

The World Economic Forum and the Global Battery Alliance emphasize the value of such cross-sector coupling. However, they also recognize broader concerns about the battery cost chain (economic dimension). For example, the production of basic raw materials for batteries has been linked to hazardous working conditions, child labor, poverty, and other problems related to social and environmental sustainability dimensions [33]. To overcome these obstacles, the European Union, for example, will propose by 2023 a new regulation to update the regulation of 2006. This new regulation will restrict not only battery producers but also companies that produce electric vehicles (EVs).

The main battery system used in EVs is lithium-ion batteries, mainly due to the relatively high-energy density and the high diversity of power/energy ratios. This system is complex and by each battery component (electrodes, separator, electrolyte, current collector, and others) there exists a large number of different materials that can be used [34]. Once that, the recycling process of this system experiences many difficulties. Different techniques (metallurgical processes, including pyrometallurgy and hydrometallurgy) have been developed and improved to recycle these materials. The use of eutectic solvents to extract active materials from the cathodes, the use of natural materials, and the use of biodegradable precursors and solvents to allow their reutilization are examples of the different approaches used to increase the life cycle of these systems and to reduce the environmental impact of materials, devices, and fabrication processes [35].

The cathode electrode is the main costly battery component and is mainly responsible for battery energy density. Cathode-active materials such as lithium nickel cobalt aluminum oxides and lithium nickel manganese oxides are used in majority of EVs. To achieve the sustainability of lithium-ion batteries, one of the possible routes is the use of more sustainable materials with less toxicity (cobalt-free) as lithium iron phosphate. Furthermore, research has been focusing on the development of other systems, such as sodium, sulfur, and others, to develop batteries without lithium to overcome the problem associated with lithium extraction [36]. Together with direct recycling, different approaches, such as second-use applications, can avoid battery accumulation in landfills. Second-use application is based on giving another use to a battery that no longer has the needed capacity in the application that requires less capacity. This approach will not only prolong the life of spent batteries but also avoid the extraction, synthesis, purification, and modification of materials to fabricate a new battery, increasing the sustainability of battery systems [37].

Hydrogen (H_2) is the most exciting example of a circular economy across industries, a clean fuel with no direct emissions of harmful pollutants or greenhouse gases. However, the industrial demand for hydrogen is currently almost completely covered by fossil fuels and is therefore one of the most important causes of CO_2 emissions. The perspective of a hydrogen-powered future is based on the production of hydrogen from low-carbon energy sources such as renewable generation (green hydrogen) or from natural gas (blue hydrogen) [38].

Hydrogen could play an important role in a sector-coupled circular economy, offering a solution to decarbonize a wide range of sectors, including transport, heating, chemicals and iron and steel, and production/transformation [39]. Hydrogen and its derivatives such as synthetic fuels or chemicals offer an opportunity to replace fossil hydrocarbons, which can reduce emissions [39]. In this scope, electricity and hydrogen are complementary and both will be needed to achieve climate neutrality and contribute to climate protection and sustainability. Since hydrogen can be generated by capturing excess production from solar and wind power, it is also a leading contender for a storage solution for renewable energy generation.

1.4 Conclusions and Outlook

Energy sustainability approaches and circular economy are nowadays topics with increasing relevance since energy consumption increases year after year. Thus, it is essential to be aware in which way this growing energy consumption will impact the ecological footprint. The transition to clean energy generation/storage and urgent financial/environmental considerations are taking place toward a sustainable energy system to guarantee the future quality of life.

Environmental sustainability is the responsibility to preserve natural resources and protect global ecosystems to support health and well-being, now and in the future. Life cycle assessment is a commonly used tool for evaluating the environmental impact. The less damage is caused to the environment, the more positively this criterion is fulfilled. Opportunities and challenges of the different energy technologies must be faced and explored. Some energy technologies, such as photovoltaic, have the possibility of recycling up to 95% of their materials; however, challenges such as the presence of hazardous substances including cadmium, arsenic, lead, and antimony should be overcome. In wind turbine technology, 90% of components could be recycled, although the huge size makes transportation costs prohibitive for long-distance hauls. Energy storage systems have the opportunity for all metals from a battery to be recycled. Nevertheless, increasing the economic profitability and economic efficiency of the recycling process can be challenging, due to fluctuating material values.

Raw materials recovery is essential, although recovering these materials and reintroducing them into production cycles present some challenges. The first is the processing obstacles due to the use of composite materials, the presence of hazardous substances, and low concentrations of valuable/scarce elements. Furthermore, the devices and systems are typically not designed to facilitate end-of-life/recyclability aspects. Nowadays, market conditions do not properly price using virgin materials vs. recycled ones, associated with logistical issues due to the remote locations, size, and safety requirements of energy infrastructure. The inclusion of materials with high recycling characteristics and less harmful to nature must be the way to overcome such obstacles. The standardization of devices, materials, and processing will simplify the later stages of recycling allowing the interconnection of raw materials to different applications.

Efforts to overcome these barriers are being implemented by different governments/authorities in different countries of the world. Sustainability is ruled by three main dimensions: economic, social, and environmental. These three dimensions are strongly interlinked and equally interdependent. It is important to keep the balance between these dimensions in order to maintain, for example, economic growth without compromising the natural environment, resources, and community health. Life cycle costing, social life cycle assessment, and life cycle assessment are necessary tools for a better assessment of these dimensions.

Acknowledgments

The authors thank the Fundação para a Ciência e Tecnologia (FCT) and COMPETE 2020 for financial support under the framework of Strategic Funding grants UIDB/04650/2020, UID/FIS/04650/2020, UID/EEA/04436/2020, and UID/QUI/00686/2020 and under projects, MIT-EXPL/TDI/0033/2021, POCI-01-0247-FEDER-046985, 2022.03931.PTDC, and PTDC/FIS-MAC/28157/2017. The authors also thank the FCT for the investigator contracts CEECIND/00833/2017 (RG) and 2020.04028.CEECIND (C.M.C.). Financial support from the BIDEKO Project, funded by MCIN/AEI, NextGenerationEU, PRTR, and Basque Government Industry Department under the ELKARTEK program are acknowledged.

References

1 Morseletto, P. (2020). *Targets for a circular economy. Resources, Conservation and Recycling* 153: 104553.

2 Ahmed, Z., Mahmud, S., and Acet, D.H. (2022). *Circular economy model for developing countries: evidence from Bangladesh. Heliyon* 8 (5): e09530.

3 Brydges, T. (2021). *Closing the loop on take, make, waste: Investigating circular economy practices in the Swedish fashion industry. Journal of Cleaner Production* 293: 126245.

4 Geissdoerfer, M., Savaget, P., Bocken, N.M.P., and Hultink, E.J. (2017). *The circular economy – a new sustainability paradigm? Journal of Cleaner Production* 143: 757–768.

5 Invernizzi, D.C., Locatelli, G., Velenturf, A. et al. (2020). *Developing policies for the end-of-life of energy infrastructure: coming to terms with the challenges of decommissioning. Energy Policy* 144: 111677.

6 New to circular economy overview. ellenmacarthurfoundation.org (accesed 19 January 2022).

7 Pearce, D.W.T.R.K. (1990). *Economics of Natural Resources and the Environment.* Baltimore: Johns Hopkins University Press.

8 Circular economy – Overview. https://ec.europa.eu/eurostat/web/circular-economy (accesed 19 January 2022).

9 Bleischwitz, R., Yang, M., Huang, B. et al. (2022). *The circular economy in China: achievements, challenges and potential implications for decarbonisation. Resources, Conservation and Recycling* 183: 106350.

10 Betancourt Morales, C.M. and Sossa, J.W.Z. (2020). *Circular economy in Latin America: a systematic literature review. Business Strategy and the Environment* 29 (6): 2479–2497.

11 United States Environmental Protection Agency (2021). National Recycling Strategy: Part One of a Series on Building a Circular Economy for All, EPA 530-R-21-003, U.S. EPA Office of Resource Conservation and Recovery.

12 Deutz, P. (2020). Circular economy. In: *International Encyclopedia of Human Geography*, 2nde (ed. A. Kobayashi), 193–201. Oxford: Elsevier.

13 Report of the World Commission on Environment and Development: Our Common Future-A/42/427 Annex-UN Documents: Gathering a body of global agreements. www.un-documents.net (accesed 19 January 2022).

14 Desing, H., Widmer, R., Beloin-Saint-Pierre, D. et al. (2019). *Powering a sustainable and circular economy—an engineering approach to estimating renewable energy potentials within earth system boundaries. Energies* 12 (24): 4723.

15 Net zero no more: We are entering the era of honest electricity. Canary Media. https://www.canarymedia.com/articles/clean-energy/net-zero-no-more-we-are-about-to-enter-the-era-of-honest-electricity (accessed 19 January 2022).

16 Enosi raises $1.5 million for clean energy tech, as demand for renewables ramps up. SmartCompany. https://www.smartcompany.com.au/startupsmart/news/enosi-powertracer-1-5-million-clean-energy-renewables/ (accessed 19 January 2022).

17 Gallagher, J., Basu, B., and Browne, M. (2019). et al, *Adapting stand-alone renewable energy technologies for the circular economy through eco-design and eecycling. Journal of Industrial Ecology* 23 (1): 133–140.

18 Sustainability in the energy sector. www.pwc.de/en/sustainability/sustainability-in-the-energy-sector.html (accessed 13 June 2022).

19 Thomas, A. Updated: potential first for UK in plans to reuse off shore platforms for wind farms. Energy Voice. https://www.energyvoice.com/oilandgas/north-sea/206114/dno-in-talks-to-reuse-uknorth-sea-platforms-to-support-wind-farms (accessed 1 July 2022).

20 Ibbetson, C. Engineers convert old oil rigs into wave energy sites. New Civil Engineer. https://www.newcivilengineer.com/latest/engineers-convert-old-oil-rigs-into-wave-energy-sites-03-04-2019 (accessed 1 July 2022).

21 Consortium aiming to re-purpose Danish North Sea fields for CO_2 storage. www.offshore-mag.com/home/article/14178029/consortium-aiming-to-repurpose-danish-north-sea-fieldsfor-carbon-dioxide-storage. (accessed 5 July 2022).

22 Maersk Drilling pursuing net-negative emissions technology. www.offshore-mag.com/drilling-completion/article/14180573/maersk-drilling-pursuing-netnegative-carbon-dioxideemissions-technology. (accessed 5 July 2022).

23 Net Power website, https://netpower.com. Patel, Sonal, "300-MW Natural Gas Allam Cycle Power Plant Targeted for 2022," Power magazine, https://www.powermag.com/300-mw-natural-gas-allam-cycle-power-plant-targeted-for-2022. (accessed 5 July 2022).

24 Clean Energy Systems website, "Zero Emissions Power Generation," https://www.cleanenergysystems.com/zero-emissions-power-generation. (accessed 5 July 2022).

25 Alves Dias, P., Pavel, C., Plazzotta, B., and Carrara, S. (2020). *Raw Materials Demand for Wind and Solar PV Technologies in the Transition Towards a Decarbonised Energy System.* Publications Office of the European Union.

26 Fortescue, June 2020 Quarterly Production Report analyst call. www.fmgl.com.au/docs/default-source/announcements/investor-and-analyst-call-transcript-(june-2020-quarterly-production-report).pdf?sfvrsn=299b584_8. (accessed 16 June 2022).

27 Rio Tinto, news release. www.riotinto.com/news/releases/2020/Rio-Tinto-to-build-first-solar-plant-in-Western-Australia-to-power-iron-ore-mine. (accessed 16 June 2022).

28 BHP, Climate Change Report 2020. www.bhp.com/investor-centre/climate-change-2020. (accessed 16 June 2022).

29 Shell, news release. www.shell.com/business-customers/chemicals/media-releases/2019-media-releases/shell-moerdijk-solar-farm.html. (accessed 24 June 2022).

30 Johnson Matthey, Annual Report and Accounts 2018. https://matthey.com/en/investors/report-archive/annual-report-2018. (accessed 24 June 2022).

31 The International Renewable Energy Agency (IRENA) and the International Energy Agency Photovoltaic Power Systems Programme, End-of-Life Management: Solar Photovoltaic Panels. https://www.irena.org/-/media/Files/IRENA/Agency/Publication/2016/IRENA_IEAPVPS_End-of-Life_Solar_PV_Panels_2016.pdf. (accessed 28 June 2022).

32 Costa, C.M., Barbosa, J.C., Castro, H. et al. (2021). *Electric vehicles: to what extent are environmentally friendly and cost effective? – comparative study by european countries. Renewable and Sustainable Energy Reviews* 151: 111548.

33 Global Battery Alliance and World Economic Forum, A Vision for a Sustainable Battery Value Chain in 2030. www3.weforum.org/docs/WEF_A_Vision_for_a_Sustainable_Battery_Value_Chain_in_2030_Report.pdf. (accessed 28 June 2022).

34 Costa, C.M., Barbosa, J.C., Gonçalves, R. et al. (2021). *Recycling and environmental issues of lithium-ion batteries: advances, challenges and opportunities. Energy Storage Materials* 37: 433–465.

35 Yuan, Z., Liu, H., Yong, W.F. et al. (2022). *Status and advances of deep eutectic solvents for metal separation and recovery. Green Chemistry* 24 (5): 1895–1929.

36 Tian, Y., Zeng, G., Rutt, A. et al. (2021). *Promises and challenges of next-generation "Beyond Li-ion" batteries for electric vehicles and grid decarbonization. Chemical Reviews* 121 (3): 1623–1669.

37 Tao, Y., Rahn, C.D., Archer, L.A. et al. (2021). Second life and recycling: energy and environmental sustainability perspectives for high-performance lithium-ion batteries. *Science Advances* 7 (45): eabi7633.

38 Zhang, C., Cao, X., Bujlo, P. et al. (2022). *Review on the safety analysis and protection strategies of fast filling hydrogen storage system for fuel cell vehicle application. Journal of Energy Storage* 45: 103451.

39 Arsad, A.Z., Hannan, M.A., Al-Shetwi, A.Q. et al. (2022). *Hydrogen energy storage integrated hybrid renewable energy systems: a review analysis for future research directions. International Journal of Hydrogen Energy* 47 (39): 17285–17312.

2

Reactive Metals as Energy Storage and Carrier Media

Hüseyin Ersoy[1], *Manuel Baumann*[1], *Marcel Weil*[1,2], *Linda Barelli*[3], *and Stefano Passerini*[2]

[1] *Institute for Technology Assessment and System Analysis (ITAS), Karlsruhe Institute of Technology (KIT), 76133, Karlsruhe, Germany*
[2] *Helmholtz Institute Ulm for Electrochemical Energy Storage (HIU), Karlsruhe Institute of Technology (KIT), 89081, Ulm, Germany*
[3] *Dipartimento di Ingegneria, Università degli Studi di Perugia (UNIPG), 06123, Perugia, PG, Italy*

2.1 Introduction

An entirely decarbonized energy system is indispensable to overcoming the impacts of climate change and accomplishing the global emission reduction targets [1]. A continuous supply of carbon-free or -neutral electricity is essential for the entire energy system. This is mostly dependent on the availability of sustainable energy storage technologies and fuels required due to the fluctuating energy provision from renewables. Furthermore, storage technologies, i.e., pumped hydro storage (PHS), compressed air energy storage, and Power-to-X (e.g. hydrogen, chemicals, and fuels) are the proposed main storage options for fulfilling the mid- to long-term storage demand [2]. On the other hand, batteries are the most appealing technologies for the compensation of the short-term fluctuations (excess/deficit of the renewable generation) on the generation with fast-response behavior, not to mention their significance for electric mobility [3]. The current rapidly increasing demand for batteries will accelerate the depletion of metal resources with concerns regarding material scarcity and supply risks [3]. Especially, the use of critical materials such as lithium (Li), cobalt (Co), nickel (Ni), and natural graphite (C) in batteries has promoted the investigation of alternative sustainable energy carriers, particularly for the stationary energy storage demand [4–6]. All these concerns highlight the need for a re-established metal economy to reduce the reliance on critical materials and maintain their efficient use [7]. This can only be rendered possible by avoiding the additional primary demand and keeping the materials in a closed-loop cycle for as long as possible. As a potential solution for introducing recyclable energy carriers and storage mediums, earth-abundant reactive metals and their compounds are viable alternatives with satisfactory energy material properties [8, 9]. Metals are

Sustainable Energy Storage in the Scope of Circular Economy: Advanced Materials and Device Design, First Edition. Edited by Carlos Miguel Costa, Renato Gonçalves, and Senentxu Lanceros-Méndez.

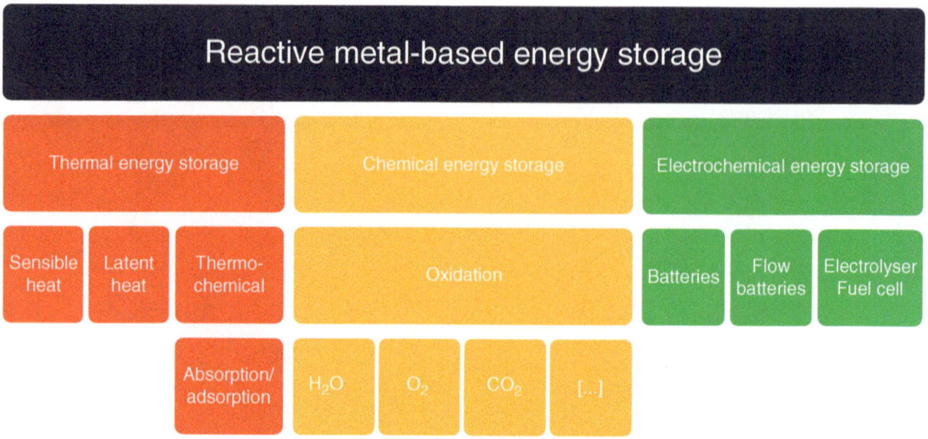

Figure 2.1 Reactive metal-based energy storage methods. *Source:* Adapted from Refs. [10–12].

dense energy carriers enabling thermal (sensible and latent heat), and chemical (thermochemical, electrochemical, etc.) energy conversion through various pathways (Figure 2.1).

Metals and their oxides demonstrate very high thermal energy storage potential both in sensible and latent form for the direct storage of heat. When longer storage durations (e.g. seasonal and annual) and higher storage densities are targeted, reactive metal-based chemical energy storage technologies possess promising properties including direct combustion of metals, thermochemical heat storage, and use of metals as anode material in mechanically rechargeable batteries for direct electrical energy storage. Metal combustion yields a significant amount of heat and metal oxides as a combustion product desirably in the solid phase. Solid combustion products can be easily captured and recycled to their parent metals via well-established and existing reduction processes which in theory can be carried out without increasing the CO_2 intensity of the metal [9]. Also, including their water-splitting ability metals provide various alternative pathways for an extensive spectrum of applications (industrial energy storage, load shifting, mobility energy demand management, etc.). An overview of the recent advances for direct electrical conversion in reactive metal-based batteries is also given. This chapter attributes particular importance to reactive metal-based chemical energy storage and their use as an energy carrier and storage medium for a sustainable circular economy. Metals and semi-metals as well as their compounds (hydrides, hydroxides, etc.) are analyzed and their potential energy storage applications are discussed.

2.2 Significance of a Circular Metal Economy for the Energy Transition

Metals are essential permanent materials that can be kept in a circular economy almost indefinitely for a sustainably developing global economy [13, 14] (Figure 2.2). The supply of affordable and clean energy is considered as a majorly contributing solution for a large spectrum of social, economic, and technological problems [15]. In addition, availability and

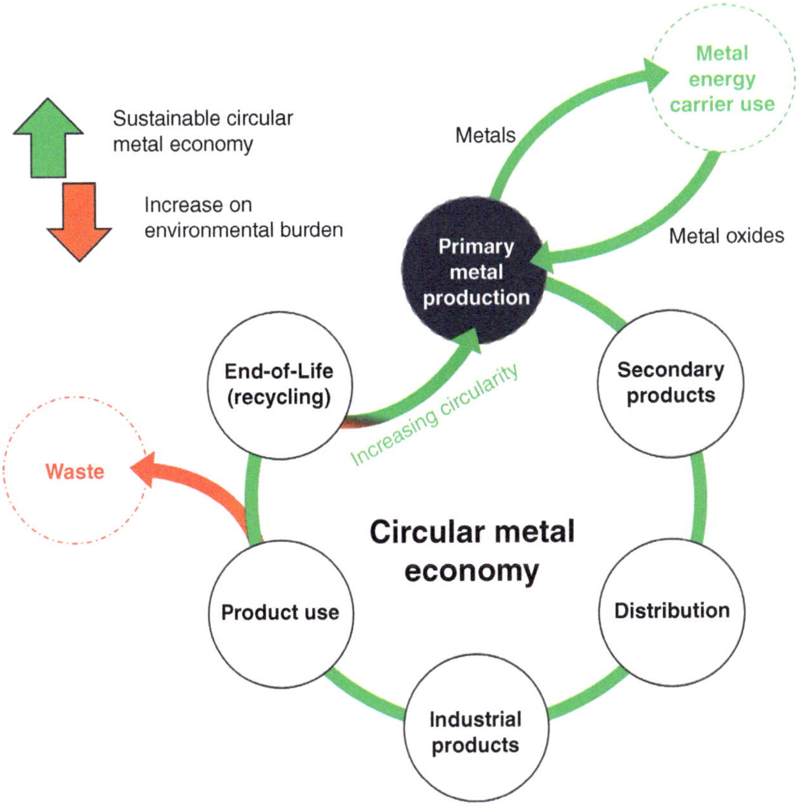

Figure 2.2 Life cycle stages of metals in a circular economy.

a secure supply of functional metals are important pillars for a global sustainable transition of energy systems [16]. Furthermore, considering the significance of a specific metal for a product or process, the named technology can be a crucial component for another technology or product highlighting this intersectional relationship. Wind turbine generators are a good example to explain this co-dependency based on the high supply risk of niobium (Nb), dysprosium (Dy), neodymium (Nd), and praseodymium (Pr), which are key materials for permanent magnet synchronous generators (PMSGs) (neodymium–iron–boron [NdFeB]) representing a challenge for circularity [14]. The aforementioned metals belong to the class of light rare earth elements (LREE) and heavy rare earth elements (HREE), thus, the criticality of these metals would put the production of highly efficient PMSG wind turbines at risk [17]. As a result, low penetration of wind-based electricity will reduce the carbon-free electricity supply significantly (based on PMSG wind turbines), increasing the CO_2 intensity of all commercial products due to the increased cost of wind turbines [18].

Moreover, metals are considered as almost fully recyclable materials if proper recycling mechanisms (including collection, dismantling, and sorting) are well integrated into the life cycle and efficient recycling processes are introduced with lower energy demand in comparison to the primary material extraction process [19]. Among commonly used energy materials, Li has been one of the key energy carriers, especially for battery energy storage

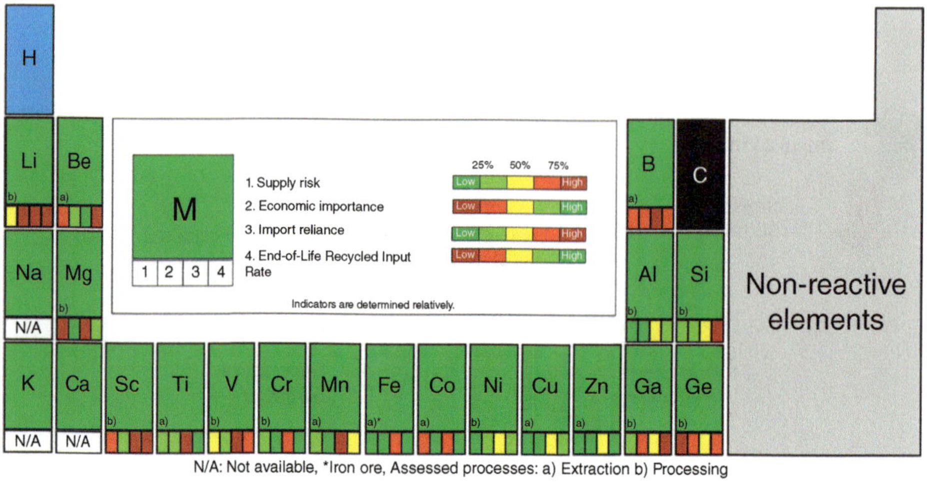

Figure 2.3 Suitable reactive metals/semi-metals (green), with European Union supply risk, economic importance, import reliance, and end-of-life recycle input rate indicators. *Source:* Adapted from Ref. [23].

and mobility applications. However, a detailed study conducted by Moreau et al. (2019) asserts that if half of the intraday energy storage demand is maintained with Li-ion batteries (LiBs) using NMC (nickel–manganese–cobalt) chemistries, the impact on resource depletion is anticipated to be very high, especially for Co, Ni, and Li [20]. This finding supports the need for the introduction of an efficient recycling process for metals used in batteries. However, there are earth-abundant metals with low supply risk (e.g. aluminum (Al), iron (Fe), sodium (Na), etc.) that can be used more intensively in energy technologies, in fact, these are highly functional metals that can deliver large energy densities for energy storage applications [8, 21, 22]. When examining the periodic table many potential metals (Figure 2.3) can be identified as reactive energy carriers, besides hydrogen (H), carbon (C), and non-reactive (under atmospheric conditions) elements. To address metal energy carriers with a high application potential, a brief screening can be made using some fundamental constraints. The following properties are essential for the proposed metal: high reactivity, non-toxic element or oxides, and they must be economical.

Metals such as Be, Cr, Mn, and Co can be excluded as a pure energy carrier due to the toxicity risk of the elements/oxides [24–27]. Sc, V, Ga, and Ge are functional metals that are too expensive to be used as pure energy carriers. As a result, Li, B, Mg, Al, Si, Fe, and Zn can be identified as optimal pure reactive energy carriers at first glance [8, 9]. Furthermore, some of these metals can be used as energy carriers benefiting from their fully established oxide reduction processes (e.g. carbothermic reduction, electrolytic reduction).

2.3 Energy Carrier Properties of Reactive Metals

Combining all the requirements of a sustainable and circular global economy, the use of pure metals as energy carriers can potentially be a direct and innovative solution to meet energy storage demands [8, 28]. Metals may enable the continuous development of

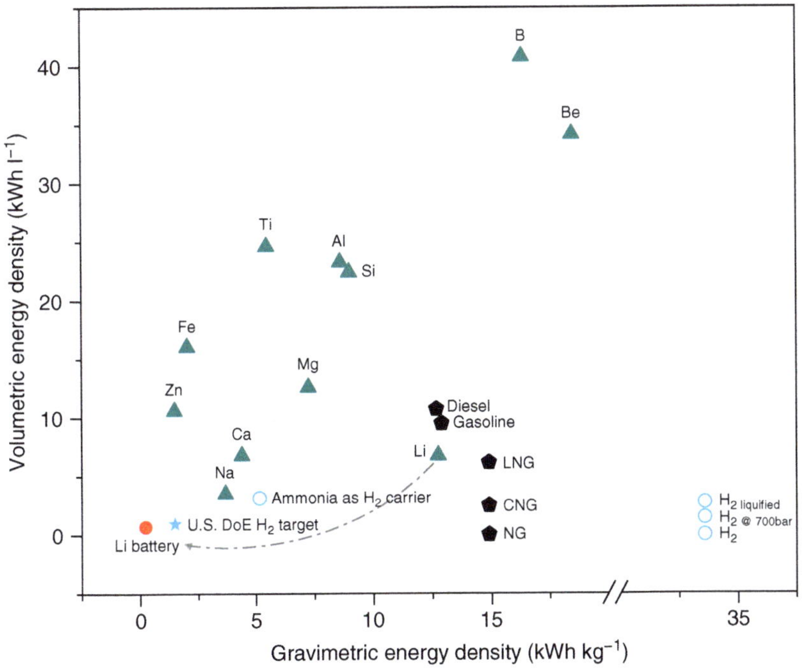

Figure 2.4 Specific and volumetric energy density of conventional energy carriers and metals. *Source:* Adapted from Ref. [8].

the carbon-free energy system if efficient utilization paths are developed and sustainable material recovery processes are established [7]. The high volumetric energy density of pure metals (see Figure 2.4) allows the storage of energy in various transformable forms. Metals and metal compounds (e.g. oxides, hydroxides, and hydrides) have high theoretical thermal and electrochemical energy or H_2 storage capacities [10]. Metal-based energy materials provide high flexibility and versatility in terms of energy conversion.

In comparison with conventional fuels (i.e. diesel, gasoline, or LNG) metal energy carriers offer advantages due to their energy-dense and reactive behavior, storability, abundance, production in large quantities, and in particular recyclability. The same applies to battery technologies, as illustrated in Figure 2.4, for instance, LiBs as state-of-the-art provide significantly lower energy densities when compared to pure Li metal but their high technological readiness level (TRL) and efficiency are ideal for mobile and stationary applications [29]. Furthermore, considering the use of metals in various industries, integrating the metals into the entire chain can support an established fully circular economy. Thus, reactive metals can be treated as transportable fuel commodities or anode materials for batteries. Their relatively easy and low-effort closed-loop recycling can raise the value of abundant metals. Considering their transportability and easy storability in an inactive form (e.g. aluminum ingots) to avoid explosion hazards, energy carrier application of metals can cover any storage purpose starting from the short-term to seasonal or even annual energy storage durations.

2.4 Potential Reactive Metal Energy Carrier and Storage Applications

In principle, the use of metals as energy carrier hinges upon two mechanisms: endothermic–exothermic reversible reactions and electrochemical energy conversion. Hence, endothermic reactions represent the energy conversion via heat absorption, exothermic reactions stand for the exploitation of the stored energy as heat [30]. This can be carried out via either the absorption–adsorption reaction where energy is stored in molecular bounds or via the reduction–oxidation principle. Reduction of the metals would refer to the metal production (i.e. reduction of Al_2O_3 to Al) while oxidation is the path where energy is released and the metal is converted to its oxide [11]. Batteries are electrochemical devices employing the reduction-oxidation (redox) principle in metals; in this case, the energy is transferred via charge-carrying electrons between cathode and anode reversibly passing through the electrolyte to form a storage cycle [12]. Storage cases are classified under three categories based on the stored energy form in order to discuss the potential reactive energy carrier and storage applications as thermal, chemical, and electrochemical energy storage categories. The given energy storage methods are able to provide power in the range of 10 kW to 1 GW in various forms with storage durations spanning from intraday to long-term as illustrated in Figure 2.5.

2.4.1 Metals as Thermal Energy Carriers

Metals and metal salts are the most promising storage mediums to store thermal energy in the form of heat for thermal energy storage applications for direct use and electricity generation. The most common heat storage and transfer medium has been, and is water due to its availability, high specific heat capacity, and cost-efficient supply. Nevertheless, for

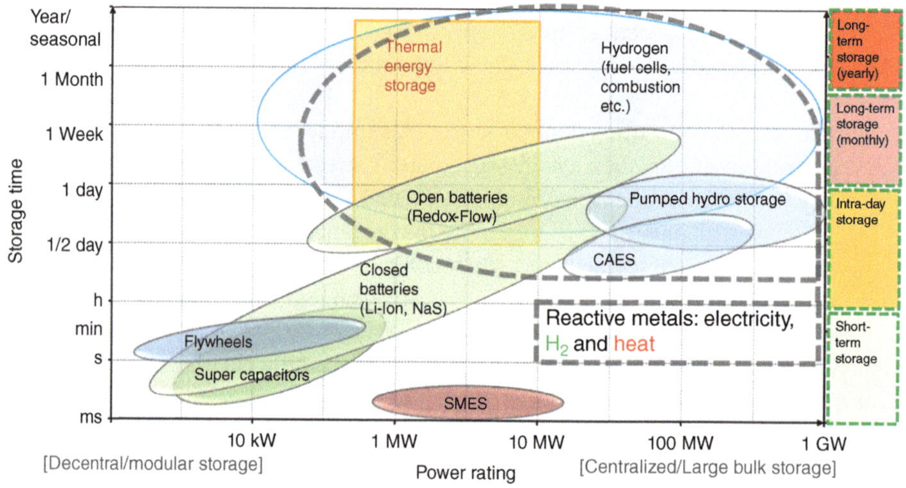

Figure 2.5 Power rating vs. storage time of various energy technologies including reactive metals. *Source:* Adapted from Refs. [22, 31–35].

large-scale heat storage applications, the demand for volumetric energy-dense storage materials is growing for use in industrial thermal energy storages and long-term utilization of renewable heat, e.g., in solar thermal power plants. The fluctuating behavior of renewables requires innovative approaches also for the efficient use of heat. Hence, the availability of energy-dense carriers with techno-economically feasible utilization pathways has been the focus of the last decades [36]. Depending on the application case, the heat demand can be maintained via sensible, latent, and thermochemical heat carriers. These differ depending on the form of heat storage and transfer path. The sensible heat transfer takes place under temperature differences, and they are the cheapest and most mature heat carriers using molten salts, ceramics, rocks, sand, etc., but they require high insulation to avoid convective heat losses through surfaces. Metals such as molten Na, Fe, Al, and Li are the metal sensible heat carriers used in practical applications [37]. Besides pure metals, also metal oxides and alloys are highly performant sensible heat storage materials with large energy densities (see Table 2.1). Among previously considered pure metals Al provides the highest volumetric energy density while aluminum oxide and magnesia are the metal oxides with the largest availability.

The latent heat carriers predicate on the phase change, so the heat is stored in the form of chemical energy. Correspondingly, latent heat carriers are named as phase-changing material (PCM) as shown in Table 2.2. This aspect makes PCMs more efficient as the heat loss is significantly lower in comparison with sensible heat carriers. The disadvantages are

Table 2.1 Reactive metal-based sensible heat storage mediums.

Storage medium	Empirical formula	Specific energy (kJ kg^{-1})	Energy density (MJ m^{-3})
Aluminum oxides	Al_2O_3	294	1146
Sintered bauxite particles	Al_2O_3 (40-60%)	385	770
Alumina (99.5%)	Al_2O_3	280	1108
Aluminum	Al	315	852
Magnesia	MgO	329	1172
Magnesia fire bricks	MgO	420	1260
Sodium liquid metal	Na	455	437
Sodium chloride	NaCl	315	680
Pure iron	Fe	157	556
Cast iron-I	Fe-alloys	294	2322
Cast iron (BSgrade3D)	Fe-alloys	189	1474
Cast steel	Fe-alloys	210	1638
Silicon	Si	250	575
Silica fire bricks	SiO_2	350	637
Copper	Cu	133	1190

Note: BSgrade3D: British Standards stock grade of 3D.
Source: Adapted from Refs. [10, 38].

Table 2.2 Reactive metal-based latent heat storage mediums.

Name	Empirical formula	Specific energy ($kJ\,kg^{-1}$)	Energy density ($MJ\,m^{-3}$)
Aluminum	Al	397	1,072
Aluminum nitrate (sol)	$Al(NO_3)_3$	131	166
Magnesium chloride	$MgCl_2$	452	967
Magnesium nitrate hexahydrate	$Mg(NO_3)_2{\bullet}6H_2O$	156	251
Magnesium chloride hexahydrate	$Cl_2Mg{\bullet}6H_2O$	167	252
Sodium chloride (sol)	NaCl	222	252
Sodium nitrate	$NaNO_3$	172	358
Sodium chloride	NaCl	492	1,063
Disodium phosphate dodecahydrate	$HNa_2O_4P{\bullet}12H_2O$	280	415
Sodium thiosulfate pentahydrate	$Na_2O_3S_2{\bullet}5H_2O$	198	339
Sodium acetate trihydrate	$CH_3COONa{\bullet}3H_2O$	245	334
Calcium chloride (sol)	$CaCl_2$	283	316
Potassium nitrate	KNO_3	266	504
Potassium fluoride	KF	452	1071
Potassium carbonate	K_2CO_3	236	540
Lithium nitrate	$LiNO_3$	360	706
Lithium hydride	LiH	2582	2040
Silicon	Si	1800	4140
Gallium	Ga	80	472
Indium	In	28	205
Titan nitride	TiN	61	443
Zinc	Zn	108	771

Source: Adapted from Refs. [10, 42].

mainly the high cost of the storage materials, low technology maturity, and corrosion hazards as most of the PCMs are salts and organics. They perform well when combined with sensible heat carriers. In particular, Ca, Na, K, and Li-based metal salts are commonly used state-of-the-art latent heat solar thermal energy storage materials [39]. The combined heat transfer use of abundant metals can be considered as a more cost-efficient alternative especially for Al compared to Si, and Li. Si is another dense material with high abundancy in the earth's crust, the fact remains that its supply is risky and Si is of high strategic importance for semi-conductor manufacturers (i.e. for solar panels and silicon carbide (SIC) semiconductors for power electronics), not to mention its prospective use in batteries as an anode material [40, 41].

Thermochemical energy storage methods are the most advanced option with significantly higher energy storage densities. The energy storage principle relies on absorption and adsorption, referring to thermal charging and discharging, respectively. The adsorption of the material yields solid oxides and CO_2, H_2O, and H_2 in gas phases. The reaction products must be stored, which requires the storage of the gas phases which is a costly complex process compared to sensible and latent heat storage methods. The applications are usually employing Mg, Na, and Ca-based metal carbonates, hydroxides, and hydrides (see Table 2.3). Especially, metal hydrides embody volumetrically larger amounts of energy due to the higher H_2 concentration of the compounds. Hence, this can also be a potential path for direct conversion of heat to H_2 or the other way around. For instance, the given reaction $Mg_2FeH_{6(s)} + \Delta H \leftrightarrow 2Mg_{(s)} + Fe_{(s)} + 3H_{2(g)}$ reduces the dimagnesium iron(II) hydride to Mg and Fe while releasing approximately 150 kg of H_2 per cubic meter of Mg_2FeH_6. This amount corresponds to ~21 GJ/5.8 MWh (based on HHV) thermal energy

Table 2.3 Reactive metal-based thermochemical heat storage mediums.

Storage medium	Empirical formula	Specific energy ($kJ\,kg^{-1}$)	Energy density ($MJ\,m^{-3}$)
Magnesium hydride	MgH_2	2810	4074
Magnesium hydroxide	$Mg(OH)_2$	1390	3252
Magnesium sulfate heptahydrate	$MgSO_4$•$7H_2O$	2500	3325
Magnesium chloride hexahydrate	$MgCl_2$. . .•$6H_2O$	1720	2003
Magnesium hydroxide (sorption)	$Mg(OH)_{2(s)}$	1340	1396
Magnesium hydride (sorption)	MgH_2	2880	2088
Dimagnesium iron(II) hydride (sorption)	$Mg_2FeH_{6(s)}$	2106	5768
Magnesium nickel hydride (sorption)	$Mg_2NiH_{4(s)}$	1160	3142
$NaMg_2H_{3(s)}$	$NaMg_2H_{3(s)}$	1721	1721
$NaMgH_2F_{(s)}$	$NaMgH_2F_{(s)}$	1416	1968
Sodium phosphide dibasic pentahydrate	Na_2HPO_3. . .•$5H_2O$	3840	3563
Calcium carbonate	$CaCO_{3(s)}$	1764	2491
Calcium hydroxide	$Ca(OH)_{2(s)}$	1406	1640
Calcium hydride	$CaH_{2(s)}$	3587	7374
Calcium hydride	H_2Ca	4280	7276
Calcium hydroxide	$Ca(OH)_2$	1470	3292
Calcium chloride (sol)	$CaCl_2 + H_2O$	910	426
Lithium chloride	$LiCl + H_2O$	4390	1439

Source: Adapted from Refs. [10, 42].

storage in a 1 m^3 volume of the metallic compound while the required liquified H_2 storage volume for such a capacity is ~2.13 m^3 [43]. The mentioned theoretical potential makes such a metal hydride interesting as an H_2 storage medium due to the availability and low cost of its constituents (Mg and Fe), however, the major challenge for enabling its use is related to its synthesis. Mainly, Mg_2FeH_6 synthesis is limited due to the miscibility gap between Mg and Fe (under thermodynamically favorable conditions) and long synthesis reaction time [44]. Similar issues are also observed for other metal hydrides in thermochemical H_2 storage applications.

Overall, metals are theoretically ideal candidates due to their physicochemical properties for use in thermochemical or combined thermal energy storage applications. Particularly, abundant metals in a pure state for all thermal energy storage pathways (sensible, latent, or thermochemical). Thus far, few literature studies for solar thermal energy storage applications have been conducted [39, 45–49].

2.4.2 Combustible Metal Fuels, and Hydrogen Carriers

One of the most promising methods is the direct combustion of metals due to the ease of capturing the solid combustion products (i.e. metal oxides) and recyclability. Combustible metals were initially considered as fuels for solid-propellant rockets, additives in liquid hydrocarbon slurry fuels, air-breathing ramjet engines, and water-breathing propulsion systems [50–52]. However, the use of such large energy-dense materials has not been considered for energy storage purposes. Theoretically, the combustion of metals with air, water, nitrogen oxides, fluorine, and carbon dioxide yields high energy. An overview of the metal oxidation reactions with various oxidizers is given in Table 2.4.

Considering various oxidizers, the oxidation of metals with air and water (i.e. dry and wet combustion, respectively) are the most feasible [8, 21, 55–58]. The use of metals in the Power-to-X context has been proposed as an innovative alternative where primary renewable energy is used for metal oxide reduction to produce pure metals (Figure 2.6). Then, the produced metals can be oxidized with water or air for re-electrification, and formed metal oxides can be kept in the cycle infinitely. When metals are oxidized with water, the reaction yields a significant amount of heat and H_2. The released heat can be utilized via heat engines, and H_2 can be supplied for direct use or converted into electricity using fuel cells.

Table 2.4 Metal oxidizer types and reactions.

Oxidizer type	Reaction
Oxygen	$xM + y/2\ O_2 \rightarrow M_xO_y + \text{heat}$
Water	$xM + yH_2O \rightarrow M_xO_y + yH_2\ \text{heat}$
Carbon dioxide	$xM + yCO_2 \rightarrow M_xO_y + yCO + \text{heat}$
Carbon dioxide	$xM + 2CO_2 \rightarrow M_xCO_3 + CO + \text{heat}$
Nitrous oxide	$xM + yN_2O \rightarrow M_xO_y + yN_2 + \text{heat}$
Fluorine	$xM + yF_2 \rightarrow M_xF_y + \text{heat}$

Source: Adapted from Refs. [8,28,52–54].

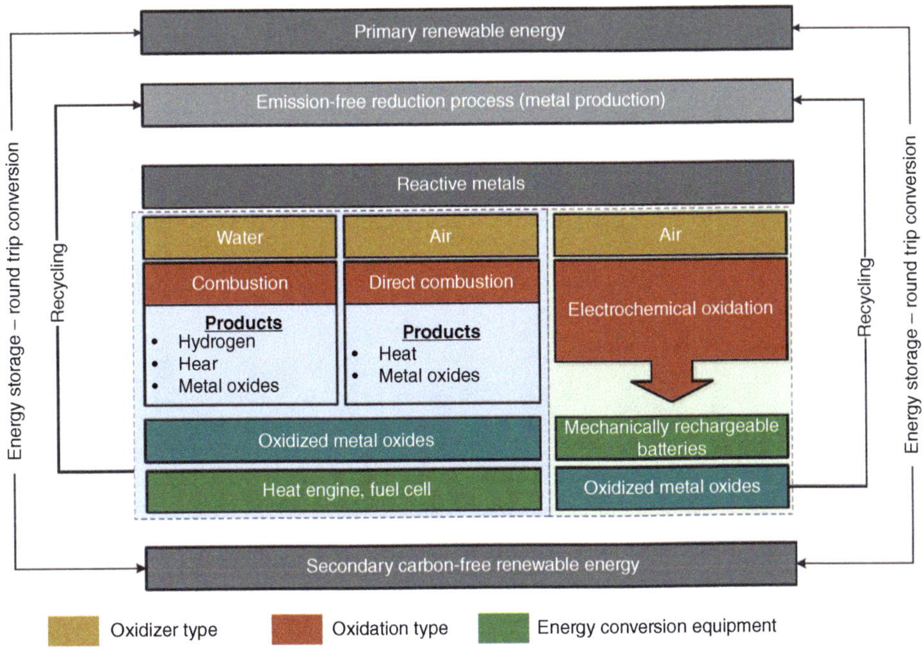

Figure 2.6 Metal production and utilization as a secondary renewable energy storage supply medium.

Additionally, the re-electrification can be carried via electrochemical conversion by integrating a mechanically rechargeable battery system (e.g. a metal-air battery) into the Power-to-X. In this case, pure metals are used in a primary battery as an electrode material, and oxidized metals after discharge must be recycled via a reduction process as in the metal combustion path. Depending on the operational requirements, efficiency, and cost-effectiveness the latter can also provide a feasible conversion path for using metals as energy carriers (see Section 4.3).

If air is chosen as an oxidizer, the stored energy is released as heat and it can be used either for thermal applications or converted into electricity. All the given metals in Figure 2.7 are combustible metals under theoretical considerations of their reactive nature. However, metals such as Zr, Ti, Si, and Li may not be suitable as a storage medium due to operational constraints (e.g. geographical availability), high demand in other industries, and low production rates due to practical reasons (mentioned in Chapter 2). For a more insightful comparison, there are other criteria to consider such as environmental sustainability, economics, and operability of the supply chain. Setting such constraints already concretizes the fact, Al, Fe, and Mg are the most competitive metals based on operational reasons [28]. The remaining materials are Be and B, the first is highly toxic and carcinogenic, while the latter faces supply concerns in Europe, and its oxides are toxic to the environment [23, 25].

From a thermophysical perspective to identify suitable metals for combustion, two thermophysical parameters are important, these are the boiling and oxide volatilization temperatures. This is mainly due to "Glassman's criterion," which stipulates a volatilization

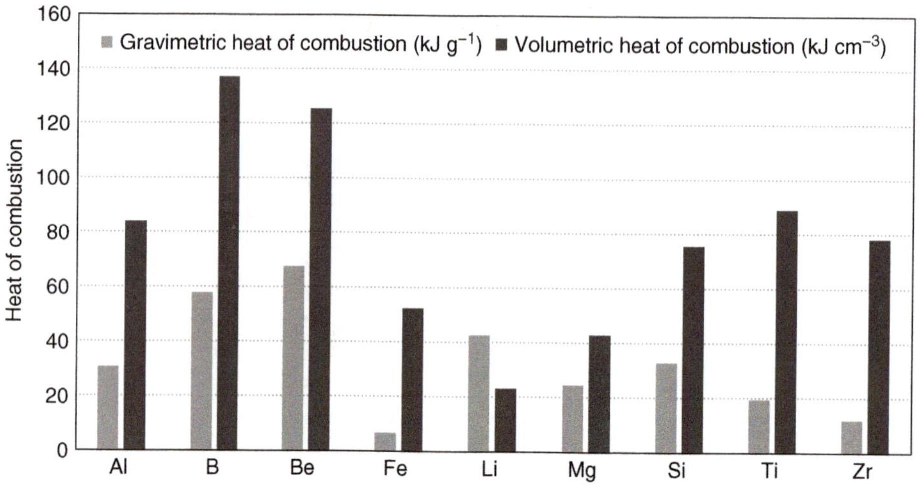

Figure 2.7 Theoretical gravimetric and volumetric energy density of different metals. *Source:* Adapted from Ref. [59].

temperature higher than the boiling point of the metal for vapor phase homogenous combustion [52]. Considering Glassman's criterion, combustible metals can be identified based on their boiling temperature at which combustion products dissociate into the gas phase molecules.

Taking the boiling and volatilization temperatures in Table 2.5 into account; Al, Be, Cr, Fe, Hf, Li, Mg, and Ti can be identified as so-called combustible metals in vapor-phase diffusion flames homogenously. On the other hand, B and Zr are the metals that demonstrate heterogeneous combustion. Furthermore, the combustion of Cr, Fe, Hf, and Ti will not yield stable combustion behavior due to a very small difference between boiling point

Table 2.5 Thermophysical properties of selected metals and metal oxides.

Metal	Boiling temperature (K)	Oxide	Volatilization temperature (K)	$\Delta H_{f,298\,K}$ (kJ mol^{-1})	ΔH_{vol} (kJ mol^{-1})	$H_{T,vol} - H_{298\,K} + \Delta H_{vol}$ (kJ mol^{-1})
Al	2791	Al_2O_3	4000	−1676	1860	2550
B	4139	B_2O_3	2340	−1272	360	640
Be	2741	BeO	4200	−608	740	1060
Cr	2952	Cr_2O_3	3280	−1135	1160	1700
Fe	3133	FeO	3400	−272	610	830
Hf	4876	HfO_2	5050	−1088	1014	1420
Li	1620	Li_2O	2710	−599	400	680
Mg	1366	MgO	3430	−601	670	920
Ti	3631	Ti_3O_5	4000	−2459	1890	2970
Zr	4703	ZrO_2	4280	−1097	920	1320

Source: Ref. [60]/With permissison of Elsevier.

and volatilization temperature. Hence, due to the thermophysical and operational advantages explained above, Al and Mg are potentially beneficial abundant metals for homogenous gas-phase combustion [50, 61–63]. Although, the homogenous direct oxidation of metals with air/oxygen results in gas combustion products, condensation, and separation of the metal oxides is a must to ensure their recyclability.

On the other hand, the use of water as an oxidizer is another high-potential path for the metal combustion process which presents an opportunity for H_2 storage/supply. The wet combustion yields the same theoretical energy densities, but a fraction of the energy is released in the form of H_2. This is supposedly a solution for avoiding the issues encountered for H_2 storage that increases the cost, and energy conversion efficiency of the common practice H_2 storage technologies [64]. Metals such as B, Mg, Al, Cr, and Mn demonstrate high H_2 storage density on a gravimetric and volumetric basis (Figure 2.8). Furthermore, among the earth-abundant metals Al and Mg result in outstanding H_2 release rates with high yields at varying temperatures [59].

Al provides a very high potential for a business case since it is produced in large quantities and its availability is ensured with proven reserves [23, 65, 66]. Aluminum is produced via an electrolytic reduction process (Hall-Héroult process) [67] which enables production using only electricity. Hence, a storage cycle benefiting from this aspect can be designed

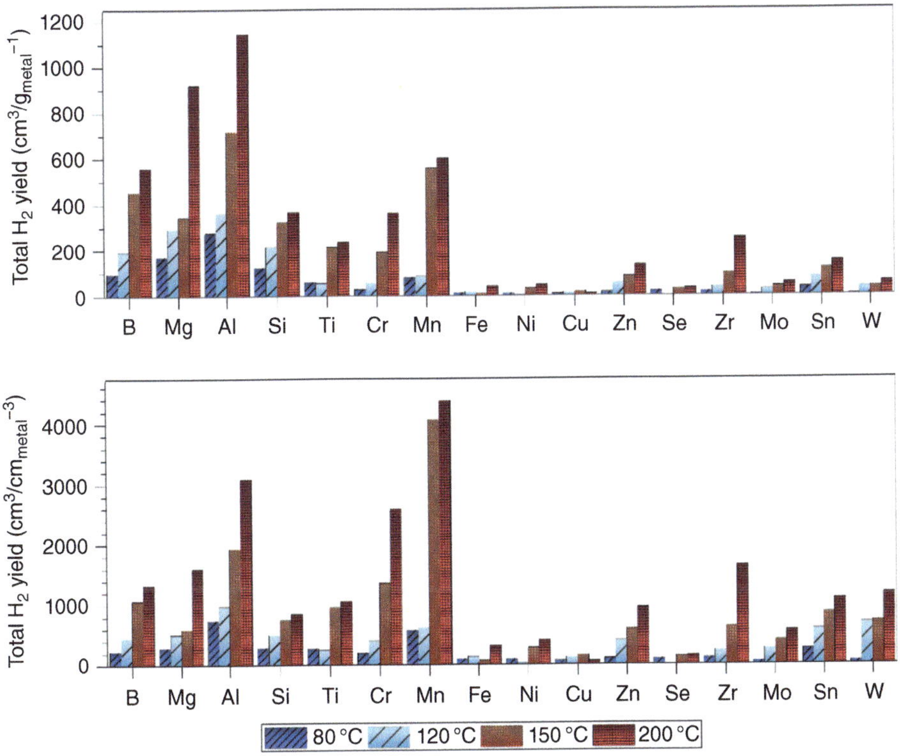

Figure 2.8 H_2 yield of metal combustion based on mass and volume. *Source:* Adapted from Ref. [59].

using the excess renewable electricity generation for Al production (Power-to-Al) and storing energy within Al. Moreover, several literature studies are available for the demonstration of such a wet combustion plant for electricity generation and H_2 supply [21, 22, 49, 68, 69]. The operation principle predicates on the supply of Al from carbon-free Al smelters and the combustion of Al in the proposed plant designs. The resulting heat can be converted into electricity using thermal energy conversion equipment while H_2 can be supplied as it is or used for electricity generation via fuel cells. The combustion reaction depending on the reaction stoichiometry yields various aluminum oxides (Al_2O_3, $Al(OH)_3$, or $AlO(OH)$), among potential oxides Al_2O_3 (alumina) at elevated temperatures, proves to offer the best reaction yield and kinetics [70]. The primary material of the Hall-Héroult process is alumina which facilitates the direct recycling of solid transportable combustion products [71]. Noting that, the Hall-Héroult process needs to be retrofitted to reduce energy consumption and direct greenhouse gas (GHG) emissions [72, 73]. This can be achieved through the introduction of inert anodes to substitute conventional carbon anodes which are consumed therefore contributing to CO_2 emissions.

In summary, the use of combustible energy-dense metals has enormous potential to function as a renewable energy storage vector, but their applicability is highly restricted by the operability of the supply chain as in the case of Mg, Mn, and Cr. In Europe due to supply concerns the circular use of these metals for energy storage purposes is turbulent. Nevertheless, Al is the ideal metal as a circular energy carrier for mid- to long-term energy storage durations (daily, seasonal, or annual). The demand for long-term energy storage vectors is expected to grow rapidly. To illustrate the volumetric energy advantage of Al, a 7.5 m thick Al layer stored beneath a soccer field itself enables the storage of 1 TWh_e energy (based on 81% electrical energy conversion efficiency) [22].

2.4.3 Reactive Metal-Based Electrochemical Energy Storage

In the battery industry metals such as Li, Co, and Ni are in high demand for use in battery cell production. An overview of a simplified classification of electrochemical energy storage technologies is given in Table 2.6. LiBs are advanced electrochemical energy storage technologies that perform well with round-trip efficiencies of around 95% and they are convenient for large power applications and also in mobility applications due to their relatively high specific energy densities [74]. This elevated demand for LiBs is mostly driven by developments in electric vehicle technology.

Currently, the practical performance and energy densities of batteries are advantageous for the stationary energy storage demand and for residual load management. However, the future of mobility requires more energy-dense (especially, gravimetric) batteries and the development of electric vehicle technology strongly depends on the availability of materials necessary for LiBs. Mostly the material criticality and energy density concerns are the driving forces to advance the research on LiBs. In this respect, alternative batteries are considered as employing abundant metals with high specific energies such as magnesium-ion, lithium–sulfur, nickel–zinc, a combination of existing cathode chemistries with silicon anodes (i.e. Li-ion with Si anodes), sodium-metal, aluminum-air, aluminum-ion, and solid-state batteries. Since the approach of this chapter focuses on reactive abundant metals in the circular economy frame, the scope herein is post-lithium-ion technologies.

Table 2.6 Classification of electrochemical energy storage technologies.

Electrochemical energy storage technologies	Flow batteries	True/classical	Same species in both half cell (e.g. All vanadium)
			Different species in each half cell (e.g. Fe–Cr)
		Hybrid	Solid reactant or product at one electrode (e.g. Zn–Br_2)
			Both electrodes involve solid phases (e.g. soluble Pb)
			H_2 or O_2 (e.g. Zn–air)
	Rechargeable Batteries	Lead-based	Flooded
			VRLA
		Lithium-based	Lithium-polymer
			Lithium-ion • Metal Oxides (LCO, NMC, NCA, LTO, etc.) • Iron phosphate (LFP)
		Nickel-based	Nickel–cadmium
			Nickel–iron
			Nickel–hydrogen
			Nickel metal hydride
			Nickel–zinc

The primary aim is substituting Li with other monovalent abundant metals such as Na^+ and K^+. Later on, multivalent metals such as Mg, Ca, Zn, or Al benefit from the accumulated technological development progress of the LiBs [75]. (see abundancy, reduction potential, and energy density of alternative metals in Figure 2.9) However, Na-ion batteries are the only batteries that pass TRL-5 (demonstration of concept) [75, 78, 79]. The Na-ion technology was very close to commercialization and in 2021 Chinese manufacturer Contemporary Amperex Technology Co. Limited (CATL) announced their first generation Na-ion battery with the aim of full commercialization by 2023. According to their announcement, their Na-ion battery performs with superior operation at low temperatures, system integration efficiency, fast charging properties, proving similar lifetime and safety features with respect to lithium–iron–phosphate (LFP) batteries [80]. Not to mention its reported prospective environmental sustainability competitiveness in the literature [81, 82]. From this aspect, Na-ion batteries seem to be a strong candidate to play an important role in battery energy storage applications.

Potassium-ion batteries (KIBs) were initially determined to be suitable due to its abundancy and potentially low cost [83]. Limitations stem from the difficulties to intercalate relatively large-sized K ions with favorable electrochemical properties to develop efficient electrode manufacturing processes [84]. Another abundant metal with a reasonable cost with respect to its energy density is Mg-based batteries, their technological development journey strongly depends on the development of suitable anode–cathode pairs, as well as electrolytes [85]. Also, considering the onerous Mg import reliance on China from a sustainability and environmental perspective means that despite Mg's high technical

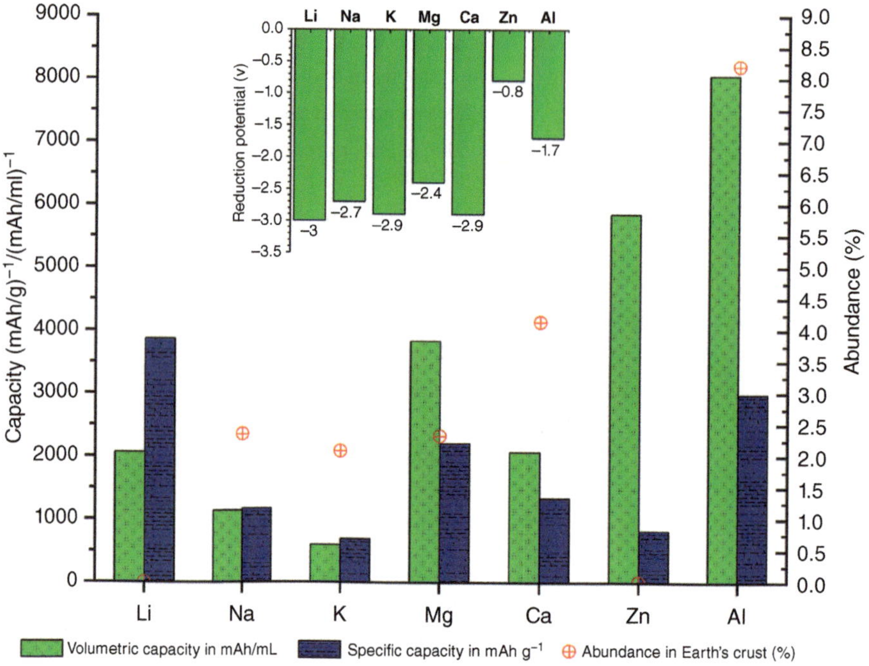

Figure 2.9 Volumetric and specific capacity of metals, and abundancy in the Earth's crust. *Source:* The volumetric and specific capacity of the metals are calculated based on their theoretical electrical properties, and the abundancy estimations are made based on [76]. The inset plot shows the reduction potential of the selected metals. *Source:* Adapted from Refs. [76, 77].

potential, it is not the most suitable option [86–88]. Ca possesses the lowest TRL as it has been recently recognized as a viable metal for battery applications. The advantages of Ca-ion batteries (CiBs) can be enumerated as a high potential to achieve a theoretical energy density of the metal, natural abundance in carbonate form, low ion mobility, and size [89]. From a technical perspective, CiBs are identified with optimal operation at elevated temperatures and poor reversibility which requires further research on the development of suitable cathodes and electrolytes for their electrochemical stability [90]. Some breakthrough advancements are reported regarding the development of electrolytes with high Coulombic efficiency and operation at room temperatures which are currently in the proof-of-concept stage [91].

Zn-based batteries are of interest due to their inherent safety and the fact that they are environmentally friendly. The development of aqueous Zn-based batteries mainly aims to achieve higher voltages and develop electrolytes that do not dissolve the electrode material. Proposed Zn-ion, dual-ion hybrid Zn, and Zn–air batteries have been investigated with various cathode materials (Mn-, Ni-, Co-, V-, air, PBA-based materials, etc.) [92]. In applications, alkaline rechargeable zinc batteries (ARZBs) are available on the market and with room for improvement [93]. Moreover, Ni–Zn and Zn–Co batteries are prospective Zn-based chemistries for stationary energy storage and mobility applications, respectively. Aqueous zinc-ion batteries (AZiBs), zinc-air batteries, and hybrid alkaline zinc batteries are candidates for mobility applications due to their high specific energy [93]. Advances in high performance, long cycling stability, and manufacturability are a necessity for Zn-based batteries to be commercially viable [94].

Al provides specific energies comparable to Li with higher volumetric energy densities than all the considered metals. It is highly abundant (more than 8% in the Earth's crust, see Figure 2.9) and produced in large quantities worldwide. Moreover, it has around 35% recycled material content, and globally varying 42–70% end-of-Life recycling input rate (EoL-RIR) (EoL-RIR, Li: <1%, Zn: 35–60%, Mg: 39%, Fe: 50–90%) which lowers the substantial demand on bauxite [95]. The use of Al as electrode material in batteries dates back to the 1850s, its application in Zn-based primary cells. Under aqueous and non-aqueous electrolyte classification as rechargeable batteries, the non-aqueous ionic liquids offer high reversible stripping and deposition process of Al therefore they have the potential to be implemented into numerous applications [77]. This important advancement was accomplished with the use of $AlCl_3$/1-ethyl-3-methylimidazolium chloride (EMIC) ionic liquid electrolyte [96]. Following this achievement, 98% of Coulombic efficiencies are reached with the use of three-dimensional graphitic-foam cathodes yielding ultrafast charging rates [97]. However, ongoing studies have not extended the desired theoretical specific energy ranges to ensure the future of Al-ion batteries.

Metal–air batteries can theoretically outperform non-aqueous batteries in high-power applications; metals such as Li, Na, Zn, Mg, and Al have been considered as suitable anode materials [98–105]. Due to the irreversibility of the electrochemical process, mechanically rechargeable metal-air batteries have been utilized with proven high energy densities achieved by Phinergy's Al–air battery [106]. The proposed operation requires the replacement of the unit and recovery of the Al from the formed Al oxides in the electrolyte. The use of mechanically rechargeable metal-air batteries is also a promising option for long-term stationary energy storage applications.

2.5 Economic and Environmental Implications of Reactive Metals

Material prices and the environmental sustainability of the produced storage technologies undoubtedly influence their feasibility and application potential in energy storage projects [107, 108]. Besides the supply of materials with low risk, their sustainable manufacturing, presence in the market with affordable prices, and recyclability in a circular economy are the main requisites for future energy carriers. The use of such technologies is unreachable without providing economically feasible business cases. Considered metal energy carriers display varying price ranges with Fe being the cheapest with an average price of 0.46 € kg^{-1}. However, considering the energy content B becomes the cheapest with 0.05 € kWh^{-1} followed by Si (0.2 € kWh^{-1}), Al (0.21 € kWh^{-1}), Fe (0.22 € kWh^{-1}), and Mg (0.33 € kWh^{-1}) based on the current commodity prices [109–111] (see Figure 2.10). The scope of this chapter is to highlight the energy storage aspects of the reactive metals, hence, the material cost on an energy content basis is more relevant. To provide a deeper understanding of the economics, the energy conversion efficiency is also an important factor.

Hence, the provided overview aims to indicate the theoretical economic potential of the metals. Correlating all the indicators, Al and Fe are ideal candidates for energy applications

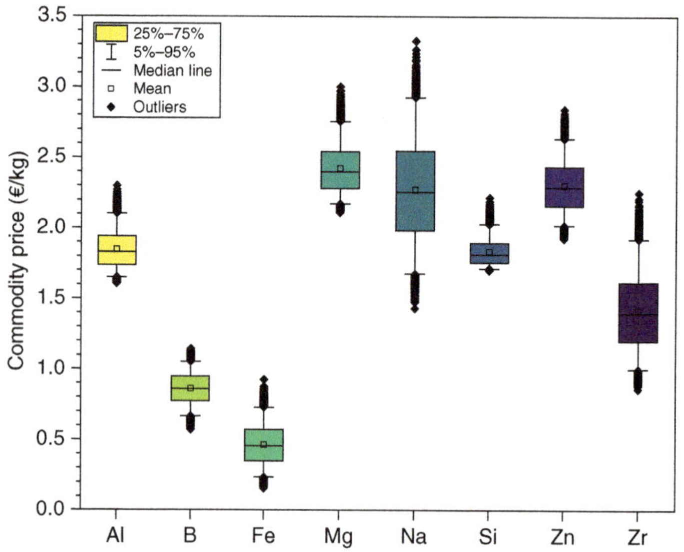

Figure 2.10 Commodity prices (2010–2020 timeframe) of the selected metals. *Source:* Adapted from Refs. [109–111].

for the integration of the energy generation sector into the circular metal economy. This is because their long-term availability is ensured and they are produced in large quantities at affordable prices worldwide. The remaining criterion in this case should be their environmental sustainability. From the resource efficiency perspective, Al and Fe deliver additional benefits, as the formed metal oxides can be recycled using well-established smelting processes. A comparison of different metals and intermediate products is provided in Figure 2.11. Here a simple approach is used by selecting "market for****" processes from the ecoinvent database V 3.7.1 using International Reference Life Cycle Data System (ILCD) midpoint indicators [112]. The functional unit for impact calculation is 1 kg of metal.

These processes represent the average consumption mix for a selected product within a certain geographical area. Note that impacts for example ZrO_2 are related to intermediate products not including further steps, for example, Van-Arkel–de-Boer-process for producing high-purity metals, such as pure Zr or boron [113]. On the contrary, for Li or Al these steps have been considered, but do not contain steps to e.g. shred and passivate Al for wet combustion. The results are only indicative and only partially comparable and are thus indicated separately. In addition, a recent study exhibits that the life-cycle assessment (LCA) of metals which is based on average inventory data from the mining industry does not show the potential broad bandwidth of life-cycle impacts caused by mining and refining [114]. Nevertheless, some first implications can be derived from this comparison. There are several (in total 12) impact categories (e.g. climate change, ozone depletion, photochemical ozone formation) covered within the ILCD impact method. Here only the global warming potential in kg CO_2-eq. is shown and discussed in the following [115]. This impact category summarizes greenhouse gases, such as CO_2, N_2O, CH_4, or hydrocarbons, that are emitted into the atmosphere mainly through the combustion of fossil fuels leading to an increase in global average temperatures. Detailed information about the impacts of different metals can be found in Harpprecht et al. (2021), Nuss and Eckelmann (2014), and Strezov et al. (2021) [116–118].

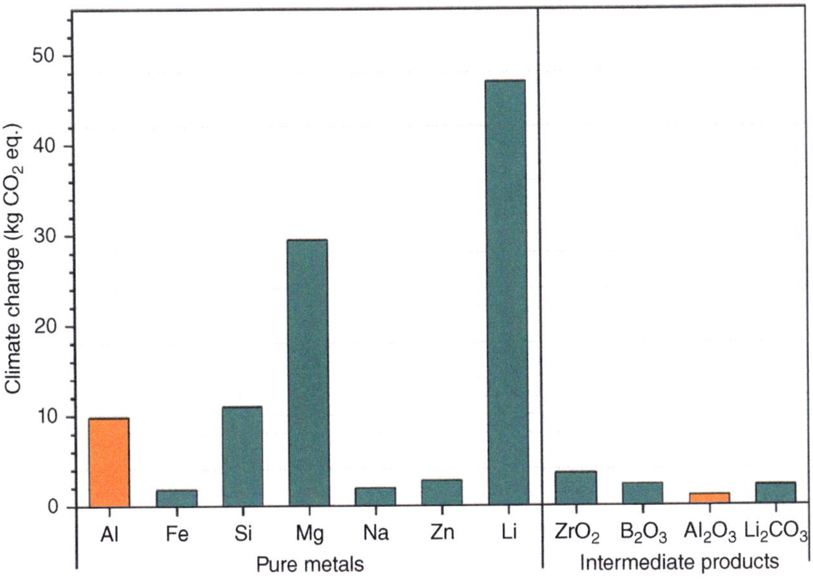

Figure 2.11 Some of the selected sustainability life-cycle sustainability indicators.

Also, Mg as an abundant reactive metal has a high impact on climate change compared to other prospective metal energy carriers. This is mainly due to the dominance of China in Mg production and the high penetration of coal-based electricity in the Chinese electricity mix [119]. Regarding the climate change impact of Al, Na, and Fe there is a high possibility to make these metals carbon-free or -neutral with the energy transition. Electrolytic reduction of the iron ore is currently a method on laboratory scale demonstration toward carbon-free Fe production. Nevertheless, the common practice iron-making process is a carbothermic reaction that yields direct CO_2 emissions for the reduction of the iron ore where the impact of carbon-free electricity is currently insignificant. Metallic sodium production is a very energy-intensive process, requiring full decarbonisation of the electricity source [120]. Additionally, storage of Na metal is not that easy due to its high reactivity with oxygen and water. Al is presently produced employing anodes made of carbon resulting in direct CO_2 emissions and the formation of perfluorocarbons (PFCs, mainly CF_4 [$GWP_{100,CF4} = 7380$] and C_2F_6 [$GWP_{100,C2F6} = 12400$] equivalent of CO_2 according to IPCC 6th assessment report) [121]. As a solution to avoid direct CO_2 emissions inert anode technology is an encouraging prospect that is anticipated to be implemented in the near future [72]. Most of the CO_2 emissions in the primary Al production result from electricity, hence, it has a high potential for being carbon-free or -neutral (as indicated in Figure 2.11 for Al_2O_3 and Al).

From the environmental perspective, Na has a low impact which can be taken as an indication of the environmental sustainability of future Na-ion batteries, but there are other challenges to be tackled (e.g. energy demand during cell manufacturing and the electrolyte salts) [81]. Impacts for lithium carbonate (Li_2CO_3) production for, e.g. LiB production is dependent on the production path (brine evaporation or mining and beneficiation of spodumene) [122]. The impact on metallic Li production, for example, Li-metal batteries is the potentially highest among the selected alternatives [123]. The magnitude of the impacts results from LiCl-electrolysis and the corresponding electricity mix. In conclusion, abundant reactive metals are prospective candidates for energy-based applications, but some of them are restricted by the current technological developments [8, 68]. Significant reduction of environmental impacts can be achieved through adequate recycling strategies and the use of renewables, which also allow the reuse of high-grade metals in the frame of a circular economy [124].

2.6 Conclusion and Outlook

The aim of this chapter is to highlight the potential of earth-abundant metals as an energy carrier in storage applications. Various factors (technical performance, economics, environmental sustainability, etc.) influence their ability to serve as feasible energy carriers. Although their use is currently limited due to the inexistence of suitable technologies, current advancements in thermal energy storage designs and electrochemical energy storage concepts mean they have scope for commercialization. This is mostly driven by the need for substituting fossil fuels coinciding with the demand for sustainable energy carriers as well as techno-economic and socio-economic constraints related to resources. Reactive metal energy carrier applications offer great potential for the transition to a circular economy. One important aspect is the high flexibility of energy storage in terms of energy forms and accelerated performance development of the energy conversion technologies (e.g. fuel cells, new combustion technologies, sustainable batteries) which aim to increase the overall conversion efficiency.

Metals with high TRL potential and prospectively improved environmental sustainability are identified; namely Al, Fe, and Na are the most suitable metals (in terms of cost, availability, energy density, and global warming potential) for use as an energy carrier. The future of electrochemical energy storage is anticipated to shift from Li to Na as an abundant and more sustainable substitute. On the other hand, Al and Fe can be identified as ideal energy carriers for the long-term energy storage demand, due to availability, large-scale industrial production, high thermal energy storage properties, and/or large H_2 storage capacities. Besides the low TRL thermal energy storage technologies and emerging batteries, direct combustion of Al and Fe is identified as a promising Power-to-X path based on the recently proposed high-performance concepts. Notably, Al is suitable for wet combustion because of its high heat and H_2 storage capacities, it is the ideal medium for providing renewable energy in various energy storage cases (stationary, seasonal, annual, etc.) and specified cases where heat, electricity, and H_2 are required simultaneously.

A combined approach with a focus on empirical studies on the use of abundant reactive metals should be promoted other than the current theoretical research approaches. An example is the ongoing work funded by the European Union at TU Eindhoven regarding the use of Fe as an energy carrier. Furthermore, the resulting demand for research is centered around techno-economic feasibility, environmental sustainability, and societal aspects. An asset to be included is the integration of the life cycle within the technology development time course to ensure sustainable concept advancements align with the circular economy approaches. Hence the eventual introduction of such energy carriers is essential for securing a sustainable global energy transition.

Aknowledgements

The authors acknowledge the support of the European Commission under the project STORIES (GAP-101036910). H.E., M.B., and S.P. acknowledge the funding from KIT Future Fields programme.

References

1 European Commission (2019). *Secretariat-General, Communication from the Commission to the European Parliament, the European Council, the Council, the European Economic and Social Committee and the Committee of the Regions The European Green Deal*. Brussels: European Commission https://eur-lex.europa.eu/legal-content/EN/TXT/?uri=CELEX:52019DC0640.

2 IEA (2020). *Innovation in Batteries and Electricity Storage | A Global Analysis Based on Patent Data*. Paris: IEA.

3 EnergyWatchGroup (EWG) (2018). *Energy Transition in Europe Across Power, Heat, Transport and Desalination Sectors – Key Findings*. Berlin: EnergyWatchGroup (EWG).

4 Weil, M., Peters, J., and Baumann, M. (2020). *The Material Basis of Energy Transitions*, 71–89. Elsevier.

5 Vaalma, C., Buchholz, D., Weil, M., and Passerini, S. (2018). *Nature Reviews Materials* 3: 18013.

6 Weil, M., Ziemann, S., and Peters, J. (2018). *Behaviour of Lithium-Ion Batteries in Electric Vehicles* (ed. G. Pistoia and B. Liaw), 59–74. Cham: Springer International Publishing.

7 European Union (2020). *Circular Economy Action Plan - For a Cleaner and More Competitive Europe*. Brussels: European Commission.

8 Julien, P. and Bergthorson, J.M. (2017). *Sustainable Energy & Fuels* 1: 615.

9 Bergthorson, J.M. (2018). *Progress in Energy and Combustion Science* 68: 169.

10 Kalaiselvam, S. and Parameshwaran, R. (2014). *Thermal Energy Storage Technologies for Sustainability: Systems Design*. Amsterdam: Assessment and Applications, Elsevier/AP.

11 Sterner, M. and Stadler, I. (ed.) (2019). *Handbook of Energy Storage: Demand, Technologies, Integration*. Berlin, Heidelberg: Springer.

12 Huggins, R. (2016). *Energy Storage: Fundamentals, Materials and Applications*. Cham: Springer International Publishing : Imprint: Springer.

13 European Recycling Industries Confederation (2021). *EuRIC Circular Metals Strategy*. EURIC.

14 Bobba, S., Carrara, S., Huisman, J. et al. (2020). *A Foresight Study*. Luxembourg: European Union.

15 EERA (2021). *White Paper on the Clean Energy Transition*. Brussels: European Energy Research Alliance.

16 IEA (2021). *The Role of Critical Minerals in Clean Energy Transitions*. Paris: IEA.

17 Alnasir, Z. and Kazerani, M. (2014). *2014 IEEE 27th Canadian Conference on Electrical and Computer Engineering (CCECE)*, 1–8. Toronto, ON, Canada: IEEE.

18 Barthelmie, R.J. and Pryor, S.C. (2021). *Climate* 9: 136.

19 European Recycling Industries Confederation (2020). *Metal Recycling Factsheet*. Brussels: EURIC.

20 Moreau, V., Dos Reis, P.C., and Vuille, F. (2019). *Resources* 8: 29.

21 Shkolnikov, E.I., Zhuk, A.Z., and Vlaskin, M.S. (2011). *Renewable and Sustainable Energy Reviews* 15: 4611.
22 Barelli, L., Baumann, M., Bidini, G. et al. (2020). *Energy Technology* 8: 2000233.
23 European Commission (2020). *Critical Raw Materials Resilience*. European Commission.
24 O'Neal, S.L. and Zheng, W. (2015). *Current Environmental Health Reports* 2: 315.
25 Strupp, C. (2010). *The Annals of Occupational Hygiene* https://doi.org/10.1093/annhyg/meq073.
26 Sun, H., Brocato, J., and Costa, M. (2015). *Current Environmental Health Reports* 2: 295.
27 Leyssens, L., Vinck, B., Van Der Straeten, C. et al. (2017). *Toxicology* 387: 43.
28 Bergthorson, J.M., Goroshin, S., Soo, M.J. et al. (2015). *Applied Energy* 160: 368.
29 Weil, M. and Tübke, J. (2015). *TATuP* 24: 4.
30 Dincer, I. and Rosen, M.A. (2011). *Thermal Energy Storage: Systems and Applications*. Chichester: Wiley.
31 Baumann, M., Weil, M., Peters, J.F. et al. (2019). *Renewable and Sustainable Energy Reviews* 107: 516.
32 Ram, M., Bogdanov, D., Aghahosseini, A. et al. (2018). *Global Energy System Based on 100% Renewable Energy – Energy Transition in Europe Across Power, Heat, Transport and Desalination Sectors. Study by LUT University and Energy Watch Group*. Berlin: Lappeenranta.
33 Cattaneo, B. (2020). Growth of solar energy is not in line with the needs for ambitious energy transition, can be found under https://ec.europa.eu/jrc/en/news/growth-solar-energy-not-line-needs-ambitious-energy-transition.
34 Energy Storage - Technology Options (n.d.). can be found under http://www.esru.strath.ac.uk/EandE/Web_sites/14-15/Industrial_autonomy/2_1_technology_options.html (accessed 26 November 2021).
35 Muehlenpfordt, J. (2020). Time series. Package version 06 October 2020. https://doi.org/10.25832/TIME_SERIES/2020-10-06.
36 Potrč, S., Andreja, N., Čuček, L., and Zdravko, K. (2021). *Chemical Engineering Transactions* 88: 1153.
37 Wyman, C., Castle, J., and Kreith, F. (1980). *Solar Energy* 24: 517.
38 Siegel, N.P. (2012). *WENE* 1: 119.
39 Bauer, T., Pfleger, N., Laing, D. et al. (2013). *Molten Salts Chemistry*, 415–438. Elsevier.
40 Pushpakaran, B.N., Subburaj, A.S., Bayne, S.B., and Mookken, J. (2016). *Renewable and Sustainable Energy Reviews* 55: 971.
41 Ashuri, M., He, Q., and Shaw, L.L. (2016). *Nanoscale* 8: 74.
42 Clifford, K.H. and Ambrosini, A. (2020). *2020 U.S. DOE Energy Storage Handbook*. U.S. Department of Energy.
43 Niaz, N.A., Ahmad, I., Khalid, N.R. et al. (2013). *Journal of Nanomaterials* 2013: 1.
44 Dubrovinskaia, N., Dubrovinsky, L., Kantor, I. et al. (2005). *Physical Review Letters* 95: 245502.
45 Bauer, T., Steinmann, W.-D., Laing, D., and Tamme, R. (2012). *Annual Review of Heat Transfer* 15: 131.
46 Laube, T., Marocco, L., Niedermeier, K. et al. (2020). *Energy Technology* 8: 1900908.
47 Niedermeier, K., Marocco, L., Flesch, J. et al. (2018). *Applied Thermal Engineering* 141: 368.

48 Niedermeier, K., Flesch, J., Marocco, L., and Wetzel, T. (2016). *Applied Thermal Engineering* 107: 386.

49 Haller, M.Y., Carbonell, D., Dudita, M. et al. (2020). *Energy Conversion and Management: X* 5: 100017.

50 Yetter, R.A., Risha, G.A., and Son, S.F. (2009). *Proceedings of the Combustion Institute* 32: 1819.

51 Markstein, G.H. (1963). *AIAA Journal* 1: 550.

52 Ross, H.D. (2014). *Microgravity Combustion: Fire in Free Fall*. London: Elsevier Science.

53 Corcoran, A., Mercati, S., Nie, H. et al. (2013). *Combustion and Flame* 160: 2242.

54 Schiemann, M., Bergthorson, J.M., Scherer, V. et al. (2018). *Clearwater Clean Energy Conference 2018*. Clearwater, FL: Coal Technologies Associates.

55 Dirven, L., Deen, N.G., and Golombok, M. (2018). *Sustainable Energy Technologies and Assessments* 30: 52.

56 Vlaskin, M., Shkolnikov, E., Lisicyn, A. et al. (2010). *International Journal of Hydrogen Energy* 35: 1888.

57 Nie, H., Zhang, S., Schoenitz, M., and Dreizin, E.L. (2013). *International Journal of Hydrogen Energy* 38: 11222.

58 Auner, N. and Holl, S. (2006). *Energy* 31: 1395.

59 Yavor, Y., Goroshin, S., Bergthorson, J.M., and Frost, D.L. (2015). *International Journal of Hydrogen Energy* 40: 1026.

60 Glassman, I., Yetter, R.A., and Glumac, N. (2015). *Combustion*. Waltham, MA: Academic Press.

61 Dreizin, E.L. (2003). *Combustion, Explosion, and Shock Waves* 39: 681.

62 Dreizin, E.L. (2000). *Progress in Energy and Combustion Science* 26: 57.

63 Yang, S., Jang, Y.-H., Kim, C.H. et al. (2010). *Powder Technology* 197: 170.

64 Satyapal, S. (2020). U.S. Department of Energy Hydrogen and Fuel Cells Program: 2019 Annual Merit Review and Peer Evaluation Report; April 29 - May 1, 2019, Arlington, Virginia, United States.

65 U.S. Geological Survey (2021). Mineral commodity summaries 2021. U.S. Geological Survey, 200 p, https://doi.org/10.3133/mcs2021.

66 World Aluminium (2021). Primary Aluminium Production, can be found under https://www.world-aluminium.org/statistics/.

67 Balomenos, E., Panias, D., and Paspaliaris, I. (2011). *Mineral Processing and Extractive Metallurgy Review* 32: 69.

68 Trowell, K.A., Goroshin, S., Frost, D.L., and Bergthorson, J.M. (2020). *Applied Energy* 275: 115112.

69 Franzoni, F., Milani, M., Montorsi, L., and Golovitchev, V. (2010). *International Journal of Hydrogen Energy* 35: 1548.

70 Petrovic, J. and Thomas, G. (2011). *Reaction of Aluminum with Water to Produce Hydrogen-2010 Update*. Washington DC.: Office Of Energy Efficiency And Renewable Energy (EERE).

71 Haller, M.Y., Amstad, D., Dudita, M. et al. (2021). *Renewable Energy* 174: 879.

72 He, Y., Zhou, K., Zhang, Y. et al. (2021). *Journal of Materials Chemistry A* 9: 25272.

73 Coursol, P., Dufour, G., Coté, J. et al. (2012). *JOM* 64: 1326.

74 Kosonen, A., Koponen, J., Ahola, J., and Peltoniemi, P. (2015). *2015 17th European Conference on Power Electronics and Applications (EPE'15 ECCE-Europe)*, 1–10. Geneva: IEEE.

75 Fichtner, M. (2022). *Batteries & Supercaps* 5 (2): 1–9.

76 Rumble, J.R., Lide, D.R., and Bruno, T.J. (ed.) (2018). *CRC Handbook of Chemistry and Physics: A Ready-Reference Book of Chemical and Physical Data*. Boca Raton London New York: CRC Press, Taylor & Francis Group.

77 Elia, G.A., Marquardt, K., Hoeppner, K. et al. (2016). *Advanced Materials* 28: 7564.

78 Chen, L., Fiore, M., Wang, J.E. et al. (2018). *Advanced Sustainable Systems* 2: 1700153.

79 Chayambuka, K., Mulder, G., Danilov, D.L., and Notten, P.H.L. (2020). *Advanced Energy Materials* 10: 2001310.

80 Hu, Y.-S. and Li, Y. (2021). *ACS Energy Letters* 6: 4115.

81 Peters, J.F., Baumann, M., Binder, J.R., and Weil, M. (2021). *Sustainable Energy & Fuels* https://doi.org/10.1039.D1SE01292D.

82 Peters, J., Buchholz, D., Passerini, S., and Weil, M. (2016). *Energy & Environmental Science* 9: 1744.

83 Xu, Y.-S., Duan, S.-Y., Sun, Y.-G. et al. (2019). *Journal of Materials Chemistry A* 7: 4334.

84 Ahmed, S.M., Suo, G., Wang, W.A. et al. (2021). *Journal of Energy Chemistry* 62: 307.

85 Liu, F., Wang, T., Liu, X., and Fan, L. (2021). *Advanced Energy Materials* 11: 2000787.

86 Montenegro, C.T., Peters, J.F., Baumann, M. et al. (2021). *Journal of Energy Storage* 35: 102053.

87 Bautista, S.P., Weil, M., Baumann, M., and Montenegro, C.T. (2021). *Energy Technology* 9: 2000964.

88 Montenegro, C.T., Peters, J.F., Zhao-Karger, Z. et al. (2019). *Energy and Environment Series* (ed. M. Fichtner), 309–330. Cambridge: Royal Society Of Chemistry.

89 Gummow, R.J., Vamvounis, G., Kannan, M.B., and He, Y. (2018). *Advanced Materials* 30: 1801702.

90 Nielson, K.V. and Liu, T.L. (2020). *Angewandte Chemie, International Edition* 59: 3368.

91 Stievano, L., de Meatza, I., Bitenc, J. et al. (2021). *Journal of Power Sources* 482: 228875.

92 Mallick, S. and Raj, C.R. (2021). *ChemSusChem* 14: 1987.

93 Shi, Y., Chen, Y., Shi, L. et al. (2020). *Small* 16: 2000730.

94 Hosseini, S., Masoudi Soltani, S., and Li, Y.-Y. (2021). *Chemical Engineering Journal* 408: 127241.

95 Graedel, T.E., Allwood, J., Birat, J.-P. (et al.) (2011)., United Nations Environment Programme, Working Group on the Global Metal Flows, Recycling Rates of Metals: A Status Report.

96 Jayaprakash, N., Das, S.K., and Archer, L.A. (2011). *Chemical Communications* 47: 12610.

97 Lin, M.-C., Gong, M., Lu, B. et al. (2015). *Nature* 520: 324.

98 Elia, G.A., Hassoun, J., Kwak, W.-J. et al. (2014). *Nano Letters* 14: 6572.

99 Grande, L., Paillard, E., Hassoun, J. et al. (2015). *Advanced Materials* 27: 784.

100 Geng, D., Ding, N., Hor, T.S.A. et al. (2016). *Advanced Energy Materials* 6: 1502164.

101 Hartmann, P., Bender, C.L., Vračar, M. et al. (2013). *Nature Materials* 12: 228.

102 Elia, G.A., Hasa, I., and Hassoun, J. (2016). *Electrochimica Acta* 191: 516.

103 Drillet, J.-F., Adam, M., Barg, S. et al. (2019). *ECS Transactions* 28: 13.

104 Neburchilov, V., Wang, H., Martin, J.J., and Qu, W. (2010). *Journal of Power Sources* 195: 1271.

105 Li, W., Li, C., Zhou, C. et al. (2006). *Angewandte Chemie, International Edition* 45: 6009.

106 Aluminiumleader (2015). Aluminium-Air: The Future of Electric Cars, can be found under https://aluminiumleader.com/focus/aluminium_air_the_future_of_electric_cars/.

107 Parra, D., Swierczynski, M., Stroe, D.I. et al. (2017). *Renewable and Sustainable Energy Reviews* 79: 730.

108 Tan, K.M., Babu, T.S., Ramachandaramurthy, V.K. et al. (2021). *Journal of Energy Storage* 39: 102591.

109 Durchschnittspreise mineralischer Rohstoffe bis 2019 (2019), can be found under https://de.statista.com/statistik/daten/studie/260427/umfrage/durchschnittspreise-ausgewaehlter-mineralischer-rohstoffe/ (accessed 26 November 2022).

110 Boron | Price | per kg | USD (2019) can be found under https://www.pharmacompass.com/price/boron (accessed 26 November 2022).

111 Preise für strategische Metalle im Juni 2020 | Institut für Seltene Erden und Metalle (2020) can be found under https://institut-seltene-erden.de/preise-fuer-strategische-metalle-im-juni-2020/ (accessed 26 November 2022).

112 European Commission (2011). *Joint Research Centre, Institute for Environment and Sustainability, International Reference Life Cycle Data System (ILCD) Handbook General Guide for Life Cycle Assessment: Provisions and Action Steps*. Luxembourg: Publications Office.

113 van Arkel, A.E. and de Boer, J.H. (1925). *Zeitschrift für Anorganische und Allgemeine Chemie* 148: 345.

114 Pell, R., Tijsseling, L., Goodenough, K. et al. (2021). *Nature Rev Earth Environment* 2: 665.

115 Hauschild, M., Goedkoop, M., Guinee, J., et al. (2011). Recommendations for Life Cycle Impact Assessment in the European context - based on existing environmental impact assessment models and factors (International Reference Life Cycle Data System - ILCD handbook). EUR 24571 EN. Luxembourg (Luxembourg): Publications Office of the European Union; 2011. JRC61049.

116 Harpprecht, C., Oers, L., Northey, S.A. et al. (2021). *Journal of Industrial Ecology* https://doi.org/10.1111/jiec.13181.

117 Nuss, P. and Eckelman, M.J. (2014). *PLoS One* 9: e101298.

118 Strezov, V., Zhou, X., and Evans, T.J. (2021). *Scientific Reports* 11: 10116.

119 Ehrenberger, S., Dieringa, H., Friedrich, H.E. (2013). Life cycle assessment of magnesium components in vehicle construction. Project Report. Deutsches Zentrum für Luft- und Raumfahrt. 109 S. https://elib.dlr.de/87332/

120 Stern, A.G. (2017). *Recent Improvements of Power Plants Management and Technology* (ed. A.B. Nikolic and Z.S. Janda). InTech.

121 Forster,P., Storelvmo, T., Armour, K., et al. (2021). The Earth's Energy Budget, Climate Feedbacks, and Climate Sensitivity. In Climate Change 2021: The Physical Science Basis. Contribution of Working Group I to the Sixth Assessment Report of the Intergovernmental Panel on Climate Change, IPCC.

122 Manjong, N.B., Usai, L., Burheim, O.S., and Strømman, A.H. (2021). *Batteries* 7: 57.

123 Zhang, X., Han, A., and Yang, Y. (2020). *Journal of Materials Chemistry A* 8: 22455.

124 Ziemann, S., Müller, D.B., Schebek, L., and Weil, M. (2018). *Resources, Conservation and Recycling* 133: 76.

Part II

Sustainable Materials for Batteries and Supercapacitors

3

Lithium-Ion Batteries: Electrodes, Separators, and Solid Polymer Electrolytes

Manuel Salado[1], Renato Gonçalves[2], Carlos Miguel Costa[3,4,5], and Senentxu Lanceros-Méndez[1,6]

[1] *BCMaterials, Basque Center for Materials, Applications and Nanostructures, UPV/EHU Science Park, 48940, Leioa, Spain*
[2] *Center of Chemistry, University of Minho, 4710-057, Braga, Portugal*
[3] *Physics Centre of Minho and Porto Universities (CF-UM-UP), University of Minho, Campus de Gualtar, Braga, 4710-057, Portugal*
[4] *Laboratory of Physics for Materials and Emergent Technologies, LapMET, University of Minho, Campus de Gualtar, Braga, 4710-057, Portugal*
[5] *Institute of Science and Innovation for Bio-Sustainability (IB-S), University of Minho, 4710-057, Braga, Portugal*
[6] *IKERBASQUE, Basque Foundation for Science, 48009, Bilbao, Spain*

3.1 Introduction

The internet of things (IoT) concept represents a revolution not only in the industrial sector but also in the daily lifestyle, with rechargeable batteries representing one of the cornerstones and a limiting factor in its widespread implementation. According to the current battery requirements (e.g. energy and power density, technology maturity), lithium-ion batteries (LIBs) are the most suitable commercial technology [1]. Generally, the different parts of a LIB include an anode, a cathode, current collectors, and a separator. The anode consists of a carbonaceous material to incorporate the lithium ions during the oxidation, such as graphite, and a polymer binder (e.g. polyvinylidene fluoride, PVDF) [2]. Whereas the cathode is made of lithium transition metal oxides, such as $LiCoO_2$ (LCO), $LiMn_2O_4$ (LMO), $LiFePO_4$ (LFP), and $Li[Ni_xMn_yCo_z]O_2$ (NMC), a carbon conductive agent, and a polymer binder [3]. Further, both electrodes (anode and cathode) are coated on Cu and Al current collectors, respectively [4]. At a glance, there are some key elements necessary for LIBs, including lithium, nickel, cobalt, manganese, and graphitic carbon. More specifically, Figure 3.1 depicts a general scheme of a commercial LIB, specifying the amount of the different components needed to fabricate a LIB [5]. It is worth mentioning that the cathode electrode entails one third of the total material amount of the battery.

However, this technology still presents some limitations in the scope of the ongoing expansion of electric vehicles (EVs) (Figure 3.2a), such as cost, safety issues (e.g. thermal runaway due to electrolyte degradation/evaporation), or decrease performance over time. Among them, aging is one of the most critical issues, as batteries are degraded, whether

Sustainable Energy Storage in the Scope of Circular Economy: Advanced Materials and Device Design, First Edition. Edited by Carlos Miguel Costa, Renato Gonçalves, and Senentxu Lanceros-Méndez.

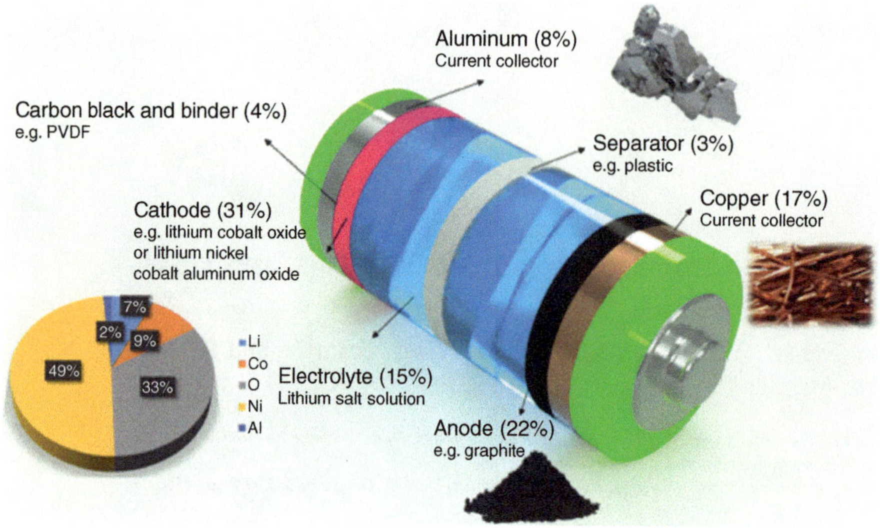

Figure 3.1 Amount (wt%) of raw materials required for the fabrication of a LIB.

they are in use or not [10]. In order to overcome these limitations, different strategies (e.g. solid-state electrolytes, metal–sulfur or metal–air batteries, among others) are being studied [11]. Additionally, independently from the future strategy selected, it will be a strong increment in the consumption of some scarce materials (e.g. cobalt, manganese, nickel, etc.) used mainly in the fabrication of the cathode in the actual battery design [12].

As it can be observed in Figure 3.2b, the key materials used in the fabrication of current cathodes for LIBs, are concentrated in certain countries, which represent an increasing concern because of the possibility of supply disruptions caused by government policies or socio-political instabilities. Specifically, the main bottleneck related to the large production of the batteries entails the presence of cobalt, which increases considerably the cost of the cathode material. One of the main reasons is that the expensive element is mined primarily in the Democratic Republic of the Congo (Figure 3.2c), giving rise to human rights abuses (e.g. artisanal miners and children dedicated to scavenge for rocks containing cobalt in the discarded by-products). Additionally, excessive and uncontrolled mining together with an absence of proper battery disposal and recycling systems contribute to water and soil pollution (Figure 3.2d) with the accumulation of heavy metals and the potential risks to human health [13–15]. Bearing this in mind and considering the potential impacts (e.g. emissions from and to the environment) of large-scale battery production, use, and disposal, recycling is a needed approach to be implemented in the scope of life-cycle assessment (LCA). Recently, Porzio et al. [16] presented a complete study focusing on the common practices in LIB LCAs. The authors analyzed them from the raw material extraction to battery end-of-life (EoL) and recycling, including battery manufacturing as well. As a conclusion, it highlighted the importance of a consensus in the parameters (e.g. critical raw material availability, geographic distribution of near- and long-term sources, and estimated improvement in extraction/processing methods) to be evaluated in order to compare different energy storage technologies. Typically, the quantification of the environmental

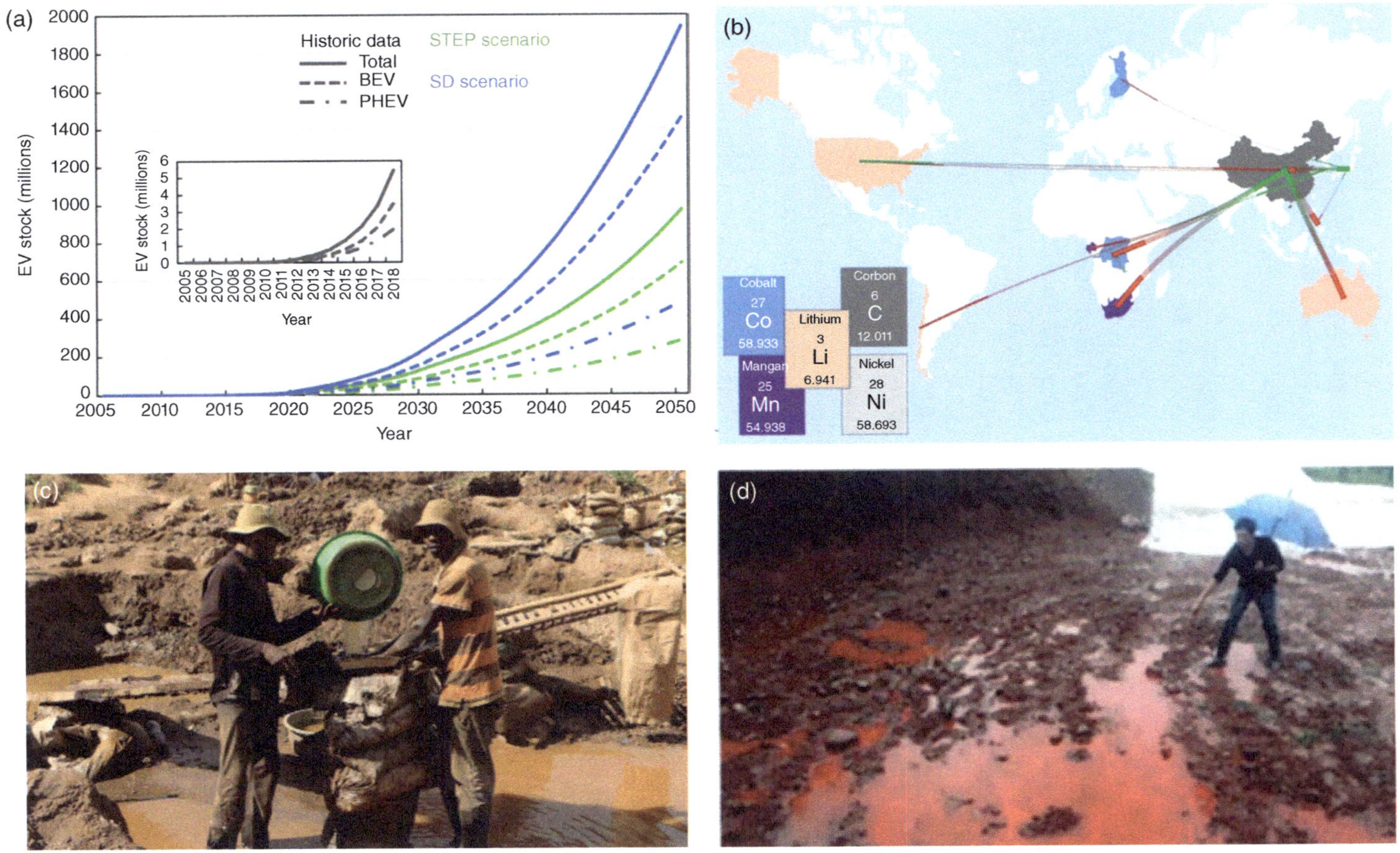

Figure 3.2 (a) Global EV stock development projected until 2050 depending on Stated Policies (STEP) or Sustainable Development (SD) scenarios. The inset represents the evolution of the EV stock of battery electric vehicles (BEV) and plug-in hybrid electric vehicles in the period (2005–2018). *Source:* Ref. [6]/Springer Nature/CCBY-4.0. (b) Geographic distribution of the main elements used in the cathode fabrication. *Source:* Reproduced with permission from [7]. Copyright © 2017 Elsevier Inc. (c) Diggers washing in a ore located at Kambove district of DR Congo. *Source:* Panella et al. [8]/With permission of John Wiley & Sons. (d) Environmental effects of water and soil pollution derived from mining and battery fabrication processes. *Source:* Negi et al. [9]/With permission of Springer Nature.

impacts is achieved considering different parameters such as global warming (GW), fine particulate matter formation (FPMF), terrestrial acidification (TA), mineral and fossil resource scarcity, and human toxicity (HT). Among them, GW and HT are the most sensible in the scope of public concern on well-being and climate change [17]. Accordingly to several reports [18–20], significant efforts have to be addressed toward policy-making independently of the geographical areas to guarantee that the sourcing of raw materials is achieved in an ecological and responsible manner.

Recycling and reusing are necessary strategies to strengthen the sustainability of battery supply chains and reduce primary resource requirements [21]. For that, integrate circular economy strategies into international policy will help to meet the climate and Sustainable Development Goals (specifically SDG11 "Sustainable Cities and Communities" and SDG13 "Climate change") [22]. For example, the company "Redwood Materials" can recover approximately 95% of the raw materials used to fabricate LIBs. These valuable materials cannot just go directly back into the supply chain to make batteries for new EVs but also reduce the need to extract them from the environment [23].

As it can be observed in the insert chart of Figure 3.1, the cathode contains the majority of the critical materials in a specific ratio (e.g. $LiNi_{0.8}Co_{0.15}Al_{0.05}O_2$). Consequently, most of the developed technologies for LIBs recycling are focused on cathode materials. Nevertheless, there exist several methods comprising the recycling of all the components of the battery (e.g. direct recycling) [24–33]. Generally, the estimation of the quantity of material contained in a full battery is complicated and uncertain, as manufacturers do not share their specific designs and formulations. Therefore, a preliminary classification of the different components is required. Besides, safety issues concerning the composition of the cathode and the electrolyte need to be taken into account during processing. Table 3.1 summarizes the different methods of recovering metal components from the battery. Initially, a pre-treatment (usually a mechanical process) is used to separate the different components, which can also reduce the energy consumption in the subsequent recycling processes (e.g. in the hydrometallurgical, direct, or biological treatments). Although several benefits of recycling are identified, such as the conservation of non-renewable resources, prevention of pollution, or reduction of the solid wasted that ends up in landfills, a careful selection of the recycling method needs to be addressed, in particular, to avoid the emission of hazardous by-products potentially harmful for living beings as well as polluting the environment. In addition, current methods require a huge energy investment and do not guarantee good quality products [46].

One of the main concern of the different methods is the presence of impurities introduced during the recycling processes that could lessen the performance of recycled materials due to the complex composition (especially of the cathode). For instance, Beak et al. [47] investigated the precipitation behavior of impurities present in the purified leachate used to regenerate cathode material. In particular, the authors concluded that the presence of non-metallic elements (such as F or S) has a detrimental influence on the structural properties and consequently the reversible electrochemical behavior. Nonetheless, a recent study [48] proved that recycled cathode materials (in particular $LiNi_{1/3}Mn_{1/3}Co_{1/3}O_2$ (NMC111)) behave better than the control sample, with lifetimes of 11 600 cycles (at 1 Ah). The authors suggested that the improvement in the performance was attributed to an optimized microstructure with a larger surface area that facilitates Li diffusion and reduces

Table 3.1 Advantages and drawbacks of the different methods of recovering metal components from the battery.

Methods	Characteristics	Advantages	Drawbacks	References
Pre-treatments	The disassembly of the cell is usually conducted manually with knives and saws, in order to extract cathode and anode sheets. Then, crushing (e.g. two-blade rotor crusher or hammer crushing) and grinding	—	—	[34, 35]
Pyrometallurgical process	After the initial mechanical process, this process consists of two consecutive heat treatment steps; first, electrolytes and organic solvent are removed from the batteries at low temperature (100–550 °C), and then the temperature is increased above 1500 °C to produce Co, Ni, and Cu alloys. Lithium, however, is retained in the slag (Li_2O or Li_2CO_3) and has to be extracted and recovered by other processes.	• Can be processed along with other types of waste (e.g. integrated smelters and refineries (ISR)). • Short process time.	• Not suitable for cathodes with cobalt-rich chemistries. • High energy consumption. • Production of hazardous by-products (e.g. HF from the PVDF binder).	[36, 37]
Hydrometallurgical process	Usually, chemical methods are used to achieve material recovery by leaching in acid (inorganic acid (H_2SO_4, HNO_3, HCl, H_3PO_4, HF), organic acid [citric acid ($C_6H_8O_7$), oxalic acid ($H_2C_2O_4$), D,L-malic acid ($C_4H_5O_6$), L-tartaric acid ($C_4H_6O_6$)] or mixed acid), turning metal elements into ions. Consecutively, these ions are separated by solvent extraction, ion exchange, and chemical precipitation.	• Low waste gas emission. • Low energy consumption due to the low-temperature requirement. • Faster kinetics, flexible, and selective.	• Not suitable for cathodes with cobalt-rich chemistries. • Volume of solvents required. • Use of the highly toxic N-methyl-2-pyrrolidone (NMP) as a solvent for dissolution. • Long process time.	[38, 39]

(Continued)

Table 3.1 (Continued)

Methods	Characteristics	Advantages	Drawbacks	References
Direct recycling	A difference from the previous methods, this process does not use chemical methods. Once the cells are dismantled, they are treated with supercritical CO_2, the temperature and pressure are then lowered and the electrolyte can be regenerated.	• Cathode materials can be reincorporated into a new cathode electrode with minimal changes to the crystal morphology. • Reduction of the time-consuming and expensive purification steps. • All battery components can be recovered and re-used after further processing.	• Different processes are needed depending on the cathode formulations. • The state of health of the battery has a strong influence on the efficiency of the process.	[40–42]
Biological recycling	As a novel strategy, the use of microorganisms to digest metal oxides from the cathode selectively and to reduce these oxides to produce metal nanoparticles has attracted recent attention.	• Eco-friendly and economical method. • Higher recovery yield. • Promising technology.	• Difficulty in microorganism reproduction/culture. • Slow kinetics and selective recovery. • Bacteria toxicity	[43–45]

mechanical stress. Another strategy toward enhancing the sustainability of Li^+-ion batteries entails the re-design of the current materials and components. Cobalt-free cathodes (such as $LiFePO_4$) would reduce production costs and energy consumption involved in their extraction, refining, and processing. In addition, alkali-metal elements (e.g. Na and K) or bivalent metals (Ca or Mg)-based anode batteries, present several advantages in terms of safety, cost as well as worldwide abundance. For instance, one of the most promising technology consists of Li–sulfur batteries due to their "almost limitless" abundance and the lack of derived health risks. Besides, with an outstanding theoretical specific capacity of $1675\,mAh\,g^{-1}$ ($11 \times LiCoO_2$ [$\sim 140\,mAh\,g^{-1}$] cathode) and the recent progress in the stabilization of irreversible relocation of polysulfides (polysulfide shuttle) with the addition of saccharide-based binder [49], would pave the way its development as an efficient energy storage technology. Electrolytes, usually LIBs are carbonate-based organic solvents that also entail safety concerns even when they can be synthesized from renewable sources. Consequently, different strategies of solid-state electrolytes (solid-polymer, gel-like, inorganic, or composite polymer electrolyte) have been widely investigated in order to improve battery performance and to overcome the current problems of electrolyte degradation.

In this chapter, a general overview of the components and applications of current LIBs is presented (Section 3.2), followed by a compilation of the sustainable materials available for the different components of the battery (electrodes, separator, and novel strategies such as solid polymer electrolytes) (Section 3.3). Finally, Section 3.4 gathers the main conclusions as well as an outlook on the next generation of more environmentally friendly energy storage systems.

3.2 Lithium-Ion Batteries

LIBs dominate the energy storage market due to their excellent characteristics for portable electronic devices. Nevertheless, their application to EVs entails a strong investment to design, efficient and safe batteries. In this section, a detailed description of the current components of LIBs is provided, together with the main drawbacks that they present when upscaling is envisaged.

3.2.1 Electrodes

Figure 3.1 shows that cathode and anode correspond to 50% of the total weight of the cell. Consequently, analyzing and redesigning the chemistry of these components is the first approach in order to reduce the amount of critical materials used in current systems. Intercalation cathode materials (Figure 3.3a) are divided into three categories depending on the crystal structure and therefore the insertion pathways of lithium ions. Layered oxides (e.g. LCO), especially $LiNi_{1-x-y}Mn_xCo_zO_2$ (NMC), exhibit suitable properties, such as cell nominal voltage, high capacity, and rate capability, that cannot be outperformed by other intercalation material [56]. However, as it was previously discussed, the presence of cobalt in the composition entails several economic and socio-economic concerns. For that reason, the main companies (e.g. Tesla, Toyota, and BMW) from the automotive industry sector are working with their battery suppliers to reduce the cobalt content in batteries and

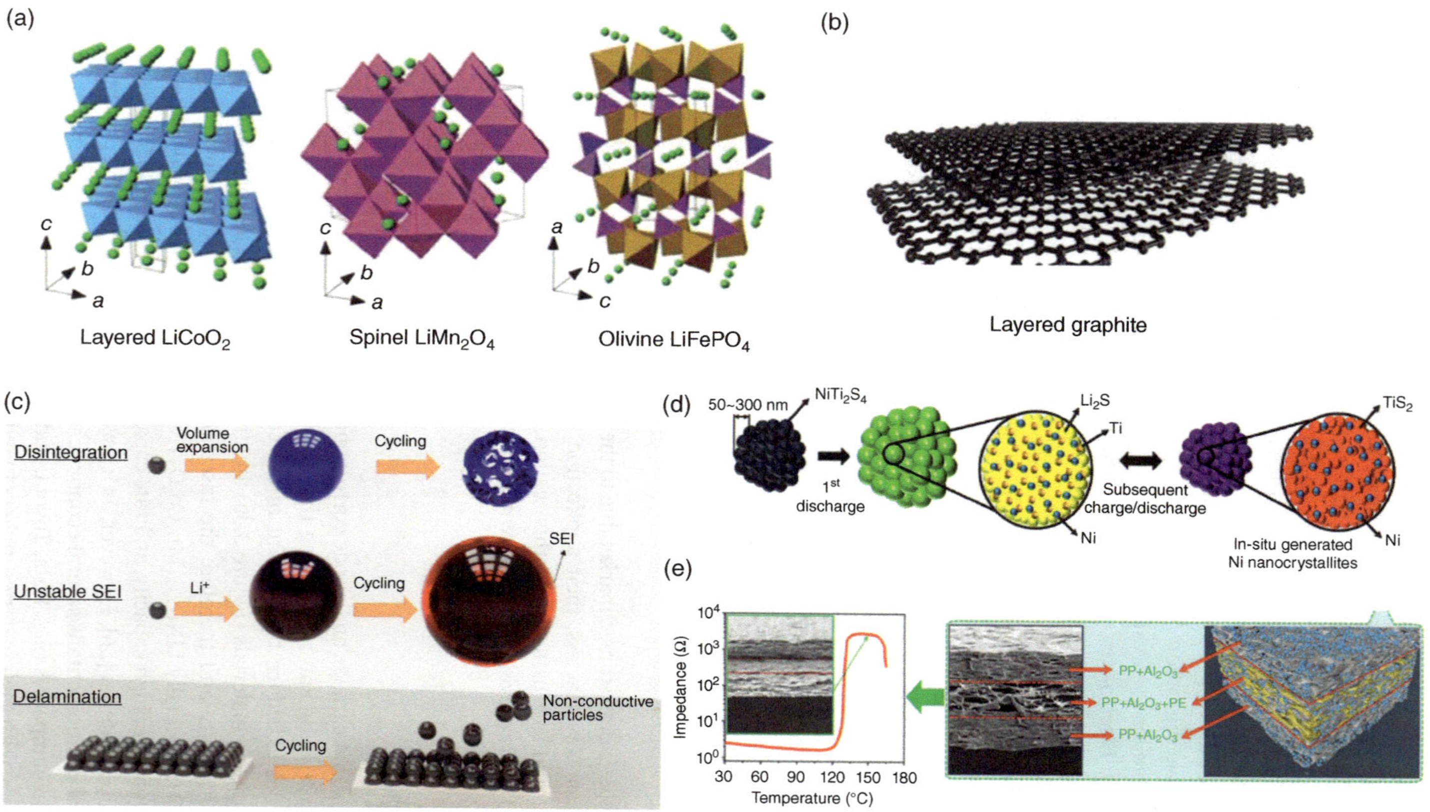

Figure 3.3 (a) Crystal structures of different metal-oxides cathode materials. *Source:* Adapted from Refs. [50, 51]. (b) The two-dimensional structure of graphite material for stable anode material. *Source:* Figure adapted from [52]. (c) Different anode degradation processes of alloy-type material. *Source:* Sun et al. [53]/With permission of Elsevier. (d) Schematic illustration for discharged and charged states of Ni–Ti_2S_4. *Source:* Ding et al. [54], Reproduced with permission from American Chemical Society. (e) (Left) Impedance variations as a function of temperature for SS||SS symmetric cells and (right) cross-section SEM images and pore structure diagnosis of trilayer (polypropylene [PE]–Al_2O_3–polyethylene [PP]) separators. *Source:* Ding et al. [55]/With permission of American Chemical Society.

even replace it entirely with a nickel alternative [57]. Unfortunately, cobalt is an essential element in order to thermally and chemically stabilize the structure formed along with nickel and manganese. It is already proved that the lack (or deficiency) of cobalt in the chemical composition, entails a segregation/dissolution of the manganese that provokes, first a poisoning of the anode and second a change of the crystal structure from layered to spinel structure [58]. Cathodes of the polyanion class with an olivine-type crystal structure (e.g. LFP) are considered a transitory solution for its use in EVs due to their high thermal stability, allowing to reduce the use of cobalt. Nevertheless, the lack of excellent electronic properties in comparison to layered or spinel cathodes makes it necessary to optimize the size of the small particles coated with carbon to minimize the drop in the volumetric energy density [59].

Although layered expanded graphite ($372\,mAh\,g^{-1}$) is widely applied in standard anodes for LIBs (Figure 3.3b) due to its low cost, safety, and lower material expansion (13%), the need to increase the energy density makes the research community investigate new insertion materials to store more Li ions. Among them, lithium is the best candidate with a specific capacity of 3862 followed by silicon (Si), aluminum (Al), and tin (Sn) with 3600, 2235, and 990 ($mAh\,g^{-1}$), respectively. Unfortunately, alloy-type materials suffer from a complete structural change as the metal ions go inside them. Therefore, a volume expansion from less than 1% (Li) to 250% (Sn) or even 600% (Al) is observed [53, 60]. Consequently, stable long-term performance is limited due to a decay of specific capacity during cycling that is attributed to stress formation (delamination), mechanical fracture (disintegration), and instable solid electrolyte interphase (SEI) boosted by the volume changes (Figure 3.3c).

Bearing this in mind, anode materials need to be designed to take on those challenges and to be reversible in and out over a few thousand cycles. Recently, Ni_2–TiS_2 composite was synthesized by a facile mechanical milling method. The presence of Ni nanocrystals induces an improvement of the electrical conductivity as well as reduces the volume change (Figure 3.3d). As a result, it shows good performance with a reversible capacity of $635\,mAh\,g^{-1}$ after 50 cycles at a current density of $1\,A\,g^{-1}$. A similar idea has been followed by Wang et al. [61] where a ceramic-based electrode was engineered for rechargeable batteries. The strategy consisted of the anchoring of Ni nanoparticles on carbon and then coated perovskite $Li_{0.33}La_{0.56}Ti_{0.9}Ni_{0.1}O_{3-\delta}$ particles. The coating layer acted as a buffer layer that controls the volume expansion to ~4%. In addition, the improved electrical (enhanced by oxygen vacancies) and ionic ($10^{-3}\,S\,cm^{-1}$ at RT) result in an excellent rate performance of $352\,mAh\,g^{-1}$ after 250 cycles at $0.2\,A\,g^{-1}$.

3.2.2 Separator

Although the separator traditionally has the only function to avoid electrical short-circuit as it does not take part in redox reactions, the numerous applications of LIBs make necessary the introduction of new properties in order to overcome not only safety issues but to improve the kinetics and electrochemical performance. The ideal properties of a separator entail inherent electrical insulation and chemical inertness (0–5 V), a porous structure (40–60%; <1 μm pore size), and high wettability. All these properties determine the ion transfer route and interface processes and consequently the power and energy of the battery. Further, functional separators have been actively studied and developed. For instance,

Ding et al. [55] designed a trilayer (PE–Al_2O_3–PP) film separator with shutdown functionality (Figure 3.3e). The excellent pore size distribution and the presence of Al_2O_3 coating on the top surface and the pores improve the mechanical properties, thermal resistance (130 °C) as well as enhance the wettability (electrolyte uptake (121.8%)). Accordingly, to the presence of Ni in the anode electrode, 2D nickel hydroxide nanosheets (NHNs) were investigated as a filler in a PVDF matrix [62]. It was stated that the presence of NHNs significantly enhances the thermal stability (150 °C) and improves electrolyte uptake (327.6%) and ionic conductivity (1.5 $mS\,cm^{-1}$) with just a 3 wt% of filler loading. Future perspective in the design of novel separator materials implies the presence of sustainable materials as bio-polymers [63, 64], including cellulose-based [65, 66] composite separators (see Section 3.3.2).

3.2.3 Electrolyte

Current electrolytes for commercial LIBs must be electrochemically (2–4.5 V) and thermally (−30 to 100 °C) stable, ionically conductive ($>10^{-3}\,S\,cm^{-1}$), and should present low toxicity and low manufacturing cost. Furthermore, the electrolyte composition plays a critical role during the cell aging processes as it has an important impact on the appearance of side reactions and consequently degradation of the SEI or cathode electrolyte interface (CEI) [67, 68]. Recently, temperature stability is one of the main safety concerns as the standard electrolytes using ethylene carbonate–ethyl methyl carbonate (EC/EMC) as solvents present a flammable nature. The incorporation of specific additives allows enhancing the formation of a stable CEI on the cathode interface and therefore allows working at high temperature and high voltage. For example, with just the addition of 0.1 wt% of dopamine to the electrolyte composition, it avoided oxidative decomposition when the cell operates at high voltage (up to 4.5 V) and 60 °C [69]. It was suggested the formation of a robust polydopamine-based passivating layer protects the cathode and leads to a higher discharge capacity (Figure 3.4a). Similar conclusions were obtained when tripropargyl phosphate [71] and *N,N*-dimethylformamide (DMF) [72] were added to the electrolyte composition. When batteries were cycled at 60 °C, cells exhibited excellent performance after the addition of the different additives, with 141.6 $mAh\,g^{-1}$ after 200 cycles (in the case of TTP) and 125.8 $mAh\,g^{-1}$ after 150 cycles (in the case of DMF).

On the other hand, electrolytes that exhibit excellent performance at low temperatures (<0 °C) have not been widely investigated even when it is paramount for EVs (e.g. EVs will require them to function during winter just as reliably as during summer) or expanding their role in the aerospace area. In this sense, a new electrolyte solvent, isoxazole (IZ), was proposed [70] to improve the ionic conductivity (27.5 $mS\,cm^{-1}$ at 20 °C) and the thermal resistance (with a high boiling point >95 °C and low melting point −67 °C) avoiding the use of toxic additives (Figure 3.4b). In the same direction, localized high-concentration electrolytes (LHCEs), in particular 1,1,2,2-tetrafluoroethyl methyl ether, was proposed to fabricate a high-voltage battery (3–4.9 V vs. Li/Li^+) that could operate with a relatively high current density (≥0.1 C) at a low temperature (−50 °C). $LiNi_{0.5}Mn_{1.5}O_4||Li$ battery configuration delivered an outstanding 80% of its RT discharge capacity at 0.2 C [73].

Nonetheless, the development of solid-state electrolytes will boost the energy density, long cycling life as well as will minimize safety issues [74]. Ceramic-based

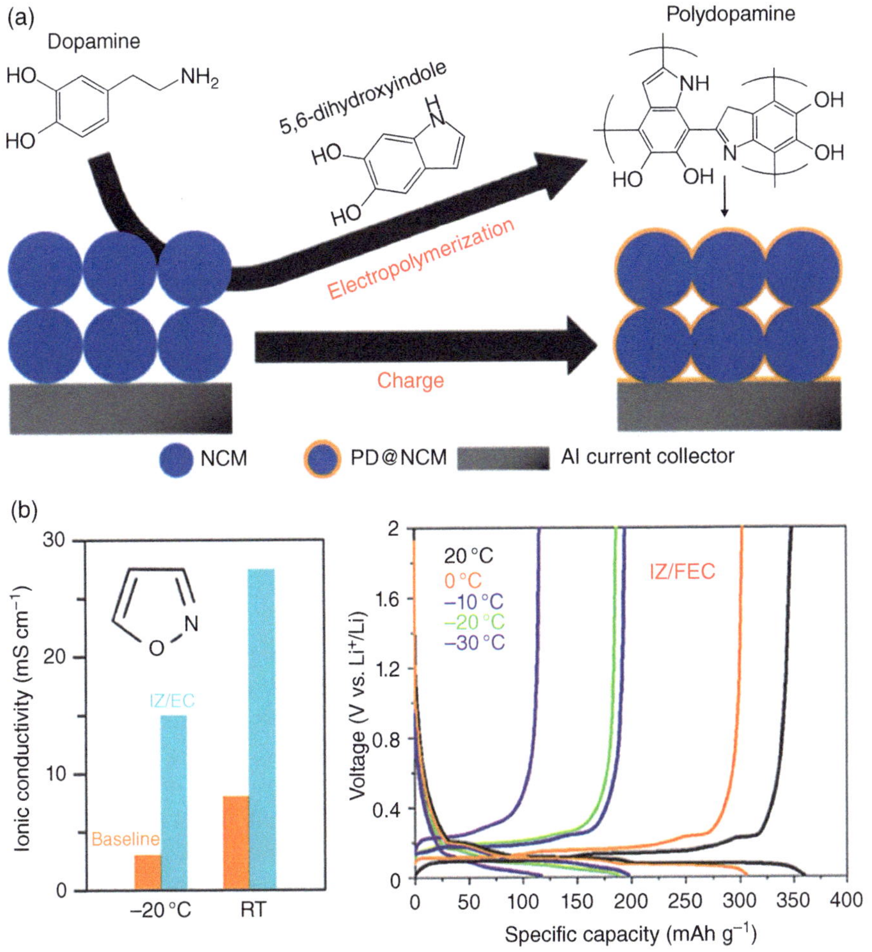

Figure 3.4 (a) Schematic representation of the passivating layer formed at the cathode surface. *Source:* Reproduced with permission from [69]. Copyright © 2016, American Chemical Society. (b) (Left) Comparison of the ionic conductivity of standard electrolyte composition (1 M $LiPF_6$ in EC/EMC (3/7 wt%)) and 1 M $LiPF_6$ in EC/IZ (1/10 vol%) electrolytes at room temperature and −20 °C and (right) voltage profiles of the Li/graphite cell using 1 M lithium difluoro(oxalato)borate in IZ/FEC (10/1, vol%) at various temperatures at the C/10 rate. *Source:* Reproduced with permission from [70]. Copyright © 2021, American Chemical Society.

solid-electrolytes present a lack of an intimate interface contact with the electrodes that causes low ionic conductivity and poor compatibility. On the other hand, gel-like electrolytes require the deposition of a thick layer due to the low mechanical resistance to the growth of dendrites [75]. Bearing this in mind, solid-polymer (see Section 3.3.3) electrolytes represent excellent candidates that offer good interface contact as well as good mechanical properties. As an example, ultrahigh elastic polyurethane-based electrolytes exhibit high ionic conductivity ($2.86 \times 10^{-4}\,S\,cm^{-1}$) and deliver a specific capacity of $150\,mAh\,g^{-1}$ at 0.2 C [76].

3.3 Sustainable Materials for Li-Ion Batteries

Sustainability in the LIBs field is related to more sustainable materials and/or synthesis processes for their fabrication, based on environmentally friendlier processes and green reagents.

3.3.1 Electrodes

In the electrodes, sustainability focuses on the active materials and polymeric binders, as well as on the respective solvent. In the case of polymeric binders, the tendency is to replace the common polymer binder used, poly(vinylidene fluoride), PVDF, for natural polymers and biopolymers [77]. Furthermore, PVDF is typically dissolved by *N*-methyl-2-pyrrolidone (NMP) which leads to a negative environmental impact and health problems, being possible to be replaced by *N,N'*-dimethylpropyleneurea [78] or DMF [79].

For the polymer binder of silicon anodes, the use of a new binder based on cross-linked dextrin in an aqueous solution has been reported, showing excellent cycling stability and superior rate capability when compared to commercial carboxymethyl cellulose binder [80]. In addition, other binders based on latex [81] and epoxy, a polyisocyanate-based (ICN) binder or polyacrylic acid [82], have been used with $LiNi_xCo_yMn_zO_2$ cathodes.

Also for the manufacture of LiNi0.5Mn1.5O4 (LMNO) electrodes, two different binders have been used to replace PVDF: pullulan (PU) and the bifunctional electronically conductive poly(3,4-ethylenedioxythiophene)–poly(styrenesulfonate), with excellent cycling behavior [83].

Regarding the active material for anode, the widely used ones are based on carbonaceous materials and silicon where the focus is on sustainable sources and green synthesis.

Through the biomass fermentation treatment, a novel hierarchical porous silicon/N-doped carbon composite (Si/N-PC) has been produced for anode electrodes as shown in Figure 3.5a) [84]. This new anode electrode shows a reversible discharge capacity of $696.1\,mAh\,g^{-1}$ for more than 300 cycles at a current density of $1\,A\,g^{-1}$, the fabrication process being sustainable, scalable, and low cost [84].

Furthermore, for the anode electrode, the solid residue from the hydrothermal liquefaction (HTL) of wet biomass waste has been converted to hard carbon (HC@WA) or converted through KOH activation to a carbon with a higher specific surface area (HC@AA), as shown in Figure 3.5b, and used in LIBs applications [85]. This material shows a discharge capacity of $319\,mAh\,g^{-1}$ with good cycling stability over many cycles and represents a new approach to zero-waste, considering circular economy approaches [85].

Also, to reduce the cost and improve the cycling behavior, an Al foil has been developed with the addition of 1% silicon ($Al_{99.0}Si_{1.0}$), the battery performance is shown in Figure 3.5c. This anode shows a stable cycle life (>100 cycles) and excellent coulombic efficiency (>99.5%) [86].

In addition, a new anode based on the three-dimensional (3D) sandwich-like hollow silicon nanoarray/graphene (hollow-Si-nanoarray/graphene) architecture has been produced as shown in Figure 3.5d [87]. This anode is characterized by a discharge capacity value of $700\,mAh\,g^{-1}$ at 5C after 500 cycles. This architecture presents high flexibility, a good electronically conductive network, lithium-ion pathways, and an efficient accommodation for volume change during discharge/charge [87].

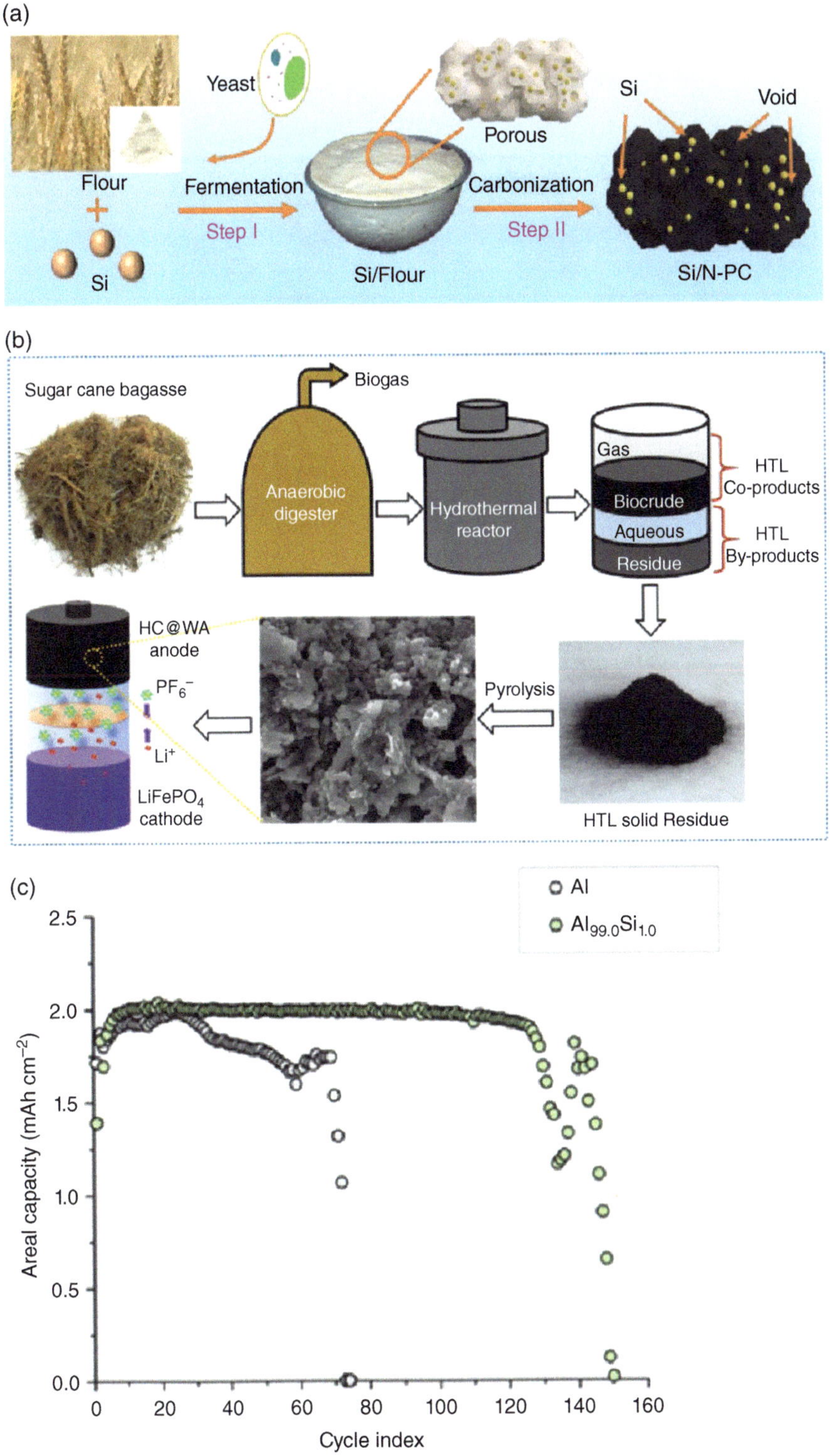

Figure 3.5 (a) Schematic representation of the fabrication process of Si/N-PC. *Source:* Wang et al. [84]/With permission of Elsevier, (b) schematic representation of the development of carbon composite material. *Source:* Sharma et al. [85]/With permission of Elsevier, (c) cycling performance of $Al_{99.0}Si_{1.0}$ vs. Al anodes. *Source:* Sharma et al. [86]/With permission of American Chemical Society, (d) schematic representation of the 3D Sandwich-Like Hollow-Si-Nanoarray/Graphene Architecture fabrication. *Source:* Huang et al. [87]/With permission of American Chemical Society, and (e) recycling and fabrication procedure of two wastes for $LiFePO_4$/C. *Source:* Lan et al. [88]/With permission of Springer Nature.

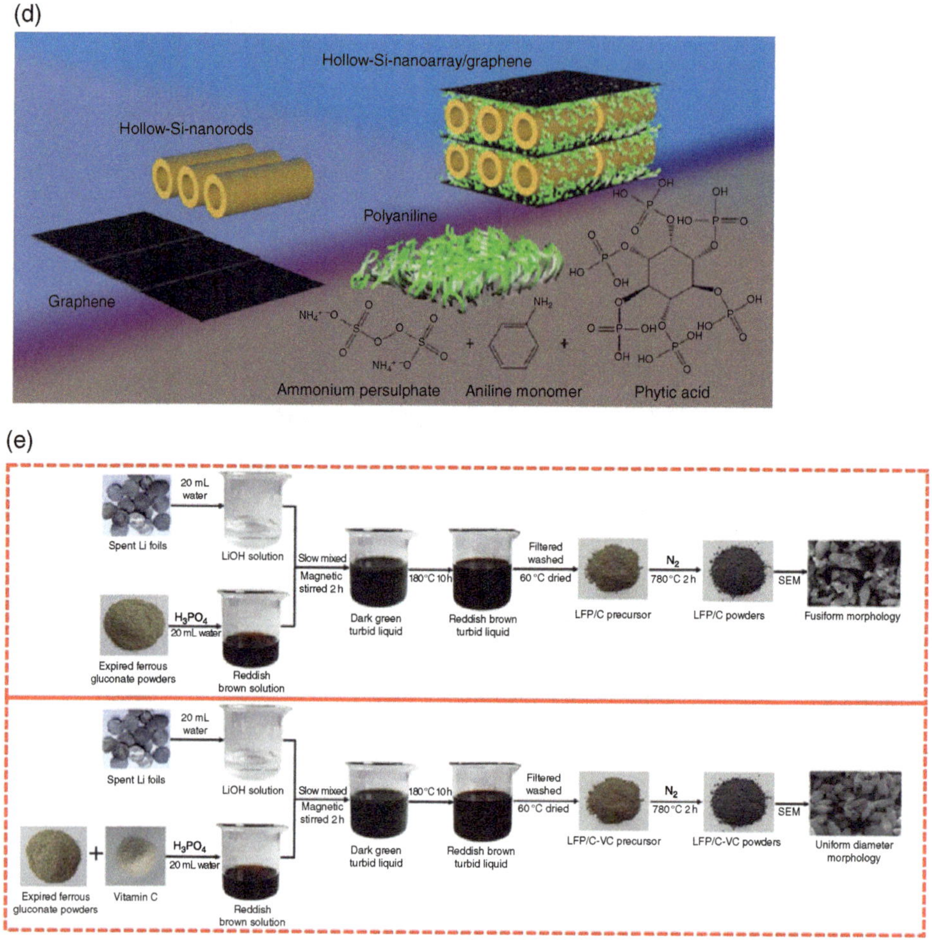

Figure 3.5 (Continued)

For the sustainable development of the cathode electrode and taking into account that carbon-coated lithium iron phosphate ($LiFePO_4$/C) is widely used, this active material has been extracted and reused from two wastes (expired ferrous gluconate, spent Li foils), and the positive role of vitamin C (VC) as seen in Figure 3.5e has been demonstrated [88]. It is observed that carbon content prevented the oxidation of Fe(II) and inhibited the formation of the insulator Li_3PO_4 phase, showing a discharge capacity value of 134.6 $mAh\,g^{-1}$ at 1 C for 400 cycles, representing a reduction in the accumulation of waste based on the circular economy [88].

Another strategy for cathode electrodes is to use organic materials such as 1,3,5-tri(9,10-anthraquinonyl)benzene [89] and 11,11,12,12-tetracyano-9,10-anthraquinonedimethane (TCAQ) [90] as active material.

In particular, 1,3,5-tri(9,10-anthraquinonyl)benzene has excellent electrochemical performance with an ultra-long cycle life of 10 000 cycles and a working temperature range from −70 to 100 °C [89], whereas TCAQ active material results in 83 mAh g^{-1} at a current density of 17 mA g^{-1} [90].

Different approaches to the sustainability of electrodes have been used with excellent battery performance and low cost, these materials, and synthesis procedures being suitable for next-generation sustainable LIBs.

3.3.2 Separator

Regarding the separator component, the focus is on replacing synthetic polymers with natural polymers and biopolymers. The most widely used natural polymer is cellulose in its different derivatives and processing protocols, from solvent casting to electrospinning [66, 91].

Thus, a new porous composite membrane based on cellulose nanofibrils and laponite has been developed, as shown in the schematic representation of Figure 3.6a [92]. This membrane is characterized by a high porosity (68%), high ionic conductivity (0.977 mS cm^{-1}), and good cycling performance (92.6% capacity retention after 100 cycles at 0.5 C) when compared to commercial poly(propylene) PP separator [92].

Also, through the acetylation process, a cellulose acetate (CA) separator of cellulose nanocrystals has been developed with a discharge capacity of 132.9 mAh g^{-1} at 0.2 C, where Figure 3.6b shows the fabrication process of the separator [93].

Mesoporous cellulose nanocrystal (MCNC) membranes with specific surface areas up to 172 m^2 g^{-1} have been developed with a discharge capacity value of 85 mAh g^{-1} at a 2 C rate, as shown in Figure 3.6c [94]. Another separator produced from nanofibrillated cellulose has been prepared from kraft and unbleached bamboo pulps using different steps with an intrinsically flame-retardant property [97].

Modified natural wood films have been also applied as battery separators with excellent mechanical properties, thermal stability, and high ion conductivity which represents a sustainable approach in this field [98].

Furthermore, cyanoethyl-chitin nanofiber (CCN) separators have been developed with high ionic conductivity, excellent mechanical strength, and improved rate capability with capacity retention compared to commercial polypropylene (PP) separators [99].

A new approach for sustainable separators has been developed through the production of porous nanofiber membranes based on lignin/polyvinyl alcohol, which show high thermal stability, flame retardant properties, and high ionic transport property [100].

Considering that poly(vinylidene fluoride) and its copolymers are widely used in this field [101], and to increase its sustainability, a novel polyethylene oxide/lignocellulose (PEO/LIGC) coated dimpled electrospun P(VDF-TrFE) nanofibrous membrane has been produced with high porosity (86%) and ionic conductivity value (7 mS cm^{-1}) and better cycling behavior when compared to the commercial separators [102].

Also, silk fibroin (SF) is being used as a battery separator through different approaches, highlighting single porous membranes with different pore sizes as shown in Figure 3.6d [95], as well as hybrid membranes (Figure 3.6e) with silk sericin (SS) [96]. For SF porous membranes with different pore sizes, the electrolyte uptake percentage and ionic conductivity value are the highest for membranes with SF 250–106 μm pore size membranes. It is also observed the relationship between β-sheet conformation percentage existing in the

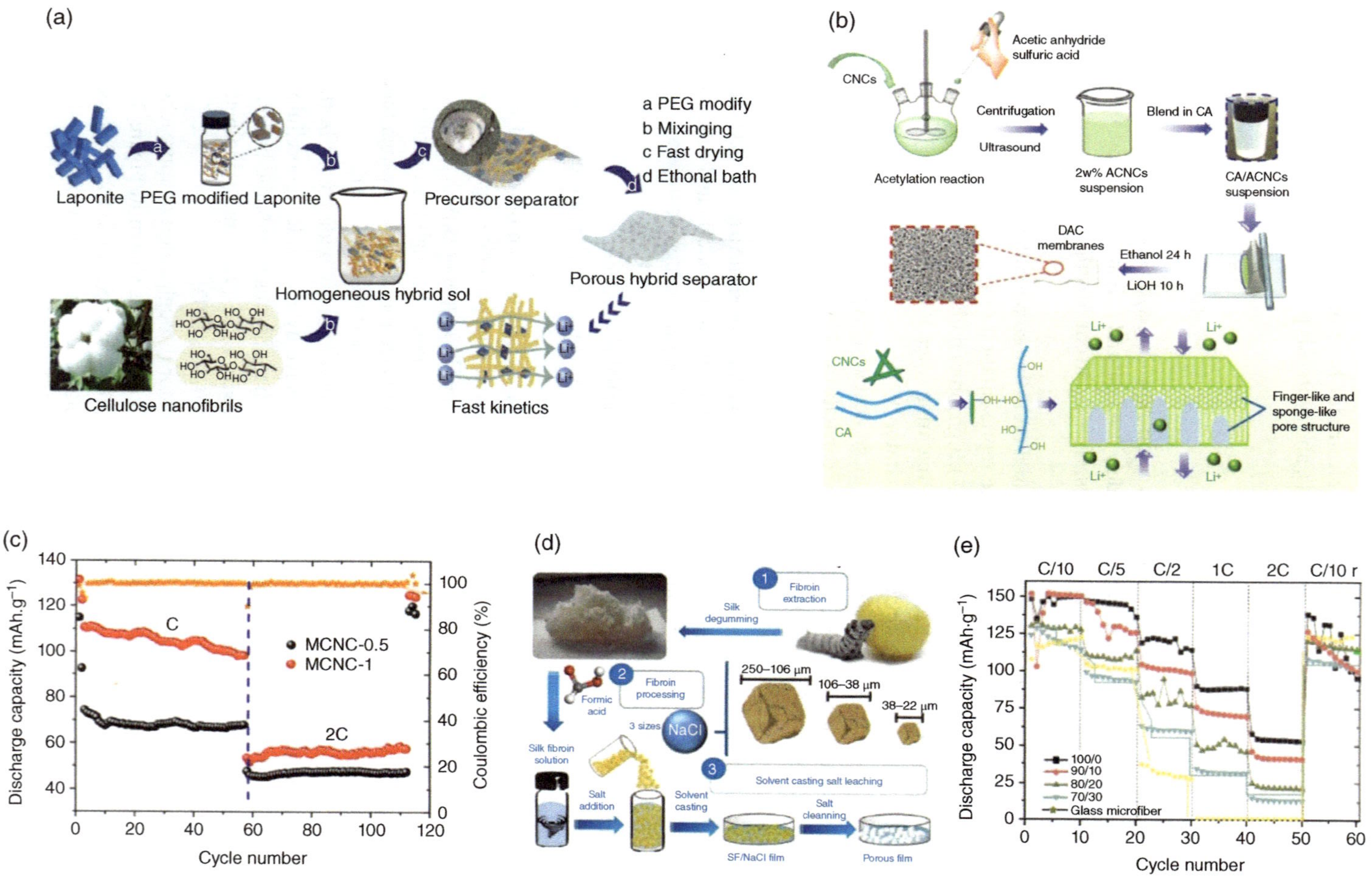

Figure 3.6 (a) Schematic representation of porous composite membranes based on CNF/Laponite. *Source:* Wang et al. [92]/With permission of Elsevier, (b) experimental process for the preparation of cellulose acetate (CA) separators. *Source:* Xu et al. [93]/With permission of John Wiley & Sons, (c) cycling battery performance using MCNC-0.5 and MCNC-1 separators. *Source:* Gonçalves et al. [94]/With permission of American Chemical Society, (d) main preparation steps for porous silk fibroin (SF) membranes. *Source:* Reizabal et al. [95]/With permission of Elsevier, and (e) rate performance as a function of cycle number for the different SF/SS separator membranes. *Source:* Reizabal et al. [96]/With permission of Elsevier.

membrane with the pore size and consequently with battery performance [95]. Also, for the SF/SS separator, it is observed the stabilizing effect of sericin on the discharge capacity for membranes containing 10 wt% of SS [96].

Thus, after highlighting their potential, naturals and biopolymers must continue to be explored for application in this field.

3.3.3 Solid Polymer Electrolytes

In order to reduce the need to use liquid electrolytes in the separator membrane and consequently increase its safety, solid polymer electrolytes emerge as an effective solution for obtaining solid-state batteries, basically composed of organic and inorganic materials [103].

Organic materials are basically constituted of different synthetic polymer matrices with dispersed fillers where their advantages are flexibility and low toxicity and their disadvantages are the low ionic conductivity value and high interface resistance. In relation to inorganic materials, they are based on ceramic materials, which have the advantages of a wide electrochemical window, and high ionic conductivity at >100 °C, but present low mechanical properties and poor contact with electrodes as disadvantages [104].

To solve these disadvantages of both materials, the development of solid polymer electrolytes (SPEs) is a possible solution.

In SPEs, the polymer matrices widely used are poly(vinylidene fluoride) (PVDF), poly(vinylidene fluoride-*co*-hexafluoropropylene) (PVDF-HFP), PEO, poly(acrylonitrile), and UV-curable polymers such as polyurethane acrylate, among others, where the tendency is to replace these polymers by natural and biopolymers [104]. This trend is based on poly(ethylene glycol) methacrylate, cellulose, as well as other polymer matrices with different ionic liquids and lithium salts (lithium bis(trifluoromethanesulfonyl)imide (LiTFSI)), but without the adequate ionic conductivity value and consequently poor battery performance.

These new SPEs still do not present adequate performance, their behavior depends on different parameters, such as lithium salt dissociation and ion-dipole interactions between ion and polymer matrix [105]. In addition, some SPEs based on natural polymers present battery performance at high temperatures (>50 °C) which limits their applicability [106, 107]. For the next generation of SPEs, these materials with natural polymers should also present smart functionalities such as shutdown function and/or flame retardancy.

More research efforts are needed for the field of sustainable SPEs for LIBs, in order to achieve a new generation of sustainable and environmentally friendly materials with improved ionic conductivity value, battery capacity, and improved safety.

3.4 Conclusions and Outlook

With the improvement of the quality of social human life, there is an increasing energy demand in which efficient energy storage devices are essential. This demand is due to the increasing use of small and portable devices, and EVs, among others, and must be accompanied by a reduction in its dependence on fossil fuels to achieve a sustainable world. Improving the ecological footprint is key in a circular economy approach to reduce the environmental impact of the used technologies. Considering that the widely used energy storage systems are LIBs, they must be increasingly based on sustainable materials and processes for the different components (electrodes, separators, and solid polymer electrodes) to obtain sustainable batteries in the scope of the circular economy.

There are several studies on sustainable approaches for electrodes (anode and cathode) and separators. Regarding electrodes, these approaches are basically centered on active materials and polymer binders through natural materials and/or green synthesis. Another sustainable approach used for active materials for the electrodes is recycling to reduce the waste and improve the reuse of the materials with excellent conversion rate and battery performance which reduces environmental impacts related to raw materials extraction. For polymer binders and separators, the highlight is the replacement of synthetic polymers with natural polymers.

The solid polymer electrolytes of sustainable materials are being developed but still do not present the needed characteristics to be implemented in high-performance batteries.

Thus, more efforts are needed to replace different battery components with materials based on sustainable materials, sustainable processes, and green syntheses with special attention to solid polymer electrolytes, in order to develop a new generation of sustainable solid-state batteries to balance of reducing hazards issues and improve safety and battery performance.

In summary, it is concluded that additional and complementary steps are needed on sustainable approaches for the different LIBs components through new materials and processes in order to reduce the environmental impact related to the increase of the implementation of those batteries in fast-growing areas such as portable devices and EVs.

Acknowledgments

The authors thank the Fundação para a Ciência e Tecnologia (FCT) and COMPETE 2020 for financial support under the framework of Strategic Funding grants UIDB/04650/2020, UID/FIS/04650/2020, UID/EEA/04436/2020, and UID/QUI/00686/2020 and under projects, MIT-EXPL/TDI/0033/2021, POCI-01-0247-FEDER-046985, 2022.03931.PTDC and PTDC/FIS-MAC/28157/2017. The authors also thank the FCT for the investigator contracts CEECIND/00833/2017 (R.G.) and 2020.04028.CEECIND (C.M.C.). Financial support from the BIDEKO Project, funded by MCIN/AEI, NextGenerationEU, PRTR, and Basque Government Industry Department under the ELKARTEK program are acknowledged.

References

1 Zeng, X., Li, M., Abd El-Hady, D. et al. (2019). *Commercialization of lithium battery technologies for electric vehicles. Advanced Energy Materials* 9 (27): 1900161.

2 Yoshino, A. (2021). *From polyacetylene to carbonaceous anodes. Nature Energy* 6 (4): 449.

3 Zhou, H., Xin, F., Pei, B. et al. (2019). *What limits the capacity of layered oxide cathodes in lithium batteries? ACS Energy Letters* 4 (8): 1902–1906.

4 Jin, S., Jiang, Y., Ji, H. et al. (2018). *Advanced 3D current collectors for lithium-based batteries. Advanced Materials* 30 (48): 1802014.

5 Hofmann, M., Hofmann, H., Hagelüken, C. et al. (2018). *Critical raw materials: a perspective from the materials science community. Sustainable Materials and Technologies* 17: e00074.

6 Xu, C., Dai, Q., Gaines, L. et al. (2020). *Future material demand for automotive lithium-based batteries. Communications Materials* 1 (1): 99.

7 Olivetti, E.A., Ceder, G., Gaustad, G.G. et al. (2017). *Lithium-ion battery supply chain considerations: analysis of potential bottlenecks in critical metals. Joule* 1 (2): 229–243.

8 Panella, C. and De Putter, T. (2022). The invisible lives of cobalt miners and green traceability discourses. *Anthropology Today* 38: 13–16. https://doi.org/10.1111/1467-8322.12753.

9 Negi, S., Batoye, S., Singh, K., and Waraich, J.S. (2021). Environmental pollution, its causes and impact on ecosystem. In: *New Frontiers of Nanomaterials in Environmental Science* (ed. R. Kumar, R. Kumar and G. Kaur). Singapore: Springer https://doi.org/10.1007/978-981-15-9239-3_1.

10 Han, X., Lu, L., Zheng, Y. et al. (2019). *A review on the key issues of the lithium ion battery degradation among the whole life cycle. eTransportation* 1: 100005.

11 Jin, D., Park, J., Ryou, M.-H. et al. (2020). *Structure-controlled Li metal electrodes for post-Li-Ion batteries: recent progress and perspectives. Advanced Materials Interfaces* 7 (8): 1902113.

12 Kwade, A., Haselrieder, W., Leithoff, R. et al. (2018). *Current status and challenges for automotive battery production technologies. Nature Energy* 3 (4): 290–300.

13 Melchor-Martínez, E.M., Macias-Garbett, R., Malacara-Becerra, A. et al. (2021). *Environmental impact of emerging contaminants from battery waste: a mini review. Case Studies in Chemical and Environmental Engineering* 3: 100104.

14 Afolayan, A.O. (2018). *Accumulation of heavy metals from battery waste in topsoil, surface water, and garden grown maize at omilende area, Olodo,*Nigeria. *Global Challenges* 2 (3): 1700090.

15 Mrozik, W., Rajaeifar, M.A., Heidrich, O. et al. (2021). *Environmental impacts, pollution sources and pathways of spent lithium-ion batteries. Energy & Environmental Science* 14 (12): 6099–6121.

16 Porzio, J. and Scown, C.D. (2021). *Life-cycle assessment considerations for batteries and battery materials. Advanced Energy Materials* 11 (33): 2100771.

17 da Silva Lima, L., Quartier, M., Buchmayr, A. et al. (2021). *Life cycle assessment of lithium-ion batteries and vanadium redox flow batteries-based renewable energy storage systems. Sustainable Energy Technologies and Assessments* 46: 101286.

18 Xia, X. and Li, P. (2022). *A review of the life cycle assessment of electric vehicles: considering the influence of batteries. Science of the Total Environment* 814: 152870.

19 Salado, M., Lizundia, E., Oyarzabal, I. et al. (2022). *The role of critical raw materials for novel strategies in sustainable secondary batteries. Physica Status Solidi* 219 (15): 2100710.

20 Shafique, M. and Luo, X. (2022). *Environmental life cycle assessment of battery electric vehicles from the current and future energy mix perspective. Journal of Environmental Management* 303: 114050.

21 Pinegar, H. and Smith, Y.R. (2019). *Recycling of end-of-life lithium ion batteries, Part I: commercial processes. Journal of Sustainable Metallurgy* 5 (3): 402–416.

22 Giannetti, B.F., Agostinho, F., JJC, E. et al. (2020). *Cleaner production for achieving the sustainable development goals. Journal of Cleaner Production* 271: 122127.

23 R. Materials, Recycle with us and help build a clean energy future. https://www.redwoodmaterials.com (accessed 26 November 2021).

24 Sumitomo Metal Mining, C., *Development of a New Recycling Process for Lithium Ion Secondary Batteries and Operation of a Pilot Plant Sumitomo*. 2019.

25 Diekmann, J., Hanisch, C., Loellhoeffel, T. et al. (2016). *Ecologically friendly recycling of lithium-ion batteries-the lithorec process. ECS Transactions* 73 (1): 1.

26 Diekmann, J., Hanisch, C., Froböse, L. et al. (2016). *Ecological recycling of lithium-ion batteries from electric vehicles with focus on mechanical processes. Journal of the Electrochemical Society* 164 (1): A6184.

27 Sloop, S.E. and R. Parker, *System and method for processing an end-of-life or reduced performance energy storage and/or conversion device using a supercritical fluid*. 2011, Patent US8067107B2.

28 Sloop, S.E., *Reintroduction of lithium into recycled battery materials*. 2014.

29 Sloop, S.E. and M. Allen, *Recycling and reconditioning of battery electrode materials*. 2016.

30 Sloop, S.E., *Recycling positive-electrode material of a lithium-ion battery*. 2017.

31 Tedjar, F. and J.-C. Foudraz, *Method for the mixed recycling of lithium-based anode batteries and cells*. 2010.

32 McLaughlin, W. and T.S. Adams, *Li reclamation process*. 1999.

33 Sommerville, R., Zhu, P., Rajaeifar, M.A. et al. (2021). *A qualitative assessment of lithium ion battery recycling processes. Resources, Conservation and Recycling* 165: 105219.

34 Siqi, Z., Guangming, L., Wenzhi, H. et al. (2019). *Recovery methods and regulation status of waste lithium-ion batteries in China: a mini review. Waste Management & Research* 37 (11): 1142–1152.

35 Vanderbruggen, A., Salces, A., Ferreira, A. et al. (2022). *Improving separation efficiency in end-of-life lithium-ion batteries flotation using attrition pre-treatment. Minerals* 12 (1): 72.

36 Makuza, B., Tian, Q., Guo, X. et al. (2021). *Pyrometallurgical options for recycling spent lithium-ion batteries: a comprehensive review. Journal of Power Sources* 491: 229622.

37 Zhou, M., Li, B., Li, J. et al. (2021). *Pyrometallurgical technology in the recycling of a spent lithium ion battery: evolution and the challenge. ACS ES&T Engineering* 1 (10): 1369–1382.

38 Liang, Z., Cai, C., Peng, G. et al. (2021). *Hydrometallurgical recovery of spent lithium ion batteries: environmental strategies and sustainability evaluation. ACS Sustainable Chemistry & Engineering* 9 (17): 5750–5767.

39 Asadi Dalini, E., Karimi, G., Zandevakili, S. et al. (2021). *A review on environmental, economic and hydrometallurgical processes of recycling spent lithium-ion batteries. Mineral Processing and Extractive Metallurgy Review* 42 (7): 451–472.

40 Xu, P., Yang, Z., Yu, X. et al. (2021). *Design and optimization of the direct recycling of spent Li-ion battery cathode materials. ACS Sustainable Chemistry & Engineering* 9 (12): 4543–4553.

41 Gao, H., Tran, D., and Chen, Z. (2022). *Seeking direct cathode regeneration for more efficient lithium-ion battery recycling. Current Opinion in Electrochemistry* 31: 100875.

42 Lu, Y., Han, X., and Li, Z. (2021). *Enabling intelligent recovery of critical materials from li-ion battery through direct recycling process with internet-of-things. Materials* 14 (23): 7153.

43 Roy, J.J., Cao, B., and Madhavi, S. (2021). *A review on the recycling of spent lithium-ion batteries (LIBs) by the bioleaching approach. Chemosphere* 282: 130944.

44 Roy, J.J., Rarotra, S., Krikstolaityte, V. et al. (2021). *Green recycling methods to treat lithium-ion batteries E-waste: a circular approach to sustainability. Advanced Materials* 34: 2103346.

45 Jegan Roy, J., Srinivasan, M., and Cao, B. (2021). *Bioleaching as an eco-friendly approach for metal recovery from spent NMC-based lithium-ion batteries at a high pulp density. ACS Sustainable Chemistry & Engineering* 9 (8): 3060–3069.

46 Kim, S., Bang, J., Yoo, J. et al. (2021). *A comprehensive review on the pretreatment process in lithium-ion battery recycling. Journal of Cleaner Production* 294: 126329.

47 Beak, M., Park, J., Park, S. et al. (2022). *Understanding the effect of nonmetallic impurities in regenerated cathode materials for lithium-ion battery recycling by tracking down impurity elements. Journal of Hazardous Materials* 425: 127907.

48 Ma, X., Chen, M., Zheng, Z. et al. (2021). *Recycled cathode materials enabled superior performance for lithium-ion batteries. Joule* 5 (11): 2955–2970.

49 Huang, Y., Shaibani, M., Gamot, T.D. et al. (2021). *A saccharide-based binder for efficient polysulfide regulations in Li-S batteries. Nature Communications* 12 (1): 5375.

50 Whittingham, M.S. (2014). *Ultimate limits to intercalation reactions for lithium batteries. Chemical Reviews* 114 (23): 11414–11443.

51 Masquelier, C. and Croguennec, L. (2013). *Polyanionic (phosphates, silicates, sulfates) frameworks as electrode materials for rechargeable Li (or Na) batteries. Chemical Reviews* 113 (8): 6552–6591.

52 Sun, Y.W., Liu, W., Hernandez, I. et al. (2019). *3D strain in 2D materials: to what extent is monolayer graphene graphite? Physical Review Letters* 123 (13): 135501.

53 Sun, L., Liu, Y., Shao, R. et al. (2022). *Recent progress and future perspective on practical silicon anode-based lithium ion batteries. Energy Storage Materials* 46: 482–502.

54 Kim, H.-H., Kim, K.-H., Lee, J. et al. (2021). *Electrochemical properties and reaction mechanism of $NiTi_2S_4$ ternary metal sulfide as an anode for lithium ion battery. ACS Sustainable Chemistry & Engineering* 9 (29): 9680–9688.

55 Ding, L., Yan, N., Zhang, S. et al. (2022). *Low-cost mass manufacturing technique for the shutdown-functionalized lithium-ion battery separator based on Al_2O_3 coating online construction during the β-iPP cavitation process. ACS Applied Materials & Interfaces* 14 (5): 6714–6728.

56 Akhilash, M., Salini, P.S., John, B. et al. (2021). *A journey through layered cathode materials for lithium ion cells – from lithium cobalt oxide to lithium-rich transition metal oxides. Journal of Alloys and Compounds* 869: 159239.

57 Seck, G.S., Hache, E., and Barnet, C. (2022). *Potential bottleneck in the energy transition: The case of cobalt in an accelerating electro-mobility world. Resources Policy* 75: 102516.

58 Park, K.-Y., Zhu, Y., Torres-Castanedo, C.G. et al. (2022). *Elucidating and mitigating high-voltage degradation cascades in cobalt-free $LiNiO_2$ lithium-ion battery cathodes. Advanced Materials* 34 (3): 2106402.

59 Chen, S.-P., Lv, D., Chen, J. et al. (2022). *Review on defects and modification methods of $LiFePO_4$ cathode material for lithium-ion batteries. Energy & Fuels* 36 (3): 1232–1251.

60 Wang, C., Yang, C., and Zheng, Z. (2022). *Toward practical high-energy and high-power lithium battery anodes: present and future. Advanced Science* 2105213.

61 Wang, J., Wang, M., Xiao, J. et al. (2022). *A microstructure engineered perovskite super anode with Li-storage life of exceeding 10,000 cycles*. *Nano Energy* 94: 106972.

62 Gao, Z., Wen, R., Deng, H. et al. (2022). *Composite membrane of poly(vinylidene fluoride) and 2D Ni(OH)$_2$ nanosheets for high-performance lithium-ion battery*. *ACS Applied Polymer Materials* 4 (2): 960–970.

63 Jeong, J., Lee, J., Kim, J. et al. (2021). *A biopolymer-based functional separator for stable Li metal batteries with an additive-free commercial electrolyte*. *Journal of Materials Chemistry A* 9 (12): 7774–7781.

64 Lizundia, E. and Kundu, D. (2021). *Advances in natural biopolymer-based electrolytes and separators for battery applications*. *Advanced Functional Materials* 31 (3): 2005646.

65 Jo, J.H., Jo, C.-H., Qiu, Z. et al. (2020). *Nature-derived cellulose-based composite separator for sodium-ion batteries*. *Frontiers in Chemistry* 8.

66 Lizundia, E., Costa, C.M., Alves, R. et al. (2020). *Cellulose and its derivatives for lithium ion battery separators: a review on the processing methods and properties*. *Carbohydrate Polymer Technologies and Applications* 1: 100001.

67 Pender, J.P., Jha, G., Youn, D.H. et al. (2020). *Electrode degradation in lithium-ion batteries*. *ACS Nano* 14 (2): 1243–1295.

68 Hu, D., Su, Y., Chen, L. et al. (2021). *The mechanism of side reaction induced capacity fading of Ni-rich cathode materials for lithium ion batteries*. *Journal of Energy Chemistry* 58: 1–8.

69 Lee, H., Han, T., Cho, K.Y. et al. (2016). *Dopamine as a novel electrolyte additive for high-voltage lithium-ion batteries*. *ACS Applied Materials & Interfaces* 8 (33): 21366–21372.

70 Tan, S., UND, R., Shadike, Z. et al. (2021). *Novel low-temperature electrolyte using isoxazole as the main solvent for lithium-ion batteries*. *ACS Applied Materials & Interfaces* 13 (21): 24995–25001.

71 Wang, C., Hu, Q., Hao, J. et al. (2021). *The electrolyte additive effects on commercialized Ni-rich $LiNi_xCo_yMn_zO_2$ $(x + y + z = 1)$ based lithium-ion pouch batteries at high temperature*. *ACS Applied Energy Materials* 4 (3): 2292–2299.

72 You, L., Duan, K., Zhang, G. et al. (2019). *N,N-dimethylformamide electrolyte additive via a blocking strategy enables high-performance lithium-ion battery under high temperature*. *The Journal of Physical Chemistry C* 123 (10): 5942–5950.

73 Feng, T., Yang, G., Zhang, S. et al. (2022). *Low-temperature and high-voltage lithium-ion battery enabled by localized high-concentration carboxylate electrolytes*. *Chemical Engineering Journal* 433: 134138.

74 Zhou, L., Zuo, T.-T., Kwok, C.Y. et al. (2022). *High areal capacity, long cycle life 4 V ceramic all-solid-state Li-ion batteries enabled by chloride solid electrolytes*. *Nature Energy* 7 (1): 83–93.

75 Manafi, P., Nazockdast, H., Karimi, M. et al. (2021). *A study on the microstructural development of gel polymer electrolytes and different imidazolium-based ionic liquids for dye-sensitized solar cells*. *Journal of Power Sources* 481: 228622.

76 Zheng, T., Cui, X., Chu, Y. et al. (2022). *Ultrahigh elastic polymer electrolytes for solid-state lithium batteries with robust interfaces*. *ACS Applied Materials & Interfaces* 14 (4): 5932–5939.

77 Costa, C.M., Lizundia, E., and Lanceros-Méndez, S. (2020). *Polymers for advanced lithium-ion batteries: state of the art and future needs on polymers for the different battery components*. *Progress in Energy and Combustion Science* 79: 100846.

78 Gören, A., Mendes, J., Rodrigues, H.M. et al. (2016). *High performance screen-printed electrodes prepared by a green solvent approach for lithium-ion batteries. Journal of Power Sources* 334: 65–77.

79 Sliz, R., Valikangas, J., Silva Santos, H. et al. (2022). *Suitable cathode NMP replacement for efficient sustainable printed Li-ion batteries. ACS Applied Energy Materials* 5 (4): 4047–4058.

80 Huang, J., Liu, B., Zhang, P. et al. (2021). *A low-cost and sustainable cross-linked dextrin as aqueous binder for silicon anodes in lithium-ion batteries. Solid State Ionics* 373: 115807.

81 Sandaruwan, R.D.L., Wang, B.-Y., SADR, M. et al. (2022). *Boosting the performances of water-processable $LiNi_{0.5}Co_{0.2}Mn_{0.3}O_2$ cathode with conventional white latex as binder ingredient. Chinese Journal of Analytical Chemistry* 50 (7): 100101.

82 Radloff, S., Scurtu, R.G., Hölzle, M. et al. (2022). *Water-based $LiNi_{0.83}Co_{0.12}Mn_{0.05}O_2$ electrodes with excellent cycling stability fabricated using unconventional binders. Journal of the Electrochemical Society* 169 (4): 040514.

83 Brilloni, A., Marchesini, F., Poli, F. et al. (2022). *Performance comparison of LMNO cathodes produced with pullulan or PEDOT:PSS water-processable binders. Energies* 15 (7): 2608.

84 Wang, T., Wu, F., Yang, W. et al. (2022). *Facile fabrication of hierarchical porous silicon/N-doped carbon composites via biomass fermentation treatment for high-performance lithium-ion batteries. Journal of Alloys and Compounds* 898: 162781.

85 Sharma, I., ADK, D., Pham, H.D. et al. (2022). *Zero-waste: Carbon and SiO_2 composite materials from the solid residue of the hydrothermal liquefaction of anaerobic digestion digestate for Li-ion batteries. Sustainable Materials and Technologies* 31: e00364.

86 Sharma, S.S., Crowley, P.J., and Manthiram, A. (2021). *Aluminum–silicon alloy foils as low-cost, environmentally friendly anodes for lithium-ion batteries. ACS Sustainable Chemistry & Engineering* 9 (43): 14515–14524.

87 Huang, R.-A., Guo, Y., Zhang, Z. et al. (2022). *Insight into the self-assembled three-dimensional sandwich-like hollow silicon nanoarray/graphene lithium storage architecture by sonication-assisted functionalization. Energy & Fuels* 36 (6): 3283–3292.

88 Lan, J., Hou, H., Huang, B. et al. (2022). *The positive role of vitamin C in spindle-like $LiFePO_4/C$ cathode derived from two wastes. Ionics* 28 (4): 1583–1593.

89 Li, M., Yang, J., Shi, Y. et al. (2022). *Soluble organic cathodes enable long cycle life, high rate, and wide-temperature lithium-ion batteries. Advanced Materials* 34 (5): 2107226.

90 Shi, W., Wang, J., Zhang, X. et al. (2022). *11,11,12,12-tetracyano-9,10-anthraquinonedimethane as a sustainable cathode for room temperature all solid-state lithium battery. International Journal of Energy Research* 46 (6): 7686–7693.

91 Zhou, S., Nyholm, L., Strømme, M. et al. (2019). *Cladophora cellulose: unique biopolymer nanofibrils for emerging energy, environmental, and life science applications. Accounts of Chemical Research* 52 (8): 2232–2243.

92 Wang, Z., Hu, M., Yu, X. et al. (2021). *Uniform and porous nacre-like cellulose nanofibrils/nanoclay composite membrane as separator for highly safe and advanced Li-ion battery. Journal of Membrane Science* 637: 119622.

93 Xu, P., Zhang, D., Shao, Z. et al. (2021). *Cellulose acetate-based separators prepared by a reversible acetylation process for high-performance lithium-ion batteries. Journal of Applied Polymer Science* 138 (30): 50738.

94 Gonçalves, R., Lizundia, E., Silva, M.M. et al. (2019). *Mesoporous cellulose nanocrystal membranes as battery separators for environmentally safer lithium-ion batteries. ACS Applied Energy Materials* 2 (5): 3749–3761.

95 Reizabal, A., Gonçalves, R., Fidalgo-Marijuan, A. et al. (2020). *Tailoring silk fibroin separator membranes pore size for improving performance of lithium ion batteries. Journal of Membrane Science* 598: 117678.

96 Reizabal, A., Fidalgo-Marijuan, A., Gonçalves, R. et al. (2022). *Silk fibroin and sericin polymer blends for sustainable battery separators. Journal of Colloid and Interface Science* 611: 366–376.

97 Tong, C., Zhang, S., Zhong, T. et al. (2021). *Highly fibrillated and intrinsically flame-retardant nanofibrillated cellulose for transparent mineral filler-free fire-protective coatings. Chemical Engineering Journal* 419: 129440.

98 Yang, Y., Li, N., Lv, T. et al. (2022). *Natural wood-derived free-standing films as efficient and stable separators for high-performance lithium ion batteries. Nanoscale Advances* 4 (7): 1718–1726.

99 Zhang, T.-W., Chen, J.-L., Tian, T. et al. (2019). *Sustainable separators for high-performance lithium ion batteries enabled by chemical modifications. Advanced Functional Materials* 29 (28): 1902023.

100 Uddin, M.-J., Alaboina, P.K., Zhang, L. et al. (2017). *A low-cost, environment-friendly lignin-polyvinyl alcohol nanofiber separator using a water-based method for safer and faster lithium-ion batteries. Materials Science and Engineering B* 223: 84–90.

101 Costa, C.M. and Lanceros-Mendez, S. (2021). *Recent advances on battery separators based on poly(vinylidene fluoride) and its copolymers for lithium-ion battery applications. Current Opinion in Electrochemistry* 29: 100752.

102 Bicy, K., Mathew, D.E., Stephen, A.M. et al. (2022). *Sustainable lithium-ion battery separators derived from polyethylene oxide/lignocellulose coated electrospun P(VDF-TrFE) nanofibrous membranes. Surfaces and Interfaces* 29: 101716.

103 Costa, C.M., Lee, Y.-H., Kim, J.-H. et al. (2019). *Recent advances on separator membranes for lithium-ion battery applications: From porous membranes to solid electrolytes. Energy Storage Materials* 22: 346–375.

104 Barbosa, J.C., Gonçalves, R., Costa, C.M. et al. (2022). *Toward sustainable solid polymer electrolytes for lithium-ion batteries. ACS Omega* 7 (17): 14457–14464.

105 Zhou, Q., Ma, J., Dong, S. et al. (2019). *Intermolecular chemistry in solid polymer electrolytes for high-energy-density lithium batteries. Advanced Materials* 31 (50): 1902029.

106 Chen, Y., Wang, Y., Li, Z. et al. (2021). *A flame retarded polymer-based composite solid electrolyte improved by natural polysaccharides. Composites Communications* 26: 100774.

107 Li, C., Huang, Y., Chen, C. et al. (2022). *A high-performance solid electrolyte assisted with hybrid biomaterials for lithium metal batteries. Journal of Colloid and Interface Science* 608: 313–321.

4

Solid Batteries Chemistries Beyond Lithium

Mary York[1], Karl Larson[1], Kailot C. Harris[1], Eric Carmona[1], Paul Albertus[1], Rosy Sharma[2], Malachi Noked[3], Ela Strauss[4], Heftsi Ragones[5], and Diana Golodnitsky[6]

[1] *Department of Chemical and Biomolecular Engineering, Maryland Energy Innovation Institute, University of Maryland, 1206B Engineering Lab Building, UMD College Park, 20740, USA*
[2] *Department of Chemistry, Indian Institute of Technology (Banaras Hindu University), Varanasi, Uttar Pradesh, 221005, India*
[3] *Department of Chemistry, Bar-Ilan University, Ramat Gan, 52900, Israel*
[4] *Israel Science Foundation, Jerusalem, 9104001, Israel*
[5] *Faculty of Engineering, Holon Institute of Technology, Holon, 5810201, Israel*
[6] *School of Chemistry, Tel Aviv University, Tel Aviv, 6997801, Israel*

4.1 Introduction

Over the past decades, our increasing knowledge of environmental processes has led to a growing awareness of anthropogenic climate change and its alarming consequences for future generations. Unfortunately, it is not enough to reduce the horrendous carbon footprint of common energy facilities and transportation by transition toward the production of clean energy and usage of emission-free transportation. Typically, solar or wind energy rely on suitable energy conversion and storage technologies, either as a result of intermittent energy production of renewable energy sources or of clean mobility needs. Electrochemical energy storage in the form of rechargeable batteries is the most efficient solution for various kinds of applications, including mobile and stationary storage systems. According to the United States Geological Survey [1], in 2020, over 70% of the annual production of lithium was used for batteries. Annual production in 2020 was about 85 000 metric tons of lithium, with three countries (Australia, Chile, and China) providing more than 85% of that annual production. Identified lithium reserves (i.e. lithium content that can be economically extracted at current prices) are 21 million metric tons. In addition, lithium resources (i.e. lithium content in a form that may be economically extracted at the some point) are 86 million metric tons. Of those resources, about 50 million metric tons may be found in Bolivia, Argentina, and Chile. The current

Mary York, Karl Larson, Kailot C. Harris, and Eric Carmona contributed equally.

Sustainable Energy Storage in the Scope of Circular Economy: Advanced Materials and Device Design, First Edition. Edited by Carlos Miguel Costa, Renato Gonçalves, and Senentxu Lanceros-Méndez.

price of lithium carbonate, the lithium salt often used in the synthesis of cathode active materials, is about 8000US$/metric ton. The greatest short-term constraint for lithium supply is likely in scaling up production to meet the 5 to 10 times projected increase in annual Li-ion production over the next 15 years. Research in "beyond lithium" rechargeable electrochemical energy storage systems based on alternative anodes has been the subject of much research in the past decade, of which there are many extensive reviews on the topic. A range of metal and alloy anodes includes sodium, potassium, and magnesium and more are under consideration.

Sodium-ion batteries (NaBs) are gaining increasing attention since metallic sodium and sodium salts are less expensive and more abundant than lithium metal and lithium salts used in lithium-ion batteries (LIBs). Sodium salts (including NaCl in the ocean) are more widely distributed and are of greater abundance and lower cost. In the case of sodium carbonate alone, world production was about 55 million metric tons in 2020, nearly a thousand times that of lithium carbonate. Reserves of sodium carbonate are more than 25 billion metric tons, and the price in 2020 was about 150US$/ton (lower than lithium carbonate by more than a factor of 50) [1]. Other sodium salts (especially NaCl) are even more widely available and cost less than sodium carbonate. Metal-oxide cathodes, for sodium systems with the composition Na_xMO_2, where M is a transition metal (e.g. Co, Mn, V, Ni, etc.), have been widely researched. Polyanion phosphate cathodes have also seen widespread development in sodium batteries because of their high theoretical capacities and operating voltage (154 $mAh\,g^{-1}$ at 2.9 V vs. Na/Na^+ for $NaFePO_4$), excellent thermal stability, and tunable characteristics for improved intercalation rates.

Rechargeable potassium-ion batteries (KIBs) are promising low-cost alternatives to lithium-ion technology for large-scale applications and potential high-voltage energy storage systems. One of the crucial advantages of potassium is its relatively high terrestrial reserves in the Earth's crust (2.58 wt% vs. 0.0017 wt% of Li), which makes it a sustainable option to share the load of LIBs. The standard redox potential of potassium is very close to that of lithium in nonaqueous electrolytes, enabling its operation as an anode in high-voltage-battery systems, thus providing high-power density. Faster ion-transport kinetics of K^+-based electrolytes attributed to lower Stokes radius of solvated K^+ ion is another aspect to be considered when seeking improved rate performance (Figure 4.1). Unlike sodium, potassium forms stable intercalation compounds with graphite. Thus, graphite can be used as the anode for KIB, just as in Li-ion batteries, meaning that the production line of the current Li-ion batteries can be adjusted with that of K-ion batteries. On the other hand, the heavy mass and bulky size of K^+ ions result in compromised energy density and uncontrollable volume expansion that need to be addressed. Furthermore, because of the inherently high reactivity of potassium compared to other alkali metals, the safety concerns of KIBs remain a significant challenge, particularly with the use of highly flammable nonaqueous organic electrolytes.

Among multivalent metals, the closest to realization as anode in a rechargeable "rocking-chair" battery is magnesium [3, 4]. Magnesium, as compared to lithium, has a lower impact on the environment (extraction, disassembly, and recyclability) in terms of abundance and very high volumetric specific capacity (2080 $mAh\,cm^{-3}$ for Li and 3800 $mAh\,cm^{-3}$ for Mg). In addition, magnesium has an advantage over sodium and

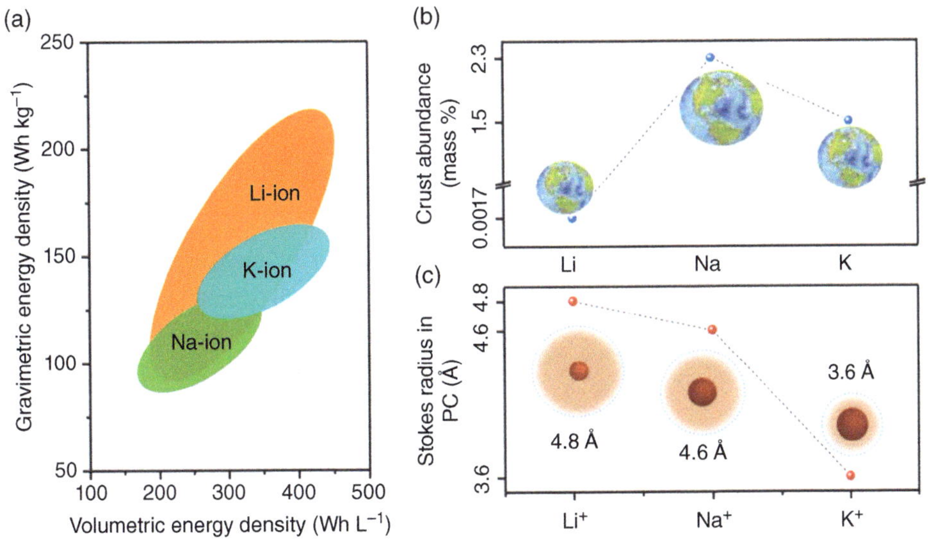

Figure 4.1 (a) Comparison of LIB, SIB, and PIB in terms of energy density. (b) Abundance of lithium, sodium, and potassium metal in earth's crust (wt%). (c) Stokes radii of Li^+, Na^+, and K^+ in propylene carbonate (PC). *Source:* Reproduced with the permission from [2]/American Association for the Advancement of Science.

potassium in terms of safety (higher redox potential and lower reactivity allowing handling in ambient air) and specific gravimetric and volumetric capacity. Although Mg batteries have been under investigation for several decades, their practical application is restricted as a result of the absence of suitable high-voltage cathode material that can overcome the sluggish diffusion of the Mg^{2+} ion into the host. To overcome the sluggish mobility of magnesium ions, V_2O_5 nanowires and spherical particles have been considered as promising cathodes. Metal sulfides, MS_2 (M = Ti, Zr, Hf, Nb, Ta, Mo, W, V), providing an operating voltage of 1.8 V have been suggested as possible cathode materials for an Mg/MS_2 battery.

The discovery and development of nonflammable alkali-ion and alkaline-earth-ion conducting solid electrolytes have paved the way for solid-state batteries (SSBs), which have become the focus of research by both scientific and industrial communities because of their high safety and energy density. SSBs are considered to be ecofriendly systems because they use thermally and electrochemically stable solid electrolytes instead of liquid organic electrolytes to conduct ions. They have advantages in power and energy per given weight and volume [5]. When combined with energy harvesting, SSBs can leave a device in the field almost indefinitely. SSBs work over a very broad range of temperatures, making it possible to use them for ruggedized applications.

Extensive research has been carried out recently by numerous scientific and industrial communities to speed up innovation in this field, and although noticeable achievements have been made, the large-scale application of beyond-lithium SSBs still faces numerous problems ranging from fundamental understanding to industrial manufacturing.

This chapter addresses recent studies, challenges, and perspectives of sodium, potassium, and magnesium SSBs.

4.2 Brief Overview of Solid Alkali-Ion and Alkaline-Earth-Ion Electrolytes

The solid electrolyte lies at the heart of the SSB concept and is an ionic conductor with negligible electronic conductivity. High-performance electrolytes, compatible with anode and cathode materials, are key to the success of an advanced, beyond-lithium battery. Unlike liquid-electrolyte systems, the performance of a solid electrolyte dominates the overall performance of a battery. However, no single current material can satisfy the requirements for solid electrolytes – high ionic conductivity, combined with good mechanical properties and flexibility, wide electrochemical-stability window, interface compatibility, and good electrode-adhesion properties.

4.2.1 Types of Solid Electrolytes

Usually, two types of solid electrolytes are explored: inorganic ceramic electrolytes and polymer electrolytes. Inorganic solid-state electrolytes (SSEs), also known as inorganic solid electrolytes (ISEs), are characterized according to their inorganic chemical species, such as oxides, sulfides, phosphates, and thiophosphate [6–8]. The ionic conductivity of a solid ceramic is described by the activated hopping of the cation from one occupied lattice site to an adjacent empty lattice site [8]. The ion-transport mechanism in ISEs is determined by the crystal structures that allow facile diffusion of one mobile species along topologically favored diffusion pathways. Thus, researchers are constantly seeking materials that can accommodate large amounts of mobile alkali-metal ions and have low ion-transport energy barriers (i.e. high ion conductivities), which are facilitated by high concentration of microscopic defects or disorder. A perfect crystal of an ionic compound would be an insulator. The structures that permit rapid ion transport are generally disordered, channeled, or layered [9–13]. Figure 4.2a illustrates the typical crystallographic structures of Si-doped NASICON, $Na_{1+x}Zr_2Si_xP_{3-x}O_{12}$ (NZSP) ($0 \leq x \leq 3$). Varying with the concentration of Si ions, the structure of NASICON will transform from rhombohedral to monoclinic structure within the concentration range of $1.8 \leq x \leq 2.2$ [10, 12]. For both structures, the sodium ions migrate through the triangular "bottleneck" areas consisting of three oxygen atoms in PO_4/SiO_4 and ZrO_6 clusters (see Figure 4.2b). The sodium-ion diffusion energy barrier within NZSP is strongly influenced by the sizes of these bottlenecks, which determine relatively low total ionic conductivity at room temperature (RT) (<0.1 mS cm^{-1}) [13]. Some materials with open one-, two-, or three-dimensional (3D) channels can achieve high ionic conductivities because mobile cations such as Li^+, Na^+, K^+, and Mg^{2+} can quickly diffuse along these channels. Glass and glass ceramic sodium-ion sulfides (Na_3PS_4, Na_3SbS_4, $Na_{11}Sn_2PS_{12}$, and $Na_{2.88}Sb_{0.88}W_{0.12}S_4$) have been found to exhibit high ionic conductivities of more than 1 mS cm^{-1} at room temperature [14]. As in the case of sodium-ion conductors, the diffusion of cations in $Mg_{0.5}Zr_2(PO_4)_3$ electrolyte occurs across the channels formed by three PO_4 tetrahedra and three ZrO_6 octahedra forming a 3D network of tunnels [15]. In SSBs, sluggish charge transfer across the SE/electrode interface is typically observed. It is caused by high and unstable interfacial resistance and is usually the limiting factor in the poor battery performance. The origins of the interfacial resistance are found in one or more

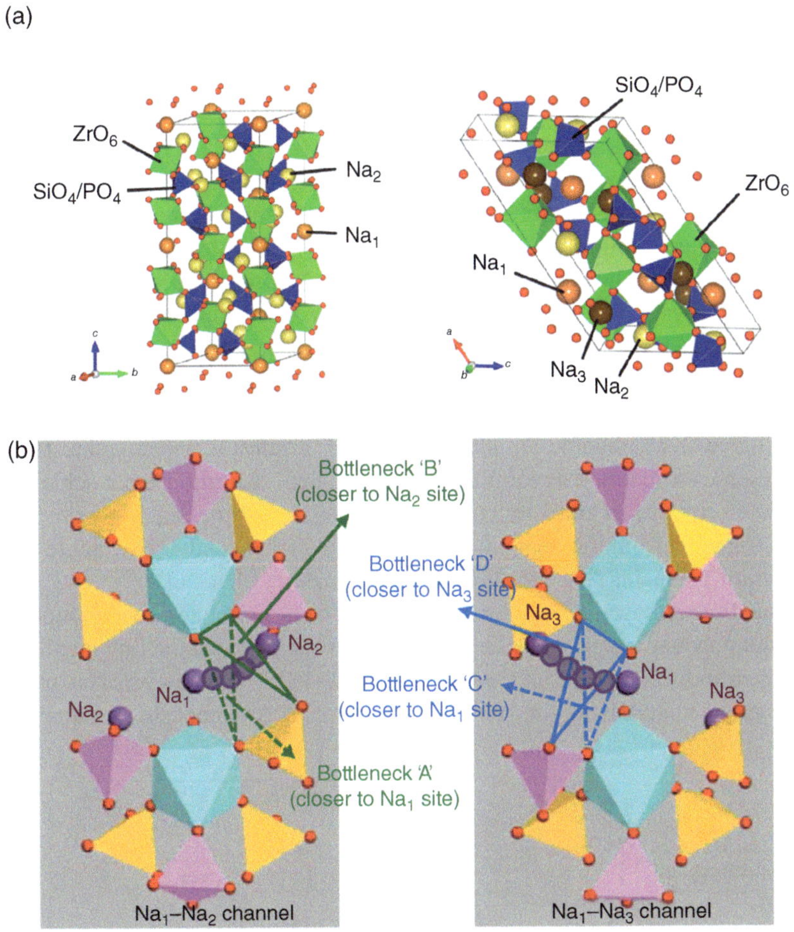

Figure 4.2 (a) Crystal structure of the NASICON-type $Na_{1+x}Zr_2Si_xP_{3-x}O_{12}$ (NZSP), rhombohedral (left) and monoclinic (right) structure. *Source:* Reproduced from Ref. [10]/With permission of American Chemical Society. (b) Illustration of four different types of bottlenecks in sodium migrating pathways in the monoclinic NZSP. *Source:* Reproduced from Ref. [12]/With permission of American Chemical Society.

factors [16]: poor physical interfacial contact, mechanical failure of the contact with volume variations, formation of lithium/sodium-depleted space-charge layers due to the large chemical potential difference between ceramics and electrode materials, and degradation at the interface that is caused by mutual diffusion of elements and/or reactivity. To improve contacts with electrodes, it was suggested [17] to form periodic misfit dislocations reconciling structural differences. Such epitaxy can tolerate large lattice mismatch, and thus may occur between many-layered electrodes and solid electrolytes. Moreover, it had been originally expected that solid ceramic electrolytes would be impervious to lithium dendrite-induced failure by virtue of their mechanical rigidity. However, recent reports have demonstrated the ability of alkali metals to penetrate solid materials [16].

Solid-polymer electrolytes (SPEs) for high-density rechargeable batteries have been investigated for decades because of their improved thermal stability compared to liquids and their lower density and cost compared to inorganic solids. Polymer electrolytes have multiphase structures at microscopic and/or macroscopic levels. Wright and coworkers were the first to measure the ionic conductivity of poly(ethylene oxide) (PEO)-salt complexes [17]. Intensive research has been carried out by Armand et al. and Bruce et al. [18, 19]. To date, PEO-based polymer electrolytes have been regarded as one of the most suitable electrolytes for lithium batteries. They consist of intricately distributed amorphous and various crystalline complexes of PEO and Li^+, which makes ion transport very complex [20–22]. A variety of relevant transport mechanisms includes the intrachain hopping of cations through the formation of a weak coordination shell between Li^+ ions and ether oxygens (EO), free ion motion along percolating channels in PEO melt, interchain ion hopping between different molecular chains or segments of PEO, etc. (Figure 4.3). Most researchers agree that high ionic conductivity requires ultrafast segmental relaxation and, according to some estimates, the required rate of segmental relaxation is not achievable in dry polymer electrolytes at ambient temperatures [18–24]. Although it is possible to decouple the ion transport from the segmental dynamics, the ionic conductivity achieved at ambient temperature remains far below the required ~1 $mS\,cm^{-1}$ [24–26].

To overcome the disadvantages of neat ceramic and polymer electrolytes, solid composite electrolytes with polymer as a host and ceramic fillers as the main conducting medium have been recently introduced by several research groups [27–35]. When powdered ceramic and polymer electrolyte materials are used for the preparation of the composite, the quality

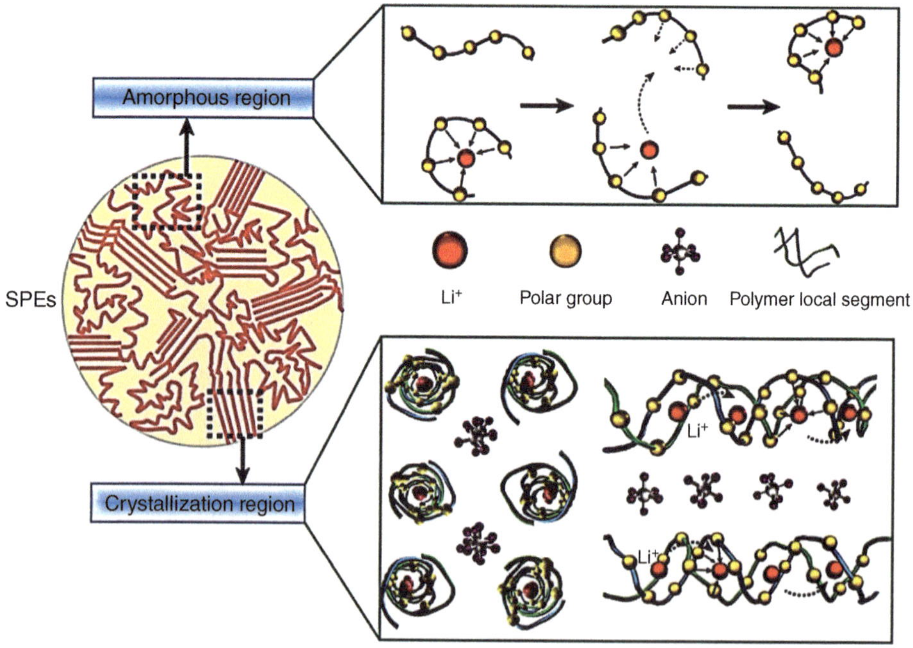

Figure 4.3 Schematic representation of ion-transport mechanism in amorphous and semicrystalline polymer electrolytes. *Source:* Chen et al. [20]/With permission of Royal Society of Chemistry.

of interfacial wetting, mutual adhesion, the microscopic porosity, and a rough surface topography of ceramics become the key considerations. Ceramic-in-polymer electrolytes with relatively low ceramic loading have higher room-temperature conductivity of ~10^{-5} S cm^{-1}, compared to polymer electrolytes without a ceramic filler (~10^{-7} S cm^{-1}). The common explanation of this phenomenon is that the addition of a few percent of ceramic particle fillers hinders polymer crystallization, enabling segmental relaxation in amorphous parts and, therefore, higher ionic conductivities and low and stable solid–electrolyte interphase (SEI) [36]. Polymer-in-ceramic composite electrolytes contain high concentrations of ceramics (ion-charge carriers) with a minimal amount of polymer. The major ion-conduction path is assumed to be through the ceramic material and through its interface, yielding high values of conduction. For example, Goodenough et al. reported that the highest conductivity of about 10^{-4} S cm^{-1} at 55 °C was obtained for an 80 wt% $Li_{6.4}La_3Zr_{1.4}Ta_{0.6}O_{12}$ (LLZTO) in PEO, i.e. in a polymer-in-ceramic electrolyte [32, 35].

An additional method for enhancing the ionic conductivity of a polymer electrolyte is by the introduction of plasticizers, like ionic liquids (ILs), which are molten organic salts at room temperature. ILs are characterized by low vapor pressure and reduced flammability, while maintaining high thermal and electrochemical stability. Quasi-solid electrolytes plasticized by IL are termed ionogel polymer electrolytes (IGPEs) [37, 38]. IGPEs can not only effectively eliminate the leaking risk of liquid electrolytes but also address the issues of inferior interface contact and low ionic conductivity of all SSEs.

4.2.2 Insights and Developments Regarding Metal Dendrites in Solid Electrolyte Systems

The discovery and development of nonflammable inorganic alkali-ion conducting solid electrolytes for higher energy density batteries has led to extensive studies of dendrite formation and propagation of metal SSBs [39–47]. These results emphasize the importance of reducing the aspect ratio of boundaries and defects in SEs and minimizing nonuniform current densities at these defects [47]. Dendrite formation is found as a common failure mechanism for solid-electrolyte/metal-anode batteries. Dendrites may penetrate the electrolyte and contact the cathode, leading to short-circuit and potential thermal runaway. Understanding the relation between dendrite formation mechanisms, electrode and electrolyte properties, and operating conditions (current density, stack pressure, temperature, etc.) is the key to developing strategies for suppressing dendrite growth and enabling practical alkali-metal batteries.

There are several hypotheses concerning dendrite formation and propagation mechanisms, which are common for lithium and sodium batteries. Deposition into electrolyte surface flaws or grain boundaries can be accelerated by current converging, causing pressure buildup at crack tips and subsequent electrolyte fracture and metal penetration. In addition, increased electronic conductivity at grain boundaries has been posited to provide a supply of electrons, allowing for metal deposition within the electrolyte rather than at the initial electrolyte/electrode interface. These mechanisms may occur independently or in conjunction depending on the electrolyte and operating conditions of the cell.

Operation at low current densities, below a "critical current density" (CCD), i.e. the current density at which dendrite initiation and propagation begins, and which also depends on other testing conditions, such as the duration of plating and previous stripping steps [48–51], leads to uniform current distributions, preventing dendrite initiation and propagation. Electrode, electrolyte, and interfacial properties can affect the CCD. Figure 4.4 shows mechanisms and highlights key properties affecting dendrite initiation and propagation. As mentioned earlier, other mechanisms are also hypothesized (e.g. Li deposition within a bulk of solid electrolyte), but in our view, dendrite propagation in brittle ceramic solid electrolytes via stress-driven crack opening has the strongest current empirical basis.

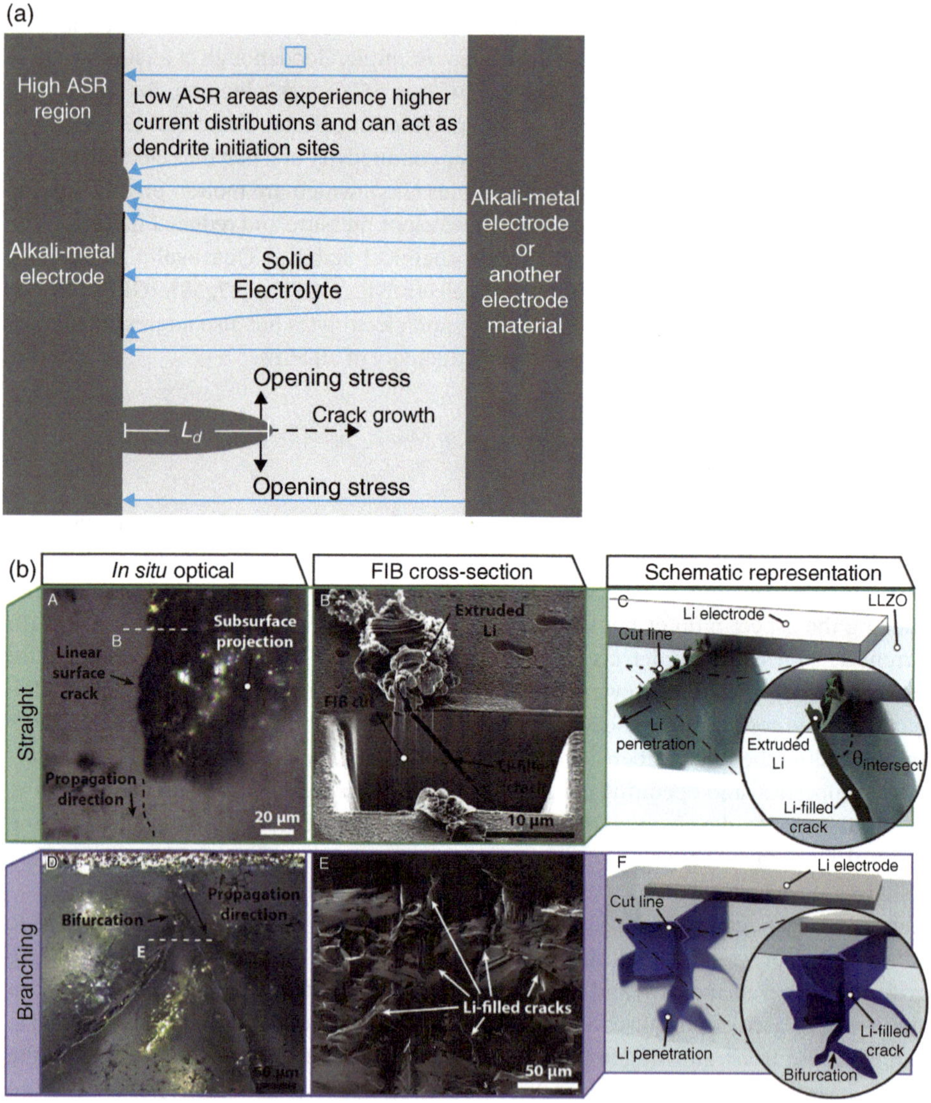

Figure 4.4 (a) Dendrite initiation and propagation schematic at a solid electrolyte/alkali metal interface. (b) Optical images ([A],[D]), SEM images ([B], [E]), and schematics ([C],) of straight Li-penetration type ([A]–[C]), branching type ([D]–[F]). *Source:* Kazyak [39]/With permission of Elsevier.

In addition to the electrolyte properties mentioned earlier, such as the ionic conductivity, κ, the electronic conductivity, σ, the Young's Modulus, E, the fracture toughness, K_{1c}, the variations of these properties within grain boundaries are extremely important for dendrite formation. Higher ionic conductivity will lead to more uniform current distribution according to the principles of primary and secondary current distributions with a varying interelectrode distance. Low electronic conductivity is required to prevent metal plating inside the electrolyte. If stresses generated at the crack tip during plating cause a stress intensity factor exceeding the fracture toughness, dendrite propagation will continue as indicated in Figure 4.4. Electrode/electrolyte interfaces are prone to defects and contaminants. Defects include surface roughness, incipient cracks, grain boundaries, among others and can lead to poor interfacial contact and inhomogeneous current distributions, both of which facilitate dendrite initiation. Surface contaminants can create interfaces with varying area-specific resistance (ASR), leading to inhomogeneous current distributions as shown in Figure 4.4. Additionally, other operating parameters, such as stack pressure and temperature, can influence plating stability. Stack pressure can help ensure contact between electrodes and the electrolyte. Mechanical properties such as E and creep properties as well as electrochemical properties such as ionic conductivity and ASR are functions of temperature, meaning that CCD is a function of temperature [52]. There have been numerous studies on dendrite formation on sodium- and lithium-metal electrodes. Investigation of potassium dendrites has been less researched, and there are some studies investigating other metal electrode dendrites such as Ca, Mg, Zn, and Ag [53–57].

4.2.2.1 Metal Growth Through Na Ceramic Solid Electrolytes

Short circuits caused by dendrite growth in molten-Na and room-temperature Na batteries with Na-β″-alumina electrolytes were first investigated by Armstrong et al. at 1970–1980s [58–63]. They theorized that the propagation mechanism for the dendrites was based on the expansion of micro-cracks due to Poiseuille pressure. Based on acoustic emission probe tests, Worrell and Redfern [60] concluded that electronic conduction in the electrolyte was associated with the presence of Na dendrites causing crack growth and that mechanical deterioration of the electrolyte was dependent on current density. Virkar et al. found that nonuniform current density in Na-β″-alumina at room temperature can exist at blocked or unwetted areas at the Na/Na-β″-alumina interface [61]. In addition, models were created that assessed crack origin and propagation and their relation to thermodynamics, mechanics, electrochemistry, and material properties [59, 62, 63]. More recently, there has been renewed investigation into dendrite formation in room-temperature sodium-metal SSBs [64–68]. Li et al. [65] established that the electron configuration in the carbon nanotubes induces instability in Na–Na dimers, preventing clustering that leads to dendrite formation and favoring evenly distributed Na adatoms at the surface of the current collector. Wang et al. demonstrated that doping of the ceramic electrolyte NASICON can lead to improved ionic conductivity and density and lowered electronic conductivity [67].

For brittle ISEs, propagation of dendrites, regardless of morphology, requires electrolyte fracture. Various groups have developed electrochemical–mechanical models detailing the stresses generated during lithium deposition and the resulting fracture mechanics of the electrolyte driven by dendrite growth [45, 69–72]. Fracture occurs when the internal stresses reach the maximum strength the solid bulk electrolyte can support, either within a bulk phase (e.g. a grain) or at a boundary between bulk regions (e.g. a grain boundary). Dendrite propagation should cause mode I fracture, which is characterized by crack opening via tensile stresses normal to the crack plane. Porz et al. developed a simple model, considering only linear elastic mechanics, for crack extension and dendrite propagation [45]. On the basis of their criteria for crack extension, the stress required for crack extension is inversely proportional to the square of the flaw length, meaning that the overpotential required for crack extension decreases with increasing flaw length. Barroso-Luque et al. developed a model for the coupled transport, plastic deformation, and elastic response due to metal deposition into a solid electrolyte [70]. The authors calculated order-of-magnitude predictions for current focusing, strain rates, and pressure buildup on the basis of a semi-analytical solution of the governing equations. An electroplastic screening length, based on metal-mechanical properties and electrolyte ionic conductivity, was found as a critical length scale. Barai et al. developed a finite-element model that assesses the major current path resulting from mechanical stress and fracture in grain boundaries [69]. This work points out the fact that Li can deposit preferentially at grain boundaries rather than grain interiors because of the stress-induced current density. The enhanced deposition at the mechanically softer grain boundary can lead to dendrite growth.

It was shown in the past that molten Li and Na batteries exhibit current densities several orders of magnitude greater than those observed in all-SSBs of the same chemistry. Richman et al. demonstrated in 1975 that molten sodium coupled with Na-β″-alumina electrolytes had an experimental CCD on the order of 1 $A\,cm^{-2}$, a figure well beyond what is desired for today's commercial applications (tens of $mA\,cm^{-2}$) [59]. In comparison, sodium coupled with Na-β″-alumina at room temperature has a CCD on the order of 1–10 $mA\,cm^{-2}$, significantly lower than those observed with molten sodium [52, 73]. It is of note that the CCDs observed in Na/Na-β″-alumina cells are roughly one order of magnitude greater than those observed in similar Li/LLZO cells at the same temperature. Bay et al. [52] found that the diffusivity of Na in the bulk is smaller than that of Na^+ in Na-β″-alumina, by only a factor of 10, whereas the diffusivity of Li in the bulk is smaller than of Li^+ in LLZO by a factor of 100. In addition, the self-diffusivity of Na is roughly 100× that of Li at room temperature. It should be noted that sodium is a softer metal than lithium, with a room-temperature yield strength approximately one-quarter that of lithium. Temperature influences the interfacial ASR, ionic conductivity, and the mechanical properties of liquid metals, which drastically differ from those of the metal at room temperature. Determining which properties or combination of these properties enables stable cycling at high current densities could direct future work on room-temperature dendrite-free metal plating and stripping.

Dendrite formation in SPEs is substantially different from that in brittle ceramic-solid electrolyte and does not follow the mechanism outlined in Figure 4.4. In particular, SPEs do not exhibit the stress-driven crack-opening mechanism, as they do not generally undergo brittle fracture. SPEs are typically ductile materials that can undergo elastic and plastic deformation, and dendrites can penetrate these materials by mechanical deformation of the SPE or through pores [74–76]. Brissot et al. demonstrated experimentally that dendritic growth can occur at current densities near the limiting current, where the ionic concentration near the electrode decreases to zero, and can also occur at low current densities, likely as a result of local inhomogeneities. Experiments have shown that dendrite prevention in polymer electrolytes can be achieved with the use of mechanical stresses; however, a clear understanding of how stresses suppress dendrite growth remains elusive, and the suppression does not enable the current densities and per-cycle areal capacities desired for many applications [77]. In a seminal work, Newman and Monroe assessed the effect of linear elastic mechanics and resultant stresses on plating stability at the polymer–electrolyte/solid–metal–electrode interface [78]. On the basis of their analytical model with prestressed conditions for the lithium and polymer electrolyte, they determined that for a polymer electrolyte with a Poisson's ratio similar to that of PEO, and dendrites should be suppressed if the electrolyte's shear modulus is twice that of alkali metal. In their model, it was concluded that the mechanical states of both the electrode and electrolyte influence the current density at the interface. These authors carefully pointed out the limitations of their model, which limit its applicability to practical cell conditions, which is often not appreciated. The electrochemical–mechanical model of Barai et al. [79] studied the effect of altering the SPE's elastic modulus and yield strength. The authors concluded that increasing the yield strength of the polymer electrolyte can help stabilize dendritic protrusions by promoting plastic deformation of the lithium before the polymer electrolyte has yielded. Research efforts to mitigate dendrite growth in sodium- and potassium-metal batteries [80] have been directed at improving the mechanical properties of these electrolytes to aide in mechanical suppression of dendrites, as well as increasing the ionic conductivity to achieve parity with conventional liquid electrolytes (10^{-3} to 10^{-2} S cm^{-1} at 25 °C) and improving the cation transference number in order to reduce concentration polarization.

As mentioned above, in addition to ISEs and SPEs, solid composite electrolytes combine ISEs with SPE in an attempt to combine the advantages of each material and address their individual shortcomings [81]. Composite electrolytes face issues with interfacial stability. Obstacles can include the interfaces between the anode and electrolyte, cathode and electrolyte, and also the polymer and ceramic. Composite electrolytes typically have difficulty maintaining interfacial stability during extended plating and stripping cycling as well as at high current densities, with failure mechanisms associated with each of the material classes present. These interfacial issues have been demonstrated to lead to dendrite growth. Volume change of the metal electrode as well as the cathode material has been shown to cause interfacial instability. Improvement of the ionic conductivity and interfacial stability are required for solid composite electrolytes to successfully suppress dendrites.

4.3 Solid-State Sodium-Ion Batteries

Just as a principal reason for investigating solid-state sodium batteries (SSNaB) is the ability to use sodium metal. Sodium-based battery would, at least for the sodium content of the battery itself, be ideal in terms of diverse supply chains and low cost. Sodium metal and its salts are of greater abundance and lower cost than lithium. A disadvantage of sodium-based vs. lithium-based cells is its lower specific capacity (3860 mAh g^{-1} for lithium metal, vs. 1165 mAh g^{-} for sodium metal [82]) and the lower cell potential for sodium-based cells. The difference in standard potential for Li vs. Na is about 0.3 V, which corresponds to about 10% "loss" of the cell potential. Assuming the same cathode properties (mAh/g, density, thickness, porosity, etc.), solid electrolyte properties, current collectors, and loading (3.5 mAh cm^{-2}), a cell with Na metal would have a specific energy and energy density about

15–20% lower than a comparable Li-metal cell. Additional drawbacks to the use of sodium include its low melting point of 97 °C, its high reactivity in air, and, like lithium metal, formation and growth of dendrites in both liquid- and solid-electrolyte systems [31]. In addition, the atomic radius and mass of sodium ion are greater than those of lithium ion (0.98 Å and 23 g mol^{-1} vs. lithium's 0.69 Å and 6.9 g mol^{-1}) and can affect ionic conductivity, cathode-specific capacity, and other properties [83]. Despite these disadvantages, sodium batteries are of the utmost importance since the Na-metal cell's performance would be comparable to today's best Li-ion cells (~275 Wh kg^{-1} and ~650 Wh L^{-1}).

To date, the principal commercial deployment of Na-based batteries with a solid electrolyte is in molten-sodium cells such as NGK's Na/S cells, which are currently sold for stationary-energy-storage applications. Indeed, given the strong presence of lithium ion in current automotive applications, a likely role for future SSNaBs is in stationary-storage applications, allowing lithium resources to be principally used for the more demanding (in terms of energy content) transportation sector [84]. Increasing SSNaB performance requires addressing similar challenges facing solid-state lithium batteries, including interfacial stability and contact (at both electrodes), processing solid materials into a high-performance porous electrode, stable metal-anode cycling, and others.

4.3.1 Solid Electrolytes for Sodium Batteries

An ideal SSNaB contains a solid electrolyte with high room-temperature ionic conductivity, interfacial compatibility with sodium metal, a wide electrochemical stability window, and strong mechanical properties in order to withstand volume changes and ensure proper contact with the anode [31, 83]. Figure 4.5 provides measured ionic conductivities for

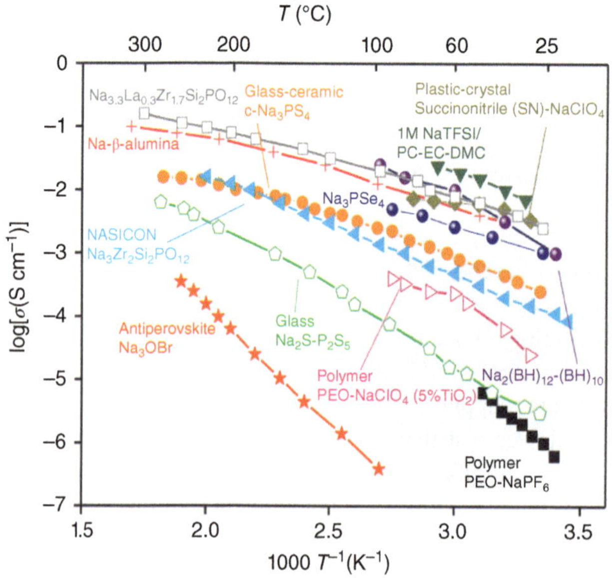

Figure 4.5 Measured ionic conductivities of various Na-ion-conducting solid-state electrolytes. *Source:* Reproduced from Ref. [31]/With permission of [Elsevier].

various solid-state electrolyte (SSE) chemistries over a range of temperatures. While most SSEs exhibit sufficient mechanical strength, side reactions that may damage the mechanical integrity of the SSE occur with the sodium metal because of thermodynamic instabilities [82]. As a result, engineering the interface between sodium metal and the SSE is a primary area of research interest [85]. The characteristics and limitations of various SSE chemistries will be discussed for ISEs and SPEs.

Similar to lithium-ion ceramic conductors, the principal migration mechanism of Na^+ in the crystalline ISEs mainly can be divided to three types: (i) vacancy diffusion mechanism: the mobile charged species migrate or hop from one vacant site to the adjacent vacant site. (ii) Direct interstitial mechanism: the mobile species with a small size can hop from one interstitial site to the unoccupied neighboring interstitial site. (iii) Correlated interstitial mechanism: upon migration interstitial Na^+ can enter into the adjacent site by substituting the adjacent lattice ion. Therefore, the charge carrier concentration, diffusion type, defect concentration, and migration path greatly influence the thermodynamics and kinetics of Na^+ conduction in the crystal structures of the ISEs [86]. In recent decades, the widely studied ISEs for sodium-based batteries are mainly focused on the oxide-based SSEs (Na–β"-alumina electrolyte and sodium superionic conductor electrolyte [NASICON]) and sulfide-based SSEs, which are characterized by high room-temperature ionic conductivity (~10^{-3} S cm^{-1}). These two oxides have been successfully deployed in SSBs achieving more than 10 000 cycles. However, they are also known to have poor interfacial compatibility with some electrodes, resulting in poor ion conduction at the electrode–electrolyte interface [82]. Sulfides can be more conductive than oxides because of the greater polarizability of sulfide than that of oxide, but sulfides can also be more reactive [82, 86]. Sulfide solid electrolytes, such as Na_3PS_4 or $Na_2S–P_2S_5$, exhibit high ionic conductivities, as shown in, but are susceptible to decomposition by sodium metal, impeding their practical application until a stable SEI can be formed [87].

Polymer-based electrolytes are a promising route to enable SSNaBs because they typically exhibit low flammability, high flexibility, and form robust interfacial contact with anodes. A major advantage of polymer electrolytes over inorganic ionic conductors is ease of processing, including the ability to process into thin films (e.g. <20 microns) and roll-to-roll processing. However, SPEs are not yet suitable for SSNaB application because they suffer from relatively low ionic conductivities at room temperature (e.g. 10^{-6} to 10^{-8} S cm^{-1}), as shown in Figure 4.5. SPEs also exhibit electrochemical instabilities that inhibit cycling performance [82]. Polyethylene oxide (PEO) is the most commonly used SPE material and is often paired with common Na-ion battery salts such as $NaClO_4$ or $NaPF_6$. Researchers have found limitations to PEO as a host material, such as its low oxidation potentials and insufficient mechanical properties at room temperature [88–90], as a majority of ionic transport occurs when the polymer is in the amorphous state, but polymer electrolytes tend to be crystalline at ambient temperature. A temperature of 60–80 °C is typically required to achieve reasonable ionic conductivities [86, 91]. PEO alternatives such as polyacrylonitrile (PAN) and polyvinyl alcohol (PVA) are also explored as promising solid-electrolyte alternatives for sodium batteries [92, 93]. Overall, a major challenge for SPEs is their low room-temperature ionic conductivity. Methods to increase the amorphous phase include, but are not limited, to blending, copolymerization, cross-linking, and doping additives [94].

In [38], an ionogel electrolyte has been developed based on investigations of phase behavior, morphology, and ion dynamics of a binary salt mixture of an organic ionic plastic crystal *N*-ethyl-*N*-methylpyrrolidinium bis(fluorosulfonyl)imide (C2mpyrFSI), with NaFSI supported on a mat of electrospun poly (vinylidene fluoride) nanofibers. It was found that there is no significant change of the conductivity in a composite electrolyte with poly(vinylidene fluoride (PVDF)-base composite ion gel electrolyte compared to the C2mpyrFSI-NaFSI mixture at lower temperatures. The conductivity for ionogel electrolyte at 35 mol% IL was $3\times10^{-3}\,S\,cm^{-1}$ at room temperature with a transference number of 0.44 at 50 °C. Na symmetric cells were demonstrated to cycle with high stability. In particular, a low polarization potential of less than 20 mV is demonstrated after plating/stripping 500 cycles (1000 hours) at $0.1\,mA\,cm^{-2}$, 50 °C, indicative of a stable solid–electrolyte interface layer formed on the Na-metal.

4.3.2 Anode Materials for Solid-State Sodium Batteries

Sodium metal is reactive with most liquid organic electrolytes [95–97]. The use of a solid-state electrolyte may help overcome these challenges, as well as stabilizing the anode by its passivation. In addition to the use of a sodium-metal foil, modifications of the sodium-metal anode include 3D structural nanoengineering to accommodate the volume change of sodium metal and stabilizing pristine sodium metal by alloying with other elemental metals [98, 99]. Pursuing a sodium-metal foil anode over a porous structure has advantages, such as higher theoretical energy density and lower cost and includes the ability to use aluminum current collectors rather than copper, as Na and Al do not alloy [100]. The high resistivity and poor stability of the sodium SEI present a challenge [101]. The low melting point of sodium metal is a possible safety hazard, but some researchers have taken advantage of this property, reporting good stability with molten sodium or liquid alloys [102, 103]. Additional anode candidates for SSNaBs are shown in Figure 4.6b.

Hard carbon (HC) is an attractive electrode material for NaB because it has a low potential (~0.1 V vs. Na/Na^+) and it is relatively inexpensive, nontoxic, lightweight, and abundant [100–103, 106]. However, because of sodium's greater ionic radius than that of lithium and its inability to form binary compounds with graphite, sodium does not intercalate suitably into graphite, and this results in poor rate capability, cyclability, and reversible capacity [107, 108]. Efforts to improve HC have focused on increasing the interstitial space for the larger Na^+ ion via manipulation of carbon nanotubes, expansion techniques, or doping [109]. Researchers observed a total capacity of $850\,mAh\,g^{-1}$ when doping graphene foam with nitrogen [110] and thermally expanded graphite reaches a reversible capacity of $284\,mAh\,g^{-1}$ [108, 111], although these values are often for small material quantities that may not translate into practical electrodes.

Conversion anodes store charge not by intercalating or inserting ions into a host matrix, but rather by undergoing an electrochemical reaction such as

$$M_aX_b + (b.n)Na^+ + (b.n)e^- \leftrightarrow aM + bNa_nX$$

where *M* is a metal with an oxidation state of *b*, *X* is an anionic species with an oxidation state of *a*, and *n* is the stoichiometric ratio of *Na* to *X* in the charged anode. Metal oxides

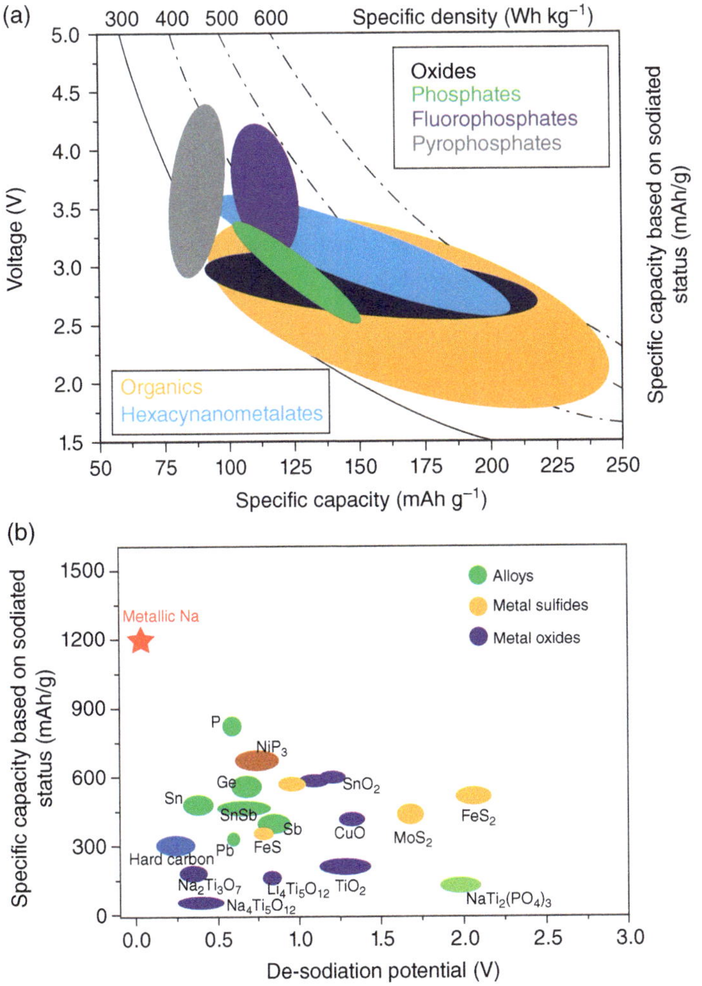

Figure 4.6 (a) Operating voltage (vs. Na/Na$^+$) vs. specific capacity for select Na-ion cathode material classes. *Source:* Reproduced from Ref. [104]/With permission of [Wiley]; (b) theoretical specific capacities and de-sodiation potentials of candidate anodes for solid-state Na batteries. *Source:* Reproduced from Ref. [105]/With permission of Elsevier.

are favored among conversion anodes for their intermediate voltage and moderate volume expansion (compared to sulfides and phosphides). Metal sulfides show promise through their high reversible capacities and moderate voltages. This is attributed to the weaker metal/sulfide bond, allowing for faster kinetics of conversion [112]. Compounds such as FeS_2 have displayed a high reversible capacity of 630 mAh g^{-1} at room temperature, while composites such as Ni_2S_3 have a lower reversible capacity (255 mAh g^{-1}) but a high initial capacity (~450 mAh g^{-1}) [113, 114]. Metals such as molybdenum, for example, in the form of MoS_2, require precise tuning for consistent capacity, but can be improved with carbon coatings [115]. Finally, metal phosphides show the greatest capacity among these conversion anode materials. Assuming complete sodiation to Na_3P, the theoretical specific capacity of P is 2596 mAh g^{-1}, and the redox potential of P is ~0.4 V vs. Na/Na^+, making it an attractive anode material. However, phosphides also undergo the largest volume change [108]. The performance of many of these candidate anode materials is displayed in Figure 4.6b

Na alloy anodes are attractive for SSBs because they inherit the high specific capacity of sodium metal while increasing the chemical compatibility of the anode with SSEs. Alloying Na with Sn can also decrease the ASR between the anode and SSE by >60× and improve sluggish Na^+ diffusion kinetics responsible for interfacial contact instabilities. Other sodium alloys, such as Na–Si and Na–Ti, have shown good cycling stability, which is attributed to improved Na^+ diffusion at the solid–solid interface [101, 116, 117]. The increased stability of Na–Sn alloys may also be due to the higher potential of these alloys, although an exact mechanism has yet to be determined.

Anode materials and electrolytes must be chosen with consideration for one another and the subsequent SEI, the study of which was pioneered by Peled [118]. The SEI is a passivating film formed instantaneously upon contact of the alkali metal with electrolyte. It consists of insoluble and partially soluble reduction products of electrolyte components. The thickness of the freshly formed layer is determined by the electron-tunneling range. SEI determines the safety, power capability, morphology of alkali-metal deposits, shelf life, and cycle life of the battery. For high performance of the battery, the SEI must be an electronic resistor in order to avoid SEI thickening leading to high internal resistance, self-discharge, and low faradaic efficiency. To eliminate concentration polarization and to facilitate the lithium dissolution-deposition processes, the cation transport number should be close to unity. To reduce overvoltage, the SEI should be highly ion conductive. In the case of the rechargeable battery, it is very important that there be uniform morphology and chemical composition in order to ensure homogeneous current distribution. The SEI must be both mechanically stable and flexible. Good adhesion to the anode is important as well. Practical primary or secondary alkaline or alkaline-earth batteries can be made only if the dissolution or corrosion of the anode can be stopped. Therefore, the electrolyte must be designed to contain at least one material and/or sacrificial additives that react rapidly with the alkali-metal anode to form an insoluble mechanically robust SEI [119]. SEI failure leads to the exposure of electrode-active material to the unstable electrolyte, thus consuming more electrolyte and creating new SEI films. Since sodium metal is easily contaminated and damaged by the crossover of the by-products that inevitably originate from the battery cathode, introduction of solid electrolyte, which may serve as an artificial barrier to reduce the detrimental anode/electrolyte interfacial reactions is considered as a promising approach to promote long-term cyclability [100, 120].

4.3.3 Cathode Materials for Solid-State Sodium Batteries

The development of SSNaBs can take advantage of cathode materials that are similar to their liquid-electrolyte counterparts, and a high-level overview of the key metrics of potential (vs. Na/Na^+) and specific capacity is provided in Figure 4.6a below. Research pushing the boundaries of the cathode performance of liquid Na-ion cells likely aids the development of cathodes for all SSNaB [82, 85, 121], and the development of some cathode materials is inspired by analogous LIB materials. Sodium's greater ionic radius and mass reduce its practical capacity in direct analogy with Li-ion materials (e.g. Li_xCoO_2 (~175 mAh g^{-1}) compared to Na_xCoO_2 (~70.4 mAh g^{-1} at 2.0–3.5 V)[122, 123]. Thus, the need to explore beyond Li-ion-based cathode materials has evolved and the field now includes a variety of materials ranging from transition-metal layered materials to phosphates. However, continued development of these materials must consider suitable solid electrolyte and anode materials to ensure stable cycling of the full cell. Poor interfacial contact and chemical/electrochemical stability of the cathode with the SE typically lead to capacity decay and eventual cell failure [124]. Research focused on improving interface contact has enlisted the use of wetting agents, mixed cathode materials, SE-coated cathodes, and more [82].

Metal oxides have been researched for applications in both Li- and Na-based batteries with the composition Na_xMO_2 where M is a transition metal (e.g. Co, Mn, V, Ni, etc.). This class of cathode materials can achieve a reversible cycling capacity greater than 200 mAh g^{-1} at close to 3 V vs. Na/Na^+[104, 125]. In addition, the operating potential can be tuned to desired voltages through the substitution of different transition metals and by varying the sodium content [126, 127]. Lithium substitution in the synthesis of these materials has led to higher operating voltages of 3.4 V vs. Na/Na^+ [128] However, during charging and discharging these materials can undergo numerous transition states, often leading to structural instabilities [125]. For example, $Na_{0.44}MnO_2$ exhibits seven intermediate phases in its voltage profile [122].

Polyanion phosphate cathodes have also seen widespread development in sodium batteries [104] because of their high theoretical capacities and operating voltage (154 mAh g^{-1} at 2.9 V vs. Na/Na^+ for $NaFePO_4$), excellent thermal stability, and tunable characteristics for improved intercalation rates, all of which attract researchers [129, 130]. In addition, sodium phosphate has been utilized as a surface coating for metal-oxide cathodes in order to improve cathode/electrolyte interfacial stability and charge-storage capacity [131]. Unfortunately, these cathode materials suffer from innately poor electronic conductivity. Attempts to improve this shortcoming involve the application of a carbon coating, reduction of particle size, and the tuning of surface morphology [132, 133]. Pyrophosphates are promising cathode materials with a wide redox potential range of 2.0 to more than 4.0 V vs. Na/Na^+ with an operating capacity of 86 mAh g^{-1} at 1 C rate for $Na_2FeP_20_7$ [104]. Fluorophosphates provide an even wider electrochemical-stability window, though because of their poor electronic conductivity, they exhibit lower capacities compared to metal oxides [104, 134].

Metal–organic cathode materials such as hexacyanometalates have demonstrated capacities on a par with metal-oxide cathodes, with some approaching 140 mAh g^{-1}[135, 136]. Prussian-blue materials are able to achieve higher capacities and voltage capabilities (up to 200 mAh g^{-1} at 3.5 V vs. Na/Na^+) because of their wider ion channels that can easily accommodate sodium ions [137]. However, Prussian-blue materials often have poor capacity retention, which has been attributed to synthesis issues resulting in crystal lattice defects. While methods are being developed to decrease the amount of crystal defects, Prussian-blue cathode materials require further particle optimization and surface treatment in order to improve cyclability and longevity [125].

All-cathode materials for SSNaB require stability and intimate contact with the solid electrolyte. Poor wetting, as well as chemical and electrochemical instabilities, has been

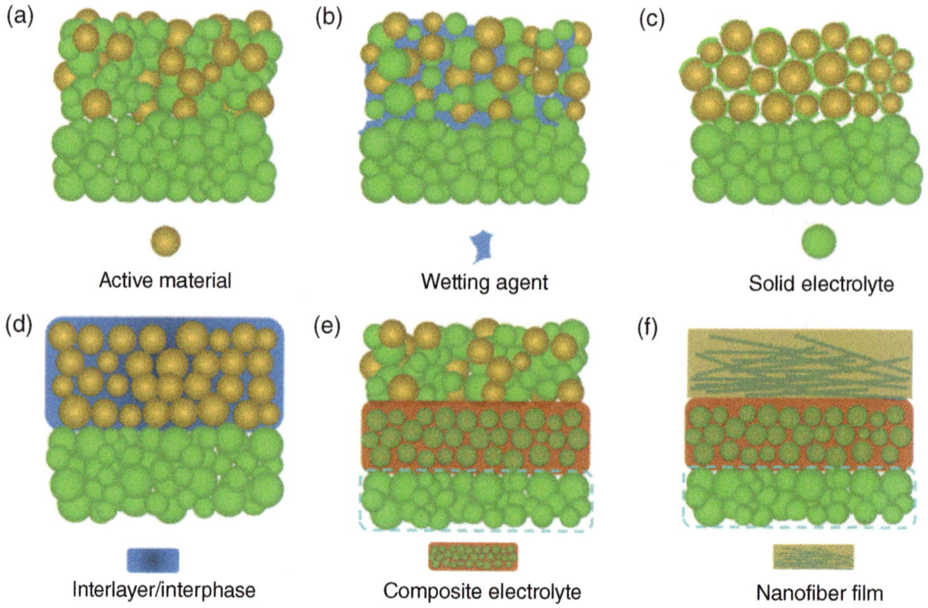

Figure 4.7 Summary of techniques to improve cathode active material mechanical contact, wetting, and electrochemical stability. (a) Composite cathode, (b) wetting agent, (c) coating by electrolyte, (d) adding interlayer, (e) composite electrolyte, and (f) nanofiber film of cathode. *Source:* Reproduced from Ref. [31]/With permission of [Elsevier].

mitigated by a variety of techniques as shown in Figure 4.7a–f. Composite cathodes consisting of both solid sulfide electrolytes and cathode-active material have been shown to provide reasonable cycling [87, 138]. Coating the cathodes with a layer of solid electrolyte or an insertion material (e.g. sodium phosphate) is also an effective technique for stabilizing metal-oxide cathodes and enhancing their longevity [131, 139, 140]. Glass-ceramic cathodes are yet another option for improving contact with the SE. Deposition of the viscous cathode material directly onto the electrolyte allows intimate contact to be formed by the fluid flow followed by crystallization [141, 142]. As the development of cathode materials for SSNaB continues into the next generation, special attention should be paid to ensuring proper interfacial electronic, mechanical, and chemical stabilities.

4.3.4 Solid-State Sodium Battery, Full-Cell Results

Na metal full cell using a $NaFePO_4$ cathode (1.75–4.0 V) and ionogel electrolyte [38] showed a good capacity retention with Coulombic efficiency close to 100% at various C rates between C/20, C/10, and C/5, achieving 120 mAh g^{-1} at C/20. Table 4.1 provides a summary of several SSNaB full-cell results, including the key cell materials, the cycles achieved cycling details, and comments. Zheng et al. [143] pursued fabrication and development of full cells utilizing SPEs. SPEs are easier to synthesize and handle than Na^+-conducting ceramics. The network-based hybrid polymer was grown to form a flexible, transparent film. The SPE in this study was first used in a symmetric cell; galvanostatic cycling at 80 °C showed stable cycling at 0.1 mA cm^{-2} for 5150 hours and at 0.5 mA cm^{-2} 3550 hours were achieved without short circuiting. The cumulative amount of charge passed was

Table 4.1 Representative full cell – components and cycling results.

Source	Anode	Electrolyte	Cathode	Cycle#	Final capacity, areal capacity (mAh g^{-1} active cathode mass, mAh cm^{-2})	C rate	Temperature (°C)	Comments
Zheng et al. [143]	Na metal	Solid polymer electrolyte (SPE)	δ-$Na_xV_2O_5$	50	123, 0.008		80	V cathode is more expensive and less abundant than Li
Song et al. [144]	Na metal	$Na_{3.1}Zr_{1.95}M_{0.05}Si_2PO_{12}$ (M = Mg, Ca, Sr, Ba	S–C composite	100	~170/ 0.28	1	20	Complex synthesis of electrolyte. Zr is more expensive and less abundant than Li
Yue et al. [139]	Na–Sn–C composite	Na_3PS_4	SE-coated Mo_6S_8	500	52, 0.25		60	
Yang et al. [145]	Na metal	PDA-zn0.1	FeS_2	300	131, 0.13	0.5	60	
Yao et al. [146]	Na metal	SPE	NVP@C	740	110, 0.21	0.5	60	Expensive cathode
Ruan et al. [147]	Na metal	NZSP ceramic	NVP mixed with NZSP	50	62.23, NA		70	Expensive cathode, SE
Wang et al. [148]	KTP	PEONa	KPBA	6000	~50, 0.75	20	80	—

Abbreviations: Polydopamine (PDA), $KTi_2(PO_4)_3$ (KTP), $Na_3Zr_2Si_2PO_{12}$ (NZSP), $Na_3V_2(PO_4)_3$ (NVP), Potassium Prussian Blue Analogues (KPBA).

6390 C cm^{-2}. Zheng et al. also built a full cell with a sodium metal anode and δ-$Na_xV_2O_5$ cathode. They found an initial discharge capacity of 305 mAh g^{-1} of active cathode material and a capacity fade over 50 cycles to 40.3% of the initial discharge. They attributed capacity fade to degradation of the vanadium oxide and not to the SPE [143]. However, the use of vanadium in the cathode-active material negates one of the benefits of switching from a lithium-based to a sodium-based cell, as vanadium is both more expensive and less abundant than the lithium salts used to make LIBs. Recent full-cell work has also used ceramic superionic conductors for improved performance. Song et al. [144] report the synthesis and application of a NASICON-like material that showed stability at room temperature with a sodium metal anode and a sulfur cathode. The cell showed an initial discharge capacity of 527 mAh g^{-1} at 2.83 μA cm^{-2} under room temperature. This cell faded over ten cycles to 160 mAh g^{-1} with the loss attributed to sulfur degradation. Through a comparison of the solid-state cell to a liquid cell with a sodium-metal anode, sulfur cathode, and ethylene carbonate (EC)+ propylene-carbonate (PC) liquid electrolyte, Song et al. showed superior capacity retention by the solid cell with only a slightly smaller initial capacity (170 mAh g^{-1} [solid] to 180 mAh g^{-1} [liquid]) and improved capacity retention (88% [solid] and 36% [liquid])[144]. In Peta et al. [149], pristine $NaPF_6$:$P(EO)_{16}$ and a composite solid electrolyte containing TiO_2 nanowires were investigated as candidates for rechargeable SSNaB. Comprehensive electrochemical characterizations were carried out, including ionic conductivity, transference number, and structural stability. At elevated temperatures, the specific capacity of an all-solid-state $Na_3Ti_2(PO_4)_3$ (Na/NTP) sodium battery was 110 mAh g^{-1}, higher than room-temperature cells with liquid electrolyte solutions. The authors attribute this behavior to increased conductivity of the polymer electrolyte, induced by the ceramic nanofiller, and combined with enhanced electronic conductivity of the NTP cathode. Table 4.1 also provides several other references with full-cell SSNaB results. It should be noted that most of the cells were cycled at elevated temperatures. While these temperatures are too high for automotive or personal electronics applications, for stationary storage, such temperatures are acceptable. As shown in Table 4.1, the highest areal capacity reached is only 0.28 mAh cm^{-2}, whereas the target value for a high-energy cell should ≥3 mAh cm^{-2}. From these references, it is clear that current research prioritizes capacity normalized to the active mass of the cathode and stabilizing the cathode–electrolyte interface. While this can demonstrate active material utilization and give insight on cathode stability, there may be additional gaps when moving to cells with electrode and separator properties that reflect practical cells on the basis of Wh kg^{-1} and Wh L^{-1}. When future work shifts to a greater emphasis on cell components reflecting practical cells, authors should include full cell information (total mass, thicknesses, densities, active mass fractions, C-rate, current density, electrode area, etc.) to help researchers identify and overcome remaining gaps in cell performance [89].

4.4 Solid-State Potassium-Ion Batteries

A paradigm shift has recently been witnessed toward developing solid-state KIBs (ssKIBs). As the solid electrolytes have low flammability and leakage issues, they offer substantially improved safety and thus are explored to mitigate the challenges associated with KIBs.

Nevertheless, despite many positive attributes, the KIB technology is in its infancy. There is plenty of room to understand the fundamental mechanism, analyze the challenges, and engineer the rechargeable K-ion batteries to boost their basic and applied research. In the following subsections, the structural and electrochemical aspects associated with various ssKIB components have been discussed to provide a clearer picture of the progress made in recent years in the field of ssKIBs.

4.4.1 Solid Electrolytes for Potassium Batteries

The search for a solid-state K^+-conductor is essential for facilitating the development of safe potassium-ion electrochemical energy storage devices. As previously mentioned, the high reactivity of potassium raises severe safety concerns when using this metal with highly flammable liquid organic electrolytes. Consequently, the primary need is to design and synthesize safer solid-state K^+ conductors with appreciable conductivity at room temperature. The low melting point of potassium (63.5 °C) and poor conductivity of most available solid electrolytes at room temperature have significantly impeded the progress of ssKIB. Several new inorganic and polymer solid electrolytes with marked ionic conductivity at room temperature have been proposed in the past few years.

Inorganic potassium solid electrolytes offer higher ionic conductivities. Up to now, the highest ionic conductivity of 8×10^{-4} S cm^{-1} at room temperature has been observed with K^+ conducting β-Al_2O_3 as a solid electrolyte [150, 151]. Though many papers reported the use of β-alumina for K/S and K/O_2 systems, the high sintering temperature and cost limit its wider application. Several inorganic oxides, like $K_{0.7}Cd_{0.15}FeO_2$, $KGaO_2$, $K_{0.89}Pb_{0.1}GaO_2$, $K_{1.90}Ba_{0.05}AlO_2$, and K_2SbPO_6 [152, 153], have been proposed in the past. However, most of them exhibit the required ionic conductivity (in the range of 10^{-3} to 10^{-5} S cm^{-1}) only above 300 °C. Recently, Doux et al. proposed $K_3Sb_4O_{10}(BO_3)$ as a solid-state K^+ conductor [154]. Attributed to 1-D interconnected channels, the material presented a relatively high ionic conductivity of about 1.5×10^{-4} S cm^{-1} at 400 °C. In contrast to all the reported inorganic solid K^+ conductors, Yuan et al. [155] presented a 3D open-framework polycrystalline ferrite, $K_2Fe_4O_7$. The material exhibited ultrafast K^+ transport with an ionic conductivity of 5.0×10^{-2} S cm^{-1} at room temperature. Another antiperovskite K_3OI and boron-doped $K_{2.9}Ba_{0.05}OI$ material with an ionic conductivity of 3.5 mS cm^{-1} at a temperature of 240 °C has been recently reported [156]. The material was also implemented in a solid-state K/$K_{2.9}Ba_{0.05}OI$/K symmetrical cell cycled at 270 °C and 0.5 mA cm^{-2} current density. At the same time, Xiao et al. [157] discovered superionic K_2CdO_2 with the use of high-throughput computational tools. However, the computational calculation suggested limited ionic conductivities of the bare material. It was later attempted to improve this with Al-doping, which resulted in about 2.2×10^{-5} S cm^{-1}.

The progress in developing the room temperature solid ion conductors is at a great pace and in the same direction, in 2021, Haffner et al. [158] reported the first nonoxide solid K-ion conductor, T5 supertetrahedral phosphidosilicate,SSi_2P_3, with excellent ionic conductivity of 2.6×10^{-4} S cm^{-1} at 25 °C.

Reddy et al. [159] designed SPE by blending polyvinylpyrrolidone (PVP), PVA and $KBrO_3$ in equal weights. The proposed SPE exhibited an ionic conductivity of 5×10^{-6} S cm^{-1} at 33 °C. Recently, Fei et al. proposed a flexible SPE using polypropylene carbonate (PPC) with cellulose nanowoven backbone (PPCB-SPE) [160]. They were able to control the ionic conductivity by optimizing the potassium bis(fluorosulfonyl)imide (KFSI) wt%. A conductivity as high as $1.36\times10^{-5}\,S\,cm^{-1}$ was obtained with 18 wt% KFSI at 20 °C. The reported ionic conductivity of PPCB-KFSI SEI at 20 °C was found to be much higher than KIO_3 in PEO at 20 °C. Furthermore, the proposed SPE did not exhibit any side reactions in the potential window of 1V-4.15 V, indicating the stability of SPE in the working range. The same group later reported flexible PEO-KFSI-based SPE [161]. The ionic conductivity of PEO-KFSI-SPE touches $2.7\times10^{-4}\,S\,cm^{-1}$ at 60 °C.

Even though the research community is progressing toward all ssKIB, much more improvement in the ionic conductivity of solid electrolytes needs to be done, especially at a temperature below the melting point of potassium. In addition, the stability and interfacial reactions of the proposed and future solid electrolytes should be explored in great depth.

4.4.2 Anode Materials for Solid-State Potassium Batteries

By reason of its high specific capacity and energy density, potassium metal was the first choice as anode material for KIBs. To date, all the ssKIB reports have been based on potassium metal as the sole anode material. However, the high reactivity of potassium results in severe interfacial parasitic reactions. Thus, the nature and characteristics of SEI are some of the most significant phenomena that dictate the performance of the KIB battery. Unlike LIBs, of which the SEI has been intensively investigated and characterized, few attempts have been made to understand the SEI in the case of KIBs.

The SEI originates from the degradation of electrolytes on the electrode surface. Some researchers relate this degradation with the energy difference between the Fermi level of the anode and the electrolyte's lowest unoccupied molecular orbital (LUMO). The Fermi level of potassium metal (and all alkali-metals) is higher than the LUMO of the electrolytes. Thus, the electron is readily transferred from potassium to the electrolyte, leading to the thermodynamically-driven electrochemical reduction on the surface of the anode. Yet, the degradation products formed serve as a passivation layer that allows ion transport, but eliminates the electronic contact between electrolyte and electrode, thereby controlling any further irreversible consumption of electrolyte, which affects the coulombic efficiency of the battery. However, because of the high reactivity of potassium and the softer nature of the cation, the spontaneously formed SEI is unstable. Moreover, because of significant volume changes, nonuniform ion flux during continuous potassium-metal deposition/stripping, and poor solubility of the SEI's degradation products, dendritic growth on the anode is observed, raising severe safety concerns. Therefore, interfacial engineering for achieving a stable and uniform SEI is of utmost importance for addressing the interfacial issues in KIBs.

It is well known that the electrolyte composition primarily determines the interfacial reactions and properties of the SEI. Thus, some groups have attempted to investigate the properties of the SEI in carbonate- and ether-based solvents. Nevertheless, to the best of our knowledge, no systematic study has been carried out to study the interfacial reactions and SEI-formation mechanism with the solid electrolytes or with a variety of salts. On the basis of investigations of liquid electrolytes, it is clear that the SEI in KIBs is much more complicated than in LIBs and involves more pronounced anion degradation in several electrolyte solutions [162–167]. Therefore, the content of inorganic components of the SEI is substantially higher. In carbonate ester-based electrolytes, thick SEI rich in inorganic species like K_4CO_4, $KHCO_3$ covered by a thick organic layer has been reported. Supported by electrochemical-impedance-spectroscopy studies, substantially increased charge-transfer resistance and impeded ion diffusion was observed during battery performance, indicating that the SEI formed in KIBs has poor ionic conductivity, which inevitably results in capacity fading [162–164].

Several other artificial protection strategies to stabilize the SEI on potassium metal have been reported [165–167]. Phosphate-based organic electrolytes form thin SEI made up of KPO_3, K_2SO_4, and $K_2(SO_2)_3$, which is characterized by a coulombic efficiency of 99.6% and stable K/K+ deposition/stripping process [162–164]. Recently, a spontaneous reaction between Red P and K has been utilized to modify the potassium surface with a K_xP_y layer [166]. Similarly, an SbF_3-based protective layer on the surface of potassium to suppress the metal corrosion and facilitate charge transfer has also been reported [167]. Stabilization of potassium is attracting much research attention, and for the progress of KIBs, the formation of a stable SEI is of utmost importance. This area is still insufficiently researched and demands detailed investigation.

4.4.3 Cathode Materials and Electrochemical Performance of Solid-State Potassium Batteries

Owing to the open channel structures, multiple redox-active sites and remarkable compositional/electrochemical tenability, Prussian blue (PB) frameworks have been widely exploited as cathode materials in LIBs, SIBs, and KIBs with all types of electrolyte solvents, aqueous, organic and solid. PB and its analogues are crystalline materials from the hexacyanoferrate family with a general formula $A_xM[Fe(CN)_6]_y{\cdot}zH_2O$, where A and M are alkali metals (e.g. Li, Na, and K) and transition metals (e.g. Fe, Mn, Co, Ni), respectively, and the value of x, which varies from 0 to 2, depends on the oxidation state of M and Fe (Figure 4.8). A large number of research publications has been reported documenting the use of $KFe[Fe(CN)_6]$ (KPB) in KIBs, mainly with aqueous systems. However, the use of KPB for solid-state KIB was first reported in 2018 by Yuan et al. [155]. The authors used $KFeFe(CN)_6$ as an active cathode material, $K_2Fe_4O_7$ as the super-ionic solid electrolyte and K-foil as the anode. A discharge capacity equivalent to 87% of the KPB theoretical capacity i.e. $76\,mAh\,g^{-1}$ was delivered at $67\,mA\,cm^{-2}$ (1 C rate) current density. Impressively, the battery retained a capacity of 75% after 50 charge/discharge cycles at a rate of 10 C. In another attempt, Lu et al. [168] designed a flexible ssKIB with the use of potassium PB particles as the cathode, or the anode with KCl-PVA as the polymer-gel electrolyte. Two steps of reversible K^+ insertion/extraction were observed in the voltage profiles corresponding to $C{-}Fe^{III}/Fe^{II}$ and $N{-}Fe^{III}/Fe^{II}$ redox processes. A discharge capacity of $58\,mAh\,g^{-1}$ was demonstrated at $100\,mA\,g^{-1}$ with a remarkable capacity retention of 82% after 200 cycles.

Recently, a solid-state potassium-ion battery has been reported that is based on the promising performance of a nickel - thioxonickel compound (Ni_3S_2) in Na-ion and Li-ion batteries. A binder-free Ni_3S_2 array structure supported on nickel foam was used as the cathode and PEO-KFSI as the SPE [168]. The material exhibited a capacity of more than $300\,mAh\,g^{-1}$ at a current density of $25\,mA\,g^{-1}$ after 100 charge-discharge cycles at 55 °C. It is noteworthy

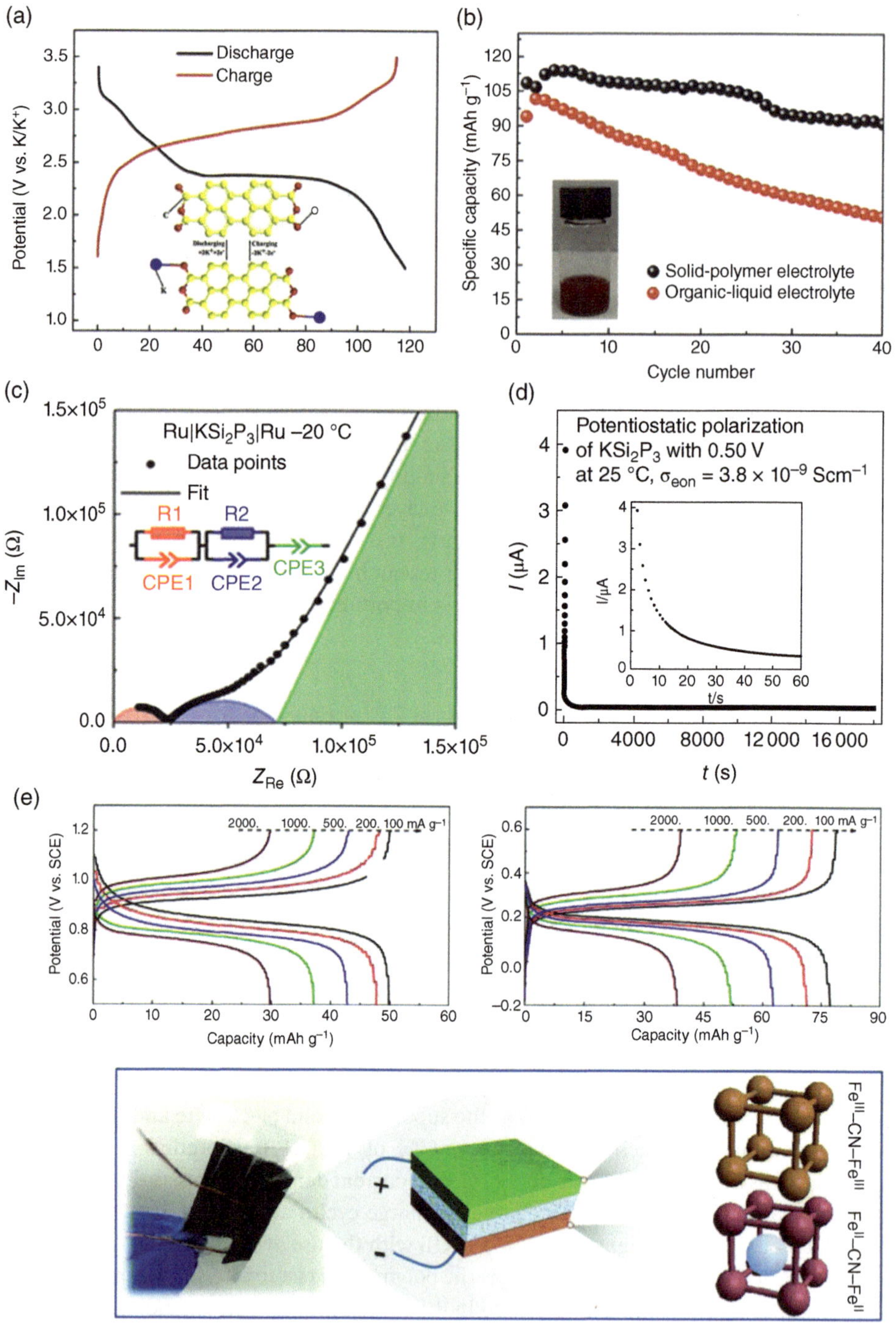

Figure 4.8 (a) Typical charge/discharge profiles of PTCDA with PPCB-SPE at a current density of 10 mA g^{-1}. The inserted picture is the schematic illustration for the proposed electrochemical reaction mechanism of PTCDA cathode in KIBs during the charge/discharge process. (b) Cycling performance of PTCDA with PPCB-SPE or organic-liquid electrolyte at a current density of 20 mA g^{-1}. *Source:* Fei et al. [161]/Reproduced with permission from Elsevier. (c) Impedance spectra of KSi_2P_3 at 20 °C. (d) Potentiostatic polarization curve over five hours showing low electronic conductivity of 3.8×10^{-9} S cm^{-1}. *Source:* Reproduced from Ref. [158]/John Wiley & Sons. (e) K-ion-storage properties of KPB@PPy@Cloth composite electrodes in the voltage window of 0.5 to 1.2 V and −0.2 to 0.6 V; and photograph and schematic illustration of the flexible solid-state symmetric K-ion cell with KPB bipolar electrodes. *Source:* Lu et al. [168]/Reproduced with permission from Royal Society of Chemistry.

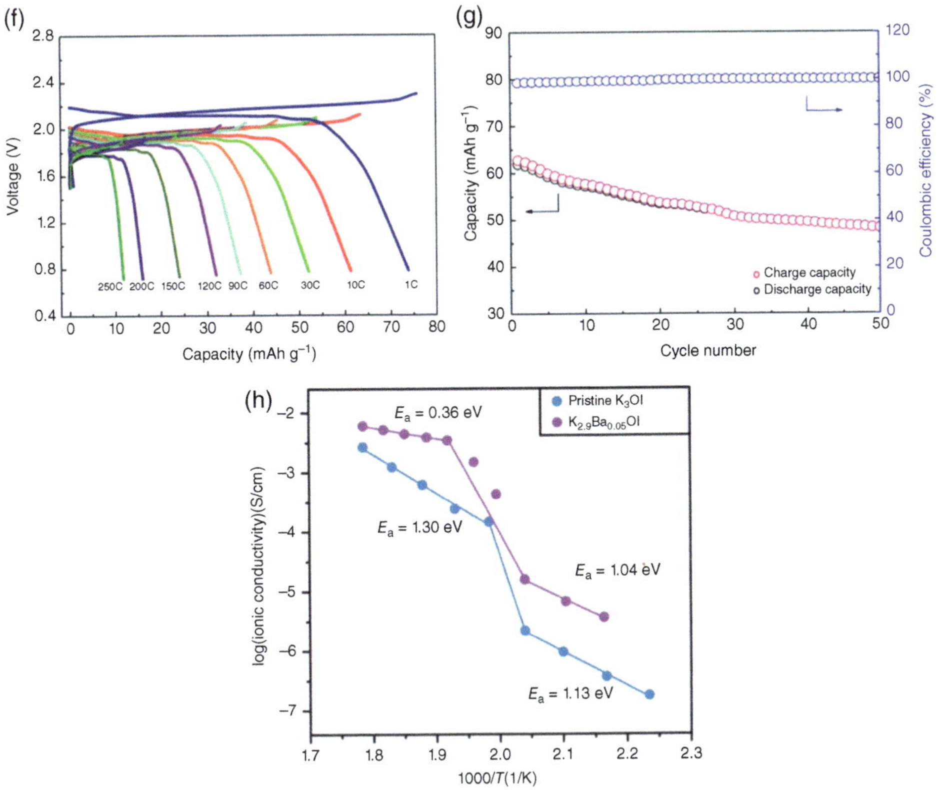

Figure 4.8 (Continued) (f) Charge and discharge curves of the K/$K_2Fe_4O_7$/KPB all-solid-state battery at different current rates. (g) Cycling performance of the K/$K_2Fe_4O_7$/KPB all-solid-state battery at a rate of 10 C. *Source:* Reproduced from Ref. [155]/With permission of Royal Society of Chemistry. (h) Arrhenius plots of the ionic conductivity of $K_{2.9}Ba_{0.05}OI$ compared with K_3OI. *Source:* Reproduced from Ref. [156]/With permission of American Chemical Society.

that the Ni_3S_2 cathode in the cell with liquid organic electrolyte suffered from a capacity fading of 90% after just 20 cycles. The stable performance of Ni_3S_2 in PEO-lithium bis(fluorosulfonyl)imide (LFSI)-SPE is ascribed to the restricted dissolution of polysulfides (PS) in the solid electrolyte.

Some research groups have recently begun exploring the subject of organic electrodes in ssKIBs. The possibility of tailoring the redox potential just by altering the synthesis route, starting precursor, or inserting electron-donating/withdrawing edge groups, offers much more freedom to custom-make organic electrodes according to the desired stability requirements, conductivity, or redox potentials. Attributed to weaker Van-der-Waals interactions between the organic layers, the interlayer spacing between the organic electrodes is suitable for storing the larger metal ions like K^+, especially in the solid state. The associated flexibility is further advantageous for accommodating the volume expansion. Among several possibilities, 3,4,9,10-perylenetetracarboxylicacid-dianhydride (PTCDA) has emerged as the possible cathode material. For KIB's, PTCDA has been investigated since 2015, when

Chen et al. [169] first demonstrated the reversible insertion-deinsertion of K^+ ions delivering a capacity of 131 mAh g^{-1} at 10 mA g^{-1} and 117 mAh g^{-1} at 50 mA g^{-1} in nonaqueous liquid electrolyte. With a capacity decay of only 0.12% after 200 cycles, PTCDA opened new avenues for redox-active organic electrodes for stable KIBs. Fei et al. [160] later reported for the first time a room-temperature solid-state KIB with the use of PTCDA as the sustainable cathode and poly(propylene carbonate) (PPC)-KFSI with cellulose nonwoven backbone as solid polymer electrolyte. He reported a discharge capacity of 113 mAh g^{-1} with a capacity fading of ~15% after 40 cycles, compared to 45.7% capacity fading witnessed with liquid organic electrolytes.

4.5 Solid-State Magnesium-Ion Batteries

Among multivalent metals the closest to realization as anode in a rechargeable "rocking-chair" battery is magnesium [3, 4]. Compared to other active metals, magnesium fits better with battery requirements: magnesium has a lower impact on the environment (extraction, disassembly, and recyclability) and it has an advantage over lithium in terms of safety, abundance and very high volumetric specific capacity (3800 mAh cm–^{3} for Mg vs. 2080 mAh cm^{-3} for Li). In addition, magnesium has an advantage over sodium and potassium in terms of safety (higher redox potential and lower reactivity, allowing handling in ambient air) and specific gravimetric and volumetric capacity.

One of the greatest challenges of rechargeable magnesium batteries (RMBs) is the development of suitable electrolytes for the efficient and reversible magnesium plating/stripping. The practical application of a Mg battery requires, in addition, well performing insertion/conversion cathode materials [170].

4.5.1 Solid Electrolytes for Magnesium-Ion Batteries

After the in-depth systematic research in the field of electrolytes for magnesium plating/stripping carried out by Gregory et al. in the 1990s [171], the scientific community concentrated efforts to find electrolyte systems with improved anodic stability (>2 V vs. Mg) and optimized conductivity for RMB applications. Aurbach's (BIU) group was the first to demonstrate a full RMB comprising a magnesium-metal foil anode, Mo_6S_8 Chevrel phase cathode and dichloro-complex organohaloaluminate liquid electrolyte (DCC) [3]. These solutions contain ether solvent and complex magnesium salt (reaction product of dibutyl magnesium and ethyl-aluminum dichloride). DCC showed high ionic conductivity and an anodic stability window of up to 2.2 V (vs. Mg). However, in spite of the high scientific contribution from this pioneer RMB prototype, its performance is insufficient to compete with LIB because of the low operating voltage (less than 1.5 V) and the low capacity of the Chevrel-phase cathode rendering the (theoretical) energy density of such an RMB far behind that of LIB.

Thus, with the general goal of improving the energy density of this RMB system, intensive research on electrolyte solutions was carried out to widen its electrochemical window and hence the cell voltage. With these goals, an all-phenyl complexes (APC) electrolyte system was developed. APC electrolyte showed an enhancement of anodic stability up to 3.2 V, potentially opening the pathway for utilization of high-voltage cathodes (e.g. V_2O_5). Unfortunately, for this type of electrolyte, several compounds detrimental to the current collector and/or to the cathode coexist. These species impede the insertion of magnesium ions into the cathode material and limit the actual battery operating voltage and cycle life. Following up the research on suitable electrolytes, a magnesium–aluminum–chloride complex (MACC) electrolyte was investigated [172]. This inorganic electrolyte has a voltage window of 3.4 V (vs. Mg) and as a result of the presence of free chloride ions, MACC electrolyte promotes the depassivation of the magnesium surface, minimizing its surface resistance. This advantage enabled reversible magnesium plating/stripping with close to 100% CE over 90 charge/discharge cycles with the use of a low-voltage Chevrel cathode material. Other researchers, like Muldoon et al. and Fichtner et al., developed new electrolyte approaches (based on recrystallization processes of Mg-complex in situ-generated electrolyte) to minimize the effect of corrosive species like chloride species and Grignard reagent and to promote a non-nucleophilic electrolyte to enable the use of conversion cathode materials like sulfur [173–175]. This new electrolyte system was based on the reaction of $HMDS_2Mg$ with $AlCl_3$. The proposed approach showed the possibility of utilizing other ethers instead of THF. Halogen-free (Grignard-free) electrolytes is another system investigated by various groups. These electrolytes are based on Mg-borohydride (relatively poor anodic stability, 2.4 V vs. Mg) or boron cluster electrolytes (anodic stability 3.2–3.8 V vs. Mg). The latter is a promising candidate because of its high anodic stability, suitability for Mg plating/stripping, noncorrosive nature (e.g. no effect on current collector), and enables the use of high-voltage cathode materials.

Despite all these efforts, significant progress toward high-energy RMBs is still hampered by the lack of a suitable electrolyte compatible with low- or high-voltage insertion and conversion cathode materials. Unfortunately, electrolyte solutions suitable for reversible magnesium anodes are not currently compatible with any relevant cathode material that may intercalate magnesium ions reversibly.

In order to overcome the problems associated with liquid electrolytes, there is a strong incentive to develop and demonstrate SSEs with good Mg^{2+} mobility and stability for RMB. The development of such an electrolyte is hostage to the relatively high charge density associated with the divalent nature of the Mg^{2+} ion. This property causes significant decrease in ionic conductivity when changing compound from its lithiated version to its magnetized parallel compound.

Early efforts for the development of a magnesium-ion solid electrolyte were focused on an oxide system, NaSICON-like phosphate systems, and borohydride materials [176, 177]. In 1987, Ikeda et al. [178] investigated Mg^{2+} conductivity in magnesium zirconium phosphate materials. $MgZr_4(PO_4)_6$ exhibited bulk ionic conductivities of 2.9×10^{-5} and $6.1\times10^{-3}\,S\,cm^{-1}$, at 400 and 800 K, respectively. Tubandt's method and electron-probe microanalysis were used to validate the fact that Mg^{2+} ions were the charge carriers in this material. Kazakos-Kijowski et al. [179] synthesized $Mg_{0.5}Zr_2(PO_4)_3$ using a sol–gel method and studied the ionic conductivities and thermal behavior of these materials. Nakano et al. [180] conducted a computational investigation of the Mg-ion conductivity and phase stability of $MgZr_4(PO_4)_6$. A schematic illustration of the monoclinic $P2_1/n$ symmetry (similar to that of β-Fe_2SO_4) crystal structure of $Mg_{0.5}Zr_2(PO_4)_3$ is shown in Figure 4.9a,b. The ionic diffusion in this electrolyte occurs across the channels formed by three PO_4 tetrahedra and three ZrO_6 octahedra forming a 3D network of tunnels. Despite the progress with the phosphate materials and the mechanistic understanding of ion pathways in the lattice, these materials exhibit high ionic conductivity only at elevated temperatures that are beyond the current target for RMB (Figure 4.9c,d).

Borohydrides were also investigated for use as Mg-ion-based SSEs. Matsuo et al. [15, 181] employed *ab initio* calculations to study the diffusion of magnesium ions in $Mg(BH_4)_2$, and they reported that Mg^{2+} ions located in tetrahedral cages surrounded by BH_4 ions had limited mobility. This low mobility is hypothesized to originate from the strong coulombic interactions of the Mg^{2+} with BH_4^- ions, and the authors suggested that partially replacing BH_4 ions by larger ions (e.g. AlH_4) will result in increased cage size and will enhance magnesium-ion mobility. Following this concept, Higashi et al. [182] investigated the difference, in terms of ionic conductivity, between $Mg(BH_4)_2$ and $Mg(BH_4)(NH_2)$.

Indeed, they showed that the larger cavities permitted lower barriers for the hopping of magnesium ions. In addition, some level of reversibility of plating/stripping of magnesium on platinum electrodes (Figure 4.10a–c) was demonstrated and with high anodic stability of this electrolyte. Roedern et al. [183] further investigated the borohydride systems. The typical crystal structure of $Mg(BH_4)(NH_2)$ is shown in Figure 4.10d–g, where BH_4 units (as

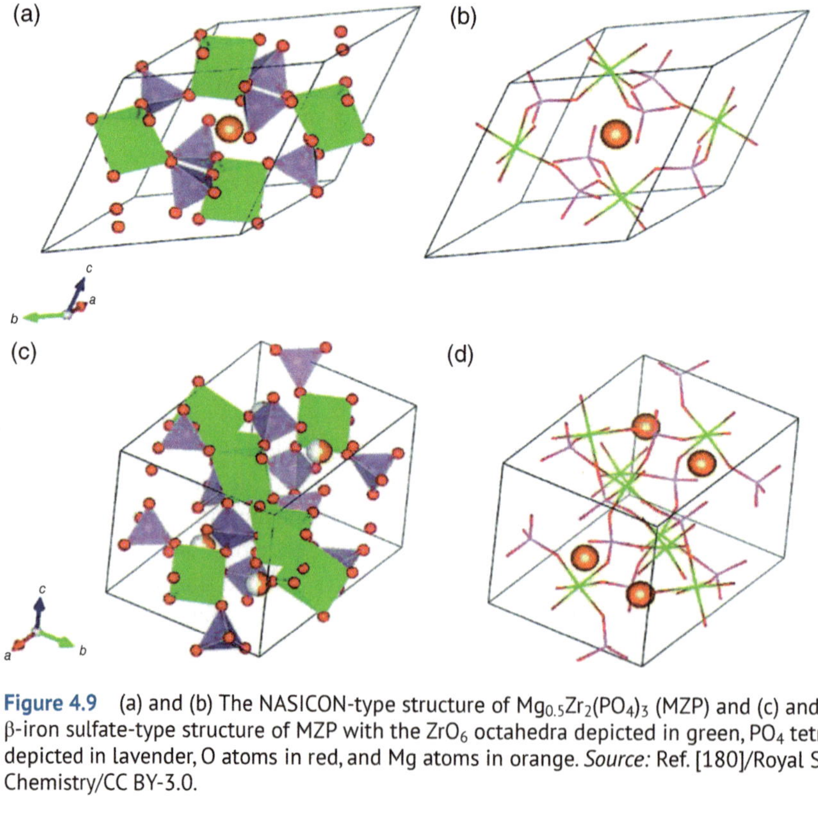

Figure 4.9 (a) and (b) The NASICON-type structure of $Mg_{0.5}Zr_2(PO_4)_3$ (MZP) and (c) and (d) the β-iron sulfate-type structure of MZP with the ZrO_6 octahedra depicted in green, PO_4 tetrahedra depicted in lavender, O atoms in red, and Mg atoms in orange. *Source:* Ref. [180]/Royal Society of Chemistry/CC BY-3.0.

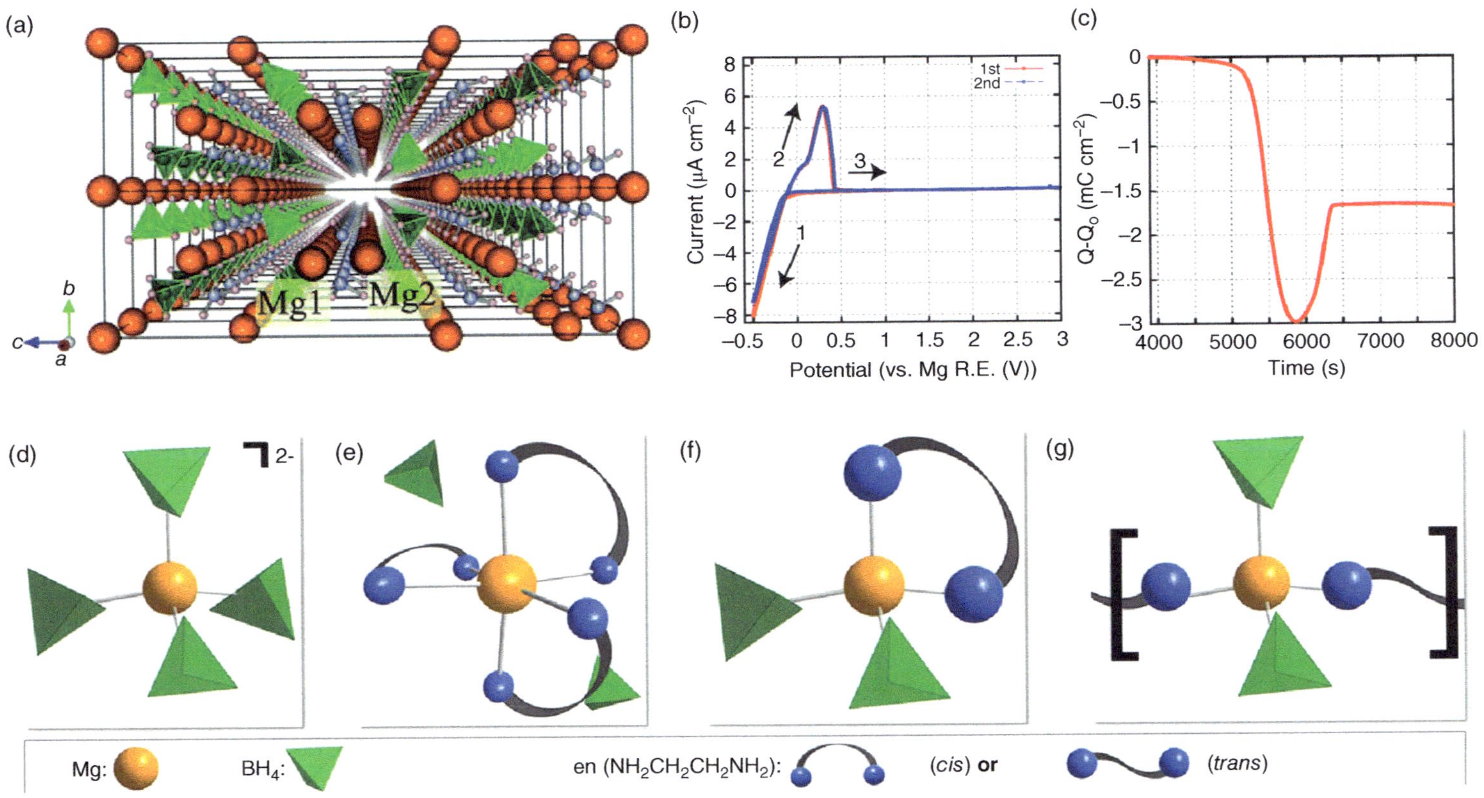

Figure 4.10 (a) Crystal structure of $Mg(BH_4)(NH_2)$ viewed along the a-axis, (b) cyclic voltammograms of $Mg/Mg(BH_4)(NH_2)/Pt$, (c) deposition/stripping charge balance (first cycle). A conceptual sketch of the local coordination of Mg^{2+} in (d) $Mg(BH_4)_2$, (e) $Mg(en)_3(BH_4)_2$, and two likely possibilities in $Mg(en)_1(BH_4)_2$ (f, g). *Source:* Reproduced from Ref. [182]/With permission of Royal Society of Chemistry and Ref. [183]/Springer Nature/CC BY-4.0.

in $Mg(BH_4)_2$) are replaced by amine groups to form a tetrahedral cage around the Mg^{2+} ion. The structure consists of alternate stacking of cations (Mg^{2+}) and anions (BH_4^- and NH_2^-) along the *c*-axis. The authors measured the ionic conductivity of 1×10^{-6} S cm^{-1} at 150 °C in $Mg(BH_4)(NH_2)$, which is three orders of magnitude higher than that of $Mg(BH_4)_2$ at the same temperature. The authors conclude that $Mg(BH_4)(NH_2)$ seems suited for application as a borohydride electrolyte in magnesium-based batteries operating above 150 °C. However, the anodic stability of this electrolyte is less than 1.5 V vs. Mg [183]. Additional progress was demonstrated with phosphate and borohydrides [176, 177], yet these pathways seem to require high temperatures, and it is difficult to see how the battery community will adjust to a solid electrolyte that operates properly only at such high temperatures.

A major breakthroughs was demonstrated by Canepa et al. [184], who reported ternary chalcogenides (e.g. $MgSc_2Se_4$) as potential solid electrolytes for RMBs. By using ab initio and first-principles calculations of the migration barriers (found to be 0.36–0.53 eV) for Mg^{2+} conduction in chalcogenide-based spinel compounds, with the general formula MgX_2Z_4 (where X=In, Y, or Sc and Z=S or Se(, Canepa et al. showed that these materials can demonstrate high ionic conductivity of 0.1 mS cm^{-1} at room temperature (Figure 4.11).

The low values of activation energy for cation migration are reported to be possible mainly because of two properties of these ternary chalcogenides. First, Mg^{2+} positioned in unfavorable coordination, specifically reduction in the coordination number of Mg^{2+} ions from 6 to 4 (from favorable octahedral to unfavorable tetrahedral sites). The second property is the dependence on high volume per anion ($O_2 > S_2 > Se_2 > Te_2$). By using Mg nuclear magnetic resonance (NMR) relaxometry and impedance spectroscopy, the authors confirmed experimentally high mobility of Mg^{2+} with a low E_a migration barrier (~370 ± 90 meV). According to the calculations, several other spinels in this family also have high Mg^{2+} mobility. These include MgY_2S_4 (~360 meV) and MgY_2Se_4 (~361–326 meV). These values of migration barriers are promising, since they are comparable even with the values of Li^+ mobility in some of the advanced garnets (~400–500 meV) and are substantially lower than other Mg-ion conductors reported before. Unfortunately, despite their promising ionic conductivity of chalcogenides, impedance spectroscopy indicated mixed conduction behavior with too high electronic conductivity, which make these materials irrelevant for battery application. Two approaches [18, 19] have been attempted to lower electronic conductivities: introduction of Se-excess phases and aliovalent doping of Sc^{3+} by Ti^{4+} and Ce^{4+}. Neither of these attempts effectively neutralized the electronic conductivity. Kundu et al. [185] synthesized modified chalcogenides by thermal treatment of the elemental (Mg, Sc, and Se) powder mix and managed to reduce the concentration of electron-conducting inclusions. This led to a significant decrease in the electronic conductivity, from ~10^{-6} to ~10^{-12} S cm^{-1}. Canepa et al. [186] showed that controlling the anion content during synthesis is an important factor in determining the defect energetics and the resultant electrical conductivity. $MgIn_2S_4$ was found to exhibit the lowest free-carrier concentration under various conditions, largely as a result of inversion within the spinel. Additionally, the authors showed that introduction of aliovalent dopants, such as Ce and Ti on Sc, may mitigate the electronic conductivity observed in $MgSc_2Se_4$. An additional strategy is to add thin coatings for the solid chalcogenide electrolyte. The coating should be thin enough, electrically insulating, and with sufficient ionic conductivity. Chen et al. [187] have recently reported the ionic transport in such potential coatings. Using a strict mobility threshold

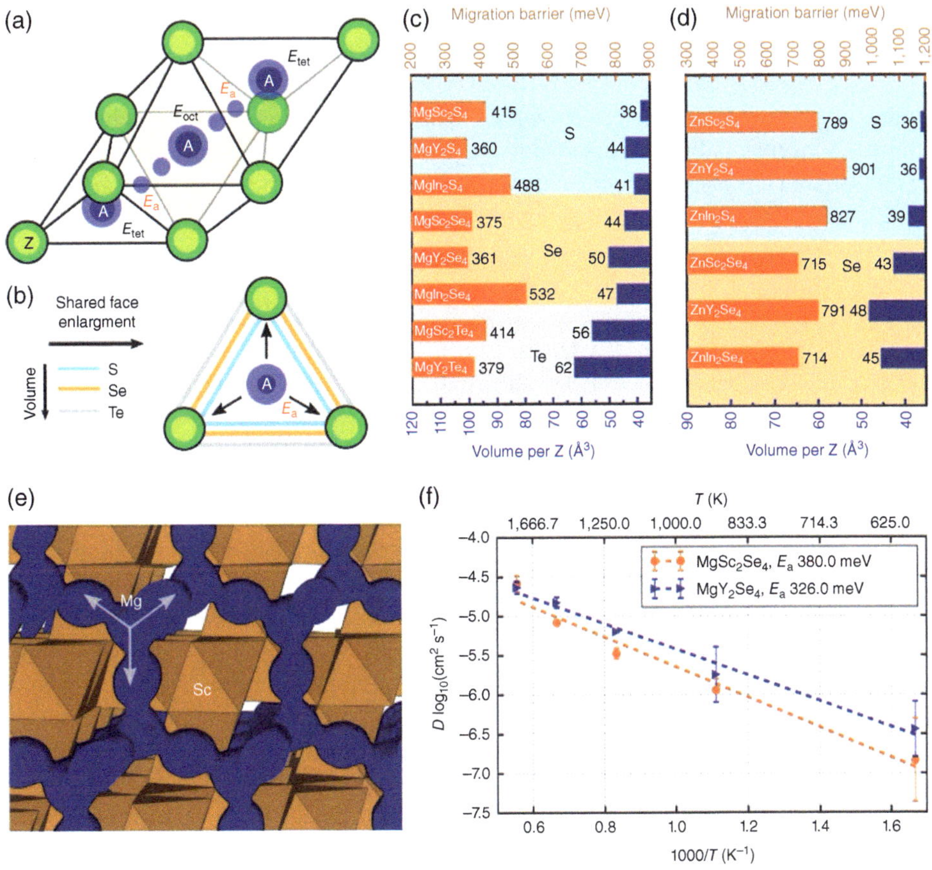

Figure 4.11 First-principles Mg and Zn migration barriers in sulfides, selenides, and tellurides AX_2Z_4 spinels (with A = Mg or Zn). (a) tet–oct–tet migration path in the AX_2Z_4 framework, with energy of the tet, oct, and transition sites indicated by E_{tet}, E_{oct}, and E_a, respectively. E_a corresponds to the migration energy. (b) Effect of the anion size on the shared (triangular) face between tet and oct sites. (c) and (d) Computed Mg and Zn migration barriers (orange bars in meV) in AX_2Z_4 spinel and volume per anion (blue bars), respectively, with X = Sc, Y, and In and Z = S, Se, and Te. (e) Probability density of Mg in $MgSc_2Se_4$ at 900 K obtained from ab initio molecular dynamic simulations (AIMDs). (f) Mg diffusivities as extrapolated from AIMD in $MgSc_2Se_4$ (orange) and MgY_2Se_4 (blue), with dashed lines and error bars indicating Arrhenius fits and SD, respectively. *Source:* Reproduced with permission from Nature 2017/Public Domain CC BY 4.0.

(600 meV), the authors allocated $MgSiN_2$, MgB_4, Mg_2 Ge, Mg_3As_2, Mg_3P_2, and $MgAl_2O_4$ as potential coatings. A more "flexible" threshold (980 meV) opened new families of potential coatings including MgH_2, Mg_3N_2, MgTe, MgSe, MgS, MgI_2, $MgBr_2$, $MgCl_2$, and $Mg(PO_3)_2$. Among the aforementioned candidates, $MgAl_2O_4$ and $Mg(PO_3)_2$ should be compatible with oxide cathodes with high anodic stability (>3 V vs. Mg), while $MgSiN_2$, MgS, MgSe, $MgBr_2$, and MgI_2 should be compatible with the magnesium-metal anode and exhibit reasonable cathodic stability (>1 V).

Similarly to lithium polymer electrolytes, earlier studies on magnesium conducting SPEs introduced PEO as a host material of magnesium salts. The ionic conductivity at room temperature is low (~10^{-9} S cm^{-1}) for PEO-based electrolytes with $MgCl_2$ and $Mg(SCN)_2$ salts. Higher room-temperature conductivity (10^{-5}–10^{-7} S cm^{-1}) was achieved with $Mg(ClO_4)_2$, $Mg(NO_3)$, and Mg(TFSI). However, extremely low magnesium transference number was measured in these electrolytes, indicating predominant anion conduction.

Shi and Vincent conducted a polarization test with PEO-$Mg(TF)_2$ and suggested that the Mg ion cannot be transported efficiently via segmental motion/site percolation due to the strong electrostatic interaction of Mg^{2+} with the ether chain oxygen of the PEO [188, 189]. PEO-PPz (Pentoxy-trifluoro-ethoxy phosphazene) with $Mg(Tf)_2$ salt showed at least one order of magnitude higher ionic conductivity over the temperature range of 50–110 °C than that of PEO and PEO-PPO electrolytes. At 50 °C, the ionic conductivity of PEO-PPz-based electrolyte was 7.4×10^{-6} S cm^{-1}, whereas 3.2×10^{-8} and 8.3×10^{-8} S cm^{-1} were reported for PEO and PEO-PPO-based electrolytes. This result was explained by the plasticizing effect of PPz in the polymer matrix [190]. In [191], it was found that up to 30 wt% of $Mg(NO_3)_2$ salt can be dissolved in the PEO-PVP (poly(vinyl pyrrolidone)) blend polymer matrix, providing the ionic conductivity of 5.8×10^{-4} S cm^{-1} at room temperature. Poly(vinyl alcohol) (PVA) and its blend with polyethylene glycol (PEG) and poly(acrylonitrile) (PAN), as well as polyethylene carbonate (PEC) and poly(vinylidene fluoride-*co*-hexafluoropropylene) PVDF-HFP, were tested as magnesium-ion hosts. PEO and poly[(styrene-4-sulfonyl trifluoromethyl sulfonyl) imide lithium] P[$(STFSI)_2Mg$] block copolymers were tested by Thelen et al. as magnesium single-ion-conducting polymer electrolytes [192]. The ionic conductivity of these electrolytes was ~3×10^{-7} S cm^{-1} at 50 °C. A comprehensive review of Park and Schaefer summarizes these data [193]. As in the lithium-ion conducting polymer electrolytes in order to increase the ionic conductivity of magnesium polymer electrolytes numerous efforts have been focused on study of the effect of type of filler and its particle size on ion conduction. Agrawal et al. [194] found that in PEO-$Mg(Tf)_2$ polymer electrolyte both ionic conductivity and magnesium transference number (1.5×10^{-5} S cm^{-1} and 0.37) were superior in the electrolytes containing TiO_2 and MgO nanofillers as compared to those with SiO_2 nanofillers and MgO microfillers.

Recently, biodegradable polymers such as agar, chitosan, cellulose acetate, pectin, and tamarind seed polysaccharide (TSP) have been considered as host materials for Mg-conducting polymer electrolytes. The conductivity of polymer electrolytes with magnesium triflate, magnesium perchlorate, and magnesium nitrate salts varies from 1×10^{-6} to 2×10^{-3} S cm^{-1}. However, these biopolymers contain high content of proton-labile functional groups such as –OH and –COOH, which makes them incompatible with a pristine magnesium anode.

4.5.2 Anode Materials for Solid-State Magnesium Batteries

A primary advantage of magnesium technology is provided by the solid magnesium anode that leads to high-energy density values, well above those of lithium-based cells. The higher reduction potential of Mg^{2+} (about 0.7 V higher than Li^+/Li) and reduced reactivity of Mg metal compared with the alkali metals provide higher electrochemical and chemical stability at the operating potential of magnesium anode. However, unlike the SEI formed on the Li anode, which is Li^+ conductive with low electronic conductivity, the SEI formed in situ on the Mg anode with the majority of electrolytes does not conduct Mg^{2+}, leading to the passivation and poor rechargeability of Mg battery. The Mg metal, which exhibits 3D crystal growth rather than one dimensional dendrites, has been expected to less likely dendritic growth than the Li metal anode. Still, Mg dendritic growth has been reported in several liquid electrolytes [193]. To overcome this obstacle, novel anode solutions have been recently proposed. Bismuth-based anodes are among the most studied alternative anodes because of the rhombohedral crystalline structure of bismuth that allows formation of alloys with high volumetric capacity [195] and superionic Mg^{2+} conductivity. However, the gravimetric capacity theoretically achievable is close to graphite, about 385 mAh g^{-1} [196] and the formation of the alloy according to the following reaction has poor reversibility.

$$2Bi + 3Mg^{2+} + 6e \rightarrow Mg_3Bi_2$$

For magnesium–tin alloys, it was found that stresses develop during dealloying of Mg_2Sn and induce pulverization of the dealloyed material and formation of nanostructured Sn [197]. A reversible capacity of 300 smAh g^{-1} was demonstrated over 150 cycles at the rate of C/5 after application of a combined sequence of regular galvanostatic cycling with an oxidative pulse to control the SEI formation. Biphasic bismuth–tin alloys were studied by Niu et al. [198]. Reversible magnetization to form Mg_3B_2 and Mg_2Sn was observed.

4.5.3 Cathode Materials and Electrochemical Performance of Magnesium Batteries

With respect to the cathode, the limitations include intrinsic challenges in three types of cathode systems, insertion, conversion, and organic cathodes. Although magnesium batteries have been studied for several decades now, their realization is restricted by the lack of suitable high-voltage cathode material that can tackle the limitations associated with the sluggish diffusion of the magnesium ion into the host [184, 199–203]. This limitation is due to the high migration barrier caused by the high charge-to-radius ratio of Mg^{2+} ions. It is expected that these limitations can be mitigated through rational engineering of cathode materials composed of morphologically tuned nanoparticles, open-tunnel structures and layered compounds [204].

The first prototype magnesium battery was demonstrated with the use of Mo_3S_4 as cathode. It exhibited excellent performance but within a low operating voltage (1.1 V vs. Mg)[3] and limited capacity. Hence, it is desirable to find a high-voltage cathode host that can reversibly intercalate the divalent Mg^{2+} ion for the realization of magnesium batteries in practical devices [205]. Previous reports on reversible Mg^{2+} intercalation hosts showed that oxides of transition metals could be promising candidates for high-voltage cathode [206–209]. Owing to the high working voltage and energy density of vanadium pentoxide (V_2O_5), it is considered extremely attractive for magnesium batteries [210, 211].

The orthorhombic V_2O_5 exhibits spin ladder configuration consisting of distorted corners and edges sharing VO_5 square pyramids connected by weak interlayer bonding, providing large interstitial space for ion intercalation (Li^+, Na^+, K^+, and Mg^{+2}). The large interstitial space arises from weak V–O interaction between three oxygen atoms (O1, O2, and O3) and vanadium atom. Previous studies confirm that lithium intercalation into an orthorhombic

V_2O_5 lattice undergoes several first-order phase transitions, of which $\alpha \leftrightarrow \epsilon \leftrightarrow \delta$ within $0 \leq x_{Li} \leq 1$ are reversible and $\delta \rightarrow \gamma \rightarrow \omega$ at $x_{Li} \geq 1$ and $x_{Li} \geq 2$ are irreversible [212–217]. The ϵ and δ phases of V_2O_5 are orthorhombic containing [VO_5] square pyramid but higher puckering within the interlayers than α phase. The irreversible γ phase belongs to the same crystal structure with [VO_5] square pyramid but even higher puckered interlayers, whereas the ω phase belongs to tetragonal structure with completely different atomic arrangement than that of orthorhombic V_2O_5 [218]. It is evident from some recent studies that orthorhombic V_2O_5 is a potential host for reversible Mg intercalation with a limit of $x_{Mg} \sim 0.5$ [204, 208, 218–221], which is significantly lower than the Li system. However, the extent of intercalation and hence the specific capacity of V_2O_5 can be increased with the use of a bilayered xerogel phase ($V_2O_5 \cdot 1.8\ H_2O$) [221, 222], which incorporates the water molecules within the interlayers of the V_2O_5 polymorph [223]. Eventually, the water molecules not only expand the interlayer spacing of V_2O_5 from 4.37–11.53 Å [223] but also coordinate to the Mg^{2+} ion, screening the charge and facilitating intercalation of Mg^{2+} ion. Tang et al. [224] reported a discharge capacity of 180 mAh.g^{-1} in surface-modified V_2O_5 aerogel. Novak et al. [225] demonstrated discharge capacities of ~170 mAh.g^{-1} with the use of $Mg(ClO_4)_2$ in water containing acetonitrile (AN) electrolyte (1 M $Mg(ClO_4)_2$ + 1 M H_2O). Yu et al. [226] showed a similar discharge capacity (~158.6 mAh g^{-1}) with $Mg(ClO_4)_2$ in water containing PC electrolyte (1 M $Mg(ClO_4)_2$ + 1.79 M H_2O/PC). However, the strategy of water containing electrolytes and aerogels is not suitable for a system with a magnesium-metal anode, therefore its utilization, even as a model system, is important scientifically but not practical.

Several researches have been carried out on V_2O_5 as cathode material for magnesium batteries with anhydrous electrolyte, but were limited in achieving stable and significant specific discharge capacity [208, 222, 227–229]. The studies gave clear evidence that the choice of nanostructured materials and thin films of V_2O_5 is a promising strategy for overcoming the sluggish mobility of magnesium ions into the cathode. Recently, Fu et al. [218] demonstrated reversible Mg intercalation in V_2O_5 nanowires with a discharge capacity of 130 mAh g^{-1} at C/20 after few activation cycles with the use of $Mg(ClO_4)_2$–AN electrolyte with a capacity retention of <30% after 100 cycles. Banerjee et al. [229] synthesized a metastable phase of V_2O_5 (ζ-phase) nanowires and demonstrated first-discharge capacity of 140 mAh g^{-1} and retained 90 mAh g^{-1} after 100 cycles, with the use of magnesium bis(trifluoromethylsulfonyl)imide (Mg-$(TFSI)_2$) in PC as electrolyte. Gershinsky et al. [208] found a reversible capacity of 150 mAh g^{-1} in monolithic V_2O_5 thin film with 20–50 nm primary particle with the use of Mg-$(TFSI)_2$ in AN. Nevertheless, thin-film methodologies are limited in the loading of active material, since this reduces the practical capacity of the cell.

Recently, Mukherjee et al. [230] demonstrated a facile and scalable approach to synthesize monodispersed orthorhombic V_2O_5 spheres and study the electrochemical activity of this material in dry $Mg(ClO_4)_2$-AN electrolyte system. The V_2O_5 spheres exhibited the highest reported first and second discharge capacity (>190 mAh g^{-1}) and very good long-term stability without any noticeable degradation, even upon long cycling at different current rates. Detailed structural and morphological study of phase evolution in V_2O_5 upon pre- and post-galvanostatic cycling predicts the potential use of this material as a cathode in next-generation magnesium batteries.

Arthur et al. [231] managed to test the high-voltage cathode in full RMB for ten stable charge/discharge cycles with α-MnO_2 cathodes. It should be noted that the low cycling stability with a K-αMnO_2 cathode was found to be linked to the nature of the electrochemical reaction at this specific cathode. Unlike lithium, which can be inserted reversibly into K-αMnO_2, the discharge of the Mg battery induces a conversion reaction with the formation of two wide bandgap insulators, MgO and MnO, on the cathode interface that blocked the reversibility and limited the cycle life of the battery.

Another intriguing approach to realizing high-energy-density RMB cells is to couple the magnesium anode with a conversion-reaction electrode, such as sulfur [174]. With this kind of cathode, the redox reaction may not rely on solid-state diffusion; hence, it is expected to react electrochemically much more quickly. This may avoid the prohibitively slow kinetics observed with most intercalation compounds upon interactions with magnesium ions. Furthermore, sulfur, as a cathode material, shows not only high specific capacity (1675 $mAh\,g^{-1}$) but also a higher operation voltage of magnesium cells compared to Chevrel-phase cathodes (1.7 vs. 1.2 V). Sulfur is a low-cost, highly abundant, comparatively eco-friendly and nontoxic element. It is just natural, though, to anticipate that the Mg–S systems will face similar issues that plague the Li–S systems [232, 233]. One of the greatest problems associated with development of a practical Li–S battery is the dissolution of soluble PS into the electrolyte solution and their spontaneous reduction on the metallic anode, which leads to the well-known shuttle mechanism that limits the capacity of Li–S cells [234]. The degree of solubility of the various PS in Li–S cells depends mainly on the electrolyte-solution composition [235].

Revolutionary attempts to realize Mg-S batteries were generally based on developing nonelectrophilic electrolyte solutions in order to avoid detrimental attacks by the sulfide species formed by sulfur reduction, which are highly nucleophilic [236]. Karger et al. followed this line and demonstrated an Mg–S system with extended cycle life and higher specific capacity [174]. Using dual-ion strategy, Gao et al. demonstrated further improvement in terms of cycle life and capacity maintenance [173]. The latest work has generally focused on the architecture of the cathodes and their interaction with the electrolyte solutions. Recently, we reported on nonelectrophilic electrolyte solutions that might be reasonable candidates for Mg–S systems [237, 238]. Indeed, Tao et al. studied this $Mg(TFSI)_2$-based electrolyte solution and showed reasonable behavior for the sulfur-cathode side.

An additional strategy for cathode development is based on organic electrode materials, made by preparing polymeric-backbone materials with conjugate bonding, which form electronically conducting frameworks containing high concentrations of –S–S– groups, bonded to them. These groups undergo reversible redox reactions plus interactions with metallic ions: $-S-S- + Mg^{2+} + 2e^- = -S-Mg-S-$. In a parallel approach, optimized porphyrins with ethynyl groups will be developed, which link and form alternating double bonds during the first oxidative half-cycle of the cell. This eliminates one of the greatest obstacles with organic electrodes, namely their solubility in organic solvents.

In contrast to their inorganic counterparts, organic cathodes (e.g. quinone-based materials) offer less rigid migration pathways for Mg^{2+} ions with lower migration barriers, which allow operation at higher rates. However, despite such advantages and the potential tuneability of redox properties in organic cathodes, the stable discharge capacity is usually quite below the theoretical values and the rate capability is limited. Therefore, the development of cathode materials with high capacity and cycle life in RMBs remains challenging.

Bitenc et al. [239] used an anthraquinone-based polymer, poly(anthraquinol) sulfide (PAQS), which is representative of the large family of electroactive quinones that are well known from Li-ion batteries. When used in polymerized form, they showed that its inherently high capacity of anthraquinone is stabilized without compromising the reaction kinetics. Recently, Abouzari-Lotf et al. [240] developed a copper–porphyrin with meso-functionalized ethynyl groups capable of reversible two- and four-electron storage at an extremely rapid rate (tested up to 53 C). The reversible four-electron redox process with cationic–anionic contributions resulted in a specific discharge capacity of 155 $mAh\,g^{-1}$ at the high current density of 1000 $mA\,g^{-1}$.

4.6 Specific Challenges and Future Perspectives

As battery technologies are in continuous development, and especially due to the rapid growth in vehicle electrification, which requires large (e.g. 100s of kg) battery packs, there has been a growing demand for more efficient, reliable, and environmentally friendly materials. Solid-state post-LIBs are considered to a possible next-generation energy-storage technology. One immediate advantage of these power sources over commercial LIBs is the potential of solving the resource issues facing LIBs, especially as cost-effective alternatives. The second advantage is the removal of the flammable liquid electrolytes. The solid electrolytes are more resistant to changes in temperature and physical damage, produce up to 80% less heat, and are able to handle more charge/discharge cycles before degradation makes them unusable. All these features point toward a longer battery lifetime. Other immediate gains include the removal of the membrane and casing required for a liquid electrolyte. This may reduce the weight and volume of each cell, leading to an increase in energy density of the battery. However, even if energy densities of current lithium-ion cells cannot be achieved by the solid-state post-Li-ion cells, this can be compensated by adding more solid-state cells into a battery pack because of the reduced volume of a single solid-state cell and the ability to create a bipolar design (which is not possible with liquid electrolytes).

Despite intensive research, solid-state alkali-metal and alkaline-earth-metal batteries are still in the early stages of their development. Solid electrolytes are the key components for these systems. The most widely studied sodium-conducting ISEs are Na-β-Al_2O_3, NASICON, sulfides, and antiperovskites. Some of these ISEs are reported to achieve bulk ionic conductivities approaching a few $mS\,cm^{-1}$ at room temperature. Potassium-conducting inorganic oxides have very low-room-temperature conductivity, while 3D open-framework polycrystalline ferrite, $K_2Fe_4O_7$, and phosphidosilicate, Si_2P_3, were found to enable rapid K^+ transport. The divalent nature of the Mg^{2+} ion results in a very sluggish mobility through traditional ionic conductors, although new ternary chalcogenides exhibit reasonably high ionic mobility. It is worth mentioning, however, that intrinsic defects and some routes in the synthesis of chalcogenides induce electronic conductivity, which has to be suppressed in order to make these ISEs relevant for RMBs. Another important issue to be resolved in sodium, potassium, and magnesium inorganic conductors is an additional resistance introduced by grain boundaries. This together with the fragility and the need for constant cell pressure may significantly limit practical applications of the batteries.

Researchers have pursued polymer electrolytes for use with metal anodes for decades. Polymer electrolytes have the advantage of lower cost and easier processability compared with inorganic SSEs, along with improved thermal and mechanical stability, and potential electrochemical stability compared with liquid electrolytes. The materials and methods for preparation of lithium polymer electrolytes are also used for sodium and potassium polymer electrolytes, but with little success for magnesium. Plasticizing additives are required to promote segmental motion and cation hopping in magnesium polymer electrolytes because of the strong electrostatic interaction of Mg^{2+} with the ether chain oxygen of the PEO.

Despite significant research efforts, challenges remain with solid composite electrolytes as well, and definite improvements in the room-temperature conductivity and ability to plate a metal electrode over other classes of solid electrolytes with commercially relevant specifications are not yet evident. Of particular value is tuning the morphologies of the ceramic phases of the electrolyte (e.g. nanoparticles, nanowires, and 3D frameworks) to provide easy transport of ions and improve the ionic conductivity of the composite electrolytes. These challenges, while being widely addressed in the past two decades, still have much to be investigated.

In the past decade, IGPEs and ionogel solid electrolytes have attracted significant attention in the research on lithium batteries. However, up to now, publications on the research on the use of these electrolytes in post Li-ion batteries are scarce. IGPEs can not only effectively eliminate the leakage risk of liquid electrolyte but also address the issue of the low ionic conductivity of all-SSEs.

As with lithium batteries, electrode/electrolyte interfaces in the solid post-Li-ion batteries are of considerable importance. These interfaces are prone to defects and contaminants. Defects include surface roughness, incipient cracks, grain boundaries, and surface contaminations among others and can lead to poor interfacial contact and inhomogeneous current distributions, both of which facilitate dendrite and/or void initiation. As in lithium batteries, dendrite growth through inorganic solid and polymer electrolytes has been widely observed and reported for solid-state sodium-metal electrodes. Investigation of potassium and magnesium dendrite formation has been less researched, especially against solid electrolytes. The diversity of experimental platforms, materials and preparation techniques, and testing conditions is a continuing challenge in reaching specific and quantitative conclusions about the causes and the principal mechanisms of dendrite penetrations in solid-state alkali- and alkaline-earth metal batteries. Determining which properties or combination of these properties enables stable cycling at high current densities could direct future work on room-temperature dendrite-free metal plating and stripping in solid batteries.

Considering the cathode materials, metal oxides, polyanion phosphate, and metal–organic materials have seen widespread development in sodium batteries because of their high theoretical capacities and operating voltage, excellent thermal stability, and tunable characteristics for improved intercalation rates. PB frameworks and nickel-thioxonickel compound have been exploited as cathode materials in potassium-ion batteries. Although magnesium batteries have already been studied for several decades, their realization is restricted by the lack of suitable high-voltage cathode material that can tackle the limitations associated with the sluggish diffusion of the magnesium ion into the host.

To summarize, there is still much room for improvements in solid-state alkali- and alkaline-earth-metal batteries. Despite dozens of reports up to the present, no sufficiently high conductivity and stability of solid electrolytes that facilitate reversible and efficient electrochemistry with sodium-, potassium-, and magnesium-metal anodes have been proved at room temperature. Progress is not expected simply by copying of materials approaches from the lithium-battery literature, but through design of new cation-conducting matrices. It is our belief that one of the most promising research directions toward advanced solid-alkali-metal and alkaline-earth-metal batteries is the development of composite IGPEs with high content of cation-conducting ceramics. Incorporating ion–dipole interactions induced by ILs and ceramic nanoparticles to the IGPEs is expected to result in a number of functionalities, including high-dimensional stability, high intrinsic conductivity, high transference number, and low interfacial resistance between the electrolyte and alkali/alkaline-earth metal anode. High coulombic efficiency and uniform, non-dendritic morphology at the anode upon extended cycling may be achieved by the engineering of an artificial SEI that will enable long-term stability and reversibility.

References

1 Survey, U.S.G. (2021). Mineral commodity summaries 2021. *Mineral Commodity Summaries* https://doi.org/10.3133/MCS2021.

2 Zhang, W., Liu, Y., and Guo, Z. (2019). Approaching high-performance potassium-ion batteries via advanced design strategies and engineering. *Science Advances* 5: 1–14. https://doi.org/10.1126/sciadv.aav7412.

3 Aurbach, D., Lu, Z., Schechter, A. et al. (2000). Prototype systems for rechargeable magnesium batteries. *Nature* 407: 724–727. https://doi.org/10.1038/35037553.

4 Yoo, H.D., Shterenberg, I., Gofer, Y. et al. (2013). Mg rechargeable batteries: an on-going challenge. *Energy & Environmental Science* 6: 2265–2279. https://doi.org/10.1039/c3ee40871j.

5 York, M., Larson, K., Harris, K.C. et al. (2022). Recent advances in solid-state beyond - lithium batteries. *Journal of Solid State Electrochemistry* 26: 1851–1869. https://doi.org/10.1007/s10008-022-05223-w.

6 Chen, R., Li, Q., Yu, X. et al. (2019). Approaching practically accessible solid-state batteries: stability issues related to solid electrolytes and interfaces. *Chemical Reviews* 120: 6820–6877. https://doi.org/10.1021/ACS.CHEMREV.9B00268.

7 Randau, S., Weber, D.A., Kötz, O. et al. ((n.d.).). Benchmarking the performance of all-solid-state lithium batteries. *Nature Energy* https://doi.org/10.1038/s41560-020-0565-1.

8 Funke, K. (1993). Jump relaxation in solid ionic conductors. *Solid State Chemistry* 22: 111–195.

9 Varzi, A., Raccichini, R., Passerini, S., and Scrosati, B. (2016). Challenges and prospects of the role of solid electrolytes in the revitalization of lithium metal batteries. *Journal of Materials Chemistry A* 4: 17251–17259. https://doi.org/10.1039/c6ta07384k.

10 Anantharamulu, N., Rao, K.K., Rambabu, G. et al. (2011). A wide-ranging review on Nasicon type materials. *Journal of Materials Science* 46: 2821–2837. https://doi.org/10.1007/s10853-011-5302-5.

11 Oh, J.A.S., He, L., Plewa, A. et al. (2019). Composite NASICON ($Na_3Zr_2Si_2PO_{12}$) solid-state electrolyte with enhanced Na + ionic conductivity: effect of liquid phase sintering. *ACS Applied Materials & Interfaces* 11: 40125–40133. https://doi.org/10.1021/ACSAMI.9B14986.

12 Park, H., Jung, K., Nezafati, M. et al. (2016). Sodium ion diffusion in nasicon ($Na_3Zr_2Si_2PO_{12}$) solid electrolytes: effects of excess sodium. *ACS Applied Materials & Interfaces* 8: 27814–27824. https://doi.org/10.1021/acsami.6b09992.

13 Fuentes, R.O., Figueiredo, F.M., Marques, F.M.B., and Franco, J.I. (2001). Influence of microstructure on the electrical properties of NASICON materials. *Solid State Ionics* 140: 173–179. https://doi.org/10.1016/S0167-2738(01)00701-9.

14 Tsuji, F., Nasu, A., Hotehama, C. et al. (2021). Preparation and characterization of sodium-ion conductive Na_3BS_3 glass and glass–ceramic electrolytes. *Materials Advances* 2: 1676–1682. https://doi.org/10.1039/d0ma00777c.

15 Nakano, K., Noda, Y., Tanibata, N. et al. (2019). Computational investigation of the Mg-ion conductivity and phase stability of $MgZr_4(PO_4)_6$. *RSC Advances* 9: 12590–12595. https://doi.org/10.1039/c9ra00513g.

16 Tarascon, J.-M. and Armand, M. (2001). Issues and challenges facing rechargeable lithium batteries. *Nature* 414, 359–367. https://doi.org/10.1038/35104644.

17 Fenton, D.E., Parker, J.M., and Wright, P.V. (1973). Complexes of alkali metal ions with poly(ethylene oxide). *Polymer (Guildf)* 14: 589. https://doi.org/10.1016/0032-3861(73)90146-8.

18 Armand, M.B., Chabagno, J.M., and Duclot, M.J. (1979). Polyethers as solid electrolytes. In: *Fast Ion Transp. Solids Electrodes Electrolytes, Proceedings International Conference*, 131, 136.

19 Armand, M.B., Bruce, P.G., Forsyth, M. et al. (2011). Polymer electrolytes. In: *Energy Mater* (ed. D.W. Bruce, D. O'Hare and R.I. Walton), 1–31. Chichester, UK: Wiley https://doi.org/10.1002/9780470977798.ch1.

20 Chen, R., Qu, W., Guo, X. et al. (2016). The pursuit of solid-state electrolytes for lithium batteries: From comprehensive insight to emerging horizons. *Materials Horizons* 3: 487–516. https://doi.org/10.1039/C6MH00218H.

21 Hallinan, D.T. and Balsara, N.P. (2013). Polymer electrolytes. *Annual Review of Materials Research* 43: 503–525. https://doi.org/10.1146/annurev-matsci-071312-121705.

22 Bocharova, V. and Sokolov, A.P. (2020). Perspectives for polymer electrolytes: a view from fundamentals of ionic conductivity. *Macromolecules* 53: 4141–4157. https://doi.org/10.1021/acs.macromol.9b02742.

23 Ferry, A., Edman, L., Forsyth, M. et al. (2000). NMR and Raman studies of a novel fast-ion-conducting polymer-in-salt electrolyte based on $LiCF_3SO_3$ and PAN. *Electrochimica Acta* 45: 1237–1242. https://doi.org/10.1016/S0013-4686(99)00386-2.

24 Wang, Y., Fan, F., Agapov, A.L. et al. (2014). Examination of the fundamental relation between ionic transport and segmental relaxation in polymer electrolytes. *Polymer (Guildf)* 55: 4067–4076. https://doi.org/10.1016/j.polymer.2014.06.085.

25 Ratner, M.A., Johansson, P., and Shriver, D.F. (2000). Polymer electrolytes: ionic transport mechanisms and relaxation coupling. *MRS Bulletin* 253 (25): 31–37. https://doi.org/10.1557/MRS2000.16.

26 Bresser, D., Lyonnard, S., Iojoiu, C. et al. (2019). Decoupling segmental relaxation and ionic conductivity for lithium-ion polymer electrolytes. *Molecular Systems Design & Engineering* 4: 779–792. https://doi.org/10.1039/c9me00038k.

27 Golodnitsky, D., Strauss, E., Peled, E., and Greenbaum, S. (2015). Review—on order and disorder in polymer electrolytes. *Journal of the Electrochemical Society* 162: A2551–A2566. https://doi.org/10.1149/2.0161514jes.

28 Croce, F., Persi, L., Scrosati, B. et al. (2001). Role of the ceramic fillers in enhancing the transport properties of composite polymer electrolytes. *Electrochimica Acta* 46: 2457–2461.

29 Bonizzoni, S., Ferrara, C., Berbenni, V. et al. (2019). NASICON-type polymer-in-ceramic composite electrolytes for lithium batteries. *Physical Chemistry Chemical Physics* 21: 6142–6149. https://doi.org/10.1039/C9CP00405J.

30 Blanga, R., Burstein, L., Berman, M. et al. (2015). Solid polymer-in-ceramic electrolyte formed by electrophoretic deposition. *Journal of the Electrochemical Society* 162: D3084–D3089. https://doi.org/10.1149/2.0221511jes.

31 Hou, W., Guo, X., Shen, X. et al. (2018). Solid electrolytes and interfaces in all-solid-state sodium batteries: progress and perspective. *Nano Energy* 52: 279–291. https://doi.org/10.1016/j.nanoen.2018.07.036.

32 Chen, L., Li, Y., Li, S.P. et al. (2018). PEO/garnet composite electrolytes for solid-state lithium batteries: From "ceramic-in-polymer" to "polymer-in-ceramic,". *Nano Energy* 46: 176–184. https://doi.org/10.1016/J.NANOEN.2017.12.037.

33 Menkin, S., Lifshitz, M., Haimovich, A. et al. (2019). Evaluation of ion-transport in composite polymer-in-ceramic electrolytes. Case study of active and inert ceramics. *Electrochimica Acta* 304: 447–455. https://doi.org/10.1016/j.electacta.2019.03.006.

34 Blanga, R., Golodnitsky, D., Ardel, G. et al. (2013). Quasi-solid polymer-in-ceramic membrane for Li-ion batteries. *Electrochimica Acta* 114: 325–333. https://doi.org/10.1016/j.electacta.2013.09.106.

35 Bae, J., Li, Y., Zhang, J. et al. (2018). A 3D nanostructured hydrogel-framework-derived high-performance composite polymer lithium-ion electrolyte. *Angewandte Chemie, International Edition* 57: 2096–2100. https://doi.org/10.1002/anie.201710841.

36 Peled, E. and Golodnitsky, D. (2004). SEI on lithium, graphite,disordered carbons and tin-based alloys. *Lithium-Ion Battery* 1–69. https://doi.org/10.1142/9781860946448_0001.

37 Watanabe, M., Thomas, M.L., Zhang, S. et al. (2017). Application of ionic liquids to energy storage and conversion materials and devices. *Chemical Reviews* 117: 7190–7239. https://doi.org/10.1021/ACS.CHEMREV.6B00504.

38 Malunavar, S., Wang, X., Makhlooghiazad, F. et al. (2021). Highly conductive ionogel electrolytes based on N-ethyl-N-methylpyrrolidinium bis(fluorosulfonyl)imide FSI and NaFSI mixtures and their applications in sodium batteries. *Journal of Physics Materials* 4: 044005. https://doi.org/10.1088/2515-7639/ac0800.

39 Kazyak, E., Garcia-Mendez, R., LePage, W.S. et al. (2020). Li penetration in ceramic solid electrolytes: operando microscopy analysis of morphology, propagation, and reversibility. *Matter* 2: 1025–1048. https://doi.org/10.1016/j.matt.2020.02.008.

40 Sudo, R., Nakata, Y., Ishiguro, K. et al. (2014). Interface behavior between garnet-type lithium-conducting solid electrolyte and lithium metal. *Solid State Ionics* 262: 151–154. https://doi.org/10.1016/j.ssi.2013.09.024.

41 Ren, Y., Shen, Y., Lin, Y., and Nan, C.W. (2015). Direct observation of lithium dendrites inside garnet-type lithium-ion solid electrolyte. *Electrochemistry Communications* 57: 27–30. https://doi.org/10.1016/j.elecom.2015.05.001.

42 Suzuki, Y., Kami, K., Watanabe, K. et al. (2015). Transparent cubic garnet-type solid electrolyte of Al_2O_3-doped $Li_7La_3Zr_2O_{12}$. *Solid State Ionics* 278: 172–176. https://doi.org/10.1016/j.ssi.2015.06.009.

43 Cheng, E.J., Sharafi, A., and Sakamoto, J. (2017). Intergranular Li metal propagation through polycrystalline $Li_{6.25}Al_{0.25}La_3Zr_2O_{12}$ ceramic electrolyte. *Electrochimica Acta* 223: 85–91. https://doi.org/10.1016/j.electacta.2016.12.018.

44 Aguesse, F., Manalastas, W., Buannic, L. et al. (2017). Investigating the dendritic growth during full cell cycling of garnet electrolyte in direct contact with Li metal. *ACS Applied Materials & Interfaces* 9: 3808–3816. https://doi.org/10.1021/acsami.6b13925.

45 Pөrz, L., Swamy, T., Sheldon, B.W. et al. (2017). Mechanism of lithium metal penetration through inorganic solid electrolytes. *Advanced Energy Materials* 7: https://doi.org/10.1002/aenm.201701003.

46 Swamy, T., Park, R., Sheldon, B.W. et al. (2018). Lithium metal penetration induced by electrodeposition through solid electrolytes: example in single-crystal $Li_6La_3ZrTaO_{12}$ garnet. *Journal of the Electrochemical Society* 165: A3648–A3655. https://doi.org/10.1149/2.1391814jes.

47 Westover, A.S., Dudney, N.J., Sacci, R.L., and Kalnaus, S. (2019). Deposition and confinement of Li metal along an artificial lipon-lipon interface. *ACS Energy Letters* 4: 651–655. https://doi.org/10.1021/acsenergylett.8b02542.

48 Lu, Y., Zhao, C.Z., Yuan, H. et al. (2021). Critical current density in solid-state lithium metal batteries: mechanism, influences, and strategies. *Advanced Functional Materials* https://doi.org/10.1002/ADFM.202009925/FORMAT/PDF.

49 Park, R.J.Y., Eschler, C.M., Fincher, C.D. et al. (2021). Semi-solid alkali metal electrodes enabling high critical current densities in solid electrolyte batteries. *Nature Energy* 6: 314–322. https://doi.org/10.1038/s41560-021-00786-w.

50 Kasemchainan, J., Zekoll, S., Jolly, D.S. et al. (2019). Critical stripping current leads to dendrite formation on plating in lithium anode solid electrolyte cells. *Nature Materials* 18: https://doi.org/10.1038/s41563-019-0438-9.

51 Flatscher, F., Philipp, M., Ganschow, S. et al. (2020). The natural critical current density limit for $Li_7La_3Zr2O_{12}$ garnets. *Journal of Materials Chemistry A* 8: 15782–15788. https://doi.org/10.1039/c9ta14177d.

52 Bay, M.C., Wang, M., Grissa, R. et al. (2020). Sodium plating from Na-β″-alumina ceramics at room temperature, paving the way for fast-charging all-solid-state batteries. *Advanced Energy Materials* 10: https://doi.org/10.1002/aenm.201902899.

53 Jäckle, M. and Groß, A. (2014). Microscopic properties of lithium, sodium, and magnesium battery anode materials related to possible dendrite growth. *The Journal of Chemical Physics* 141: 174710. https://doi.org/10.1063/1.4901055.

54 Hundekar, P., Basu, S., Fan, X. et al. (2020). In situ healing of dendrites in a potassium metal battery. *Proceedings of the National Academy of Sciences of the United States of America* 117: 5588–5594. https://doi.org/10.1073/pnas.1915470117.

55 Scrosati, B. and Butherus, A.D. (1972). Electrochemical properties of $RbAg_{[4]}I_{[5]}$ solid electrolyte. *Journal of the Electrochemical Society* 119: 128. https://doi.org/10.1149/1.2404147.

56 Davidson, R., Verma, A., Santos, D. et al. (2019). Formation of magnesium dendrites during electrodeposition. *ACS Energy Letters* 4: 375–376. https://doi.org/10.1021/acsenergylett.8b02470.

57 Pu, S.D., Gong, C., Gao, X. et al. (2020). Current-density-dependent electroplating in Ca electrolytes: from globules to dendrites. *ACS Energy Letters* 5: 2283–2290. https://doi.org/10.1021/acsenergylett.0c01153.

58 Armstrong, R.D., Dickinson, T., and Turner, J. (1974). The breakdown of β-alumina ceramic electrolyte. *Electrochimica Acta* 19: 187–192. https://doi.org/10.1016/0013-4686(74)85065-6.

59 Richman, R.H. and Tennenhouse, G.J. (1975). A model for degradation of ceramic electrolytes in Na-S batteries. *Journal of the American Ceramic Society* 58: 63–67. https://doi.org/10.1111/J.1151-2916.1975.TB18986.X/FORMAT/PDF.

60 Worrell, C.A. and Redfern, B.A.W. (1978). Acoustic emission studies of the breakdown of beta-alumina under conditions of sodium ion transport. *Journal of Materials Science* 13: 1515–1520.

61 Virkar, A.V., Viswanathan, L., and Biswas, D.R. (1980). On the deterioration of I′-alumina under electrolytic conditions ceramics. *Journal of Materials Science* 15: 302–308.

62 Brennan, M.P.J. (1980). The failure of beta-alumina electrolyte by a dendritic penetration mechanism. *Electrochimica Acta* 25: 621–627. https://doi.org/10.1016/0013-4686(80)87067-8.

63 Virkar, A.V., Miller, G.R., and Gordon, R.S. (1981). Comment on "the failure of beta alumina electrolyte by a dendritic penetration mechanism". *Electrochimica Acta* 26: 1023–1027. https://doi.org/10.1016/0013-4686(81)85071-2.

64 Liu, Q., Zhang, L., Sun, H. et al. (2021). In situ observation of sodium dendrite growth and concurrent mechanical property measurements using an environmental transmission electron microscopy–atomic force microscopy (ETEM-AFM) platform. *ACS Energy Letters* 6: 40. https://doi.org/10.1021/acsenergylett.0c01214.

65 Li, M., Sun, B., Ao, Z. et al. (2020). Atomic-scale identification of influencing factors of sodium dendrite growth on different current collectors. *Journal of Materials Chemistry A* 8: 10199–10205. https://doi.org/10.1039/d0ta01853h.

66 Shen, L., Yang, J., Liu, G. et al. (2021). High ionic conductivity and dendrite-resistant NASICON solid electrolyte for all-solid-state sodium batteries. *Materials Today Energy* 20: https://doi.org/10.1016/j.mtener.2021.100691.

67 Wang, X., Chen, J., Mao, Z., and Wang, D. (2022). Effective resistance to dendrite growth of NASICON solid electrolyte with lower electronic conductivity. *Chemical Engineering Journal* 427: 130899. https://doi.org/10.1016/j.cej.2021.130899.

68 Tsai, C.-L., Lan, T., Dellen, C. et al. (2020). Dendrite-tolerant all-solid-state sodium batteries and an important mechanism of metal self-diffusion. *Journal of Power Sources* 476: 228666. https://doi.org/10.1016/j.jpowsour.2020.228666.

69 Barai, P., Higa, K., Ngo, A.T. et al. (2019). Mechanical stress induced current focusing and fracture in grain boundaries. *Journal of the Electrochemical Society* 166: A1752–A1762. https://doi.org/10.1149/2.0321910jes.

70 Barroso-Luque, L., Tu, Q., and Ceder, G. (2020). An analysis of solid-state electrodeposition-induced metal plastic flow and predictions of stress states in solid ionic conductor defects. *Journal of the Electrochemical Society* 167: 020534. https://doi.org/10.1149/1945-7111/ab6c5b.

71 Klinsmann, M., Hildebrand, F.E., Ganser, M., and Mcmeeking, R.M. (2019). Dendritic cracking in solid electrolytes driven by lithium insertion. *Journal of Power Sources* 442: 227226. https://doi.org/10.1016/j.jpowsour.2019.227226.

72 Yuan, C., Lu, W., and Xu, J. (2021). Unlocking the electrochemical–mechanical coupling behaviors of dendrite growth and crack propagation in all-solid-state batteries. *Advanced Energy Materials* 11: https://doi.org/10.1002/AENM.202101807/FORMAT/PDF.

73 Spencer Jolly, D., Ning, Z., Darnbrough, J.E. et al. (2020). Sodium/Na β″ alumina interface: effect of pressure on voids. *ACS Applied Materials & Interfaces* 12: 678–685. https://doi.org/10.1021/acsami.9b17786.

74 Frenck, L., Sethi, G.K., Maslyn, J.A., and Balsara, N.P. (2019). Factors that control the formation of dendrites and other morphologies on lithium metal anodes. *Frontiers in Energy Research* 7: 115. https://doi.org/10.3389/FENRG.2019.00115/BIBTEX.

75 Brissot, C., Rosso, M., Chazalviel, J.-N. et al. (1999). Dendritic growth mechanisms in lithiumrpolymer cells. *Journal of Power Sources* 81–82: 925–929.

76 Song, J.Y., Wang, Y.Y., and Wan, C.C. (1999). Review of gel-type polymer electrolytes for lithium-ion batteries. *Journal of Power Sources* 77: 183–197.

77 Harry, K.J., Higa, K., Srinivasan, V., and Balsara, N.P. (2016). Influence of electrolyte modulus on the local current density at a dendrite tip on a lithium metal electrode. *Journal of the Electrochemical Society* 163: A2216–A2224. https://doi.org/10.1149/2.0191610jes.

78 Monroe, C. and Newman, J. (2005). The impact of elastic deformation on deposition kinetics at lithium/polymer interfaces. *Journal of the Electrochemical Society* 152: A396. https://doi.org/10.1149/1.1850854.

79 Barai, P., Higa, K., and Srinivasan, V. (2017). Lithium dendrite growth mechanisms in polymer electrolytes and prevention strategies †. *Physical Chemistry Chemical Physics* 19: 20493. https://doi.org/10.1039/c7cp03304d.

80 Yin, H., Han, C., Liu, Q. et al. (2021). Recent advances and perspectives on the polymer electrolytes for sodium/potassium-ion batteries. *Small* 17: https://doi.org/10.1002/SMLL.202006627/FORMAT/PDF.

81 Li, S., Zhang, S.Q., Shen, L. et al. (2020). Progress and perspective of ceramic/polymer composite solid electrolytes for lithium batteries. *Advancement of Science* 7: https://doi.org/10.1002/ADVS.201903088/FORMAT/PDF.

82 Zhao, C., Zhao, J., Liu, L. et al. (2018). Solid-state sodium batteries sodium-ion batteries view project inorganic crystal compounds view project solid-state sodium batteries. *Advanced Energy Materials* 8: 1703012. https://doi.org/10.1002/aenm.201703012.

83 Yu, Y., Xu, H., Wang, Z., and Shao, G. (2021). Transport of sodium ions in solids: progress in first-principle theoretical formulation of potential solid sodium-ion electrolytes. *Batteries & Supercaps* 4: 1096–1107. https://doi.org/10.1002/BATT.202000320.

84 Tarascon, J.M. (2020). Na-ion versus Li-ion batteries: complementarity rather than competitiveness. *Joule* 4: 1616–1620. https://doi.org/10.1016/J.JOULE.2020.06.003.

85 Lu, Y., Li, L., Zhang, Q. et al. (2018). Electrolyte and interface engineering for solid-state sodium batteries. *Joule* 2: 1747–1770. https://doi.org/10.1016/J.JOULE.2018.07.028.

86 Li, Z., Liu, P., Zhu, K. et al. (2021). Solid-state electrolytes for sodium metal batteries. *Energy & Fuels* 35: 9063–9079. https://doi.org/10.1021/ACS.ENERGYFUELS.1C00347.

87 Yue, J., Han, F., Fan, X. et al. (2017). High-performance all-inorganic solid-state sodium-sulfur battery. *ACS Nano* 11: 4885–4891. https://doi.org/10.1021/ACSNANO.7B01445.

88 Qi, X., Ma, Q., Liu, L. et al. (2016). Sodium bis(fluorosulfonyl)imide/poly(ethylene oxide) polymer electrolytes for sodium-ion batteries. *ChemElectroChem* 3: 1741–1745. https://doi.org/10.1002/CELC.201600221.

89 Ma, Q., Liu, J., Qi, X. et al. (2017). A new Na[$(FSO_2)(n\text{-}C_4F_9SO_2)N$]-based polymer electrolyte for solid-state sodium batteries. *Journal of Materials Chemistry A* 5: 7738–7743. https://doi.org/10.1039/C7TA01820G.

90 Moreno, J.S., Armand, M., Berman, M.B. et al. (2014). Composite PEOn:NaTFSI polymer electrolyte: Preparation, thermal and electrochemical characterization. *Journal of Power Sources* 248: 695–702. https://doi.org/10.1016/J.JPOWSOUR.2013.09.137.

91 Ponrouch, A., Monti, D., Boschin, A. et al. (2014). Non-aqueous electrolytes for sodium-ion batteries. *Journal of Materials Chemistry A* 3: 22–42. https://doi.org/10.1039/C4TA04428B.

92 Mokhtar, M., Majlan, E.H., Talib, M.Z.M. et al. (2016). A short review on alkaline solid polymer electrolyte based on polyvinyl alcohol (PVA) as polymer electrolyte for electrochemical devices applications. *International Journal of Applied Engineering Research* 11: 10009–10015.

93 Osman, Z., Md Isa, K.B., Ahmad, A., and Othman, L. (2010). A comparative study of lithium and sodium salts in PAN-based ion conducting polymer electrolytes. *Ionics (Kiel)* 16: 431–435. https://doi.org/10.1007/S11581-009-0410-9/FIGURES/5.

94 Vignarooban, K., Kushagra, R., Elango, A. et al. (2016). Current trends and future challenges of electrolytes for sodium-ion batteries. *International Journal of Hydrogen Energy* 41: 2829–2846. https://doi.org/10.1016/J.IJHYDENE.2015.12.090.

95 Seh, Z.W., Sun, J., Sun, Y., and Cui, Y. (2015). A highly reversible room-temperature sodium metal anode. *ACS Central Science* 1: 449–455. https://doi.org/10.1021/ACSCENTSCI.5B00328.

96 Cohn, A.P., Muralidharan, N., Carter, R. et al. (2017). Anode-free sodium battery through in situ plating of sodium metal. *Nano Letters* 17: 1296–1301. https://doi.org/10.1021/ACS.NANOLETT.6B05174.

97 Lee, S.E. and Tang, M.H. (2019). Reliable reference electrodes for nonaqueous sodium-ion batteries. *Journal of the Electrochemical Society* 166: 3260–3264. https://doi.org/10.1149/2.0401914jes.

98 Luo, W., Zhang, Y., Xu, S. et al. (2017). Encapsulation of metallic Na in an electrically conductive host with porous channels as a highly stable Na metal anode. *Nano Letters* 17: 3792–3797. https://doi.org/10.1021/ACS.NANOLETT.7B01138.

99 Kim, Y.J., Lee, H., Noh, H. et al. (2017). Enhancingthe cycling stability of sodium metal electrodes by building an inorganic–organiccomposite protective layer. *ACS Applied Materials & Interfaces* 9: 6000. https://doi.org/10.1021/ACSAMI.6B14437.S001.

100 Lee, B., Paek, E., Mitlin, D., and Lee, S.W. (2019). Sodium metal anodes: emerging solutions to dendrite growth. *Chemical Reviews* 119: https://doi.org/10.1021/ACS.CHEMREV.8B00642.

101 Oh, J.A.S., He, L., Chua, B. et al. (2021). Inorganic sodium solid-state electrolyte and interface with sodium metal for room-temperature metal solid-state batteries. *Energy Storage Materials* 34: 28–44. https://doi.org/10.1016/J.ENSM.2020.08.037.

102 Lu, X., Li, G., Kim, J.Y. et al. (2014). Liquid-metal electrode to enable ultra-low temperature sodium–beta alumina batteries for renewable energy storage. *Nature Communications* 51 (5): 1–8. https://doi.org/10.1038/ncomms5578.

103 Baclig, A.C., McConohy, G., Poletayev, A. et al. (2018). High-voltage, room-temperature liquid metal flow battery enabled by Na-K| K-β′-alumina stability. *Joule* 2: 1287–1296. https://doi.org/10.1016/j.joule.2018.04.008.

104 Dai, Z., Mani, U., Tan, H.T. et al. (2017). Advanced cathode materials for sodium-ion batteries: what determines our choices? *Small Methods* 1: 1700098. https://doi.org/10.1002/SMTD.201700098.

105 Zheng, X., Bommier, C., Luo, W. et al. (2019). Sodium metal anodes for room-temperature sodium-ion batteries: applications, challenges and solutions. *Energy Storage Materials* 16: 6–23. https://doi.org/10.1016/J.ENSM.2018.04.014.

106 Bommier, C., Ji, X., and Greaney, P.A. (2018). Electrochemical properties and theoretical capacity for sodium storage in hard carbon: insights from first principles calculations. *Chemistry of Materials* 31: 658. https://doi.org/10.1021/acs.chemmater.8b01390.

107 Hou, H., Qiu, X., Wei, W. et al. (2017). Carbon Anode materials for advanced sodium-ion batteries. *Advanced Energy Materials* 7: 1602898. https://doi.org/10.1002/AENM.201602898.

108 Perveen, T., Siddiq, M., Shahzad, N. et al. (2020). Prospects in anode materials for sodium ion batteries - a review. *Renewable and Sustainable Energy Reviews* 119: 109549. https://doi.org/10.1016/J.RSER.2019.109549.

109 Stevens, D.A. and Dahn, J.R. (2000). High capacity anode materials for rechargeable sodium-ion batteries. *Journal of the Electrochemical Society* 147: 1271. https://doi.org/10.1149/1.1393348/XML.

110 Xu, J., Wang, M., Wickramaratne, N.P. et al. (2015). High-performance sodium ion batteries based on a 3d anode from nitrogen-doped graphene foams. *Advanced Materials* 27: 2042–2048. https://doi.org/10.1002/ADMA.201405370.

111 Wen, Y., He, K., Zhu, Y. et al. (2014). Expanded graphite as superior anode for sodium-ion batteries. *Nature Communications* 5: https://doi.org/10.1038/NCOMMS5033.

112 Kim, Y., Ha, K.H., Oh, S.M., and Lee, K.T. (2014). High-capacity anode materials for sodium-ion batteries. *Chemistry - A European Journal* 20: 11980–11992. https://doi.org/10.1002/CHEM.201402511.

113 Kim, J.S., Cho, G.B., Kim, K.W. et al. (2011). The addition of iron to Ni3S2 electrode for sodium secondary battery. *Current Applied Physics* 11: S215–S218. https://doi.org/10.1016/J.CAP.2010.11.082.

114 Kim, T.B., Choi, J.W., Ryu, H.S. et al. (2007). Electrochemical properties of sodium/pyrite battery at room temperature. *Journal of Power Sources* 174: 1275–1278. https://doi.org/10.1016/J.JPOWSOUR.2007.06.093.

115 David, L., Bhandavat, R., and Singh, G. (2014). MoS_2/graphene composite paper for sodium-ion battery electrodes. *ACS Nano* 8: 1759–1770. https://doi.org/10.1021/NN406156B.

116 Yang, J., Gao, Z., Ferber, T. et al. (2020). Guided-formation of a favorable interface for stabilizing Na metal solid-state batteries. *Journal of Materials Chemistry A* 8: 7828–7835. https://doi.org/10.1039/D0TA01498B.

117 Rao, R.P., Zhang, X., Phuah, K.C., and Adams, S. (2019). Mechanochemical synthesis of fast sodium ion conductor $Na_{11}Sn_2PSe_{12}$ enables first sodium–selenium all-solid-state battery. *Journal of Materials Chemistry A* 7: 20790–20798. https://doi.org/10.1039/C9TA06279C.

118 Peled, E. (1979). The electrochemical behavior of alkali and alkaline earth metals in nonaqueous battery systems—the solid electrolyte interphase model. *Journal of the Electrochemical Society* 126: 2047–2051. https://doi.org/10.1149/1.2128859.

119 Peled, E., Golodnitsky, D., and Penciner, J. (2007). The anode/electrolyte interface. In: *Handbook of Battery Materials* (ed. C. Daniel and J.O. Besenhard), 419–456. Wiley https://doi.org/10.1002/9783527611676.CH18.

120 Fincher, C.D., Zhang, Y., Pharr, G.M., and Pharr, M. (2020). Elastic and plastic characteristics of sodium metal. *ACS Applied Energy Materials* 3: 1759–1767. https://doi.org/10.1021/ACSAEM.9B02225.

121 Xiang, X., Zhang, K., and Chen, J. (2015). Recent advances and prospects of cathode materials for sodium-ion batteries. *Advanced Materials* 27: 5343–5364. https://doi.org/10.1002/ADMA.201501527.

122 Islam, M.S. and Fisher, C.A.J. (2014). Lithium and sodium battery cathode materials: computational insights into voltage, diffusion and nanostructural properties. *Chemical Society Reviews* 43: 185–204. https://doi.org/10.1039/C3CS60199D.

123 Wang, L., Chen, B., Ma, J. et al. (2018). Reviving lithium cobalt oxide-based lithium secondary batteries-toward a higher energy density. *Chemical Society Reviews* 47: 6505–6602. https://doi.org/10.1039/C8CS00322J.

124 Zhang, W., De Zhao, C., and Wu, X.L. (2020). Research progresses on interfaces in solid-state sodium batteries: a topic review. *Advanced Materials Interfaces* 7: 2001444. https://doi.org/10.1002/ADMI.202001444.

125 Qian, J., Wu, C., Cao, Y. et al. (2018). Prussian blue cathode materials for sodium-ion batteries and other ion batteries. *Advanced Energy Materials* 8: 1702619. https://doi.org/10.1002/AENM.201702619.

126 Hasa, I., Buchholz, D., Passerini, S., and Hassoun, J. (2015). A comparative study of layered transition metal oxide cathodes for application in sodium-ion battery. *ACS Applied Materials & Interfaces* 7: https://doi.org/10.1021/am5080437.

127 Liu, Q., Hu, Z., Chen, M. et al. (2019). Recent progress of layered transition metal oxide cathodes for sodium-ion batteries. *Small* 15: 1805381. https://doi.org/10.1002/SMLL.201805381.

128 Kim, D., Kang, S.-H., Slater, M. et al. (2011). Enabling sodium batteries using lithium-substituted sodium layered transition metal oxide cathodes. *Advanced Energy Materials* 1: 333–336. https://doi.org/10.1002/AENM.201000061.

129 Jin, T., Li, H., Zhu, K. et al. (2020). Polyanion-type cathode materials for sodium-ion batteries. *Chemical Society Reviews* 49: 2342–2377. https://doi.org/10.1039/C9CS00846B.

130 Wang, M.-Y., Guo, J.-Z., Wang, Z.-W. et al. (2020). Isostructural and multivalent anion substitution toward improved phosphate cathode materials for sodium-ion batteries. *Small* 16: 1907645. https://doi.org/10.1002/SMLL.201907645.

131 Hyeon Jo, J., Ung Choi, J., Konarov, A. et al. (2018). Sodium-ion batteries: building effective layered cathode materials with long-term cycling by modifying the surface via sodium phosphate. *Advanced Functional Materials* 28: 1705968. https://doi.org/10.1002/ADFM.201705968.

132 Wu, H., Zou, X., and Wu, X.L. (2020). Nanoconstruction and nanoeffect of phosphate-based cathode materials for advanced sodium-ion batteries. *Nano Futures* 4: 042001. https://doi.org/10.1088/2399-1984/ABC103.

133 Gu, Z.Y., Guo, J.Z., Sun, Z.H. et al. (2020). Carbon-coating-increased working voltage and energy density towards an advanced $Na_3V_2(PO_4)_2F_3$@C cathode in sodium-ion batteries. *Scientific Bulletin* 65: 702–710. https://doi.org/10.1016/J.SCIB.2020.01.018.

134 Serras, P., Palomares, V., Goñi, A. et al. (2012). High voltage cathode materials for Na-ion batteries of general formula $Na_3V_2O_{2x}(PO_4)_2F_{3-2x}$. *Journal of Materials Chemistry* 22: 22301–22308. https://doi.org/10.1039/C2JM35293A.

135 Aubrey, M.L. and Long, J.R. (2015). A dual–ion battery cathode via oxidative insertion of anions in a metal–organic framework. *J. Am. Chem. Soc.* 137: 13594–13602. https://doi.org/10.1021/jacs.5b08022.

136 Nie, P., Shen, L., Pang, G. et al. (2015). Flexible metal–organic frameworks as superior cathodes for rechargeable sodium-ion batteries. *Journal of Materials Chemistry A* 3: 16590–16597. https://doi.org/10.1039/C5TA03197D.

137 Lee, H.W., Wang, R.Y., Pasta, M. et al. (2014). Manganese hexacyanomanganate open framework as a high-capacity positive electrode material for sodium-ion batteries. *Nature Communications* 5: https://doi.org/10.1038/NCOMMS6280.

138 Duchêne, L., Kühnel, R.S., Stilp, E. et al. (2017). A stable 3 V all-solid-state sodium–ion battery based on a closo-borate electrolyte. *Energy & Environmental Science* 10: 2609–2615. https://doi.org/10.1039/C7EE02420G.

139 Yue, J., Zhu, X., Han, F. et al. (2018). Long cycle life all-solid-state sodium ion battery. *ACS Applied Materials & Interfaces* 10: 39645–39650. https://doi.org/10.1021/ACSAMI.8B12610.

140 Banerjee, A., Park, K.H., Heo, J.W. et al. (2016). Na_3SbS_4: A solution processable sodium superionic conductor for all-solid-state sodium-ion batteries. *Angewandte Chemie, International Edition* 55: 9634–9638. https://doi.org/10.1002/ANIE.201604158.

141 Yamauchi, H., Ikejiri, J., Sato, F. et al. (2019). Pressureless all-solid-state sodium-ion battery consisting of sodium iron pyrophosphate glass-ceramic cathode and β''-alumina solid electrolyte composite. *Journal of the American Ceramic Society* 102: 6658–6667. https://doi.org/10.1111/JACE.16607.

142 Yamauchi, H., Ikejiri, J., Tsunoda, K. et al. (2020). Enhanced rate capabilities in a glass-ceramic-derived sodium all-solid-state battery. *Scientific Reports* 101 (10): 1–12. https://doi.org/10.1038/s41598-020-66410-1.

143 Zheng, Y., Pan, Q., Clites, M. et al. (2018). High-capacity all-solid-state sodium metal battery with hybrid polymer electrolytes. *Advanced Energy Materials* 8: 1801885. https://doi.org/10.1002/AENM.201801885.

144 Song, S., Duong, H.M., Korsunsky, A.M. et al. (2016). A Na(+) superionic conductor for room-temperature sodium batteries. *Scientific Reports* 6: https://doi.org/10.1038/SREP32330.

145 Yang, J., Liu, G., Avdeev, M. et al. (2020). Ultrastable all-solid-state sodium rechargeable batteries. *ACS Energy Letters* 5: 2835–2841. https://doi.org/10.1021/ACSENERGYLETT.0C01432/SUPPL_FILE/NZ0C01432_SI_001.PDF.

146 Yao, Y., Wei, Z., Wang, H. et al. (2020). Toward high energy density all solid-state sodium batteries with excellent flexibility. *Advanced Energy Materials* 10: 1903698. https://doi.org/10.1002/AENM.201903698.

147 Ruan, Y., Guo, F., Liu, J. et al. (2019). Optimization of $Na_3Zr_2Si_2PO_{12}$ ceramic electrolyte and interface for high performance solid-state sodium battery. *Ceramics International* 45: 1770–1776. https://doi.org/10.1016/J.CERAMINT.2018.10.062.

148 Wang, Z., Yang, L., Liu, J. et al. (2020). Tuning rate-limiting factors to achieve ultrahigh-rate solid-state sodium-ion batteries. *ACS Applied Materials & Interfaces* 12: 48677–48683. https://doi.org/10.1021/ACSAMI.0C15015.

149 Peta, G., Bublil, S., Alon-Yehezkel, H. et al. (2021). Toward high performance all solid-state Na batteries: investigation of electrolytes comprising $NaPF_6$, poly(ethylene oxide) and TiO_2. *Journal of the Electrochemical Society* 168: 110553. https://doi.org/10.1149/1945-7111/AC330D.

150 Yao, Y.-F.Y. and Kummer, J.T. (1967). Ion exchange properties of and rates of ionic diffusion in beta-alumina. *Journal of Inorganic and Nuclear Chemistry* 29: 2453–2475. https://doi.org/10.1016/0022-1902(67)80301-4.

151 Verma, R., Didwal, P.N., Hwang, J.Y., and Park, C.J. (2021). Recent progress in electrolyte development and design strategies for next-generation potassium-ion batteries. *Batteries & Supercaps* 4: 1428–1450. https://doi.org/10.1002/BATT.202100029.

152 Wang, E. and Greenblatt, M. (1991). Ionic conductivity of potassium phosphatoantimonates and some of their ion-exchanged analogues. *Chemistry of Materials* 3: 542–546. https://doi.org/10.1021/CM00015A034.

153 Wang, E. and Greenblatt, M. (1991). Ionic conductivities of ion-exchanged $A_3Sb_3P_2O_{14}$(A = Na, K, Rb), $A_5Sb_5P_2O_{20}$(A = Li, Na, K, Rb), and partially substituted $K_5Sb_{5-x}M_xP_2O_{20}$(M = Nb, Ta). *Chemistry of Materials* 3: 703–709. https://doi.org/10.1021/CM00016A026.

154 Doux, J.M., Leguay, L., Le Gal La, A. et al. (2018). $K_3Sb_4O_{10}(BO_3)$: a solid state K-ion conductor. *Solid State Ionics* 324: 260–266. https://doi.org/10.1016/j.ssi.2018.07.018.

155 Yuan, H., Li, H., Zhang, T. et al. (2018). A $K_2Fe_4O_7$ superionic conductor for all-solid-state potassium metal batteries. *Journal of Materials Chemistry A* 6: 8413–8418. https://doi.org/10.1039/C8TA01418C.

156 Zheng, J., Fang, H., Fan, L. et al. (2021). Antiperovskite K_3OI for K-ion solid state electrolyte. *Journal of Physical Chemistry Letters* 12: 7120–7126. https://doi.org/10.1021/ACS.JPCLETT.1C01807/SUPPL_FILE/JZ1C01807_SI_001.PDF.

157 Xiao, R., Li, H., and Chen, L. (2020). High-throughput computational discovery of K_2CdO_2 as an ion conductor for solid-state potassium-ion batteries. *Journal of Materials Chemistry A* 8: 5157–5162. https://doi.org/10.1039/C9TA13105A.

158 Haffner, A., Hatz, A.-K., Zeman, O.E.O. et al. (2021). Polymorphism and fast potassium-ion conduction in the T5 supertetrahedral phosphidosilicate KSi_2P_3. *Angewandte Chemie, International Edition* 60: 13641–13646. https://doi.org/10.1002/ANIE.202101187.

159 Reddy, C.V.S., Sharma, A.K., and Narasimha Rao, V.V.R. (2006). Electrical and optical properties of a polyblend electrolyte. *Polymer (Guildf)* 47: 1318–1323. https://doi.org/10.1016/J.POLYMER.2005.12.052.

160 Fei, H., Liu, Y., An, Y. et al. (2018). Stable all-solid-state potassium battery operating at room temperature with a composite polymer electrolyte and a sustainable organic cathode. *Journal of Power Sources* 399: 294–298. https://doi.org/10.1016/j.jpowsour.2018.07.124.

161 Fei, H., Liu, Y., An, Y. et al. (2019). Safe all-solid-state potassium batteries with three dimentional, flexible and binder-free metal sulfide array electrode. *Journal of Power Sources* 433: 226697. https://doi.org/10.1016/J.JPOWSOUR.2019.226697.

162 Liu, S., Mao, J., Zhang, L. et al. (2021). Manipulating the solvation structure of nonflammable electrolyte and interface to enable unprecedented stability of graphite anodes beyond 2 years for safe potassium-ion batteries. *Advanced Materials* 33: 2006313. https://doi.org/10.1002/ADMA.202006313.

163 Zhang, Q., Han, B., Zou, Y. et al. (2021). Enabling atomic-scale imaging of sensitive potassium metal and related solid electrolyte interphases using ultralow-dose cryo-TEM. *Advanced Materials* 33: https://doi.org/10.1002/ADMA.202102666/FORMAT/PDF.

164 Liu, S., Mao, J., Zhang, Q. et al. (2020). An intrinsically non-flammable electrolyte for high-performance potassium batteries. *Angewandte Chemie* 132: 3667–3673. https://doi.org/10.1002/ANGE.201913174.

165 Wang, S., Yan, Y., Xiong, D. et al. (2021). Towards dendrite-free potassium-metal batteries: rational design of a multifunctional 3D polyvinyl alcohol-borax layer. *Angewandte Chemie, International Edition* 60: 25122–25127. https://doi.org/10.1002/ANIE.202111753.

166 Shi, P., Zhang, S., Lu, G. et al. (2021). Red phosphorous-derived protective layers with high ionic conductivity and mechanical strength on dendrite-free sodium and potassium metal anodes. *Advanced Energy Materials* 11: 2003381. https://doi.org/10.1002/AENM.202003381.

167 Xiao, N., Zheng, J., Gourdin, G. et al. (2019). Anchoring an artificial protective layer to stabilize potassium metal anode in rechargeable K-O_2 batteries. *ACS Applied Materials & Interfaces* 11: 16571–16577. https://doi.org/10.1021/ACSAMI.9B02116/SUPPL_FILE/AM9B02116_SI_001.PDF.

168 Lu, K., Zhang, H., Gao, S. et al. (2018). High rate and stable symmetric potassium ion batteries fabricated with flexible electrodes and solid-state electrolytes. *Nanoscale* 10: 20754–20760. https://doi.org/10.1039/C8NR07268J.

169 Chen, Y., Luo, W., Carter, M. et al. (2015). Organic electrode for non-aqueous potassium-ion batteries. *Nano Energy* 18: 205–211. https://doi.org/10.1016/j.nanoen.2015.10.015.

170 Song, J., Sahadeo, E., Noked, M., and Lee, S.B. (2016). Mapping the challenges of magnesium battery. *Journal of Physical Chemistry Letters* 7: https://doi.org/10.1021/acs.jpclett.6b00384.

171 Muldoon, J., Bucur, C.B., and Gregory, T. (2014). Quest for nonaqueous multivalent secondary batteries : magnesium and beyond. *Chemical Reviews* 114: 11683–11720.

172 Canepa, P., Jayaraman, S., Cheng, L. et al. (2015). Elucidating the structure of the magnesium aluminum chloride complex electrolyte for magnesium-ion batteries. *Energy & Environmental Science* 8: 3718–3730. https://doi.org/10.1039/C5EE02340H.

173 Gao, T., Noked, M., Pearse, A.J. et al. (2015). Enhancing the reversibility of Mg/S battery chemistry through Li+ mediation. *Journal of the American Chemical Society* https://doi.org/10.1021/jacs.5b07820.

174 Zhao-Karger, Z., Zhao, X., Wang, D. et al. (2015). Performance improvement of magnesium sulfur batteries with modified non-nucleophilic electrolytes. *Advanced Energy Materials* 5: 1401155. https://doi.org/10.1002/aenm.201401155.

175 Kim, H.S., Arthur, T.S., Allred, G.D. et al. (2011). Structure and compatibility of a magnesium electrolyte with a sulphur cathode. *Nature Communications* 2: 427. https://doi.org/10.1038/ncomms1435.

176 Zhan, Y., Zhang, W., Lei, B. et al. (2020). Recent development of Mg ion solid electrolyte. *Frontiers in Chemistry* 8: 1–7. https://doi.org/10.3389/fchem.2020.00125.

177 Jaschin, P.W., Gao, Y., Li, Y., and Bo, S.H. (2020). A materials perspective on magnesium-ion-based solid-state electrolytes. *Journal of Materials Chemistry A* 8: 2875–2897. https://doi.org/10.1039/c9ta11729f.

178 Ikeda, S., Takahashi, M., Ishikawa, J., and Ito, K. (1987). Solid electrolytes with multivalent cation conduction. 1. Conducting species in MgZrPO4 system. *Solid State Ionics* 23 (1–2): 125–129. https://doi.org/10.1016/0167-2738(87)90091-9.

179 Kazakos-Kijowski, A., Komarneni, S., Agrawal, D., and Roy, R. (1988). Synthesis, crystal data and thermal stability of magnesium zirconium phosphate [$MgZr_4(PO_4)_6$]. *Materials Research Bulletin* 23 (8): 1177–1184.

180 Nakano, K., Noda, Y., Tanibata, N. et al. (2019). RSC advances computational investigation of the Mg-ion. *RSC Advances* 4: 12590–12595. https://doi.org/10.1039/c9ra00513g.

181 Matsuo, M., Oguchi, H., Sato, T. et al. (2013). Sodium and magnesium ionic conduction in complex hydrides. *Journal of Alloys and Compounds* 580: S98–S101. https://doi.org/10.1016/j.jallcom.2013.01.058.

182 Higashi, S., Miwa, K., Aoki, M., and Takechi, K. (2014). A novel inorganic solid state ion conductor for rechargeable Mg batteries. *Chemical Communications* 50: 1320–1322. https://doi.org/10.1039/c3cc47097k.

183 Roedern, E., Kühnel, R.-S., Remhof, A., and Battaglia, C. (2017). Magnesium ethylenediamine borohydride as solid-state electrolyte for magnesium batteries. *Scientific Reports* 7: 46189. https://doi.org/10.1038/srep46189.

184 Canepa, P., Bo, S.H., Sai Gautam, G. et al. (2017). High magnesium mobility in ternary spinel chalcogenides. *Nature Communications* 8: 1–8. https://doi.org/10.1038/s41467-017-01772-1.

185 Kundu, S., Solomatin, N., Kauffmann, Y. et al. (2021). Revealing and excluding the root cause of the electronic conductivity in Mg-ion $MgSc_2Se_4$ solid electrolyte. *Applied Materials Today* 23: 100998. https://doi.org/10.1016/j.apmt.2021.100998.

186 Canepa, P., Sai Gautam, G., Broberg, D. et al. (2017). Role of point defects in spinel Mg chalcogenide conductors. *Chemistry of Materials* 29: 9657–9667. https://doi.org/10.1021/acs.chemmater.7b02909.

187 Chen, T., Sai Gautam, G., and Canepa, P. (2019). Ionic transport in potential coating materials for Mg batteries. *Chemistry of Materials* 31: 8087–8099. https://doi.org/10.1021/acs.chemmater.9b02692.

188 Shi, J. and Vincent, C.A. (1993). The effect of molecular weight on cation mobility in polymer electrolytes. *Solid State Ionics* 60: 11–17. https://doi.org/10.1016/0167-2738(93)90268-8.

189 Vincent, C.A. (1995). Ion transport in polymer electrolytes. *Electrochimica Acta* 40: 2035–2040. https://doi.org/10.1016/0013-4686(95)00138-5.

190 Acosta, J.L. and Morales, E. (1998). Synthesis and characterization of polymeric electrolytes for solid state magnesium batteries. *Electrochimica Acta* 43: 791–797. https://doi.org/10.1016/S0013-4686(97)00123-0.

191 Anilkumar, K.M., Jinisha, B., Manoj, M., and Jayalekshmi, S. (2017). Poly(ethylene oxide) (PEO) – Poly(vinyl pyrrolidone) (PVP) blend polymer based solid electrolyte membranes for developing solid state magnesium ion cells. *European Polymer Journal* 89: 249–262. https://doi.org/10.1016/J.EURPOLYMJ.2017.02.004.

192 Thelen, J.L., Inceoglu, S., Venkatesan, N.R. et al. (2016). Relationship between ion dissociation, melt morphology, and electrochemical performance of lithium and magnesium single-ion conducting block copolymers. *Macromolecules* 49: 9139–9147. https://doi.org/10.1021/ACS.MACROMOL.6B01886/SUPPL_FILE/MA6B01886_SI_001.PDF.

193 Park, B. and Schaefer, J.L. (2020). Review—polymer electrolytes for magnesium batteries: forging away from analogs of lithium polymer electrolytes and towards the rechargeable magnesium metal polymer battery. *Journal of the Electrochemical Society* 167: 070545. https://doi.org/10.1149/1945-7111/AB7C71.

194 Agrawal, R.C., Sahu, D.K., Mahipal, Y.K., and Ashrafi, R. (2013). Investigations on ion transport properties of hot-press cast magnesium ion conducting Nano-Composite Polymer Electrolyte (NCPE) films: Effect of filler particle dispersal on room temperature conductivity. *Materials Chemistry and Physics* 139: 410–415. https://doi.org/10.1016/J.MATCHEMPHYS.2012.12.056.

195 Bella, F., De Luca, S., Fagiolari, L. et al. (2021). An overview on anodes for magnesium batteries: challenges towards a promising storage solution for renewables. *Nanomaterials* 11: 810–811. https://doi.org/10.3390/NANO11030810.

196 Di Leo, R.A., Zhang, Q., Marschilok, A.C. et al. (2015). Composite anodes for secondary magnesium ion batteries prepared via electrodeposition of nanostructured bismuth on carbon nanotube substrates. *ECS Electrochemistry Letters* 4: A10–A14. https://doi.org/10.1149/2.0081501EEL/XML.

197 Yaghoobnejad Asl, H., Fu, J., Kumar, H. et al. (2018). In situ dealloying of bulk Mg_2Sn in Mg-ion half cell as an effective route to nanostructured sn for high performance Mg-ion battery anodes. *Chemistry of Materials* 30: 1815–1824. https://doi.org/10.1021/ACS.CHEMMATER.7B04124/SUPPL_FILE/CM7B04124_SI_001.PDF.

198 Niu, J., Gao, H., Ma, W. et al. (2018). Dual phase enhanced superior electrochemical performance of nanoporous bismuth-tin alloy anodes for magnesium-ion batteries. *Energy Storage Materials* 14: 351–360. https://doi.org/10.1016/J.ENSM.2018.05.023.

199 Canepa, P., Sai Gautam, G., Hannah, D.C. et al. (2017). Odyssey of multivalent cathode materials: open questions and future challenges. *Chemical Reviews* 117: 4287–4341. https://doi.org/10.1021/acs.chemrev.6b00614.

200 Canepa, P., Gautam, G.S., Malik, R. et al. (2015). Understanding the initial stages of reversible Mg deposition and stripping in inorganic non-aqueous electrolytes. *Chemistry of Materials* 17: 3317–3325.

201 Sai Gautam, G., Canepa, P., Urban, A. et al. (2017). Influence of inversion on Mg mobility and electrochemistry in spinels. *Chemistry of Materials* 29: 7918–7930. https://doi.org/10.1021/acs.chemmater.7b02820.

202 Rong, Z., Malik, R., Canepa, P. et al. (2015). Materials design rules for multivalent ion mobility in intercalation structures. *Chemistry of Materials* 27: 6016–6021. https://doi.org/10.1021/acs.chemmater.5b02342.

203 Bayliss, R.D., Key, B., Gautam, G.S. et al. (2020). Probing Mg migration in spinel oxides. *Chemistry of Materials* 32: 663–670. https://doi.org/10.1021/acs.chemmater.9b02450.

204 Imhof, R., Haas, O., and Novák, P. (1999). Magnesium insertion electrodes for rechargeable nonaqueous batteries Ð a competitive alternative to lithium? *Electrochimica Acta* 45: 351–367.

205 Aurbach, D., Suresh, G.S., Levi, E. et al. (2007). Progress in rechargeable magnesium battery technology. *Advanced Materials* 19: 4260–4267. https://doi.org/10.1002/adma.200701495.

206 Yuan, C., Zhang, Y., Pan, Y. et al. (2014). Investigation of the intercalation of polyvalent cations (Mg^{2+}, Zn^{2+}) into λ-MnO_2 for rechargeable aqueous battery. *Electrochimica Acta* 116: 404–412. https://doi.org/10.1016/j.electacta.2013.11.090.

207 Livage, J. (1991). Vanadium pentoxide gels. *Chemistry of Materials* 3: 578–593. https://doi.org/10.1021/cm00016a006.

208 Gershinsky, G., Yoo, H.D., Gofer, Y., and Aurbach, D. (2013). Electrochemical and spectroscopic analysis of Mg^{2+} intercalation into thin film electrodes of layered oxides: V_2O_5 and MoO_3. *Langmuir* 29: 10964–10972. https://doi.org/10.1021/la402391f.

209 Liu, M., Rong, Z., Malik, R. et al. (2015). Spinel compounds as multivalent battery cathodes: A systematic evaluation based on ab initio calculations. *Energy & Environmental Science* 8: 964–974. https://doi.org/10.1039/c4ee03389b.

210 Gautam, G.S., Canepa, P., Malik, R., and Liu, M. (2015). First-principles evaluation of multi-valent cation insertion into orthorhombic V_2O_5. *ChemComm* 51: 13619–13622. https://doi.org/10.1039/c5cc04947d.

211 Gautam, G.S., Canepa, P., Richards, W.D. et al. (2016). Role of structural H_2O in intercalation electrodes: the case of mg in nanocrystalline xerogel – V_2O_5. *Nano Letters* 16: 2426–2431. https://doi.org/10.1021/acs.nanolett.5b05273.

212 Delmas, C., Cognac-Auradou, H., Cocciantelli, J.M. et al. (1994). The $Li_xV_2O_5$ system: an overview of the structure modifications induced by the lithium intercalation. *Solid State Ionics* 69: 257–264. https://doi.org/10.1016/0167-2738(94)90414-6.

213 Murphy, D.W., Christian, P.A., Disalvo, F.J., and Waszczak, J.V. (1979). Lithium incorporation by vanadium pentoxide. *Inorganic Chemistry* 18: 2800–2803. https://doi.org/10.1021/ic50200a034.

214 Braithwaite, J.S., Catlow, C.R.A., Gale, J.D., and Harding, J.H. (1999). Lithium intercalation into vanadium pentoxide: A theoretical study. *Chemistry of Materials* 11: 1990–1998. https://doi.org/10.1021/cm980735r.

215 Wu, Q.H., Thißen, A., and Jaegermann, W. (2005). Photoelectron spectroscopic study of Li intercalation into V 2O5 thin films. *Surface Science* 578: 203–212. https://doi.org/10.1016/j.susc.2005.01.042.

216 Rocquefelte, X., Boucher, F., Gressier, P., and Ouvrard, G. (2003). First-principle study of the intercalation process in the $Li_xV_2O_5$ system. *Chemistry of Materials* 15: 1812–1819. https://doi.org/10.1021/cm021254h.

217 Scanlon, D.O., Walsh, A., Morgan, B.J., and Watson, G.W. (2008). An ab initio study of reduction of V_2O_5 through the formation of oxygen vacancies and Li intercalation. *Journal of Physical Chemistry C* 112: 9903–9911. https://doi.org/10.1021/jp711334f.

218 Fu, Q., Sarapulova, A., Trouillet, V. et al. (2019). In operando synchrotron diffraction and in operando X-ray absorption spectroscopy investigations of orthorhombic V_2O_5 nanowires as cathode materials for Mg-ion batteries. *Journal of the American Chemical Society* 141: 2305–2315. https://doi.org/10.1021/jacs.8b08998.

219 Gregory, T.D. (1990). Nonaqueous electrochemistry of magnesium. *Journal of the Electrochemical Society* 137: 775. https://doi.org/10.1149/1.2086553.

220 Sai Gautam, G., Canepa, P., Abdellahi, A. et al. (2015). The intercalation phase diagram of Mg in V_2O_5 from first-principles. *Chemistry of Materials* 27: 3733–3742. https://doi.org/10.1021/acs.chemmater.5b00957.

221 Mukherjee, A., Sa, N., Phillips, P.J. et al. (2017). Direct investigation of Mg intercalation into the orthorhombic V_2O_5 cathode using atomic-resolution transmission electron microscopy. *Chemistry of Materials* 29: 2218–2226. https://doi.org/10.1021/acs.chemmater.6b05089.

222 Sa, N., Wang, H., Proffit, D.L. et al. (2016). Is alpha-V_2O_5 a cathode material for Mg insertion batteries? *Journal of Power Sources* 323: 44–50. https://doi.org/10.1016/j.jpowsour.2016.05.028.

223 Sa, N., Kinnibrugh, T.L., Wang, H. et al. (2016). Structural evolution of reversible Mg insertion into a bilayer structure of $V_2O_5 \cdot nH_2O$ xerogel material. *Chemistry of Materials* 28: 2962–2969. https://doi.org/10.1021/acs.chemmater.6b00026.

224 Tang, P.E., Sakamoto, J.S., Baudrin, E., and Dunn, B. (2004). V_2O_5 aerogel as a versatile host for metal ions. *Journal of Non-Crystalline Solids* 350: 67–72. https://doi.org/10.1016/j.jnoncrysol.2004.07.072.

225 Novák, P. (1993). Electrochemical insertion of magnesium in metal oxides and sulfides from aprotic electrolytes. *Journal of the Electrochemical Society* 140: 140. https://doi.org/10.1149/1.2056075.

226 Yu, L. and Zhang, X. (2004). Electrochemical insertion of magnesium ions into V_2O_5 from aprotic electrolytes with varied water content. *Journal of Colloid and Interface Science* 278: 160–165. https://doi.org/10.1016/j.jcis.2004.05.028.

227 Kim, R.H., Kim, J.S., Kim, H.J. et al. (2014). Highly reduced VO_x nanotube cathode materials with ultra-high capacity for magnesium ion batteries. *Journal of Materials Chemistry A* 2: 20636–20641. https://doi.org/10.1039/c4ta05564k.

228 Cheng, Y., Shao, Y., Raju, V. et al. (2016). Molecular storage of Mg ions with vanadium oxide nanoclusters. *Advanced Functional Materials* 26: 3446–3453. https://doi.org/10.1002/adfm.201505501.

229 Andrews, J.L., Mukherjee, A., Yoo, H.D. et al. (2018). Reversible Mg-ion insertion in a metastable one-dimensional polymorph of V_2O_5. *Chemistry* 4: 564–585. https://doi.org/10.1016/j.chempr.2017.12.018.

230 Mukherjee, A., Taragin, S., Aviv, H. et al. (2020). Rationally designed vanadium pentoxide as high capacity insertion material for Mg-ion. *Advanced Functional Materials* 30: 1–9. https://doi.org/10.1002/adfm.202003518.

231 Arthur, T.S., Zhang, R., Ling, C. et al. (2014). Understanding the electrochemical mechanism of K-αMnO_2 for magnesium battery cathodes. *ACS Applied Materials & Interfaces* 6: 7004–7008. https://doi.org/10.1021/am5015327.

232 Kolosnitsyn, V.S. and Karaseva, E.V. (2008). Lithium-sulfur batteries: problems and solutions. *Russian Journal of Electrochemistry* 44: 506–509. https://doi.org/10.1134/S1023193508050029.

233 Zhao-Karger, Z., Liu, R., Dai, W. et al. (2018). Toward highly reversible magnesium–sulfur batteries with efficient and practical $Mg[B(hfip)_4]_2$ electrolyte. *ACS Energy Letters* 3: 2005–2013. https://doi.org/10.1021/acsenergylett.8b01061.

234 Xiong, S., Xie, K., Diao, Y., and Hong, X. (2014). Characterization of the solid electrolyte interphase on lithium anode for preventing the shuttle mechanism in lithium-sulfur batteries. *Journal of Power Sources* 246: 840–845. https://doi.org/10.1016/j.jpowsour.2013.08.041.

235 Chang, D.R., Lee, S.H., Kim, S.W., and Kim, H.T. (2002). Binary electrolyte based on tetra(ethylene glycol) dimethyl ether and 1,3-dioxolane for lithium-sulfur battery. *Journal of Power Sources* 112: 452–460. https://doi.org/10.1016/S0378-7753(02)00418-4.

236 Kim, N. and Jinks-Robertson, S. (2012). Transcription as a source of genome instability. *Nature Reviews. Genetics* 13: 204–214. https://doi.org/10.1038/nrg3152.

237 Salama, M., Shterenberg, I., Shimon, L.J.W. et al. (2017). Structural analysis of magnesium chloride complexes in dimethoxyethane solutions in the context of Mg batteries research. *Journal of Physical Chemistry C* 121: 24909–24918. https://doi.org/10.1021/acs.jpcc.7b05452.

238 Shterenberg, I., Salama, M., Yoo, H.D. et al. (2015). Evaluation of $(CF_3SO_2)_2N^-$(TFSI) based electrolyte solutions for Mg batteries. *Journal of the Electrochemical Society* 162: A7118–A7128. https://doi.org/10.1149/2.0161513jes.

239 Bitenc, J., Pirnat, K., Bančič, T. et al. (2015). Anthraquinone-based polymer as cathode in rechargeable magnesium batteries. *ChemSusChem* 8: 4128–4132. https://doi.org/10.1002/cssc.201500910.

240 Abouzari-Lotf, E., Azmi, R., Li, Z. et al. (2021). A self-conditioned metalloporphyrin as a highly stable cathode for fast rechargeable magnesium batteries. *ChemSusChem* 14: 1840–1846. https://doi.org/10.1002/cssc.202100340.

5

A Rationale for the Development of Sustainable Biodegradable Batteries

Marina Navarro-Segarra[1,2] *and Juan P. Esquivel*[1,2,3]

[1] *Instituto de Microelectrónica de Barcelona, IMB-CNM (CSIC), C/ dels Til·lers sn, Campus UAB, 08193, Bellaterra, Barcelona, Spain*
[2] *BCMaterials, Basque Centre for Materials, Applications and Nanostructures, UPV/EHU Science Park, 48940, Leioa, Spain*
[3] *IKERBASQUE, Basque Foundation for Science, 48009, Bilbao, Spain*

5.1 Challenges for Powering a Digital Society

Digital technologies are generating a global transformation with economic, social, and environmental implications. Digitalization has already shown many benefits to society, it has stimulated disruptive innovation, generated new business models providing opportunities for value creation, contributed to transparency, and improved access to information. However, this digital transformation is generating a fierce debate on how to ensure a positive, sustainable, and just social impact. Furthermore, this growing trend arises profound concerns about its environmental impact [1, 2].

Big data analytics, artificial intelligence, and the Internet-of-Things (IoT) are primary enablers of the digital transformation. IoT is constituted by multiple energy autonomous systems that will be comprehensively capable of sensing, diagnosing, deciding, and actuating in a communicative and collaborative way. Although IoT is still in the early stages of growth, current estimations point to more than 9 billion connected devices around the world. This number is expected to increase exponentially, with estimates ranging from 25 to 50 billion devices in 2025 [3]. This network of devices will give rise to a new era of information access, well-being, and productivity in many sectors [4]. However, its manufacture, usage, and waste management will consume a high amount of resources.

As this global trend of society digitalization speeds up, planned obsolescence and linear economy models are consequently generating an unprecedented amount of waste electrical and electronic equipment (WEEE) or e-waste. WEEE is rapidly becoming the largest waste stream worldwide [5–7]. It has not stopped growing since 2014, leaving behind an annual generation of about 50 million metric tons, which are expected to increase up to 75 Mt in 2030, according to the latest Global e-Waste Monitor report [8]. On the other hand, the World Economic Forum estimates that the material value of our used electronic devices globally amounts to US$ 62.5 billion, three times more than the annual output of the

Sustainable Energy Storage in the Scope of Circular Economy: Advanced Materials and Device Design, First Edition. Edited by Carlos Miguel Costa, Renato Gonçalves, and Senentxu Lanceros-Méndez.

world's silver mines [9]. Thus, waste stream equally contains hazardous and valuable materials. In 2019, the documented global percentage of collected and properly recycled WEEE was 17.4%, meaning that the large majority of e-waste generated was not formally collected nor managed in a systematic manner. Europe, in particular, ranks first in terms of e-waste generation per capita with 16.2 kg (12 Mt generated in total in 2019) [8]. As depicted in Figure 5.1a, European Union (EU) member countries show a growing trend since 2014.

Moreover, the gap of data and statistics on WEEE management has already been exposed, pointing out that the lack of formally documented data implies that e-waste is mainly managed outside the official collection system. This obsolete equipment is, in some cases, shipped to developing countries,[1] where there is a complete lack of secure and sanitized facilities for e-waste recycling and material recovery [10]. The problem is even more critical when this waste is not even collected properly. Despite having a developed recycling infrastructure in high-income countries, a high number of small-size electronics are disposed of in general waste bins that usually end up in landfills where, in many cases, are incinerated. One of the most hazardous components in e-waste is batteries. Due to their chemical composition, encapsulation, and risk of explosion, batteries need special and dedicated recycling processes [11].

Due to their suitability in many areas, batteries have become today ubiquitous power sources to feed portable EEE, but they also contribute intensively to WEEE generation. Just in 2018, 191 kilotons of portable batteries were sold in the EU (see Figure 5.1b).[2] However, only about 88 kilotons of waste portable batteries and accumulators were collected for recycling, representing ca. 48% of collection rate [12]. Being the highest collection rate archived since records began. Significant efforts are being undertaken at EU level to create a sustainable framework for batteries. Most of these initiatives focus on the recovery of valuable elements, but also include the incorporation of sustainability into batteries value chain [13, 14], specially promoting environmental impact assessments [15]. The European Commission (EC) has identified batteries as a strategic value chain where

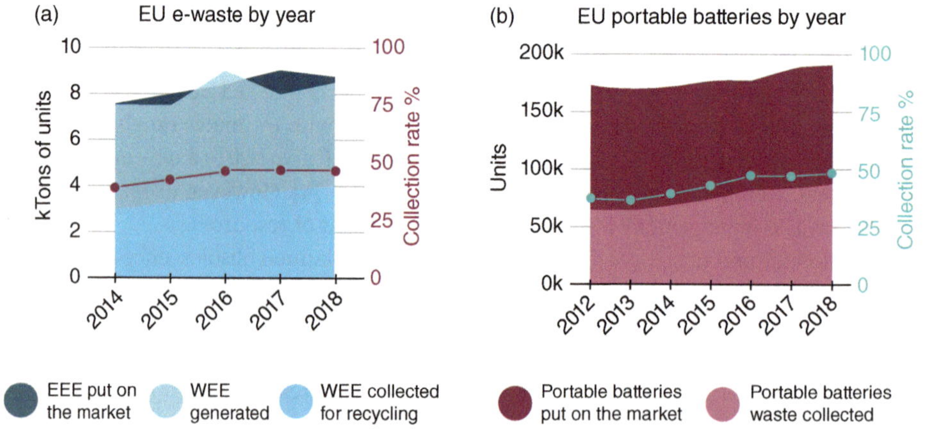

Figure 5.1 (a) EEE put on the market compared with WEEE generated and collected by year in EU in kilotons. Represented by dots, the collection rate. (b) Units of portable batteries put on the EU market in comparison with portable batteries collected for recycling by year. *Source:* Prepared by de authors based on the data supplied by Eurostat. Collection rates have been calculated as set out in the waste Electrical and Electronic Equipment (WEEE) Directive (% of the average weight of EEE put on the market in the three preceding years).

the EU must strengthen investment and innovation through an industrial policy strategy that builds an integrated, sustainable and competitive industrial base [16, 17]. Despite these valuable actions intended to improve sustainability of large-scale batteries (mostly for e-mobility and energy storage sectors), particular considerations should be made to improve portable battery formats as well. Portable batteries have shown to have a high-carbon footprint associated especially to their manufacturing and end of life (EoL). For portable batteries to stop contributing to environmental damage, and become an example for sustainable technological development, it is crucial to change the way the batteries value chain is approached.

The actual life cycle of portable batteries follows an obsolete linear economy model. Figure 5.2 gives a schematic view of the main stages of the battery life cycle, pointing out some unsustainable practices currently performed at each of the stages. Batteries depend on scarce nonrenewable materials that are extracted by intrusive methods, contributing to environmental deterioration. Concurrently, there are increasing issues related to raw materials extraction and use of EU critical raw materials (e.g. P, Co, Li), which are not sustainable in the long term [18, 19]. Their fabrication stage takes place in centralized facilities (mostly located in Asia), generating a necessity of a long-distance battery distribution around the world. Furthermore, portable batteries fabrication is usually based on standardized formats. Then, product manufacturers usually decide the battery format for their application on the basis of cost rather than optimal energy requirements. As a result, batteries are often discarded after their use without being completely depleted (sometimes almost fully charged). Finally, during their EoL, if they are collected for recycling, their complex encapsulation, together with the hazardous nature of the material they are made of, makes the recycling process really energy consuming[3] [20]. Moreover, due to the lack of local management and e-waste recycling infrastructure, battery waste sometimes needs to be exported again [10]. In a large number of cases, as it happens to e-waste, portable batteries are also improperly discarded with general waste streams, ending up in landfills or incinerated, actions that pose risks to human health [21], biodiversity, and ecosystems.

Ongoing strategies related to ecodesign, collection, repurposing, and recycling of batteries for large-scale applications [22] cannot be directly applied to portable battery formats – due to the diverse variety of use-case scenarios, their ubiquity, and that the success of the strategies depends primarily on the end-user self-responsibility and availability of dedicated collection routes. For these reasons, it is imperative to rethink the portable battery life cycle in accordance with a holistic sustainable perspective.

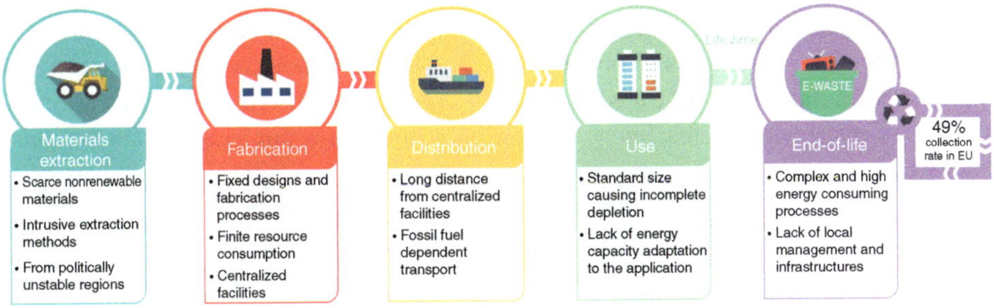

Figure 5.2 Linear life cycle of current portable batteries and practices indentified by the authors that contribute to the environmental burden at each of their stages.

This chapter presents a portable battery development approach that considers social, economic, and environmental implications. This approach aims to change the current paradigm of portable batteries to a new model in which batteries are designed to follow the life cycle of the device to be powered to minimize their environmental impact along their value chain. The ecodesign of batteries implies a detailed analysis of all the stages involved in their lifecycle, e.g. raw material selection, manufacturing process, demanded energy capacity, operational time, and disposal considerations. In this sense, batteries's life cycle has been entirely reformulated to place sustainability as core priority, by raising for instance alternative EoL scenarios; such as being capable to be biodegraded or even composted with other organic waste. First, a review of the state of the art in portable batteries with disruptive EoL is introduced (Section 5.2). Then, Section 5.3 presents the rationale behind the sustainable battery development approach theorized by the authors, in which battery life cycle stages are examined and re-thought under a doughnut economy model perspective. Finally, Section 5.4 provides a future outlook and points out shortcomings that should be overcome to ensure sustainable portable batteries development.

5.2 State of the Art of Portable Batteries with a Disruptive End of Life

In order to mitigate the associated environmental impact of EEE in general, and portable batteries in particular, researchers have focused efforts on seeking alternative end-of-life scenarios that do not require a dedicated collection route with subsequent recycling for materials recovery. In this regard, transient technology is a blooming research area that seeks materials and devices to undergo a controlled degradation process after their operational lifespan. Several recent reviews summarized the application of transiency to materials [23, 24], electronics [25–27], and battery components [28].

The reported degradation mechanisms for transient materials can be classified by the external activator that triggers the material disintegration, such as pH [29], light [30], or temperature [31]. When the degradation process is activated, a series of physical or chemical reactions occur and, after a given time, the material components transform into degradation products. The degradation mechanism depends on the nature of the material and the degrading medium.

Some academic approaches have proposed degradable batteries to satisfy the growing need for portable power sources while easing the environmental impact after disposal. In 2014, Rogers and co-workers reported a battery for medical implants made of biocompatible metal foils and polyanhydride packaging and showed its dissolution into its primary components after some hours of operation in physiological conditions (see Figure 5.3a). Later that year, the concept of transient benign batteries for *in vivo* applications was transferred to *ex vivo* environments, inaugurating in this way, a new family of so-called "*biodegradable*" batteries [32].

As the examples in Figure 5.3b,c show, magnesium has been used as primary anode material, whereas cathode electrodes are implemented with iron [34, 36], silk-fibroin-polypyrrole [37], gold [38], or molybdenum trioxide [33]. In the same way, different biopolymers are used as electrolytes and to confer the batteries with mechanical stability [39, 40]; and alternatives to metal electrodes, such as melanine [41] or activated carbon [42] have

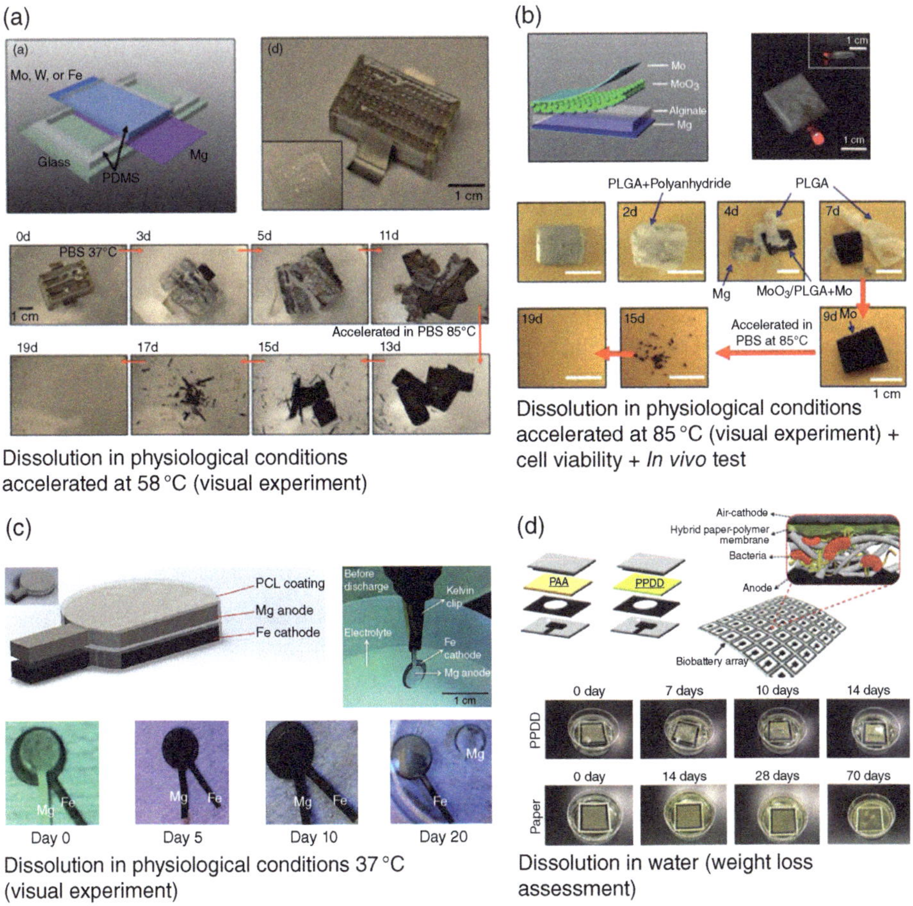

Figure 5.3 Batteries with a disruptive end of life: (a) dissolvable battery with metal foils and polyanhydride packaging. *Source:* Yin et al. [32]/With permission of John Wiley & Sons. (b) Biocompatible and dissolvable battery with magnesium anode and molybdenum trioxide as cathode. *Source:* Huang et al. [33]/With permission of John Wiley & Sons. (c) Dissolvable battery with Mg as anode and Fe as cathode. *Source:* Tsang, M et al. [34]/Springer Nature/CC BY 4.0. (d) Degradable battery with bacteria-based anode and nickel air cathode. *Source:* Mohammadifar et al. [35]/With permission of John Wiley & Sons.

also been presented by Bettinger et al. Both lithium technology and bio-based batteries – using enzymatic or microbial electrodes – have also been successfully exported to transient configurations [30, 43–45]. A remarkable microbial battery onto biodegradable paper-polymer substrates, Figure 5.3d, was reported by Choi group in 2018, showing a considerable weight loss after 70 days of degradation test [35].

However, the self-appointed "*biodegradability*" of these approaches relies on their capability to be disintegrated in tiny parts or even dissolved in ions and molecules. In these cases, biocompatibility[4] and/or bioabsorbability could be more appropriate properties to claim, since dissolvability tests are reported at physiologic conditions and, in the case of Huang et al., even demonstrated within a Sprague–Dawley (SD) rat model [33]. In fact, in literature, it is usually assumed that dissolution of biocompatible metals in river waters or soils is a sustainable EoL alternative. However, if this EoL was transferred to a battery production scale of billions, it is clear that it would suppose the loss of tons of irretrievable material together with the undesired accumulation of metal in terrestrial and aquatic ecosystems.

Therefore, it is important to point out that only when the degradation trigger is the presence of microorganisms present in natural environments, the mechanisms are enzymatic reactions executed by bacteria/fungi and the products are CO_2, H_2O, and simple molecules[5], the degradation process can be called biodegradation, according to International Union of Pure and Applied Chemistry (IUPAC)[6] [46] and regulated standards: EN 13432 European standard,[7] Organisation for Economic Co-operation and Development (OECD) Test 311, International Organization for Standardization (ISO) [49], American Society for Testing and Materials (ASTM) [50], and United States Environmental Protection Agency (US EPA) [51]. Figure 5.4 shows a schematic representation of a battery undergoing a biodegradation process. Therefore, according to regulated standards, only organic compounds can suffer biodegradability. In contrast, inorganic compounds can be considered bioabsorbable (if they can be assumed by a living organism), inert (if they do not cause an interaction), or toxic (if it causes a lethal effect at a certain concentration).

In 2015, our group reported a metal-free organic redox chemistry-based primary battery, which biodegradability was successfully tested for the first time under a standardized protocol (Figure 5.5a). Showing a biotic degradation measured by the conversion of initial total carbon mass of the material that is transformed into CO_2 and CH_4. This battery demonstrates that existing regulations to test biodegradability in materials could be adapted to assess complete devices [52]. Another approach has been recently published by M. H. Lee et al. in which a custom-made protocol to test a secondary battery biodegradability and toxicity was used (Figure 5.5b). A cytotoxicity cell assay was used to assess the toxicity of each battery component on biological systems, while at a macroscopic level, nontoxicity was also demonstrated by burying the battery and growing a plant in the same soil [53].

Batteries with a disruptive EoL represent a first attempt to minimize batteries' impact on ecosystems. However, as has been demonstrated, the lack of harmonized testing and regulations for batteries biodegradability assessment makes it difficult to ensure battery safety for terrestrial ecosystems. Furthermore, the above reviewed works do not consider the carbon footprint on the rest of the life cycle stages, rising serious doubts on the overall battery sustainability.

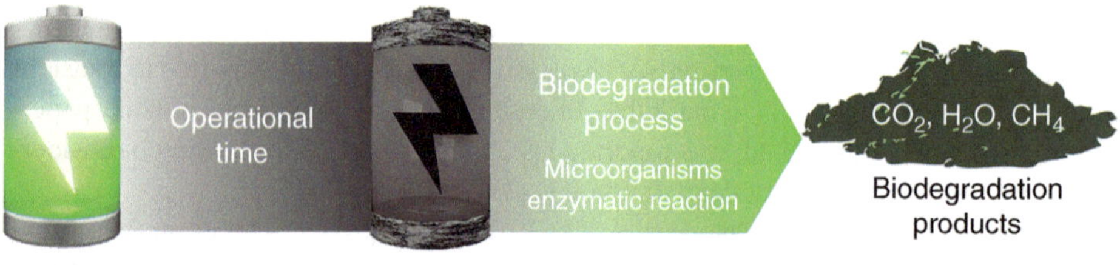

Figure 5.4 Schematic representation of the biodegradation process for a battery as established by regulated standards. Biodegradation assessment parameters, such as temperature, maximum process time, and minimum degradation percentage, are specified on the standards and will depend on the type of biodegradation process pursued.

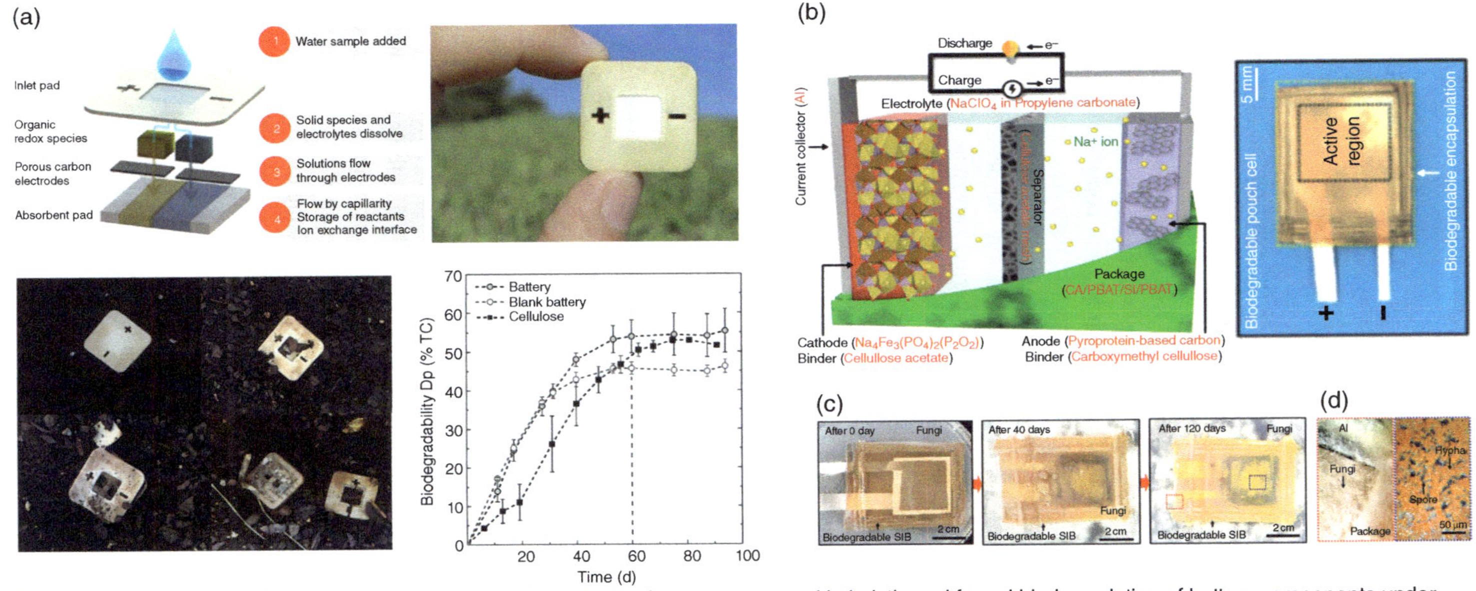

Biotic degradation test under standardized following the guideline of the OECD Test 31125 that comprises recommendations of other normalized tests (ISO, 26 ASTM, 27 US-EPA, 28)

Hydrolytic and fungal biodegradation of battery components under self-established guidelines to verify the biodegradability and the nontoxicity of battery compounds via the cytotoxicity test

Figure 5.5 (a) Biodegradable battery with redox organic-based chemistry. *Source:* Esquivel et al. [52]/Reproduced with permission from John Wiley & Sons. (b) Biodegradable secondary battery based in sodium-ion technology. *Source:* Lee et al. [53]/With permission of John Wiley & Sons.

5.3 How to Design a Truly Sustainable Battery?

In 2012, the economist and professor Kate Raworth published an Oxfam discussion paper entitled "A Safe and Just Space for Humanity" [54]. In it, Raworth presented a framework for humanity sustainable development by combining the concepts of planetary and social boundaries.

Her theory, known as *the doughnut economic model* because of its graphical representation by two concentric circles, balances essential human needs (represented as an inner boundary called "Social Foundation") and environmental degradation (depicted as an outer boundary named "Environmental Ceiling"). Between these two boundaries lies an area that, according to Raworth, represents the environmentally safe and socially just place for humanity to thrive in balance. The social foundation comprehends those resources needed – such as food, water, health care, or energy – to protect societies from critical deprivation and ensure life is lived with dignity, opportunity, and fulfillment. While the environmental ceiling sets planetary boundaries that sustain the benevolent environmental conditions achieved during the Earth's stable state period known as Holocene.[8] In order to define the dimensions of the ecological ceiling, Raworth used the planetary boundaries previously proposed by an international group of Earth system scientists led by Johan Rockström and Will Steffen, whereas the social foundation comprises 12 social dimensions derived from the priorities specified in the United Nations Conference on Sustainable Development, Rio+20.

Among others, Raworth condemns the excessive attention given to the gross domestic product (GDP) as a solo economic indicator to quantify countries' prosperity. Pointing out how such fixation on an ever-growing GDP as a goal has failed to reflect other important societal wealth metrics, such as the continuous degradation of ecosystems or the invaluable but unpaid work of carers. Ultimately, Raworth's doughnut theory evidences the necessity of providing a global perspective that takes into account all the dimensions at environmental, social and, economic level when developing a truly sustainable model.

As mentioned previously, there is a lack of legislation and highly efficient methods for portable batteries to be recovered and processed, which is leading to an irreversible ecosystem degradation. In view of this perspective, tightened environmental laws and appropriate recycling infrastructure are urgently needed. However, the development of those mechanisms is complex and sluggish, and it is very unlikely that their implementation can keep up with the increasing demand for portable power sources. In this sense, applying the doughnut economy approach to batteries development appears to be a promising way to meet the technological needs of society without compromising future generations.

Looking into Raworth's macroeconomics observations, one can find enormous similarities with the way battery development has been approached up to date. From materials screening to structures and geometries exploration, the main objectives of portable battery research have been to increase battery power, to achieve higher energy densities, and, in the case of secondary batteries, to gain in cyclability. But this excessive attention given to battery performance has overshadowed other relevant indicators. Meaningful metrics such as scarcity of raw materials, energy consumption, and complexity of manufacturing methods, equipment required, hazardousness of secondary materials involved (i.e. solvents), or recyclability of the design, suffer from a pathological lack of attention. In fact, such focus on battery performance, in terms of power and energy density, has completely ignored the indicators of the value chain that set the sustainability of the battery across technical, environmental, economic, and social dimensions.

This comprehensive understanding of sustainability suggests that, rather than traditional goal-based approaches, sustainability is a system property that requires a process-based systemic approach to plan and guide it, as described by F. Ceschin and I. Gaziulusoy [56]. For this reason, following Raworth's theory, we have drafted a doughnut-shaped model applied to battery research and development that sets a social, economic and environmental ceiling and a performance foundation. The performance foundation is composed by the six indicators researchers and companies usually use to measure commercial viability and competitiveness of a portable battery. While the ceiling has been established by the authors to ensure batteries sustainability in social, economic, and environmental dimensions (see Table 5.1). As depicted in Figure 5.6, by bringing together the two approaches a safe and sustainable space for portable batteries naturally emerges. This framework sets a new perspective on batteries development, bringing forward an opportunity to rethink their entire life cycle.

By ensuring the fulfillment of the six environmental ceiling dimensions, this change of paradigm aims to track, minimize, and, ideally, eliminate the battery environmental impact through careful device design and development. While at the same time, meeting the application minimum requirements in terms of power and operation time. Advocating for a "tailor-made" approach, the development of such batteries should be coupled with the device value chain, in such a manner that even the power source EoL could follow the device's one. However, to move into the sustainable space for batteries development is complex because the performance and the social, economic, and environmental dimensions are interconnected. For instance, a competitive manufacture might require processes that are energy intense. Hence, the production and implementation of portable batteries in a responsible way are only possible if a balance of the whole life cycle is taken into account.

This chapter proposes a research methodology that takes into account the application's energy demand as a cornerstone. This "tailor-made" approach sets a clear performance

Table 5.1 Defining a sustainable space for batteries development. The six dimensions that define the performance foundation are usually used by researchers and companies to measure commercial viability and competitiveness of a portable battery. The six dimensions of the social, economic, and environmental ceiling have been established by the authors to ensure battery sustainability across its three pillars.

Performance foundation dimensions	Social, economic, and environmental ceiling dimensions
Energy density	Materials toxicity
Power density	Ethical supply chain
Competitive manufacture	Low energy and resources consumption
Portability	Materials abundance
Operation time	Economically affordable
User-friendly	Cost-effective end-of-life

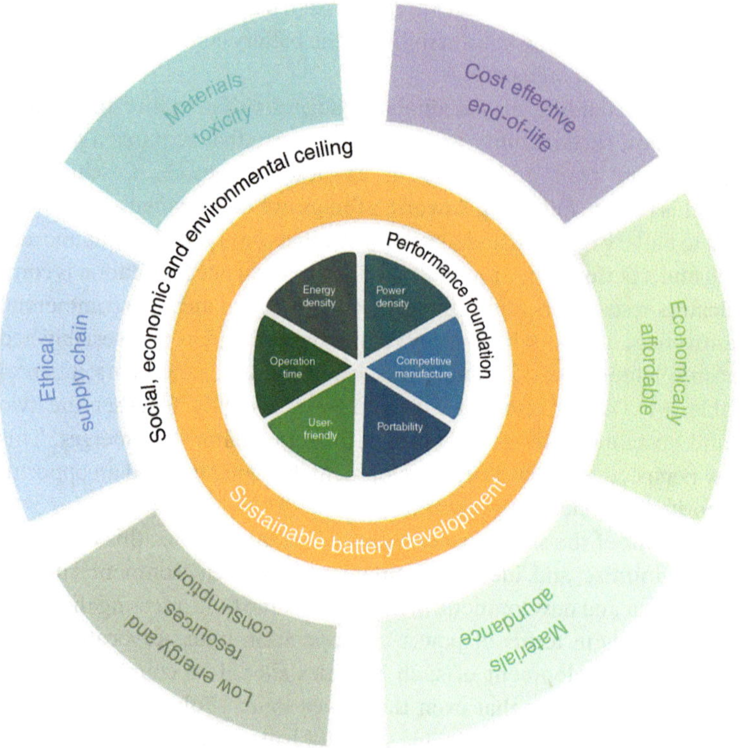

Figure 5.6 Defining a sustainable space for batteries development. The six dimensions that define the performance foundation ensure the power supply for the portable battery application. The six dimensions of the social, economic, and environmental ceiling prevent the battery to exceed the natural planetary boundaries and ensure a just and accessible technology.

foundation from the very beginning of the battery conception. From there, eco-design is used as linchpin to ensure the fulfillment of the six celling dimensions.

5.3.1 Portable Battery Development in a Doughnut Model

Ecodesign is a design for sustainability (DfS) approach that focuses on lowering the environmental impact of a product throughout the whole life cycle. From raw materials extraction to final disposal, it integrates environmental aspects into the product development process, balancing ecological and economical requirements [56].

In order to ensure that portable batteries stay within the social, economic, and environmental ceiling and performance foundation, it is essential that research and development are performed under a holistic ecodesign vision. Thus, in our proposed methodology, we have implemented three main ideas to consider at each stage of the battery life cycle, conceptualized and adapted from Raworth's theory.

1) **An integrated vision:** Societies need portable energy, and its demand is expected to significantly increase in the upcoming years. However, while creating wealth, scarce material dependence, and costly production methods could lead to inequitable access to

battery technology. Sustainability awareness at environmental, social, and economic dimensions ensures the creation of a fair access to the technology.

2) **A refocusing of priorities:** Energy demand intensification would increase raw materials and natural resources exhaustion, if not properly managed. Environmental stress must not be seen as a collateral damage, but rather as a decisive driver to guide battery development.
3) **Metrics beyond performance:** Batteries should not be ranked exclusively in terms of technical performance. A criteria to provide a holistic evaluation of a battery advantage in all dimensions must be developed by researchers and demanded by consumers. Such consensus should incorporate innovative indicators to assess battery sustainability, as its adequacy for a particular application, ease to be recycled, or efficacy in terms of material utilization.

Thus, the whole battery life cycle (i.e. materials, fabrication, distribution, application, and end of life) must be taken into account from early battery conception by the implementation of life cycle assessment (LCA). LCA is a standardized methodology for assessing the potential environmental impact associated with a product or a service over its entire life cycle. This assessment technique provides a holistic view of environmental interactions that will take into account ecological, social, and human health impacts of material consumption and environmental releases to air, water, and land associated with specific manufacturing and distribution processes on local, regional and global scales. The results from LCA would include an estimate of the carbon footprint of developed batteries to benchmark its impact based on standardized and comparable metrics [57–59]. Ecodesign of batteries would then use this output to make decisions on materials, manufacturing methods, and form factors according to the specific requirements of the application to be powered.

The following section gathers the proposed rationale pillars, a set of good practices, recommendations, and guidelines to consider in each of the battery life cycle stages that could be useful to keep the battery development within a safe space as shown in Figure 5.7.

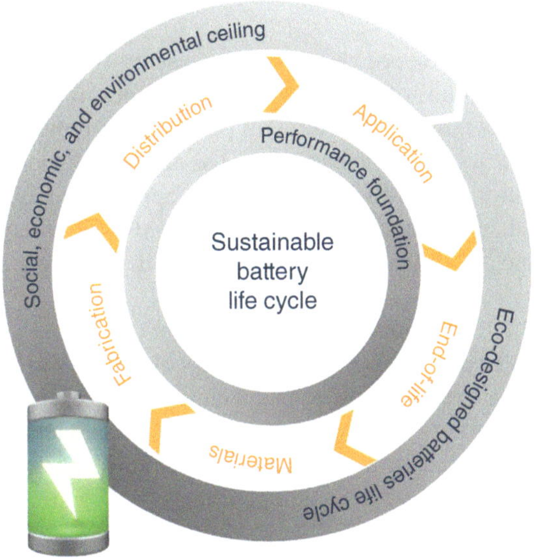

Figure 5.7 Batteries life cycle implementation within a doughnut model of sustainable development.

5.3.1.1 Materials

Conventional portable batteries have a dependence on materials with supply chain risk. Materials abundance and geographical origin is rarely prioritized against material performance, when analyzing their suitability for an application. Focusing on battery chemistries, current battery technologies rely on toxic and flammable compounds that involve aggressive and hazardous processing that produce toxic by-products and demand stringent safety measures, during their extraction, processing, and end of life.

Materials selection is the most critical step during a battery conception and design, since it largely determines the battery performance in terms of energy density and operational time. It also affects battery sustainability in terms of toxicity, cost-effectiveness, and affordability. An ethical supply chain is also largely determined by the material selection stage. There are some key points that battery materials research must assess to ensure battery performance while staying within the ceiling:

Screening of chemistries: selection of suitable chemistries based on metrics such as non-toxicity, availability, and compatibility with proposed EoL scenario. More eco-friendly and abundant chemistries already identified for large-scale applications such as Na-ion and Zn–air batteries would provide the starting point for this approach. But also the exploration of organic chemistries (e.g. quinones, vitamins, imides) could provide a feasible alternative that would reduce the current dependence on inorganic chemistries.

Optimal raw materials utilization: according to the features required by the application, the amount of electrochemically active material should be addressed. By adjusting the available energy to avoid material waste while ensuring a sufficient power supply.

Alternative materials sources: identify and explore battery materials beyond critical raw materials or materials with supply risk currently used in commercial batteries, e.g. Li metal, fluorine (Li/CF_x), manganese (Zn/MnO_2), nickel-metal hydride (NiMH). Alternative sources such as bioderived, recycled, or repurposed secondary raw materials would represent great sources. Subsequently, the exploration of novel extraction and recovery processes based on green chemistry and low-energy consuming methods must be implemented.

As mentioned above, LCA represents a key methodology specially for material selection. The results of a LCA can be used to prioritize the use of one material over another that behaves similarly in terms of performance, providing specifically environmentally conscious metrics. The LCA can be used as an iterative tool to assess different materials, while fixing others that are compulsory due to application requirements or performance reasons. All in all, materials selection must be carried out without neglecting the application power needs but prioritizing sustainability metrics that ensure the optimum selection for each application.

5.3.1.2 Fabrication and Distribution

Current manufacturing processes of batteries are complex, energy intensive, and usually centralized, mostly in Asia. Transportation of materials to the fabrication sites and then of batteries to the world contributes to a large carbon footprint. LCA studies of alkaline batteries account the majority of greenhouse gas emissions to processes related to material extraction and cell manufacturing.

Battery design should be carried out leading towards the implementation of cost and energy-efficient manufacturing methods. Defining sustainable and affordable fabrication processes will ensure the development of an accessible and inclusive technology, both at economic and social dimensions. Key points that should be address while creating/improving the manufacturing process include:

Energy-efficient fabrication: implementation of manufacturing methods based not only on the monetary cost but on the energy efficiency, leading to minimization of the carbon footprint associated.

Clean battery production: prioritize processes with the lowest amount of by-products needed and minimize waste production. Active research on reuse and reintegration methods for produced by-products or reagents is needed. Cost-effective and clean manufacturing can also be tackled from materials selection and implementation, by prioritizing those materials with lower working temperature or minimizing the use of reagents needed during material processing.

Decentralized manufacturing: efficient processes from Industry 4.0 can be implemented and reproduced locally, leading to a decentralized battery production model. The geographical democratization of battery production would avoid the creation of monopolies since batteries could be produced with available raw materials, distributed, consumed, and repurposed locally, fostering local industrial growth and competitiveness.

Implementation of best practices and policies: foster open innovation to fit with upcoming sustainable regulatory frameworks, and also promote and ensure an ethical fair trade supply chain, that avoids social and political conflicts.

The overall objective of the redefinition of batteries manufacturing routes should be introducing indicators beyond monetary cost. Energy-efficient, waste production, resources consumption metrics should be properly quantified to ensure cost-effectiveness at an environmental, social, and economic level. The implementation of such indicators should lead, ultimately, to the development of an accessible and inclusive technology.

5.3.1.3 Application

Current battery technologies have made progress in their industrialization to reduce production costs. However, this mass production strategy has ended up causing a widespread use of standard-size portable batteries in products, which have based their battery choice on price rather than power requirements. Consequently, sometimes batteries are oversized in terms of power and energy capacity (as in medical single-use tests), implying that they are disposed of before their complete depletion. In other cases, batteries are undersized, provoking excessive need of replacement and structural materials waste (as in hearing aid devices). Moreover, conventional batteries have been optimized for continuous operation and include elements to reduce leakages and overloads over the long periods of time required by non-disposable applications. These features are implemented using a wide range of materials from metallic casings, springs, rubber gaskets, paper/polymer separators, spacers, sealants, plastic/metal pouches, etc.; which are not always necessary, for instance in short-time use applications.

Technology integration with a holistic vision is only possible if, during the battery design, the application is taken into account. A tailor-made battery model would aim to ensure the following points:

Tailored design: adapting energy and power to the demands of the application would consume only the required amount of materials, and therefore efficient material utilization would be ensured. The use of additive manufacturing techniques from Industry 4.0 would enable a high level of design customization.

Operating conditions: the faradaic and energy efficiencies quantify the total amount of chemical energy stored in a battery that has been converted to useful electrical energy. These metrics strongly depend on the operating conditions in which a battery is forced to work. Careful examination of application energy demands would allow maximizing active material utilization of the battery during the operation stage.

Operating time: batteries research and development invest great efforts to ensure long and steady operational time. It is often that, while operational, novel battery technologies are discarded in their development phase due to short shelf and operational lifetimes. However, many single-use applications in sectors like portable diagnostics or precision agriculture could benefit from those short-term battery technologies.

Considering the requirements of the application from the beginning of the battery conceptualization would allow adapting battery features such as materials, geometries, operational conditions, and stored energy, which would lead to improving the overall efficiency.

5.3.1.4 End of Life

Battery recycling requires costly and complex infrastructure. Furthermore, the recycling rates through conventional waste streams are hardly met. Finding alternative EoL scenarios for batteries could help to ease the load on current recycling facilities while decreasing the environmental impact of improperly discarded batteries. When looking for an appropriate EoL, the following points may be useful:

Paired EoL: When considering the target application, the whole life cycle of the product that is being powered can be taken into consideration to match their EoL. On top of that, if the operational time of the product and the battery are very different, the usage scenario can be done considering to adequate the battery EoL to lack or existence of proper recycling facilities.

Reduce, Reuse, and Recycle: Reducing waste materials or products would undeniably reduce their impact. Similarly, reusing them, as long as the quantity of new material and energy resources needed to restore is reasonable. However, recycling implies an energy-consuming process and therefore its benefits depend on the specific product and materials to be recovered. As demonstrated by Linda Gaines, it is important to analyze deeply, and case-by-case, the pros and cons of recycling, to fully understand the involved impact tradeoffs [60].

Alternative EoL scenarios: As analyzed in Section 5.2, batteries can have EoLs as disruptive as biodegradability. Developing different paradigms for battery EoLs, and properly implementing them according to the life cycle of the application, would provide a diversity of viable waste streams that turn used batteries into valuable raw materials or a safe return into nature.

Identifying alternative EoL scenarios in which batteries consider the whole life cycle of the product they are powering can provide a minimal carbon footprint. Biodegradable batteries could eventually be processed as compost (domestic or industrial) or even undergo degradation on soil. The feasibility of this approach must include assessment of biodegradability in laboratory and field conditions, as well as the ecotoxicological impact of discarded batteries in soils for organisms. These characterizations would allow the total computation of environmental impact of the entire battery life cycle.

5.4 Global Trends and Opportunities

In this chapter, we have proposed a methodology to undertake battery development from a social, economic, and environmental sustainability standpoint, based on principles from the doughnut economics model.

The above presented approach uses ecodesign as a fundamental tool to align battery development to circular economy concepts. However, many other design for sustainability approaches could also be applied to improve the methodology and its implementation, such as green design (lowering environmental impact through redesigning individual qualities of individual products), design for sustainable behavior (DfSB) (making people adopt a desired sustainable behavior and abandon an unwanted unsustainable behavior), cradle-to-cradle design (CTC) (emphasis on a regenerative approach by the industry and closing the loops; focus on non-human species and future generations), or biomimicry design (BM) (mimicking nature in the design of forms, products, and systems by using nature as model, measure, and mentor) [56].

The given recommendations and guidelines can be implemented on any battery, but it becomes especially relevant in disruptive technologies that are introducing alternative EoL scenarios, such as biodegradability. In such cases, all materials used as electrodes, electrolytes, or structural components would be non-toxic and selected to meet the specific EoL requirements and a safe and scalable manufacturability. This set of prerequisites would lead to choose materials without supply chain risk and move toward materials obtained from a sustainable bioeconomy and recovered from recycling processes. For instance, a combination of bioderived organic species, electrolytes, biopolymers, and carbon-based electrodes comprising the batteries would result in a battery waste that can nurture soil or restore natural systems, radically changing the current battery paradigm. Energy and cost-efficient manufacturing methods from Industry 4.0 would then allow a decentralized local production and consumption model, in which batteries would not have to travel long distances to reach their destination but rather be constructed, distributed, consumed, and repurposed locally, fostering local industrial growth and competitiveness.

From a global perspective, this battery development approach is in full alignment with the Sustainable Development Goals (SDGs) defined by the United Nations [41] and not only addresses SDG7 (Affordable and Clean Energy) but also has a relevant impact on SDG9 (Industry, Innovation, and Infrastructure), SDG11 (Sustainable Cities and Communities), SDG12 (Responsible Consumption and Production), SDG13 (Climate Action), and SDG17 (Partnerships for the Goals) [61].

In this sense, the technologies created according to this approach would not only produce sustainable power source alternatives for applications in developed regions in the world but would also represent a paramount solution for low-resource settings in the world. According to the Energy Progress Report, which provides a global dashboard on progress toward Sustainable Development Goal 7 (SDG7), about two billion people – around 1 in 4 people in the world – currently live with no or limited access to electricity. Of those, 860 million people, mostly located in rural areas, live completely without electricity [62]. A recent study elaborated by SPACE10 and Quicksand explored first-hand the way families around the world live an everyday life under these conditions of limited access to electric power. Their findings confirmed that in situations of extreme energy poverty, even the first kWhs have a profound impact on living standards, providing benefits in education, livelihoods, well-being, health, and equality [63].

In particular, the biodegradable batteries developed under our proposed method would represent a suitable alternative of decentralized energy system to use in those limited-resource settings. It is especially in these regions where the lack of proper battery recycling leads to uncontrolled disposal of batteries, causing a huge impact on environment, biodiversity, and human health.

This methodology has been created from the point of life of batteries, a critical element in electronic devices and consequently in electronic waste. The same exercise can be carried out in the development of complete devices and systems. The advances in green electronics have already demonstrated the potential to generate devices that could also be biodegradable, and in this way, the whole system could then follow the same waste stream.

The authors' ambition is that the methodology proposed here becomes the starting point of an open discussion that defies unsustainable practices in battery development. Therefore, this methodology is subject to be continuously improved and updated with contributions from battery developers around the globe and turn into a beacon for the implementation of a new paradigm of portable batteries based on a holistic sustainability approach that satisfies the portable power needs of humanity without compromising our natural environment.

Acknowledgments

The authors acknowledge financial support from projects ILINK+2019 (LINKB20026), Ecotronic (RTI2018-102070-B-C21, MCIN), and BIDEKO (PLEC2021-007801), funded by MCIN/AEI/10.13039/501100011033 and European Union "NextGenerationEU"/PRTR".

Notes

1 As a part of the official WEEE management, it is usual to incorrectly categorize WEEE under the euphemism Used Electrical and Electronic Equipment (UEEE). As it has been reported by Basel Action Networks in multiple occasions, this illegal practice is used to avoid OECD waste exporting regulations and generates a non-regulated trade of WEEE to developing countries [10]. The interested reader may consult: ban.org/trash-transparency

2 75% of them non-rechargeable batteries (primary), 25% portable secondary batteries.
3 The current portable battery recycling process is focused on metal recovery, thus three metal recovery method methods have been mostly implemented for spent batteries treatment systems: hydrometallurgic, pyrometallurgical and thermal treatment.
4 Biocompatibility is define by IUPAC as "ability to be in contact with a living system without producing an adverse effect." Both, organic and inorganic materials can be considering biocompatible, however *in vitro* culture experiments or *in vivo* implantation assessments are necessaries to demonstrate biocompatibility of a material or device.
5 The above mentioned products of biodegradation are for aerobic conditions (presence of O_2), simple molecules refer to sulfides. In anaerobic conditions (absence of O_2) the products of biodegradation include CH_4.
6 IUPAC definition of biodegradation: "Breakdown of a substance catalyzed by enzymes in vitro or *in vivo*." https://doi.org/10.1351/goldbook.B00656
7 Definition of biodegradation by the European standard EN 13432 "chemical breakdown of at least 90% material into CO_2, water and minerals in six months by biological actions at 58 °C". In this standard, the material conversion to CO_2 (i.e. biodegradation process) must be evaluated accordingly to ISO14855 or ISO 17556 [47, 48].
8 Holocene is the current geological epoch, defined as an interglacial period in which planet Earth enjoys a stable state characterized by proliferation and growth of species worldwide. Researchers from different areas have alerted that current fossil fuel emissions and nature destruction activities are pushing Earth's away from Holocene preservation state. J. Hansen et al. (2013), "Assessing Dangerous Climate Change: Required Reduction of Carbon Emissions to Protect Young People Future Generations and Nature" [55].

References

1 Feroz, A.K., Zo, H., and Chiravuri, A. (2021). Digital transformation and environmental sustainability: a review and research agenda. *Sustainability* 13 (3): 1530. https://doi.org/10.3390/SU13031530.
2 World Economic Forum. (2016). Digital Transformation of Industries: Societal Implications. *World Econ. Forum White Paper*. https://www.weforum.org/reports/digital-transformation-of-industries/.
3 McKinsey Global Institute (2015). *The Internet of Things: Mapping the Value beyond the Hype Executive Summary. McKinsey Co.*
4 EPoSS (2013). Strategic Research Agenda.
5 Ilankoon, I.M.S.K., Ghorbani, Y., Chong, M.N. et al. (2018). E-waste in the international context – a review of trade flows, regulations, hazards, waste management strategies and technologies for value recovery. *Waste Management* 82: 258–275. https://doi.org/10.1016/J.WASMAN.2018.10.018.
6 Heacock, M., Kelly, C.B., Asante, K.A. et al. (2016). E-waste and harm to vulnerable populations: a growing global problem. *Environmental Health Perspectives* 124 (5): 550–555. https://doi.org/10.1289/ehp.1509699.
7 Pérez-Belis, V., Bovea, M., and Ibáñez-Forés, V. (2014). An in-depth literature review of the waste electrical and electronic equipment context: trends and evolution. *Waste Management*

& Research: The Journal for a Sustainable Circular Economy 33 (1): 3–29. https://doi.org/10.1177/0734242X14557382.

8 Forti, V., Baldé, C.P., Kuehr, R., and Bel, G. (2020). *The Global E-Waste Monitor 2020*.

9 The world's e-waste is a huge problem. It's also a golden opportunity | World Economic Forum https://www.weforum.org/agenda/2019/01/how-a-circular-approach-can-turn-e-waste-into-a-golden-opportunity/ (accessed 30 October 2021).

10 Puckett, J., Brandt, C., and Palmer, H. (2018). Holes in the circular economy: WEEE leakage from Europe. A report of the e-trash transparency project. 120.

11 A New Circular Vision for Electronics, Time for a Global Reboot | World Economic Forum https://www.weforum.org/reports/a-new-circular-vision-for-electronics-time-for-a-global-reboot (accessed 3 February 2020).

12 Waste statistics - recycling of batteries and accumulators - Statistics Explained.

13 ReCell Center https://recellcenter.org/ (accessed 31 January 2020).

14 Relib https://relib.org.uk/ (accessed 31 January 2020).

15 Liu, T., Zhang, Y., Chen, C. et al. (2019). Sustainability-inspired cell design for a fully recyclable sodium ion battery. *Nature Communications* 10 (1): https://doi.org/10.1038/s41467-019-09933-0.

16 European Battery Alliance | Internal Market, Industry, Entrepreneurship and SMEs.

17 State aid: €3.2 billion public support battery value chain.

18 European Portable Battery Association | What is a battery? https://www.epbaeurope.net/portable-batteries/portable-batteries-primary-batteries-rechargeable-batteries-consumer-batteries/ (accessed 30 October 2021).

19 European Commission (2020). *J. R. C. Critical Raw Materials for Strategic Technologies and Sectors in the EU*. Publications Office of the European Union https://doi.org/10.2873/865242.

20 European Commission - Directorate General Environment (2003). *Impact Assessment on Selected Policy Options for Revision of the Battery Directive*. European Commission.

21 Soaring e-waste affects the health of millions of children, WHO warns https://www.who.int/news/item/15-06-2021-soaring-e-waste-affects-the-health-of-millions-of-children-who-warns (accessed 30 October 2021).

22 Fichtner, M., Edström, K., Ayerbe, E. et al. (2021). Rechargeable batteries of the future—the state of the art from a battery 2030+ perspective. *Advanced Energy Materials* https://doi.org/10.1002/AENM.202102904.

23 Cao, Y. and Uhrich, K.E. (2018). Biodegradable and biocompatible polymers for electronic applications: a review. *Journal of Bioactive and Compatible Polymers* 34 (1): 3–15. https://doi.org/10.1177/0883911518818075.

24 Di Mauro, E., Rho, D., and Santato, C. (2021). Biodegradation of bio-sourced and synthetic organic electronic materials towards green organic electronics. *Nature Communications* 12 (1): 1–10. https://doi.org/10.1038/s41467-021-23227-4.

25 Li, R., Wang, L., Kong, D., and Yin, L. (2018). Recent progress on biodegradable materials and transient electronics. *Bioactive Materials* 322–333. https://doi.org/10.1016/j.bioactmat.2017.12.001.

26 Han, W.B., Lee, J.H., Shin, J.-W., and Hwang, S.-W. (2020). Advanced materials and systems for biodegradable, transient electronics. *Advanced Materials* 32 (51): 2002211. https://doi.org/10.1002/ADMA.202002211.

27 Zvezdin, A., Di Mauro, E., Rho, D. et al. (2020). En route toward sustainable organic electronics. *MRS Energy & Sustainability* 7 (1): https://doi.org/10.1557/MRE.2020.16.

28 Mittal, N., Ojanguren, A., Niederberger, M., and Lizundia, E. (2021). Degradation behavior, biocompatibility, electrochemical performance, and circularity potential of transient batteries. *Advancement of Science* https://doi.org/10.1002/ADVS.202004814.

29 Fu, K.K., Wang, Z., Yan, C. et al. (2016). All-component transient lithium-ion batteries. *Advanced Energy Materials* 6 (10): 1502496. https://doi.org/10.1002/aenm.201502496.

30 Fu, K., Liu, Z., Yao, Y. et al. (2015). Transient rechargeable batteries triggered by cascade reactions. *Nano Letters* 15: 3. https://doi.org/10.1021/acs.nanolett.5b01451.

31 Park, C.W., Kang, S.-K., Hernandez, H.L. et al. (2015). Thermally triggered degradation of transient electronic devices. *Advanced Materials* 27 (25): 3783–3788. https://doi.org/10.1002/ADMA.201501180.

32 Yin, L., Huang, X., Xu, H. et al. (2014). Biodegradable electronics: materials, designs, and operational characteristics for fully biodegradable primary batteries. *Advanced Materials* 26 (23): 3777–3777. https://doi.org/10.1002/adma.201470152.

33 Huang, X., Wang, D., Yuan, Z. et al. (2018). A fully biodegradable battery for self-powered transient implants. *Small* 14 (28): 1800994. https://doi.org/10.1002/smll.201800994.

34 Tsang, M., Armutlulu, A., Martinez, A.W. et al. (2015). Biodegradable magnesium/iron batteries with polycaprolactone encapsulation: a microfabricated power source for transient implantable devices. *Microsystems & Nanoengineering* 1: https://doi.org/10.1038/micronano.2015.24.

35 Mohammadifar, M., Yazgan, I., Zhang, J. et al. (2018). Green biobatteries: hybrid paper-polymer microbial fuel cells. *Advanced Sustainable Systems* 2 (10): 1800041. https://doi.org/10.1002/adsu.201800041.

36 She, D., Tsang, M., and Allen, M. (2019). Biodegradable batteries with immobilized electrolyte for transient MEMS. *Biomedical Microdevices* 21 (1): https://doi.org/10.1007/s10544-019-0377-x.

37 Jia, X., Wang, C., Zhao, C. et al. (2016). Toward biodegradable Mg-air bioelectric batteries composed of silk fibroin-polypyrrole film. *Advanced Functional Materials* 26 (9): 1454–1462. https://doi.org/10.1002/adfm.201503498.

38 Jia, X., Wang, C., Ranganathan, V. et al. (2017). A biodegradable thin-film magnesium primary battery using silk fibroin–ionic liquid polymer electrolyte. *ACS Energy Letters* 2 (4): 831–836. https://doi.org/10.1021/ACSENERGYLETT.7B00012.

39 Edupuganti, V. and Solanki, R. (2016). Fabrication, characterization, and modeling of a biodegradable battery for transient electronics. *Journal of Power Sources* 336: 447–454. https://doi.org/10.1016/J.JPOWSOUR.2016.11.004.

40 Xia, J., Yuan, Z., and Cai, F. (2018). Toward a biocompatible and degradable battery using a Mg-Zn-Zr alloy with β-tricalcium phosphate nanocoating as anode. *Journal of Materials Engineering and Performance* 27 (8): 4005–4009. https://doi.org/10.1007/s11665-018-3512-6.

41 Kim, Y.J., Wu, W., Chun, S.-E. et al. (2013). Biologically derived melanin electrodes in aqueous sodium-ion energy storage devices. *Proceedings of the National Academy of Sciences of the United States of America* 110 (52): 20912–20917. https://doi.org/10.1073/pnas.1314345110.

42 Kim, Y.J., Chun, S.E., Whitacre, J., and Bettinger, C.J. (2013). Self-deployable current sources fabricated from edible materials. *Journal of Materials Chemistry B* 1 (31): 3781–3788. https://doi.org/10.1039/c3tb20183j.

43 Chen, Y., Jamshidi, R., White, K. et al. (2016). Physical-chemical hybrid transiency: a fully transient li-ion battery based on insoluble active materials. *Journal of Polymer Science Part B: Polymer Physics* 54 (20): 2021–2027. https://doi.org/10.1002/polb.24113.

44 Gao, Y., Mohammadifar, M., and Choi, S. (2019). From microbial fuel cells to biobatteries: moving toward on-demand micropower generation for small-scale single-use applications. *Advanced Materials Technologies* 4 (7): 1900079. https://doi.org/10.1002/admt.201900079.

45 Zhu, Z., Kin Tam, T., Sun, F. et al. (2014). A high-energy-density sugar biobattery based on a synthetic enzymatic pathway. *Nature Communications* 5: https://doi.org/10.1038/ncomms4026.

46 Vert, M., Doi, Y., Hellwich, K.-H. et al. (2012). Terminology for biorelated polymers and applications (IUPAC recommendations 2012). *Pure and Applied Chemistry* 84 (2): 377–410. https://doi.org/10.1351/PAC-REC-10-12-04.

47 International Organization for Standardization. ISO 14855 -Determination of the Ultimate Aerobic Biodegradability of Plastic Materials under Controlled Composting Conditions — Method by Analysis of Evolved Carbon Dioxide —. *61010-1 © Iec2001* 2003, p. 13.

48 Standardization, I. O. for. ISO-17556-Plastics-—-Determination-of-the-Ultimate-Aerobic-Biodegradability-of-Plastic-Materials-in-Soil-by-Measuring-the-Oxygen-Demand-in-a-Respiro.

49 International Organization for Standardization. ISO 11734 - Water Quality - Evaluation of "Ultimate" Anaerobic Biodegradability of Organic Compounds in Digested Sludge - Method by Measurement of Biogas Production. 1995, 1–13.

50 American Society of the International Association for Testing and Materials. ASTM E1196, Test Method for Determining the Anaerobic Biodegradation Potential of Organic Chemicals, 1992.

51 United States Environmental Protection Agency (1998). *Fate, Transport and Transformation Test Guidelines OPPTS 835.3400 Anaerobic Biodegradability of Prganic Chemicals*, US Environmental Protection Agency.

52 Esquivel, J.P., Alday, P., Ibrahim, O.A. et al. (2017). A metal-free and biotically degradable battery for portable single-use applications. *Advanced Energy Materials* 7 (18): https://doi.org/10.1002/aenm.201700275.

53 Lee, M.H., Lee, J., Jung, S.K. et al. (2021). A biodegradable secondary battery and its biodegradation mechanism for eco-friendly energy-storage systems. *Advanced Materials* 33 (10): https://doi.org/10.1002/adma.202004902.

54 Raworth, K. (2012). A Safe and Just Space for Humanity: Can We Live within the Doughnut?

55 Hansen, J., Kharecha, P., Sato, M. et al. (2013). Assessing "dangerous climate change": required reduction of carbon emissions to protect young people, future generations and nature. *PLoS One* 8 (12): e81648. https://doi.org/10.1371/JOURNAL.PONE.0081648.

56 Ceschin, F. and Gaziulusoy, I. (2016). Evolution of design for sustainability: from product design to design for system innovations and transitions. *Design Studies* 47: 118–163. https://doi.org/10.1016/J.DESTUD.2016.09.002.

57 Bautista, S.P., Weil, M., Baumann, M., and Montenegro, C.T. (2021). Prospective life cycle assessment of a model magnesium battery. *Energy Technology* 9 (4): https://doi.org/10.1002/ENTE.202000964.

58 Iturrondobeitia, M., Akizu-Gardoki, O., Amondarain, O. et al. (2022). Environmental impacts of aqueous zinc ion batteries based on life cycle assessment. *Advanced Sustainable Systems* 6 (1): 2100308. https://doi.org/10.1002/ADSU.202100308.

59 Porzio, J. and Scown, C.D. (2021). Life-cycle assessment considerations for batteries and battery materials. *Advanced Energy Materials* 11 (33): 2100771. https://doi.org/10.1002/AENM.202100771.

60 Gaines, L. (2012). To recycle, or not to recycle, that is the question: insights from life-cycle analysis. *MRS Bulletin* 37 (4): 333–338. https://doi.org/10.1557/MRS.2012.40.

61 Sustainable Development Knowledge Platform https://sustainabledevelopment.un.org/index.html (accessed 30 October 2021).

62 Tracking SDG7: The Energy Progress Report 2018. Track. SDG7 Energy Prog. Rep. 2018. https://doi.org/10.1596/29812.

63 Life Without Energy: Needs, Dreams and Aspirations | SPACE10 https://space10.com/project/life-without-energy/ (accessed 30 October 2021).

6

Recent Advances of Sustainable Electrode Materials for Supercapacitor Devices

Shilpi Sengupta and Manab Kundu

Electrochemical Energy Storage Laboratory, Department of Chemistry, University Building, SRM Institute of Science and Technology, Kattankulathur, Tamil Nadu, 603203, India

6.1 Introduction

The unrelenting upsurge in the importance of renewable energy resources as well as energy storage strategies are driven by the prompt consumption of fossil fuels and subsequent very serious deleterious impacts on our environment [1]. A change from a community of "energy gatherers" those who collect fossil fuels for energy to one of "energy farmers," who utilize the energy vectors like biofuels, electricity, hydrogen, or other synthetic fuels, is inevitable. As with the development in technology electrical appliances usage has significantly increased, hence the demand for electrical energy is uninterruptedly increasing. Because of the intermittent nature of renewable energy resources, an intermediate energy storage device must be developed to maintain the regular electrical energy supply. In these circumstances, the replacement of fossil fuel power for emissions-free energy resources and electrochemical energy storage systems devices (EESD) are becoming vital storage tools to facilitate the energy transition [2]. In this pursuit, researchers and industrialists are on the hunt of finding that EESDs have a high-energy density as well as a power density without scarifying the cycle life, are economically feasible, and are eco-friendliness. In Figure 6.1, the comparative performance of conventional capacitors, supercapacitors, and batteries, with their hybrids has been presented in terms of energy and power density (Ragone plot). Moreover, it also includes fuel-consuming devices: conventional gasoline combustion engines, hydrogen fuel cells, and hydrogen combustion engines (CE). The various types of energy storage devices based on their suitability for their applications are summarized in Figure 6.2 [4].

In comparison to various energy storage systems, supercapacitors (SCs) offers more balanced energy storage performance and fill the gap between conventional capacitor (lower energy density) and batteries (lower power density) [5]. Because of the excellent electrochemical energy storage performances, e.g. high-power density, superior cycling life, and a rapid charging–discharging rate, SCs have been positioned themselves in a wide

Sustainable Energy Storage in the Scope of Circular Economy: Advanced Materials and Device Design, First Edition. Edited by Carlos Miguel Costa, Renato Gonçalves, and Senentxu Lanceros-Méndez.

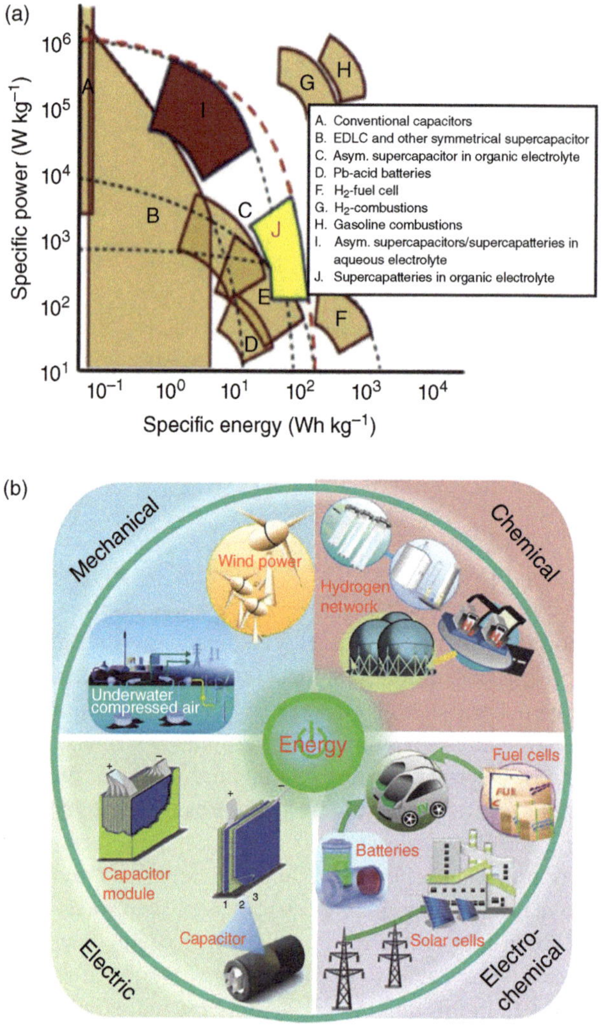

Figure 6.1 (a) The energy-power performance of conventional capacitors, batteries, and supercapacitors along with their hybrids demonstrated by Ragone plot. *Source:* Yu and Chen [3]/ with permission of Elsevier. (b) Different types of energy storage devices for sustainable applications. *Source:* Liu et al. [4]/John Wiley & Sons/CC BY 4.0.

range of application segments including hybrid cars and electric mass transit vehicles and wearable/portable electronic devices [6–8]. The energy density defines the quantitate amount of energy that can be stored/extracted, and the power density indicates how fast the device can deliver the energy.

Based on the type of electrode material, the electrical energy in SCs can be stored either through the faradic or non-faradic process at the interface between electrode and electrolyte. Depending on the charge storage behavior, supercapacitors can be divided into three categories: electric double-layer capacitor (EDLC: non-Faradaic), pseudo-capacitor (capacitive Faradaic), and battery-type supercapacitor (non-capacitive Faradic) [9].

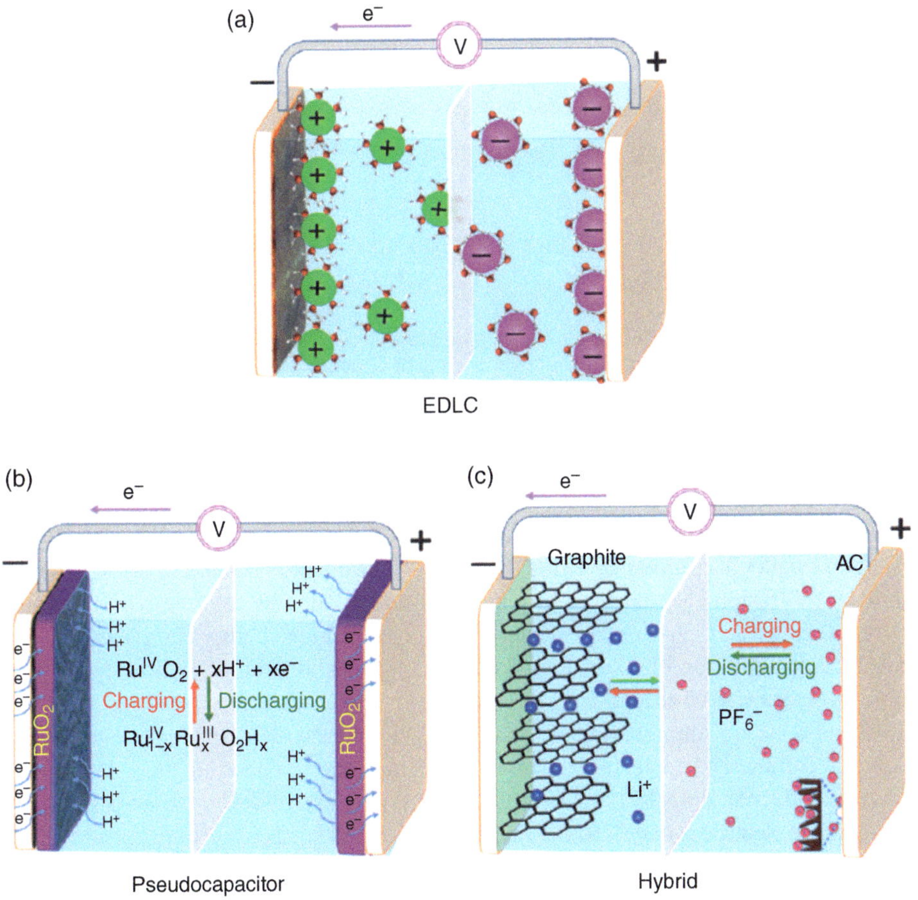

Figure 6.2 Charge storage mechanism of supercapacitors, (a) EDLC, (b) pseudocapacitors, and (c) hybrid supercapacitors.

The energy storage mechanism of supercapacitors involves two electrodes fabricated on a current collector, an electrolyte, and a separator (for solid-state devices same material can act as electrolyte and separator) which allow the ion transfer and prevent short circuits. Based on the energy storage principle, a supercapacitor can be classified as EDLCs, pseudocapacitors (PCs), and battery-type supercapacitor or hybrid supercapacitor (non-capacitive Faradic). In case of EDLC, the charge is stored through non-faradic interactions (electrostatically) means by charge separation at the electrode–electrolyte interface which is known as Helmholtz double layer. On the other hand, PCs store energy by Faradaic charge transfer via redox reactions [10, 11].

Thus, an ideal EDLC material should have a very high effective surface area. For instance, for carbon-based materials, the charge storage is based on the formation of an EDL at the active material and electrolyte interface (Figure 6.3a).

During this entire charge storage process, neither Faradaic reactions associated across the electrode/electrolyte interface nor any electrochemical reactions inside the bulk of the

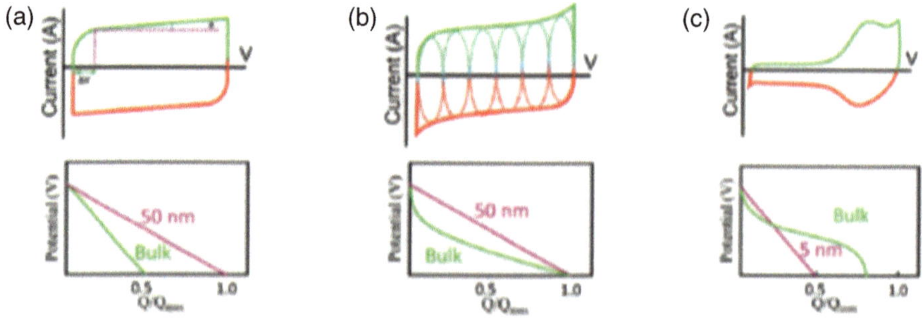

Figure 6.3 Characteristics of different ECSs storage devices: (a) EDLC, (b) Pseudocapacito, and (c) Hybrid system. *Source:* Noori et al. [12]/Royal Society of Chemistry.

electrode active material take place. Charge storage in PCs is based on Faradaic oxidation-reduction reactions at the electrode–electrolyte interface on, but no ion diffusion within the bulk of the electrode materials take place (Figure 6.3b). In this case, the diffusion length (l), is much smaller in comparison to the thickness of the diffusion layer $\left(l \ll 2\sqrt{Dt}\right)$, where D stands for diffusion coefficient and t represents time. EDL delivers high power density whereas PCs provide higher energy density. In order to achieve higher energy and power density, the combination of EDLCs and pseudocapacitors to obtain a hybrid device has become a very successful strategy.

In spite of these advantages, the low energy density of these devices is the main limitation of their commercial applications. To overcome this issue and boost up the performance of the supercapacitor, the selection of electrode material is one of the key strategies [13]. In addition to the electrode materials, the microstructure and surface morphology and other features such as surface area, surface porosity, pore size distribution, electrical/ionic conductivity, and diffusion path length are determining parameters for the performance of the SCs [14, 15].

6.2 Charge Storage Mechanism

Based on the energy storage mechanisms, electrochemical supercapacitors can be classified into electric double-layer capacitors (EDLCs, based on capacitive non-Faradaic) [16, 17] pseudocapacitors (PCs, based on non-capacitive Faradaic reaction) and non-capacitive Faradaic known as battery type [18, 19]. In case of EDLC, the formation of two charged layers takes place by electrostatically accumulating charges at the electrode/electrolyte interfaces (Figure 6.2a). The rapid adsorption and desorption of charges at the interface between the electrode and electrolyte resulting an outstanding power density [15]. Mainly carbon-based and its allotropic materials like porous carbon, activated carbon, carbon nanotubes (CNT) and doped graphene are the commonly used electrodes in the case of EDLCs owing to their good electrical conductivity, high specific surface area, cost-effectiveness, and environment-friendly [20].

On the other hand, PCs (capacitive Faradaic) store the charge by reversible and fast surface Faradaic reaction (charge-transfer reactions), through redox reaction, electrosorption/desorption, intercalation/de-intercalation, and doping/de-doping of electrolyte ions. Figure 6.2b shows the energy storage mechanism of PCs. Owing to the variable oxidation states, higher specific capacitance, low cost, tuneable morphology, abundant in nature, conductive polymers, transition metal oxides, transition metal sulfides, etc. are mainly used as electrode materials in case of PC. By a combination of electrode materials with two storage mechanisms (EDLC and PC), a hybrid capacitor can be designed. Depending on the combination of positive and negative electrode materials, capacitive non-Faradaic, capacitive Faradaic, and non-capacitive Faradaic mechanisms undergo to achieve higher energy and power densities [12].

The energy density (E) and power density (P) of electrochemical supercapacitors (ECSC) can be calculated by

$$E = \frac{CXV^2}{2} \quad P = \frac{V^2}{4R} \tag{6.1}$$

Here, C, V, and R are the capacitance in Farad, V operating potential in volts, and equivalent series resistance in Ohm, respectively.

The specific capacitance of the electrode can be expressed by

$$C = \frac{I \times \Delta t}{m \times \Delta V} \tag{6.2}$$

where I and Δt stands for discharge current in Ampere and discharge time in seconds, respectively.

Hence to achieve high energy density, it is crucial to use electrode materials having higher capacitance value and larger operating potential window. While by lowering the series resistance of the system, and extending the working potential window, power density can be enhanced. The capacitance of the supercapacitor device is solely dependent upon the kind of electrode materials used and their charge storage mechanism. Thus, the performance of ECSCs including the type of SCs, energy, and power density mainly dependent on the type of electrode materials that are being used. The detailed development of sustainable electrode materials and their charge storage mechanism related to a distinct type of SCs is described.

6.2.1 Electric Double-Layer Capacitor

In case of EDLC, the energy storage mechanism does not dependent on the charge transfer between electrode and electrolyte ions but is based on the formation of an EDL by electrostatic adsorption of the electrolyte ions at the interface of electrode and electrolyte. Thus, the key to parameter increase capacitance of EDLCs is using a highly accessible surface area and electronically conducting electrode materials. The major benefit of EDLCs is their superior power density and longer cycle life. Nanostructured carbon-based electrode materials with very high specific surface area and decent electrical conductivity are beneficial for EDLCs [21].

6.2.1.1 Activated Carbon

Among the various carbon-based electrode materials, due to the advantages of low cost, huge accessible surface area, ease of synthesis, non-toxicity, decent thermochemical stability, and surface porosity to accumulate a large number of ions at the electrodes and electrolytes interface activated carbon is widely electrode material for EDLCs [22].

To boost the performance of EDLCs with organic and aqueous electrolytes, the introduction of micropores in activated carbon has been adapted as an effective strategy. As efficient use of microporous especially at low temperatures depends on the kinds of electrolytes with different ion sizes, tailoring the pore size of activated carbons hence a high surface area to ease migration of ions and achieve a maximum specific capacitance is required. For instance, Xu et al. introduced predominantly micropores in activated carbon with numerous pores size and correlated the effect of pore size on capacitance at low temperatures [23]. It was observed that the activated carbon with the pore size of 0.9 nm (specific surface area of $2700\,m^2\,g^{-1}$) can deliver a high specific capacitance of $164\,F\,g^{-1}$ at 20 °C and only 10% of capacitance loss was exhibited at −70 °C. Whereas the specific capacitance value for activated carbon possessing a pore size of 0.7 nm (specific surface area of $2000\,m^2\,g^{-1}$) was only $12\,F\,g^{-1}$ at −70 °C which is 88% lower than at 20 °C($100\,F\,g^{-1}$). The result clearly proves that the presence of larger micropores permits quicker migration of ions by avoiding the ion trapped in the narrow pores, easing a faster charge/discharge rate of EDLCs. The existence of mesopores in activated carbon specifically for ionic liquids with long alkyl chains is advantageous and it can offer a smooth pathway for ions to easily access the electrode materials [24]. In order to control the diffusion resistance at low temperatures by facilitating the migration of ions through pores, the number of mesopores in the electrodes had been increased by using MgO-template and zeolite-template [25, 26].

Unfortunately, the volumetric capacitance will dramatically reduce in presence of large interconnected mesopores in the electrode due to the low packing density of broad pores. Therefore, developing electrode materials with a well-balanced mirco-to-mesopore ratio is highly needed to minimize diffusion resistance of the ions and also to achieve optimized gravimetric and volumetric capacitance of the EDLCs.

To exploit the beneficial features more efficiently, recently some efforts in self-supported three-dimensional activated carbon fabric have been designed [27, 28]. For example, Brandon et al. also reported binder-free activated carbon fabric having a substantial fraction of micropores which could improve the surface area of electrodes. A specific capacitance of $110\,F\,g^{-1}$ was achieved at −65 °C in $SBPBF_4$/AN electrolyte [29].

6.2.1.2 Carbon Nanotubes

CNTs as supercapacitor electrodes have been widely investigated because of their outstanding conductivity coupled with huge specific surface area, very high aspect ratio, and excellent mechanical and chemical stability [30, 31].

The electrochemical performance of CNTs can be greatly improved by creating more active sites, opening tips, and by tuning the distance between the CNTs. Zhang et al. reported a facile and effective chemical-acid-etching technique to open the tips of CNTs and also to introduce functional groups. The tip-open CNTs (functionalized and tip-open [FTO]-CNTs), demonstrated their ability to towards various types of ions (Li^+, Na^+, and Mg^{2+}) in aqueous electrolytes. In a symmetric supercapacitor device, the FTO-CNTs electrode delivered a larger areal energy density of 39 μW h cm^{-2} with a power density of 10.2 mW cm^{-2} [32].

To enhance the supercapacitive properties by minimizing the contact resistance between the active materials and the current collector, direct deposition of CNTs has been accepted as one of the effective strategies [33].

As in the low temperature, the presence of mesopores in CNTs benefits the ion accessibility to the surface of the electrode and improves the charge storing capacitive. Kang et al. [34] showed reported that mesopores in CNTs reduced internal resistance and allowed efficient ion migration to the surface of the electrode, maintaining high capacitance retention of 90% when the scan rate was increased from 5 to 40 mV s^{-1}, compared to 68% of activated carbon.

6.2.1.3 Graphene

Graphene, sp2 bonded carbon atoms made of a one-atom-thick sheet in a honeycomb crystal lattice have proven as suitable electrode materials for high-performance supercapacitors. The outstanding electrical conductivity, excellent chemical and thermal stability with very high accessible surface area (theoretical specific surface area of 2630 m^2 g^{-1}) have led it to be an attractive electrode material for supercapacitors [35, 36].

Unlike other carbon materials, it was observed that its electrochemical properties are not influenced by the distribution of pores [37, 38]. Despite these unique properties, the experimentally observed specific surface area is considerably lower than the theoretical value because of restacking of the graphene layer due to $\pi-\pi$ interactions.

To maximum utilization of intrinsic capacitance, improving the effective specific surface area of monolayer graphene is necessary. To address these stubborn issues, the introduction of other carbon-based materials like CNTs to create more active sites and prevent restacking of graphene layers had been adapted.

Zang et al. [39] reported rGO/CNTs electrode with a ratio of 5:1 which showed much improved specific capacitance of 33 mF cm^{-2} with 1-ethyl-3-methylimidazolium bis(trifluoromethylsulfonyl)imide ([Emim][TFSI]EMIMTSFI-based gel electrolyte at 40 °C, whereas the specific capacitance values for individual rGO and CNTs are only 8 mF cm^{-2} and 16 mF cm^{-2}, respectively. In this case, CNTs acted as conducting pathways and rGO delivered added active surface and the synergy effect between rGO and CNTs has made the material a very effective EDLC electrode.

6.2.1.4 Metal–Organic Frameworks (MOFs)

Recently, two-dimensional electrically conductive metal–organic frameworks (MOFs) have emerged as new electrode materials for EDLCs [40]. The key features of conductive MOFs, such as high electrical conductivities (up to 2500 S cm^{-1}), porosities (surface areas of 500 to 1400 m^2g^{-1}), and finally tunable crystalline structures make them promising materials as EDLC electrodes [41–43]. For example, MOF Ni_3(2, 3, 6, 7, 10, 11-hexaiminotriphenylene)$_2$ showed very high bulk electrical conductivity (higher than 5000 S m^{-1}) [44] whereas these values for activated carbons as well as holey graphite are ~1000 S m^{-1} [45].

Generally, in case of MOF, the transition metal M^{2+} with square planar coordination and planar conjugated organic linker molecules nodes to form $\pi-d$ conjugated two-dimensional (2D) sheets. These 2D sheets then in an eclipsed/near-eclipsed manner pile to create a

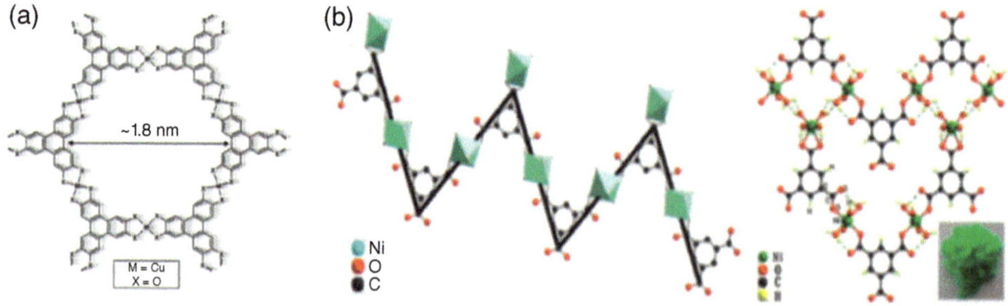

Figure 6.4 (a) Schematic demonstrating the general structure of hexasubstituted triphenylene-based conductive MOFs. The π−d conjugated 2D sheets stack to form an extended 3D honeycomb structure. This creates pores/channels that run through the material, with a pore size of 1.8 nm as calculated from the simulated structure of $Cu_3(HHTP)_2$ [46]. (b) Structural representation of the Ni-based MOF compound. *Source:* Kang et al. [47], Elsevier.

lengthy 3D honeycomb-like structure, generating pores that run through the material (Figure 6.4a). This porous structure creates a very high Brunauer–Emmett–Teller surface area. Kang et al. [47] reported a Ni-MOF electrode based on $Ni_3(btc)_2 \cdot 12H_2O$ (btc = 1,3,5-benzene tricarboxylic acid) using an easy solvothermal synthesis method. The 3D stable network structure consists of the combination of a 2D sheet motif with a one-dimensional (1D) zigzag chain building block connected by hydrogen bonds Figure 6.4b. Benefiting from the stable network structure, the Ni-MOF electrode delivered admirable specific capacitance 726 $F g^{-1}$ with an excellent capacitance retention rate of 94.6% even after 1000 cycles.

This amalgamation of very high surface area and electrical conductivity make MOFs a promising alternative to carbon-based materials for EDLCs. It has been observed that the open pores can easily host large electrolyte ions such as tetraethylammonium tetrafluoroborate ($TEABF_4$) and with their first solvation sphere in acetonitrile, which is a common organic solvent for EDLCs (ionic diameters of TEA^+, $TEA^+ \cdot 7ACN$, BF^{4-} and $BF^{4-} \cdot 9ACN$, are 0.68, 1.3, 0.46, and 1.16 nm, respectively) [48].

The easy preparation conditions of MOFs are added advantage in comparison to those followed for the synthesis of graphitic carbons, which are involved with severe pretreatment chemicals like strong bases, and carbonization at very high temperatures (<800 °C) [46].

De et al. for the first time reported conductive $Cu_3(HHTP)_2$ (HHTP = 2,3,6,7,10,11-hexa hydroxytriphenylene) nanowire arrays (NWAs) directly grown on the Cu foil and used as electrode material for symmetrical supercapacitor. The symmetrical supercapacitor delivers an admirable specific capacitance of 41.1 $\mu F cm^{-2}$ at a current density of 0.5 $A g^{-1}$, which is superior in comparison to the previously published capacitance values using various porous carbon electrodes [49].

Although too many molecular structures of MOFs were reported as electrode materials for SCs, the poor conductivity, as well as unsatisfactory molecular structures, are the main obstacle preventing from practical application.

To enhance the electrochemical performance few strategies have been adopted: (i) increase in electronic conductivity of MOFs by designing the molecular structures introducing multiple central metal ions or ligands. The introduction of multiple central metal ions or ligands generates a conductive MOF with an improved specific surface area, adaptable pores as well as crystal structure. (ii) by blending pristine MOFs with carbon-based conductive materials (such as rGO, PPy, CNTs, PANI, etc.) [50].

6.2.2 Pseudocapacitor

Unlike EDLC, the storage mechanism of pseudocapacitors is associated with the reversible faradic redox reaction, and also via adsorption or desorption, at the electrode-electrolyte interface. In presence of an applied potential, similar to the battery charge transfer takes place and the corresponding pseudocapacitance can be expressed by the equation:

$$C = \frac{dQ}{dV}$$

where dQ and dV are the charge acceptance and the potential window during charge–discharge. The pseudocapacitors have high capacitance compared to EDLC due to surface active redox reaction and higher power density than batteries as it possesses fast discharge. The electrode materials for PCs are mainly based on transition metal hydroxides, oxides, and chalcogenides (TM (OHs), TMOs, and TMChs) because of their high specific capacitance, excellent conductivity, abundant active sites, and stable structure for cyclability.

6.2.2.1 Transition Metal Hydroxides

Because of the redox-rich properties resulting in a very high capacitance, metal hydroxides ($M(OH)_2$, where M = Ni, Co, Mn, Fe, etc.) are accepted as promising pseudocapacitive materials as supercapacitor electrodes. The electrical energies stored (alkaline medium) in the metal hydroxides are mainly due to the Faradaic redox reaction:

$$M(OH)_2 + OH- \leftrightarrow MOOH + H_2O + e-$$

The similarities in crystallinity and structure of $M(OH)_2$ and MOOH materials play a vital role in achieving high and stable pseudocapacitance. In the case of iron hydroxides, the dissimilarities between oxides and hydroxides structure lead to massive rearrangement of atoms during the Faradaic redox reaction as a result it has very poor supercapacitive performance, whereas $Ni(OH)_2$ and $Co(OH)_2$ exhibit superior supercapacitive performance [51, 52].

Still, the poor electrical conductivity of the hydroxide electrode material preventing to achieve higher capacitance [53]. Guo et al. deposited $Ni(OH)_2$ directly on rGO conductive substrate and the resulting composite exhibited a very high energy density of 39.24 Wh kg^{-1} at 1962 W kg^{-1} with a capacity retains of 91.4% after 10 000 cycles at 5 A g^{-1} [54].

Han et al. reported carbon dots (CDs) in situ doped $Ni(OH)_2$ thin-film electrodes where the electrode after CDs doping electrical conductivity as well and also electrochemical activity increased dramatically [55].

Flower-like nanostructured nickel hydroxide powder directly deposited on nickel foam was compared, and it was observed that the binder-free $\alpha-Ni(OH)_2$ electrode delivered specific capacitance of 2814 F g^{-1} at the current density of 3 A g^{-1} because of its very high specific surface area (164 m^2 g^{-1}) [56].

Cobalt hydroxide is another well-studied metal hydroxide, distinctive structural property, and easy to synthesis with tailored morphology made it as an outstanding pseudocapacitive electrode material in alkaline solutions. To enhance the electrochemical properties of $Co(OH)_2$, Mn-doped Co(OH)2 nanosheets via hydrothermal decomposition was reported by Zhao et al. which displayed a high specific capacitance of $1915.88\,F\,g^{-1}$ at $1.0\,A\,g^{-1}$ and superior cycle performance of 93.5% capacitance retention over 5000 charge-discharge cycles at $10.0\,A\,g^{-1}$ [57].

To improve the electrochemical properties of transition metal hydroxides, various mixed-metal hydroxide compounds such as Ni–Al [58], Co–Ni [59, 60], Mn–Co [61], Co–Ru [62], Ni–Co–Cu [63], and Ni–Co–Zn [64] has been explored. These mixed metal hydroxides have better energy storage properties over the corresponding single-metal hydroxide as they promote integrated electrochemical performance.

6.2.2.2 Transition Metal Oxides

TMOs are used in PCs due to their exceptional electrochemical properties like extraordinary specific capacitance, chemical stability, diverse composition, tunable conductivity, and numerous crystal structure. TMOs store energy by faradic reaction and hence provide high energy density.

Among the various metal oxides, MnO_2 has been studied extensively because of its high theoretical specific capacitance value ($1370\,F\,g^{-1}$) and admirable environment-friendliness [65, 66] Shen et al. [67] reported one-dimensional MnO_2 nanotubes with surface mesoporosity and oxygen vacancies. Recently, it has been demonstrated that the mixed TMOs (with different metal cations) possess improved electrochemical activities and higher electrical conductivities compared to simple TMOs [68]. Inspired by these observations, various mixed MCo_2X_4, MMn_2X_4, MFe_2X_4, and $MMoX_4$ (M signifies transition metal: Ni, Mn, Zn, Cu, and Mg, and X present O) have been studied [69]. $NiFe_2O_4$/rGO are synthesized by hydrothermal method and this composite electrode delivered a capacitance of 584.63 F/g with a retention of 91% after 2000 cycles [70].

6.2.2.3 Transition Metal Sulfides

TMSs with reversible faradaic reactions have recently aroused attractive electrode materials for PCs and deliver amazing specific capacities. The associated advantages with TMs are their very low electronegativity, admirable electrical conductivity, and rich electrochemical redox sites resulting in high specific capacitance, in contrast with their corresponding oxide counterparts [71]. The integration of the materials with graphene will not only ease ionic transportation but also speed up the charge transfer process to promote their practical specific capacity [72] Among them, Chen et al. [73] reported CoS2@GONFs nanocomposites via a simple and facile method using reduced graphene oxide nanofibers (rGONFs) as the support materials. The nanocomposites in $6\,mol\,l^{-1}$ KOH delivered a high specific capacity of $635.8\,F\,g^{-1}$ at the applied current density of $1\,A\,g^{-1}$, which is much superior in comparison to the pure cobalt sulfides. TMS@graphene composites reported good electrical conductivity, a large specific surface area, and a reversible faradic reaction [74, 75].

To further improve the electrochemical performance of TMSs by improving the electrical conductivity mixed metal transition metal oxide has become more popular in comparison to the single metal sulfides. For instance, Liu et al. reported facile one-pot hydrothermal route to prepare ternary $Co_{0.33}Fe_{0.67}S_2$ nanoparticles in situ embedded in-between graphene nanosheets to form sandwich-like graphene/$Co_{0.33}Fe_{0.67}S_2$ composite [76]. The as-obtained composite exhibited high specific capacitance up to 310.2 C g^{-1} at 2 mV s^{-1} and outstanding rate capability with 61.8% capacitance retention at 200 mV s^{-1}. Yang et al. [77] synthesized Ni–Co–S nanoparticles supported by underlying graphene (Ni–Co–S@G) by an in situ chemical conversion route. At a current density of 1 A g^{-1}, the observed very high specific capacity of 1492 F g^{-1}, and the Ni–Co–S@G materials demonstrated a good cycling performance.

6.2.2.4 Transition Metal Diselenides

As selenium has lower ionization energy and a larger radius with a more noticeable metallic feature compared to the O and S, hence Transition metal diselenides (TMDSs) have excellent conductivity (10^{-3} S m^{-1}) therefore an attractive electrode material [78, 79].

Because of the improved electrical conductivity as well as electrochemical activity, substituting oxygen atoms in transition metal oxides with selenium has become a promising strategy.

In transition metal diselenides, the metal atom is sandwiched between the two selenide layers, and the atoms in these three layers are covalently connected, but the weak van der Waals force is the only link between the individual sheets. Therefore, extra atoms can easily intercalate in the inter-chain breaks reversibly, generating intercalation compounds or intercalates, enabling energy to be stored in the layered conductors [80].

Recently Chang et al. reported two-dimensional porous and ultrathin $NiSe_2$ nanosheets with selenium vacancies ($NiSe_2$ PNSvac) by plasma-assisted dry exfoliation. In 1 M KOH electrolyte, the observed specific capacitance of $NiSe_2$ PNSvac was 466 F g^{-1}, whereas the specific capacitance values for ultrathin $NiSe_2$ nanosheets and $NiSe_2$ particles are 328 and 251 F g^{-1}, respectively. The specific capacitance of $NiSe_2$ PNSvac's was very stable during the entire 1000 cycles, indicating its high cycling stability. The remarkable electrochemical performance of $NiSe_2$ PNSvac's can be ascribed by its high conductivity, shorter ion diffusion distance, and also large specific surface area [81].

N-doped reduced graphene oxide decorated $NiSe_2$ nanoparticles (N-rGO/$NiSe_2$) was synthesized by conversation of Ni $(OH)_2$ precursor suing solvothermal synthesis of N-rGO/$NiSe_2$ composites with varying N-rGO content. In this composite N-rGO acted as a Supporter for $NiSe_2$ nanoparticles, inhibiting them from aggregating and increasing specific surface area and electrical conductivity. The optimized N–rGO/$NiSe_2$ composite has a high specific capacitance of 2451.4 F g^{-1} at a current density of 1 A g^{-1}. In asymmetric configuration with AC as the negative electrode and the N–rGO/$NiSe_2$ composite as the positive electrode, it has a maximum energy density of 40.5 Wh kg^{-1} at a power density of 841.5 W kg^{-1}. Furthermore [82], a comparative study of directly deposited transition metal diselenides nanostructures (Co, Ni, and Mn) on nickel foam towards supercapacitors was carried out in asymmetric configuration.

The observed specific capacitance values of 148, 139, and 50 F g^{-1} for $CoSe_2$, $NiSe_2$, and $MnSe_2$, respectively, indicating the superior electrochemical performance of $CoSe_2$ over $NiSe_2$ and $MnSe_2$. Moreover, the $CoSe_2$ electrode showed better cycling stability, with capacitance retention of 98.46% upto 3000 cycles. On the other hand, $NiSe_2$ and $MnSe_2$ exhibited cycling stability of 59.44% and 86.7%, respectively. The superior performance of $CoSe_2$ was explained by the density functional theory (DFT) simulations. The quantum capacitance is greatest for $CoSe_2$, and it is reflected in electrochemical performance, $CoSe_2 > NiSe_2 > MnSe_2$.

6.3 Conclusion

Among the various electrochemical energy storage technologies, supercapacitor is of utmost encouragement for the next generation power source. This chapter has given a detailed overview of different types of supercapacitors, their working principle, advantages, and disadvantage. As the energy storage performance of any kind of supercapacitor depends on the type of electrode materials and their structural and morphological properties, the main motivation of this chapter was to highlight the progress in sustainable materials in this field and it covers various nanostructured materials for all different kind of supercapacitor. In case of EDLCs, among the various type of electrode materials MOFs are very promising for the next-generation application because of the presence of the $\pi - d$ conjugated extended 2D sheets stack forming a 3D honeycomb structure with pores/channels. Whereas, for PCs selenides with lower ionization energy and larger radius with a more noticeable metallic feature compared to their oxides and sulfides counterpart are more prospective candidates as electrode materials. Moreover, the electrochemical performance including cycle stability and rate performance of the TMDSs can be further improved by designing unique architectures and making composites with other functional conductive nanostructured materials, e.g. graphene.

References

1 Benoy, S.M., Pandey, M., Bhattacharjya, D., and Saikia, B.K. (2022). *Journal of Energy Storage* 52: 104938.
2 Balasubramaniam, S., Mohanty, A., Balasingam, S.K. et al. (2020). *Nano-Micro Letters* 12: 85.
3 Yu, L. and Chen, G.Z. (2016). *Journal of Power Sources* 326: 604.
4 Liu, J., Wang, J., Xu, C. et al. (2018). *Advancement of Science* 5: 1700322.
5 Zhang, S. and Pan, N. (2015). *Advanced Energy Materials* 5: 1401401.
6 Simon, P., Gogotsi, Y., and Dunn, B. (2014). *Science* 343: 1210.
7 Li, C., Islam, M., Moore, J. et al. (2016). *Nature Communications* 7: 1.
8 Winter, M. and Brodd, R.J. (2004). *Chemical Reviews* 104: 4245.
9 Amatucci, G., Badway, F., Du Pasquier, A., and Zheng, T. (2001). *Journal of the Electrochemical Society* 148: A930.
10 Wang, Y., Song, Y., and Xia, Y. (2016). *Chemical Society Reviews* 45: 5925.
11 Zhao, X., Sánchez, B.M., Dobson, P.J., and Grant, P.S. (2011). *Nanoscale* 3: 839.
12 Noori, A., El-Kady, M.F., Rahmanifar, M.S. et al. (2019). *Chemical Society Reviews* 48: 1272–1341.
13 Gonçalves, J.M., da Silva, M.I., Toma, H.E. et al. (2020). *Journal of Materials Chemistry A* 8: 10534.
14 Zhang, D., Zheng, L., Ma, Y. et al. (2014). *ACS Applied Materials & Interfaces* 6: 2657.
15 Abouali, S., Akbari Garakani, M., Zhang, B. et al. (2015). *ACS Applied Materials & Interfaces* 7: 13503.
16 Volfkovich, Y.M., Mikhalin, A.A., Bograchev, D.A., and Sosenkin, V.E. (2012). *Russian Journal of Electrochemistry* 48: 424.

17 Wu, M.Z. and Julia Chiang, P.C. (2004). *Electrochemical and Solid-State Letters* 7: A123.
18 Rauda, I.E., Augustyn, V., Dunn, B., and Tolbert, S.H. (2013). *Accounts of Chemical Research* 46: 1113.
19 Brezesinski, K., Wang, J., Haetge, J. et al. (2010). *Journal of the American Chemical Society* 132: 6982.
20 Borenstein, A., Hanna, O., Attias, R. et al. (2017). *Journal of Materials Chemistry A* 5: 12653.
21 Swain, N., Saravanakumar, B., Kundu, M. et al. (2021). *Journal of Materials Chemistry A* 9: 25286.
22 Hung, P., Zhang, H., Lin, H. et al. (2022). *Journal of Energy Chemistry* 68: 580.
23 Xu, J., Yuan, N.Y., Razal, J.M. et al. (2019). *Energy Storage Materials* 22: 323.
24 Ania, C.O., Pernak, J., Stefaniak, F. et al. (2006). *Carbon* 44: 3126.
25 Korenblit, Y., Kajdos, A., West, W.C. et al. (2012). *Advanced Functional Materials* 22: 1655.
26 Kado, Y., Imoto, K., Soneda, Y., and Yoshizawa, N. (2014). *Journal of Power Sources* 271: 377.
27 Zhang, Q., Rong, J.P., Ma, D.S., and Wei, B.Q. (2011). *Energy & Environmental Science* 4: 2152.
28 Markoulidis, F., Lei, C.H., and Lekakou, C. (2017). *Electrochimica Acta* 249: 122.
29 Brandon, E.J., Smart, M.C., and West, W.C. (2020). *Journal of Materials Research* 35: 113.
30 Hadjipaschalis, I., Poullikkas, A., and Efthimiou, V. (2009). *Renewable and Sustainable Energy Reviews* 13: 1513.
31 Zhang, H., Cao, G., Yang, Y., and Gu, Z. (2008). *Carbon* 46: 30.
32 Zhang, Y. and Xie, E. (2021). *Dalton Transactions* 50: 12982.
33 Reinhardt, H., Hellmann, C., Nürnberger, P. et al. (2017). *Advanced Materials Interfaces* 1: 1700508.
34 Kang, J., Atashin, S., Jayaram, S.H., and Wen, J.Z. (2017). *Carbon* 111: 338–349.
35 Hung, P.Y., Lau, K.T., Fox, B. et al. (2018). *Composites Part B: Engineering* 133: 240.
36 Miller, J.R., Outlaw, R., and Holloway, B. (2010). *Science* 329: 1637.
37 Zhang, C., Xie, Y., Zhang, C., and Lin, J. (2019). *Carbon* 153: 585.
38 Li, A., Dong, C., Dong, W. et al. (2018). *ACS Applied Materials & Interfaces* 10: 32093.
39 Zang, X., Zhang, R., Zhen, Z. et al. (2017). *Nano Energy* 40: 224.
40 Liu, J., Song, X., Zhang, T. et al. (2020). *Angewandte Chemie, International Edition* 60: 5612.
41 Ko, M., Mendecki, L., and Mirica, K.A. (2018). *Chemical Communications* 54: 7873.
42 Sun, L., Campbell, M.G., and Dinca, M. (2016). *Angewandte Chemie, International Edition* 55: 3566.
43 Dou, J.-H., Arguilla, M.Q., Luo, Y. et al. (2021). *Nature Materials* 20: 222.
44 Sheberla, D., Sun, L., Blood-Forsythe, M. et al. (2014). *Journal of the American Chemical Society* 136: 8859.
45 Xu, Y., Lin, Z., Zhong, X. et al. (2014). *Nature Communications* 5: 4554.
46 Gittins, J.W., Balhatchet, C.J., Chen, Y. et al. (2021). *Journal of Materials Chemistry A* 9: 16006.
47 Kang, L., Sun, S., Kong, L.B. et al. (2014). *Chinese Chemical Letters* 25: 957.
48 Pilon, L., Wang, H., and d'Entremont, A. (2015). *Journal of the Electrochemical Society* 162: A5158.
49 Du, X., Zhang, J., Wang, H. et al. (2020). *Materials Chemistry Frontiers* 4: 243.
50 Gao, H., Shen, H., Wu, H. et al. (2021). *Energy & Fuels* 35: 12884.
51 Gao, H., Xin, S., and Goodenough, J.B. (2017). *Chemistry* 3: 26.

52 Kim, S.W., Kim, I.H., Kim, S.I., and Jang, J.H. (2017). *Chemistry - An Asian Journal* 12: 1291.

53 Shi, F., Li, L., Wang, X.L. et al. (2014). *RSC Advances* 4: 41910.

54 Guo, B., Gao, Y., Li, Y. et al. (2022). *ACS Applied Nano Materials* 5: 7471.

55 Han, C., Si, H., Sang, S. et al. (2020). *ACS Applied Nano Materials* 12: 12106.

56 Yi, X., Sun, H., Robertson, N., and Kirk, C. (2021). *Sustainable Energy & Fuels* 5: 5236.

57 Zhao, C., Liang, S., Jiang, Y. et al. (2020). *Materials Letters* 270: 127751.

58 Wang, Y.G., Cheng, L., and Xia, Y.Y. (2006). *Journal of Power Sources* 153: 191.

59 Zhou, D., Su, X., Boese, M. et al. (2014). *Nano Energy* 5: 52.

60 Chen, J.C., Hsu, C.T., and Hu, C.C. (2014). *Journal of Power Sources* 253: 205.

61 Liu, J., Jiang, J., Cheng, C. et al. (2011). *Advanced Materials* 23: 2076.

62 Xu, J., Wang, Q., Wang, X. et al. (2013). *ACS Nano* 7: 5453.

63 Lien, C.H., Hu, C.C., Hsu, C.T., and Wong, D.S. (2013). *Electrochemistry Communications* 34: 323.

64 Wang, H., Gao, Q., and Hu, J. (2010). *Journal of Power Sources* 195: 3017.

65 Kundu, M. and Liu, L.F. (2013). *Journal of Power Sources* 243: 676.

66 Zhao, J., Zhu, B., Yang, G. et al. (2021). *Journal of Physics and Chemistry of Solids* 150: 109856.

67 Shen, M., Wang, Y., and Zhang, Y.X. (2020). *Dalton Transactions* 49: 17552.

68 Kundu, M., Karunakaran, G., Kolesnikov, E. et al. (2018). *Journal of Industrial and Engineering Chemistry* 59: 90.

69 Zhu, X. (2022). *Journal of Energy Storage* 49: 104148.

70 Askari, M.B. and Salarizadeh, P. (2020). *International Journal of Hydrogen Energy* 45: 27482.

71 Guo, M., Balamurugan, J., Thanh, T.D. et al. (2016). *Journal of Materials Chemistry A* 4: 17560.

72 Meng, X., Deng, J., Zhu, J. et al. (2016). *Scientific Reports* 6: 21717.

73 Chen, Q., Cai, D., and Zhan, H. (2017). *Journal of Alloys and Compounds* 706: 126.

74 Huang, G., Liu, H., Wang, S. et al. (2015). *Journal of Materials Chemistry A* 3: 24128.

75 Zhao, G., Li, X., Huang, M. et al. (2017). *Chemical Society Reviews* 46: 4417.

76 Liu, W., Niu, H., Yang, J. et al. (2018). *Chemistry of Materials* 30: 1055.

77 Yang, J., Yu, C., Fan, X. et al. (2016). *Energy & Environmental Science* 9: 1299.

78 Shi, X., Wang, H., Ji, S. et al. (2019). *Chemical Engineering Journal* 364: 320.

79 An, W., Gao, Y., and Liu, J. (2017). *Chemistry Select* 24: 7372.

80 Sajjad, M., Tao, R., and Qiu, L. (2021). *Journal of Materials Science: Materials in Electronics* 32: 1.

81 Chang, A., Zhang, C., Yu, Y. et al. (2018). *ACS Applied Materials & Interfaces* 10: 41861.

82 Gu, Y., Fan, L., Jian-Ling, H. et al. (2019). *Journal of Power Sources* 425: 60.

Part III

Sustainable Approaches for Fuel Cells

7

Sustainable Materials for Fuel Cell Devices

Weidong He, Shijie Zhong, Yunfa Dong, and Qun Li

National Key Laboratory of Science and Technology on Advanced Composites in Special Environments, and Center for Composite Materials and Structures, Harbin Institute of Technology, Harbin, 150080, China

7.1 Introduction

Under the background of sustainable energy rise, proton exchange membrane fuel cells (PEMFCs) have gained great advancements. The applications of PEMFCs in the areas of transportation and small-scale power generation are shown in Figure 7.1 [1].

The schematic diagram of fuel cell composition is shown in Figure 7.2 [2]. The main components of a fuel cell include proton exchange membrane, catalyst layer, gas diffusion layer, and bipolar plate. In this chapter, the contents are mainly divided into three sections.

In order to improve the power density and service life of fuel cells, each component needs to be optimized. Each section introduces one main component and its development tendency. Regarding the catalyst, the current mainstream is platinum (Pt)-based materials. It is worth noting that metal-free catalysts and metal–nitrogen–carbon catalysts are also promising candidates. The proton exchange membranes can be divided into perfluorosulfonic acid based and nonfluorinated polymer-based ones in low-temperature fuel cells. Moreover, polybenzimidazole PBI-based membrane is also a promising material for high-temperature applications. Finally, brief introductions of gas diffusion layer, bipolar plate, current collector, and sealing materials are also provided.

7.2 Catalysts

7.2.1 Introduction

Catalysts are a crucial part of fuel cells, which can reduce the activation energy for reactant molecules on the catalysts' surfaces to accelerate reactions. The catalytic activity and durability directly determine the power density and lifetime of the fuel cells. Currently, commercial fuel cell catalysts are almost platinum group metal (PGM)-based catalysts [3], which account for half the cost of the fuel cell stack due to their exorbitant price and high

Sustainable Energy Storage in the Scope of Circular Economy: Advanced Materials and Device Design,
First Edition. Edited by Carlos Miguel Costa, Renato Gonçalves, and Senentxu Lanceros-Méndez.

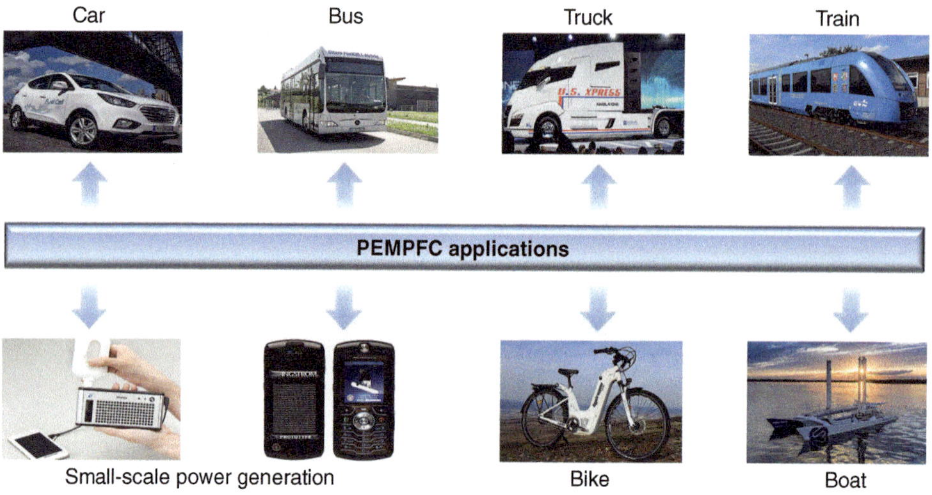

Figure 7.1 The applications of PEMFCs in the areas of transportation and small-scale power generation. *Source:* Xiong et al. [1], Elsevier.

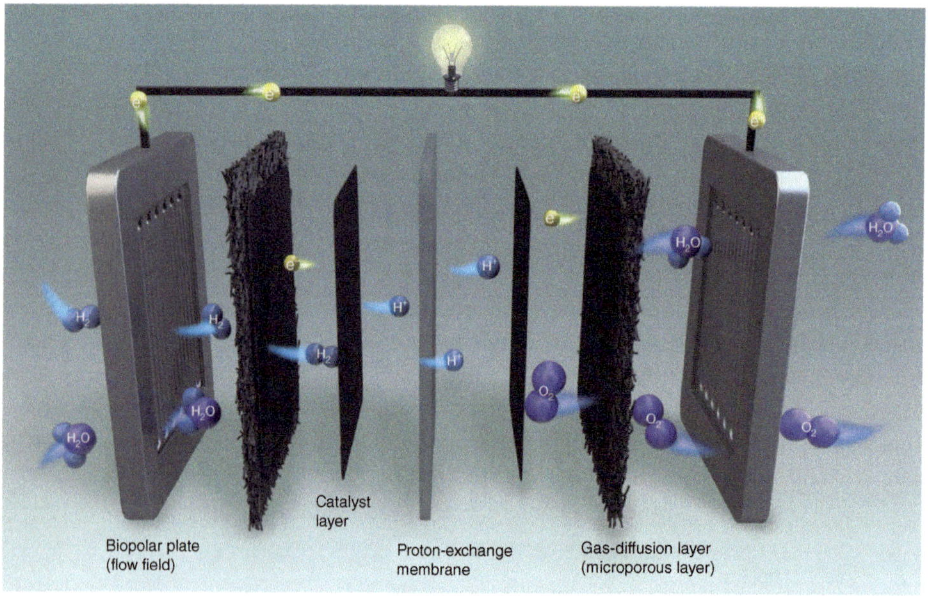

Figure 7.2 Schematic diagram of fuel cell composition. *Source:* Jiao et al. [2]/With permission of Springer Nature.

loading. The US Department of Energy (DOE) hopes that the fuel cell cost can fall to \$40 kW^{-1} by 2025 [4]. For this purpose, they set a series of targets for PGM catalysts (see in Table 7.1): the total PGM loading reaches 0.125 mg cm^{-2} (both anode and cathode included), the mass activity reaches 0.44 A mg$_{PGM}$$^{-1}$ at 0.9 V$_{IR\text{-free}}$, the performance at 0.8 V exceeds 300 mA cm^{-2}, and the loss of voltage maintains below 30 mV operating in the current density of 0.8 or 1.5 A cm^{-2} after 5000 cycles. Using PGM-free catalysts is also a

Table 7.1 DOE 2025 target of electrocatalysts.

Characteristic	Units	2025 Target
Total PGM loading	$mg_{PGM}\ cm^{-2}$	0.125
Mass activity at $0.9\,V_{IR\text{-free}}$	$A\ mg_{PGM}^{-1}$	0.44
Performance at 0.8 V	$mA\,cm^{-2}$	>300
Loss of voltage at 0.8 or $1.5\,A\,cm^{-2}$	mV	<30
PGM-free catalysts' performance at $0.9\ V_{IR\text{-free}}$	$A\,cm^{-2}$	0.044

prospective choice to reduce the cost. But these catalysts require a much thicker catalytic layer to achieve the same effect as PGM catalysts, which is unfavorable for large-scale application [5]. So that, developing a cheap, efficient, and robust catalyst for fuel cell is still a long way off.

An excellent catalyst in proton exchange membrane fuel cell must satisfy the following characteristics:

a) The catalysts must be resistant to acid attack under high temperature and high humidity to avoid being dissolved owing to strong acid environment on the surface of proton exchange membrane.
b) High electrical conductivity of catalysts is vital for charge transfer during the catalytical reactions.
c) High catalytic activity is essential for catalysts. According to Sabatier's principle, moderate adsorption bond strength between catalyst surface and reactants results in higher reaction activity.

7.2.2 PGM-Based Catalysts

PGMs like Pt, gold, ruthenium, and rhodium are often used as fuel cell catalysts due to their excellent catalytic activity, great corrosion resistance, and high electrical conductivity. In the volcano plot (shown in Figure 7.3) describing the relationship between oxygen reduction activity and oxygen-binding energy, Pt metal is closest to the peak by virtue of its unparalleled catalytic activity [6]. The first fuel cell in the space used pure Pt black. However, individual Pt atom with high surface energy leads to the agglomeration of the Pt particle, which causes significant decline of catalytic activity. To solve this problem, researchers have made great efforts to ameliorate the Pt catalysts, such as using the carriers to support Pt catalysts [7], introducing heterogeneous atoms to form ordered Pt alloy [8], and constructing core–shell structure catalysts [9].

7.2.2.1 Carbon-Based Supported PGM Catalysts

Carbon-based materials are widely used in supporting PGM catalysts because of their favorable conductivity and large surface area. Platinum-carbon (Pt/C) catalysts have been commercialized on a large scale until now. Nevertheless, once the carbon carrier is corroded or structure collapses, the catalysts will agglomerate or suffer from CO poisoning rapidly. Hence, the chemical stability of carbon materials directly affects the performance of Pt/C catalysts.

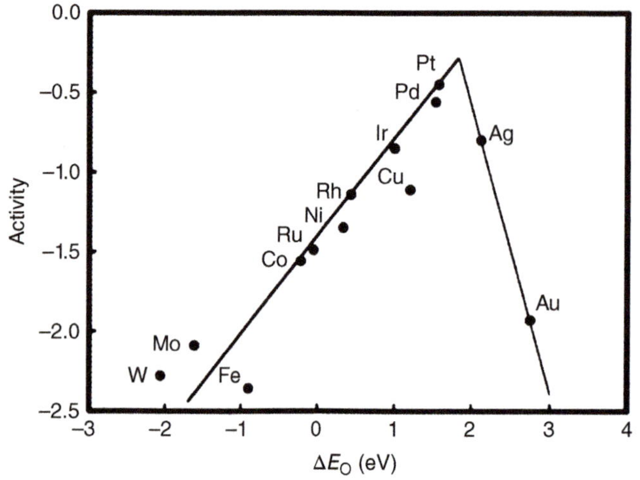

Figure 7.3 Relationship between oxygen reduction activity and oxygen-binding energy. *Source:* Norskov et al. [6]/With permission of American Chemical Society.

Amorphous carbon materials have attracted scientists' attention as their irregularity can supply large quantities of loading sites for PGM. However, the high irregularity also leads to low lifetime in fuel cell working conditions. By contrast, graphite carbon materials with more stable structures are desirable for fuel cell applications. Researches show that higher degree of graphitization is beneficial for the stabilization of Pt/C catalysts, but unfavorable for the loading of Pt atoms [10]. Functional modifications with oxygen-containing groups on the surface of graphite supporting materials are efficient for greatly improving the catalytic activity by partial loss of stability [11].

Diamond is known for being extremely robust even in nano size, because each of its atoms forms four covalent bonds with neighboring atoms. The strong covalent bonds endow it with ultrastable structure, which is suitable for supporting catalysts in fuel cells. The nanoscale diamonds produced from chemical vapor deposition are low cost and flexible for optimizing nanostructure and adjusting composition [12]. For example, boron-doped diamond is a promising catalyst support in fuel cells. The boron doping concentration directly affects the conductive properties of diamond. Optimal boron-doped diamond-supported Pt catalyst displays better performance than commercial Vulcan XC-72 and multiwalled carbon nanotube-supported catalysts [13].

Carbon nanotubes (CNTs) with unique tubular structure in carbon materials are more conducive to the dispersion of Pt atoms. Moreover, the CNTs can provide larger surface area, better electrical conductivity, and higher chemical stability than carbon black. Multiwalled CNTs have been proved to enhance oxygen reduction reaction (ORR) performance as Pt catalyst supports in fuel cells [14]. However, the research also showed that too much multiwalled CNTs would decrease the volume of primary pores and increase the volume of secondary pores in the catalyst layer. Adding about 10% multiwalled CNTs would balance the two kinds of pores volume to obtain the best fuel cell performance. Different structures of CNTs, shown in Figure 7.4, also bring different electrochemical properties. CNTs with bamboo structure have better catalytic activity benefitting from the

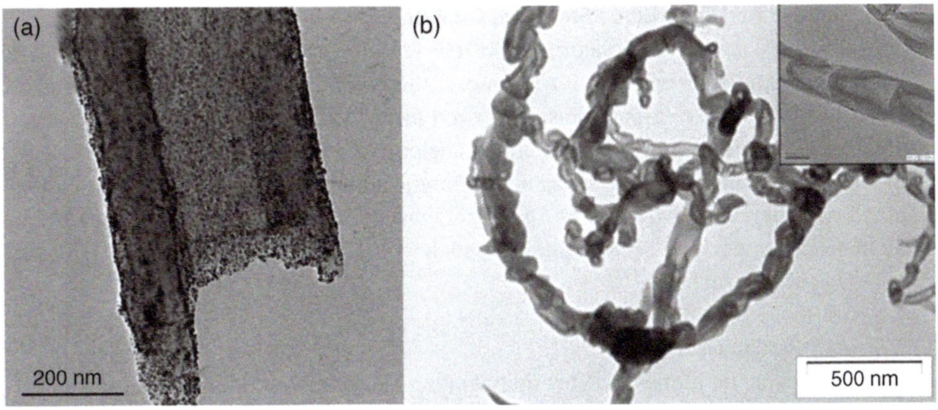

Figure 7.4 TEM images of (a) hollow-structured CNTs. *Source:* Wang et al. [15], Elsevier and (b) bamboo-structured CNTs. *Source:* Shanmugam [16], Elsevier.

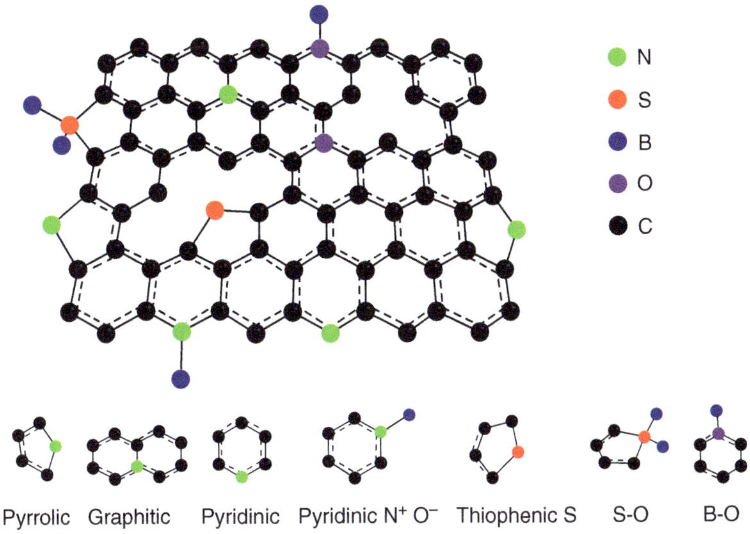

Figure 7.5 Schematic representation of heteroatom-doped carbon support materials. *Source:* Raj et al. [18]/Royal Society of Chemistry.

more defect sites at the structural edges. In contrast, hollow-structured CNTs show better durability but poorer activity [15–17]. Hence, how to balance the stability and catalytic activity of Pt/C catalysts remains a huge challenge.

Heteroatom doping in carbon support materials can significantly improve their surface area and electronic conductivity, which can enhance the Pt/C catalyst performance. Nitrogen, boron, sulfur, and oxygen-doped carbon are common forms to apply in Pt/C support materials (Figure 7.5). The nitrogen atoms possess larger electronegativity than carbon atoms, which can accelerate the electron transfer between carbon and oxygen to enhance ORR activity [19]. Meanwhile, nitrogen doping can decrease Pt particle size and prevent Pt atoms agglomerating to enlarge the catalysts' electrochemical surface area and enhance fuel cell durability [20].

7.2.2.2 Oxide-Based Supported PGM-Based Catalysts

Various transition metal oxide materials, like IrO_2 [21], TiO_2 [22], RuO_2 [23], and WO_3 [24], are stable in acidic environment. Furthermore, these materials can anchor Pt atoms on their surface to decrease Pt particle size or even form Pt single-atom catalysts. However, most of these transition metal oxide materials belong to semiconductors with a nonnegligible ohmic resistance, which goes against catalytic activity enhancement. Doping other elements or materials is proved to be an efficient way to improve catalytic performance of oxide material-supported Pt-based catalysts, such as Pt/TiON-1 [25], PtRu/WO_x [26], and Pt/$Ir_3(IrO_2)_7$ [27] in fuel cells.

7.2.2.3 Pt Alloy Catalysts

Although Pt owns the highest ORR activity among all the metals, it does not provide sufficient catalytic activity to meet realistic requirements when used alone. And single Pt catalysts can be poisoned by CO under fuel cell operating conditions. Introducing heteroatoms to form Pt alloys as fuel cell catalysts has been extensively researched. There are three effects to understand how the transition metals adjust the Pt catalysts' performance.

The facet effect shows that different facets of Pt possess different activities for ORR. Constructing Pt alloys can drive Pt to expose more active crystal faces. For example, Pt_3Ni alloy with exposure (111) facet displays record ORR-specific activity [28]. Ligand effect explains that introducing different transition metals into Pt alloys will cause different changes in d-states of Pt surface. As shown in Figure 7.6, [29] the volcano plot relates the ORR activity to the d-band center of Pt_3M (M = Ni, Co, Ti, V, Fe) catalysts. Pt_3Co and Pt_3Ni catalysts with the right d-band center are superior than other catalysts for ORR. Modulation ligands have also been shown to be effective in different ternary Pt alloys, such as PtNi/Ni–B [30] and PtNiCo [31] catalysts. Strain effect occurs when the transition metals are combined with Pt, resulting in the increase of Pt lattice spacing. The researchers have found that transition metals increasing Pt lattice spacing will decrease the O adsorption energy on the Pt surface [32], which can affect the ORR performance of Pt alloys. The strain effect not only affects the activity of the catalyst but also increases its stability. For example, PtRu/Ru (0001) catalyst displays a strong tolerance to CO benefitting from the strain effect [33].

7.2.2.4 Pt Core–Shell Structure Catalysts

Compared with Pt alloys, Pt core–shell materials possess similar effects to enhance catalytic activity of Pt-like facet effect, ligand effect, and strain effect. Expect that, they exhibit better stability and higher utilization of Pt than Pt alloys, since Pt atoms are only dispersed over the layer in core–shell structure [34]. Pt shell with appropriate shell thickness can not only protect core metal dissolution but also make full use of active sites. The $L1_0$-CoPt NP core–shell catalyst (in Figure 7.7) with about two atomic layers of Pt displays excellent membrane electrode assembly (MEA) activity and stability beyond DOE 2025 targets [35], which make core–shell structure Pt catalysts become the focus of research. However, the complex synthesis process and strict requirements on reaction conditions increase the cost and difficulty of its mass production and large-scale commercialization.

7.2.3 PGM-Free Catalysts

The most important factor limiting fuel cell development is the high cost of catalysts. Currently, the status of Pt-based catalysts is irreplaceable in fuel cells, but non-noble metal

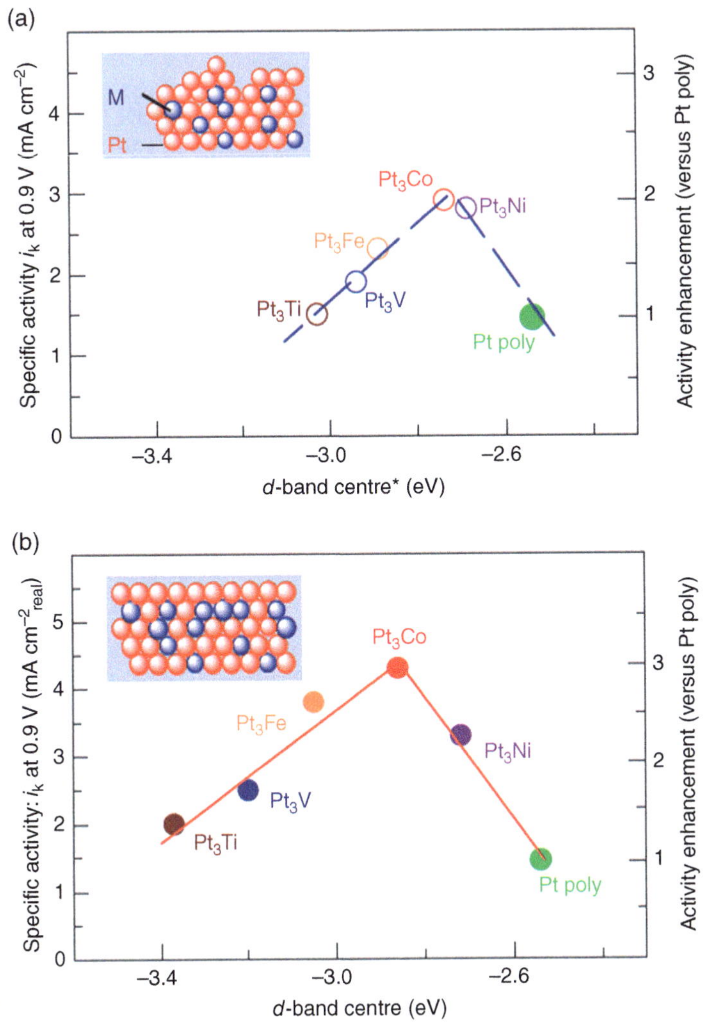

Figure 7.6 Relationships between ORR catalytic activity of Pt_3M vs. the d-band center for (a) Pt-skeleton and (b) Pt-skin. *Source:* Stamenkovic et al. [29]/With permission of Springer Nature.

catalysts are attracting more and more attention due to their low cost. As shown in Table 7.1 and Figure 7.8 [36], comparing the 2025 DOE targets with our current status, we are far from producing the satisfactory non-noble catalysts.

7.2.3.1 Metal-Free Catalysts

Carbon-based materials can be used not only as supports for PGM catalysts but also as catalysts directly due to their large surface area and high conductivity. In fuel cells, almost all metal-free catalysts are carbon-based materials. Pure carbon materials cannot promote ORR. Only by changing the electronic structure of carbon can it have catalytic activity. It is found that the catalytic activity of carbon materials can be achieved by nitrogen atom doping or nitrogen-like atom doping [37]. It is generally believed that the doping position and content of nitrogen have the greatest influence on the catalytic activity

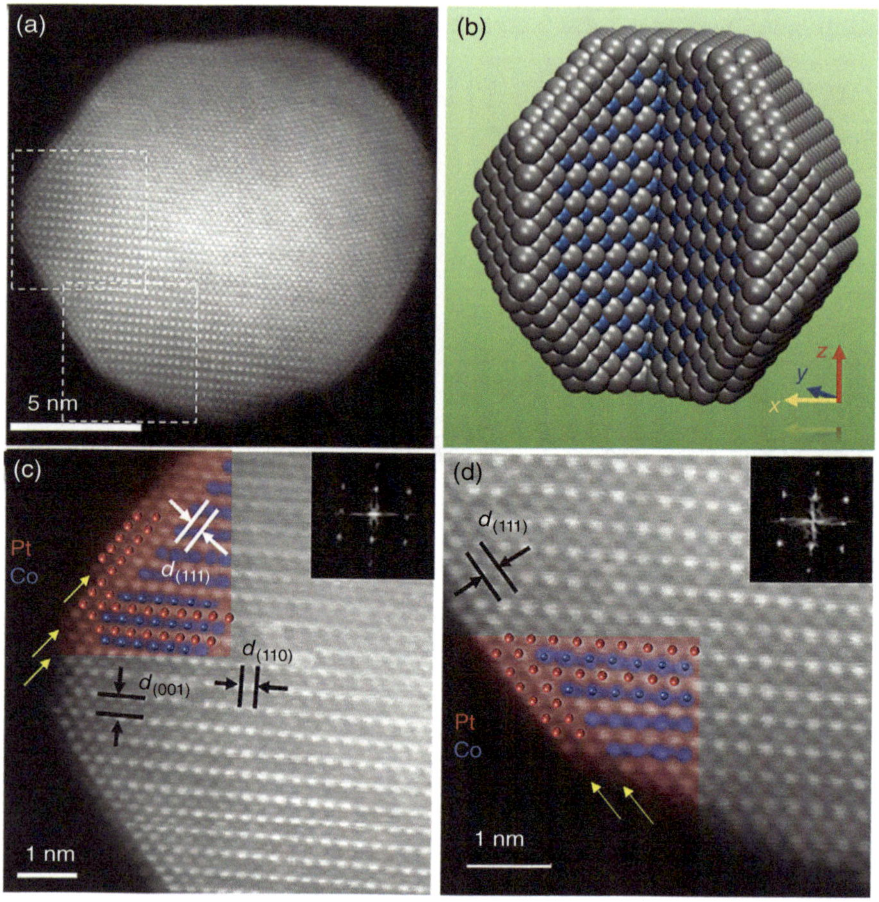

Figure 7.7 (a) STEM image and (b) schematic of $L1_0$-CoPt/Pt NPs. (c and d) Enlarged section of figure (a). *Source:* Li et al. [35]/With permission of Elsevier.

of N-doped carbon materials [38]. Additionally, B, S, and P dopings are thought to have the effect similar to that of nitrogen doping [39].

7.2.3.2 Metal–Nitrogen–Carbon Catalysts

M–N–C catalysts, especially Fe–N–C catalysts, are considered as the most promising PGM-free catalysts in fuel cells due to their outstanding catalytic activity [40]. But some researchers concern that H_2O_2 produced during fuel cell operation may produce Fenton effect with Fe, which may decompose proton exchange membrane. Fe-free M–N–C catalysts, such as Co–N–C, Zn–N–C, Cr–N–C, and Sn–N–C, get more attention [41]. The formation of M—N bond is the key factor of high catalytic activity. Hence, selecting appropriate nitrogen-containing precursors such as specific MOF materials can efficiently obtain the catalysts containing rich M–N. At present, the stability of M–N–C materials still cannot meet the commercial requirements of fuel cells, and the development and commercialization of PGM-free catalysts still face great challenges.

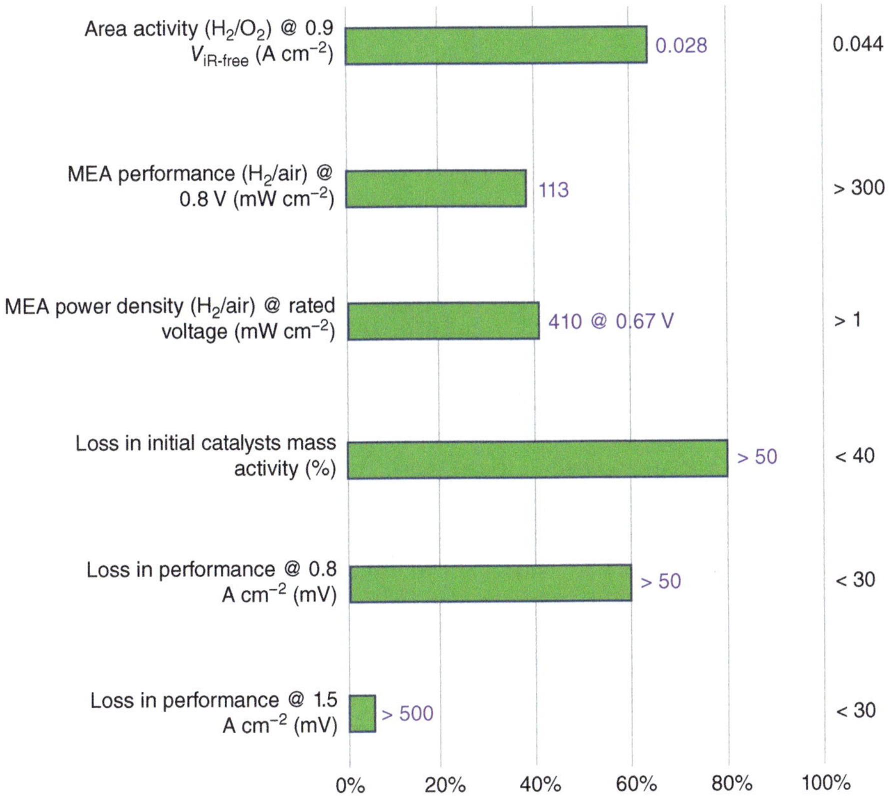

Figure 7.8 DOE 2025 targets and current status of PGM-free catalysts for transportation. *Source:* He et al. [36]/With permission of Royal Society of Chemistry.

7.3 Proton Exchange Membrane (PEM)

Proton exchange membrane (PEM) is one of the most important components in PEMFCs that works as a conductor to transport the H^+ from the anode to the cathode and barrier to prevent direct contact of fuel and oxidant. High proton conductivity is a basic requirement for ideal PEM. In addition, the material used for PEM should have low crossover of gases, extraordinary mechanical strength, and superior chemical durability. The performances of PEM are not only closely related to the membrane materials but also depend on the preparation methods. According to base materials, the PEMs can usually be classified into perfluorosulfonic acid (PFSA) membranes, sulfonated hydrocarbon polymers (SHPs), and polybenzimidazole (PBI)/H_3PO_4 membranes. There are various methods reported in recent decades to prepare high-performance PEMs including solution casting, electrospinning, electrophoretic deposition, sol–gel method, slot dye extrusion, layer-by-layer self-assembly, spin coating, and impregnation [42–45]. The PEMs obtained by different methods mentioned above generally show some good performances in some areas, while showing unsatisfied or bad performances in others.

7.3.1 PFSA and Their Composite Membranes

The most popular and widely used commercial PEM in PEMFCs is Nafion membranes based on PFSA, which is the combination of polytetrafluoroethylene (PTFE) backbone and randomly tied side chain with pendant sulfonic groups ($-SO_3H$). The Nafion membrane shows hydrophobic properties due to the nonpolar structure of PTFE backbone, while the hydrophobic PTFE provides long-term mechanical/chemical stability for membrane because of the strong bond of C–F. The hydrophilic sulfonic groups endow the membrane with good hydration in water. The proton-conducting channels in Nafion are generated among ion clusters like reverse micelle formed by sulfonic groups. It has been demonstrated that Nafion membranes have high proton conductivity of $0.2\,S\,cm^{-2}$ at 100 °C and 100% relative humidity (RH) [46]. Nevertheless, Nafion membrane has several drawbacks, such as rapid drop of proton conductivity at low humidity, severe methanol permeability, and poor mechanical strength with low thickness. To address the aforementioned problems of Nafion, various strategies have been proposed in recent years. The most impactful way to improve the performance is to regulate the micro-morphology of Nafion by polymer chemistry. Moreover, the length of the side chain of PFSA has important effects on the properties of membrane. Thus, Dow Chemical Co. explored and invented short-side-chain PFSA (SSC-PFSA) in 1980s, and the SSC-PFSA membrane applied by Ballard Power Systems exhibited considerable augmentation in fuel cell performance [47]. 3M, Solvay Specialty Polymers, and Asahi Glass further developed the preparation technology of SSC-PFSA and drastically reduced the cost of SSC-PFSA compared to Dow. The molecular structural formulas of PFSAs with different lengths of side chain are shown in Figure 7.9 [48]. SSC-PFSA provides better performance for membrane in proton conductivity at elevated temperature, degree of crystallinity, degree of hydration, and mechanical durability compared to Nafion [49]. Luo et al. revealed that fully hydrated SSC-PFSA membranes possess higher water contents (56–75 vol%), moderate λ values (15–18), and higher analytical acid concentrations (2–2.8 M)

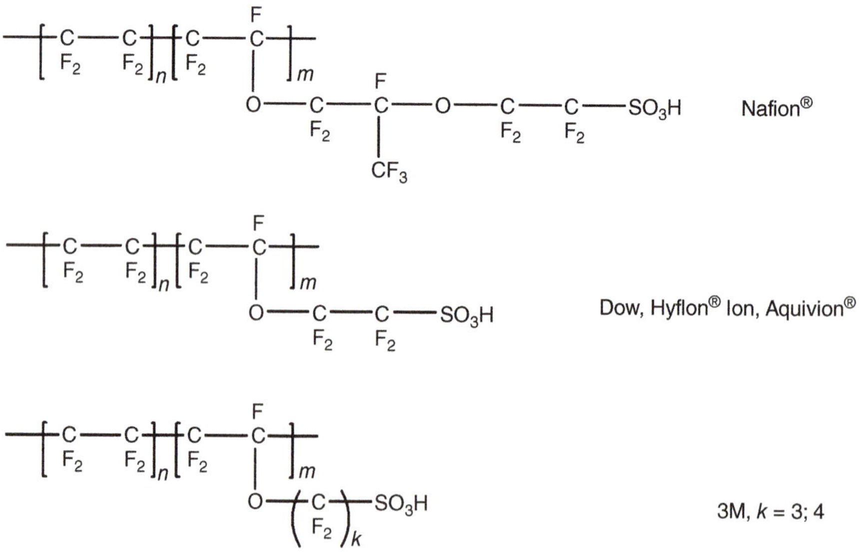

Figure 7.9 Various PFSA polymers with long- and short-side chains. *Source:* Primachenko et al. [48]/With permission of John Wiley & Sons.

compared to long-side-chain PFSA (LSC-PFSA) [50]. Molecular dynamics (MD) simulations confirm that the water molecules in SSC-PFSA membrane have easier access to the sulfonic acid groups to form a more developed water channel due to the more uniform distribution of $-SO_3H$ in SSC-PFSA membrane than Nafion ionomer [51].

PEMs based on PFSA ionomer are more suitable for low-temperature PEMFCs because they will lose water severely as the temperature exceeds the boiling point of water, resulting in the decrease of proton conductivity. Therefore, introducing inorganic fillers into PFSA membranes is an effective strategy to provide a wider temperature range of applications. Hygroscopic metal oxides (M_xO_y, M = Ti, Si, Zr, Al, Ta, etc.), clays, carbon nanomaterials, heteropoly acids (HPA), and zirconium phosphate are the most commonly used additives to improve the water retention capability and proton conductivity of PFSA membrane in elevated temperatures [42, 52]. Son et al. prepared a composite membrane by incorporating commercial silica mesostructured cellular foam (SiO_2-MSUF) into Nafion membrane [53]. Multimode atomic force microscopy and force distance analyses revealed that SiO_2-MSUF resulted in morphological changes and a phase-lag associated with variation of water uptake on Nafion membrane (Figure 7.10). The improvement of proton conductivity of

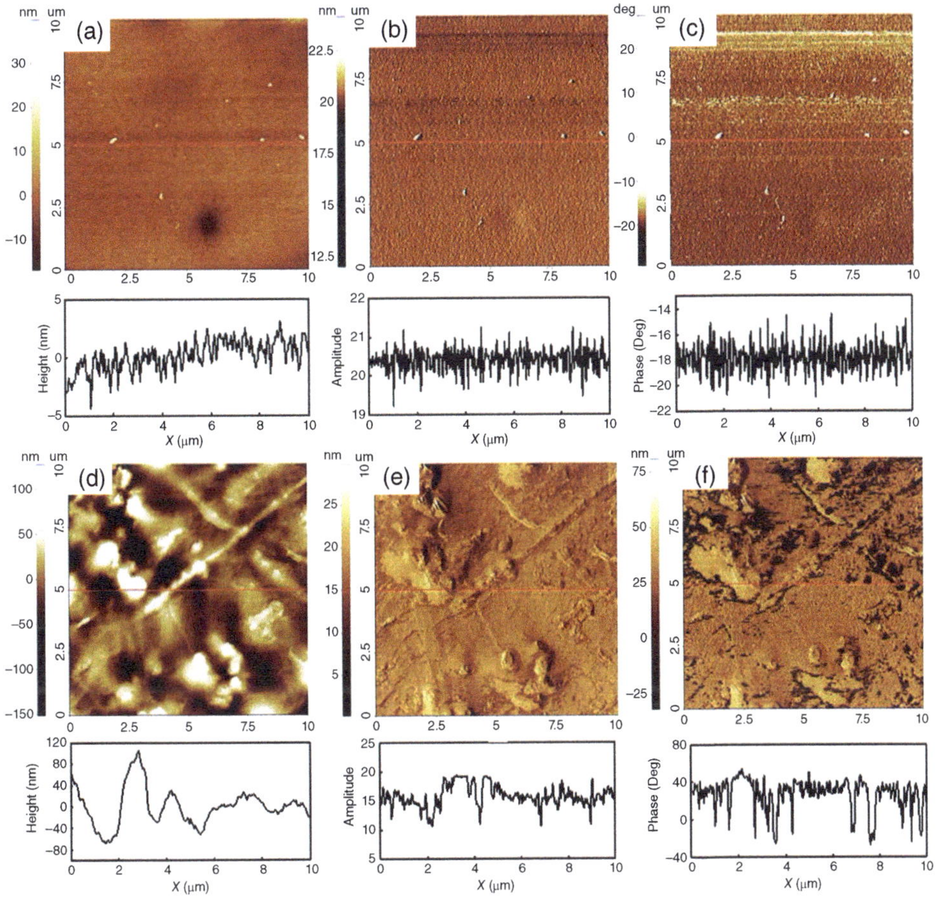

Figure 7.10 (a) Topography, (b) amplitude image, and (c) phase image of the pure Nafion and (d) topography, (e) amplitude image, and (f) phase image of the SiO_2/Nafion membrane. *Source:* Son et al. [53]/With permission of John Wiley & Sons.

SiO_2/Nafion membrane was closely associated with the separated proton-conducting structure because the hygroscopic nature of SiO_2 enhances the water uptake of membrane.

In addition, the surface functionalization of inorganic nanomaterials by grafting proton-conducting groups, such as $-SO_3H$, $-PO_3H_3$, $-NH_2$, $-CO_2H$, imparts better compatibility between inorganic fillers and PFSA and facilitates the proton conductivity [54]. These acidic group-functionalized inorganic fillers can enhance the proton transport by increasing the density of ionic groups in Nafion matrix. Vinothkannan et al. first synthesized carbon nanotubes-unzipped carbon nanotubes (CNTs-UCNTs) and then sulfonated the CNTs-UCNTs by 4-benzenediazonium sulfonate to obtain the sulfonated CNTs-sulfonated UCNTs (SCNTs-SUCNTs) hybrid, as shown in Figure 7.11a [55]. The Nafion/SCNTs-SUCNTs composite membrane showed high proton conductivity of $0.015\,S\,cm^{-1}$, power density of $0.343\,W\,cm^{-2}$, and low voltage drop of $0.15\,mV\,h^{-1}$ when working at 60 °C under 20% RH in fuel cells, respectively (Figure 7.11b,c).

PFSA blending with other polymers such as polyvinylidene fluoride (PVDF), polybenzimidazole (PBI), polyvinyl alcohol (PVA), and polyaniline (PANI) have been regarded as promising methods to improve methanol resistance and mechanical strength of Nafion.

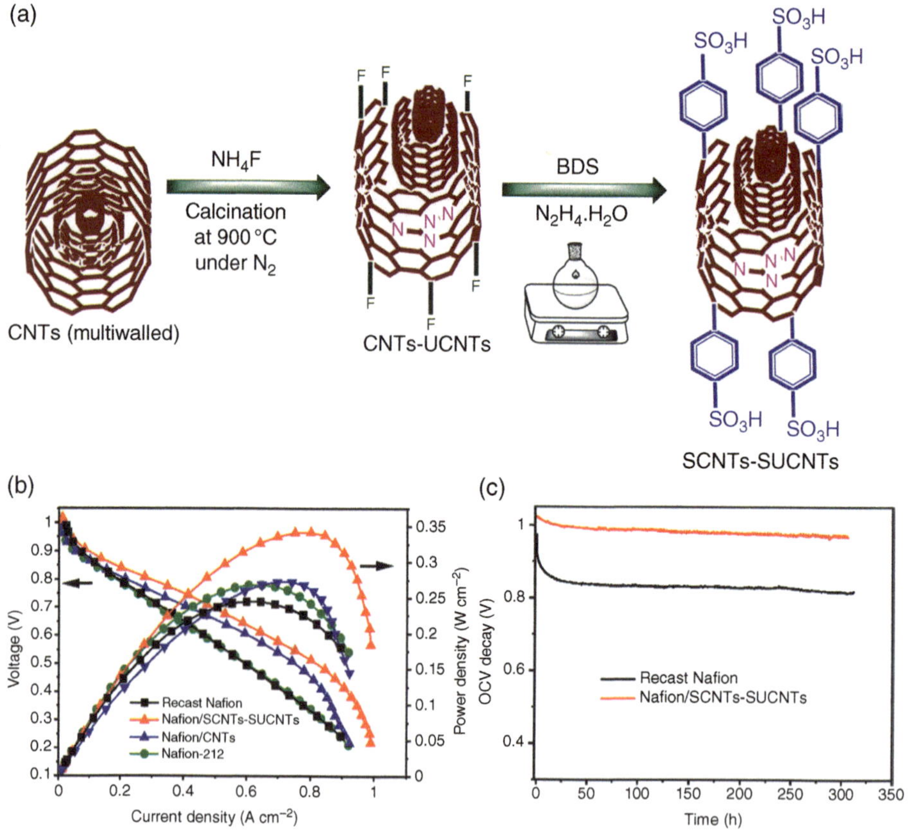

Figure 7.11 (a) Schematic illustration of synthesis of SCNTs-SUCNTs hybrid. (b) Polarization curves of MEA assembled with Nafion and its composite membranes quantified at 60 °C under 20% RH. (c) OCV degradation of Nafion and its composite membrane quantified at 60 °C under 20% RH. *Source:* Vinothkannan et al. [55]/With permisssion of Elsevier.

The polymers for blending with PFSA generally have lower methanol crossover and good compatibility with PFSA. Rao et al. studied the effects of Nafion membranes with various PVA coating thicknesses on their proton conductivity and methanol permeability [56]. The results show that the Nafion membrane with only 2 μm thick PVA coating can reduce methanol crossover through membrane by 50% due to the dense structure of PVA layer caused by hydrogen bonding in intra- and intermolecular level. Zhou et al. prepared semi-interpenetrating (SIPN) PEMs consisted of Nafion and PVDF via radiation cross-linking [57]. The SIPN PEM exhibited much better mechanical properties (53.4 MPa) and improved dimensional stability due to the PVDF molecular chains form three-dimensional (3D) cross-linked network. Moreover, introduction of these inexpensive polymers can reduce the number of PFSA, ulteriorly resulting in significant reduction in the cost of PEM.

Currently, the thickness of membrane tends to be thinner to respond to the requirements for low-cost PEM and overall transport resistances. However, thin PFSA membranes suffer from decreased mechanical properties and chemical stability. By contrast, thin perfluorinated membranes reinforced with porous polymer substrates such as PTFE, polyimide (PI), polyacrylonitrile (PAN), and PVDF exhibit enhanced mechanical and dimensional stability. Expanded PTFE (ePTFE) is the most widely used polymer as reinforcement for PFSA due to the excellent chemical and mechanical properties among these porous polymers. For example, commercial Gore-Select PEM with 10 μm applied in Toyota Mirai consists of 2.5 μm ePTFE as reinforced skeleton and PFSA containing cerium salt as proton conducting layer [58]. Durability test of fuel cell based on Gore-Select membrane shows that the car can achieve a 3000-hours real-world driving. Liu et al. further improved the durability of ePTFE-reinforced PFSA membrane by the deposition of ZrO_2 coating on polydopamine-modified ePTFE skeletons (Figure 7.12) [59]. The ZrO_2-deposited ePTFE-reinforced PEM shows a lower OCV decrease of 2.60 mV h^{-1} compared with unmodified ePTFE-reinforced PEM (3.51 mV h^{-1}).

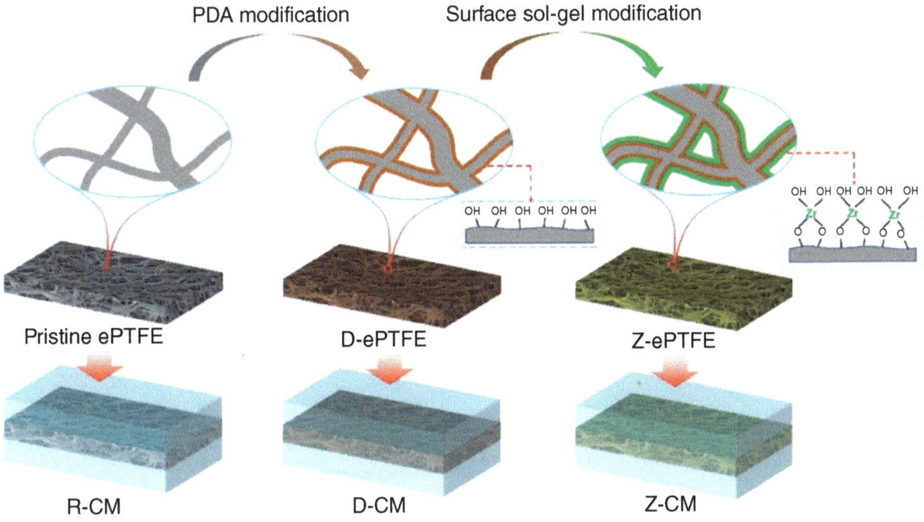

Figure 7.12 Schematic diagram of modified ePTFE skeletons and reinforced composite membranes. *Source:* Liu et al. [59]/With permission of American Chemical Society.

Another strategy to prepare reinforced PEM is the introduction of a porous nonwoven nanofiber with 3D network by electrospinning into the PFSA ionomer. Electrospinning technology can obtain soft, porous, and robust polymer nanofiber mat. The aligned nanofiber featured with high surface area, low density, and fully interconnected pores is beneficial to improve the compatibility between PFSA and polymer nanofiber mat and mechanical strength [44]. These inert substrates in PFSA also reduce the methanol permeability of PEM and dramatically keep the membrane thickness below 50 μm [60].

7.3.2 SHPs and Their Composite Membranes

SHPs and their composite membranes are considered as promising alternatives of PFSA due to their low-cost, high-water uptake, and thermal stability compared to PFSA. SHPs including nonfluorinated aliphatic and aromatic hydrocarbons are sulfonated to introduce the sulfonic acid group, which enables them to conduct proton on the hydrophobic polymer chains. SHPs with aromatic backbone have better oxidation stability due to the strong C—H bonds of the benzene ring. The typical structure of sulfonated aromatic polymer is shown in Figure 7.13 [61]. Arylene groups are connected with different linkers to form virous aromatic polymers. These repeating functional groups such as ether, ketone, sulfone, nitrile, and sulfide can form abundant polymer structures, including sulfonated poly(arylene ether sulfone) (SPAES), sulfonated poly(arylene ether ketone) (SPAEK), sulfonated poly(ether ether ketone) (SPEEK), sulfonated poly(arylene ether nitrile) (SPAEN), and sulfonated poly (sulfone) (SPSU) [61]. PEM based on SHPs should balance the high proton conductivity and long-term mechanical/chemical stability because the increase in density of $-SO_3H$ generally reduces the mechanical properties of membrane. Fang et al. prepared pure SPEEK membranes via electrophoretic deposition and studied their properties [43]. SPEEK membrane with sulfonation degree of 60% shows high proton conductivity of 0.11 $S\,cm^{-1}$ at 70 °C and 100% RH. Furthermore, the SHP membranes are widely studied in methanol fuel cells due to better resistance to methanol.

Similar to PFSA membrane, the SHP membrane is also beset by issues such as severe loss of proton conductivity and membrane degradation under high operating temperature. Therefore, all methods to modify PFSA are suitable for SHPs. For example, Huang et al.

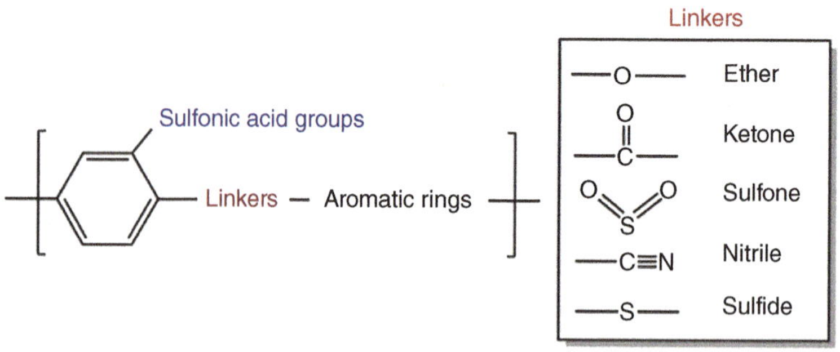

Figure 7.13 Basic chemical structure of hydrocarbon-based PEMs with aromatic backbone. *Source:* Mirfarsi et al. [61]/With permission of Elsevier.

reported an ultrathin-reinforced membrane prepared by using hexagonal boron nitride nanosheets (h-BN), SPEEK (DS = 81.4%), and ePTFE for PEMFC [62]. The BN@SPEEK/ePTFE membrane shows the highest proton conductivity (0.283 S cm^{-1} at 60 °C) compared to SPEEK/PTFE and Nafion 212.

7.3.3 PBI/H_3PO_4 Membrane

The PEM-based PFSA or SHPs are difficult to apply to high-temperature PEMFCs (HT-PEMFCs) due to the proton conduction mechanism through water. However, doping high-volatility acids into membrane is one of the effective methods to extend the operating temperature range of PEMs. Phosphoric acid (H_3PO_4, PA)-doped PBI membrane is one of the most studied high-temperature PEM due to high conductivity and excellent thermal stability at high temperature. PBI is an aromatic polymer with high tolerance to acidic or basic conditions and thermal stability ($T_g = 430$). PBI membrane owns proton conductivity without dehydration due to the presence of acid or alkaline groups as the proton donors. To improve the proton conductivity of PBI used in HT-PEMFCs, PA has been used as doping agents for various PBIs in that it can work without dependence on the RH of the environment. The high proton conductivity of PBI/H_3PO_4 membrane depends on many factors such as the content and distribution of the doped H_3PO_4. However, high amounts of PA in PBI can result in severe problems, such as reduced mechanical properties of PBI, catalyst corrosion, and leaching and condensation of PA groups. Extensive modification methods have been proposed to alleviate these problems recently, such as the introduction of hygroscopic inorganic fillers, organic frameworks, graphene oxide, and ionic liquids, which improve the acid retention capability [63]. In addition, introducing encapsulating structure into PBI/H_3PO_4 has also been proved to be an effective method to reduce PA leaching. Kannan et al. reported a PA-doped PBI membrane with three layers [64]. As shown in Figure 7.14, the middle layer contains higher PA content as an acid reservoir and two outer layers contain lower PA content as barrier layers to prevent the PA leaching into the middle layer. The multilayer structure of PBI/H_3PO_4 membrane shows much lower voltage loss rates of 2.3–4.1 μV h^{-1} compared to single-layered membranes (11–14 μV h^{-1}) at 180 °C for 10 000 hours.

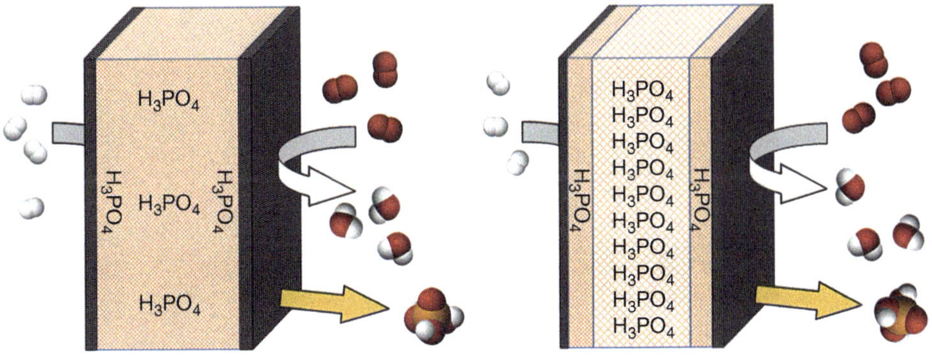

Figure 7.14 Schematic of PA-doped three-layered membrane. *Source:* Kannan et al. [64]/With permisssion of Elsevier.

7.4 The Other Components

7.4.1 Gas Diffusion Layer (GDL)

As a vital component of PEMFCs, the gas diffusion layer (GDL) has significant influence on its overall performance. It can not only support the catalyst layer and maintain the integrity of electrode but also provide the transfer area for gas, electron, heat, and water generated by reaction [65, 66]. The optimization of this component could contribute to a 10% enhancement of power density according to the estimation data [2].

An ideal GDL should possess the characteristics as following [67]:

a) A gradient pore size
b) Low resistivity
c) Dense structure along with smooth surface
d) Appropriate mechanical strength with moderate rigidity and flexibility
e) Appropriate hydrophilic/hydrophobic balance
f) High chemical and thermal stability
g) Low manufacturing cost and high performance/price ratio

The fabrication methods of gas diffusion layer include spray [68], coating [69], and electrospun [70]. The GDL can be divided into gas diffusion supporting base that affords homogenization of fuel and oxidant and the microporous layer (MPL), which mainly diminishes the contact impedance and accelerates the drainage of water. Various carbon-based raw materials or metallic materials are used to establish supporting base. Moreover, the MPL is usually made up of hybrids with carbon materials and hydrophilic or hydrophobic materials. According to the current industry development situation, carbon paper will still be the mainstream base material of GDL because of its low price, chemical stability, rigid structure, and excellent conductor of electricity [2].

Moreover, in order to comply with the requirements of industry development – improving mass transfer, promoting the water drainage, and achieving the power density of $9\,\mathrm{kW\,l^{-1}}$ – the potential approaches for further GDL design are shown in Figure 7.15 [2]. One is to establish the pore size gradient through cutting down the porosity of GDL to mitigate the scale discrepancies between the bipolar plate and

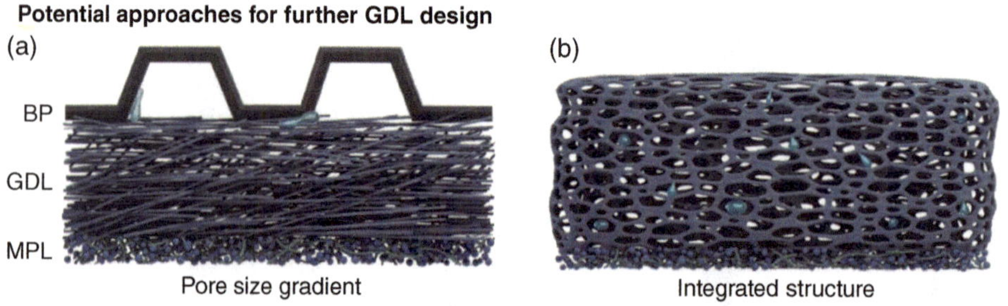

Figure 7.15 Two potential approaches for future GDL development: (a) pore size gradient, realized by manipulation of the carbon fiber arrangement. (b) Integrated BP-MEA design or GDL-less design with foam materials. *Source:* Jiao et al. [2]/With permission of Springer Nature.

membrane electrode assembly (see in Figure 7.15a). In this way, the reactant provision and water drainage can be enhanced. The other is to fabricate an integrated structure (an integrated porous bipolar plate-membrane electrode assembly design) instead of the splits system (see in Figure 7.15b). Meanwhile, the design of this concept puts forward higher requirements for the use of materials, for example, multifunctional foam material [2]. As a proof of the concept, a GDL-less structure is designed by means of utilizing corrugated mesh flow channel [71].

7.4.2 Bipolar Plate (BP)

Bipolar plate is also a critical component of the fuel cell, which functions as supporting layer, collecting current, isolating reaction medium, and provides the channels for reactant gas, reaction product, cooling liquid, and heat dissipation [72]. In general, the requirements of BP can be different under different operating conditions.

An ideal BP should possess the characteristics as following [73]:

a) High heat transfer efficiency (>20 W mK^{-1}).
b) High purity.
c) High stability with GDLs in physical and chemical properties without evident performance degradation.
d) Easiness to establish flow field with fine structure, maintain small machining error within 0.05 mm, and afford a flat surface.

The materials employed in BP are usually composite plates and metal plates. The composite plates contain carbon/carbon composite plates, thermoset-based composite plates, and thermoplastic-based composite plates. The optimization direction of BP is to decrease cost and enhance durability in the fuel cell vehicles industry particularly. The ideal cost of BP is \$3 kW^{-1} according to US DOE criterion for 2020 [2]. Regarding the environmental friendliness, the metal plates are more attractive due to the convenience of recycling and reprocessing. Moreover, the metal plates hold the merits of preparing the thinner plates and more sophisticated flow fields. The mainstream base material of metal plates is modified stainless steel due to its excellent chemical stability and superior electronic conductivity. As shown in Figure 7.16, the stainless steel (SS304L and SS430) shows superior performance compared with the graphite plate under automotive condition when iR is corrected [74]. It is gratifying that laser is considered as a promising processing method because it can fabricate the metal BP in a selective, controllable, and low-cost manner.[75]

7.4.3 Current Collector

The main function of a current collector is to conduct electrons. The main materials used in current collector are copper plates, graphite plates, and stainless steel plates coated with gold. Currently, copper plate is the mainstream base material due to its good electrical conductivity and economy effect.

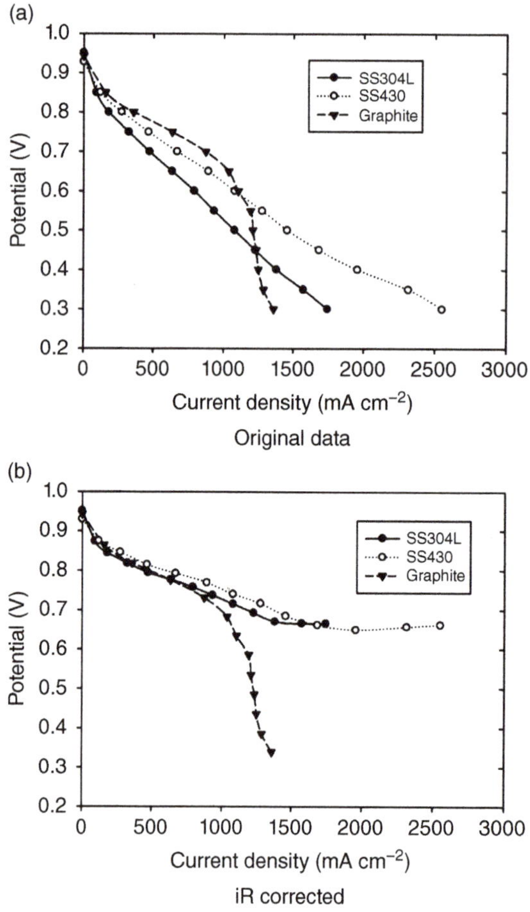

Figure 7.16 Polarization curves for serpentine flow-field plates under automotive condition: (a) original data. (b) Data from iR corrected (T_{cell} = 80 °C, anode/cathode RH = 75/0, anode/cathode stoich = 1.3/2.0, P = 274 kPa). [GDL: Toray TGP-H-060, GORE™ PRIMEA®SERIES 57 MEA (Pt: 0.1/0.4 mg cm^{-2})]. *Source:* Shimpalee et al. [74]/With permisssion of Elsevier.

7.4.4 Sealing Material (SM)

Compared with membrane electrode assembly and bipolar plate, sealing materials are an easily overlooked component, but it has a significant impact on the service life of fuel cells. The main functions of the sealing material are to seal the fuel cell, compensate for component tolerances, and keep the membrane electrode assembly in an appropriate compression state. For the practical test, the requirement for the sealing material (SM) is to guarantee the stack in a moderate compression state, avoid stack overvoltage, and minimize contact resistance in the stack. In addition, the precipitates in the SM should not cause catalyst poisoning, accelerate film aging, or block the pores in the GDL [76]. Sealing materials should also afford temperature resistance, acid resistance, moisture resistance, electrical insulation, and good shock absorption resistance [77].

Commercial sealing materials are mainly silicone rubber, fluorine rubber, and polyolefin rubber [78]. It is noteworthy that the service life of sealing materials is limited by the chemical stability of PEM, and high durability of membrane materials is beneficial to extend the life of sealing materials because of the less fluorine ions from membrane degradation [79].

References

1 Xiong, K., Wu, W., Wang, S., and Zhang, L. (2021). *Applied Energy* 301: 117443.
2 Jiao, K., Xuan, J., Du, Q. et al. (2021). *Nature* 595: 361.
3 Zhao, Z., Chen, C., Liu, Z. et al. (2019). *Advanced Materials* 31: e1808115.
4 Araya, S.S., Li, N., and Liso, V. (2022). Chapter 9 – Degradation and failure modes in proton exchange membrane fuel cells. In: *PEM Full Celles, Fundamentals, Advanced Technologies and Practical Application*, 199–222.
5 Zaman, S., Huang, L., Douka, A.I. et al. (2021). *Angewandte Chemie, International Edition* 60: 2.
6 Norskov, J.K., Rossmeisl, J., Logadottir, A. et al. (2004). *The Journal of Physical Chemistry. B* 108: 17886.
7 Chalgin, A., Song, C., Tao, P. et al. (2020). *Progress in Natural Science: Materials International* 30: 289.
8 Huang, L., Zaman, S., Tian, X. et al. (2021). *Accounts of Chemical Research* 54: 311.
9 Cai, B., Hubner, R., Sasaki, K. et al. (2018). *Angewandte Chemie, International Edition* 57: 2963.
10 Ott, S., Orfanidi, A., Schmies, H. et al. (2020). *Nature Materials* 19: 77.
11 Li, L., Wu, G., and Xu, B.Q. (2006). *Carbon* 44: 2973.
12 Spataru, N., Calderon-Moreno, J.M., Osiceanu, P. et al. (2020). *Chemical Engineering Journal* 402: 126258.
13 Halima, A.F., Rana, U.A., and MacFarlane, D.R. (2014). *Electrochimica Acta* 115: 639.
14 Kil, K.C., Hong, S.G., Park, J.O. et al. (2014). *International Journal of Hydrogen Energy* 39: 17481.
15 Wang, Y., Li, G., Jin, J., and Yang, S. (2017). *International Journal of Hydrogen Energy* 42: 5938.
16 Shanmugam, S. and Gedanken, A. (2006). *Electrochemistry Communications* 8: 1099.
17 Shao, Y., Liu, J., Wang, Y., and Lin, Y. (2009). *Journal of Materials Chemistry* 19: 46.
18 Raj, C.R., Samanta, A., Noh, S.H. et al. (2016). *Journal of Materials Chemistry A* 4: 11156.
19 Xiao, F., Xu, G.L., Sun, C.J. et al. (2020). *Nano Energy* 77: 105192.
20 Mardle, P., Ji, X., Wu, J. et al. (2020). *Applied Catalysis. B, Environmental* 260: 118031.
21 Du, J., Quinson, J., Zhang, D. et al. (2021). *ACS Catalysis* 11: 820.
22 Antoniassi, R.M., Quiroz, J., Barbosa, E.C.M. et al. (2021). *ChemCatChem* 13: 1931.
23 Zhang, Z.Q., Liu, J.P., Wang, J. et al. (2021). *Nature Communications* 12: 5235.
24 Stewart, D.W.G., Scott, K., Wain, A.J. et al. (2020). *ACS Applied Materials & Interfaces* 12: 37079.
25 Dhanasekaran, P., Selvaganesh, S.V., and Bhat, S.D. (2016). *Journal of Power Sources* 304: 360.
26 Wang, Q., Wang, G., Sasaki, K. et al. (2013). *Journal of Power Sources* 241: 728.
27 Kong, F.D., Zhang, S., Yin, G.P. et al. (2012). *Journal of Power Sources* 210: 321.
28 Stamenkovic, V.R., Fowler, B., Mun, B.S. et al. (2007). *Science* 315: 493.
29 Stamenkovic, V.R., Mun, B.S., Arenz, M. et al. (2007). *Nature Materials* 6: 241.
30 He, D.P., Zhang, L.B., He, D.S. et al. (2016). *Nature Communications* 7: 12362.
31 Zhao, Z.P., Feng, M., Zhou, J.H. et al. (2016). *Chemical Communications* 52: 11215.
32 Nilekar, A.U. and Mavrikakis, M. (2008). *Surface Science* 602: L89.

33 Diemant, T., Bergbreiter, A., Bansmann, J. et al. (2010). *ChemPhysChem* 11: 3123.
34 Gong, S., Zhang, Y.X., and Niu, Z. (2020). *ACS Catalysis* 10: 10886.
35 Li, J., Sharma, S., Liu, X. et al. (2019). *Joule* 3: 124.
36 He, Y., Liu, S., Priest, C. et al. (2020). *Chemical Society Reviews* 49: 3484.
37 Zhang, W., Sun, H., Zhu, Z. et al. (2020). *Renewable Energy* 146: 2270.
38 Yang, Z., Nie, H., Chen, X.A. et al. (2013). *Journal of Power Sources* 236: 238.
39 Ren, X., Liu, B., Liang, X. et al. (2021). *Journal of the Electrochemical Society* 168: 044521.
40 Chen, G., Liu, P., Liao, Z. et al. (2020). *Advanced Materials* 32: 1907399.
41 Zhao, C.X., Li, B.Q., Liu, J.N., and Zhang, Q. (2021). *Angewandte Chemie, International Edition* 60: 4448.
42 Haider, R., Wen, Y., Ma, Z.F. et al. (2021). *Chemical Society Reviews* 50: 1138.
43 Fang, F., Zhao, S., Zhang, W. et al. (2018). *Journal of Membrane Science* 565: 179.
44 Kallem, P., Yanar, N., and Choi, H. (2018). *ACS Sustainable Chemistry & Engineering* 7: 1808.
45 Zhang, H., Huang, H., and Shen, P.K. (2012). *International Journal of Hydrogen Energy* 37: 6875.
46 Andrews, J. and Doddathimmaiah, A.K. (2008). *Materials for Fuel Cells*, 344. Elsevier.
47 Li, J., Pan, M., and Tang, H. (2014). *RSC Advances* 4: 3944.
48 Primachenko, O.N., Marinenko, E.A., Odinokov, A.S. et al. (2020). *Polymers for Advanced Technologies* 32: 1386.
49 Kwon, S.H., Kang, H., Sohn, Y.J. et al. (2021). *Scientific Reports* 11: 8702.
50 Luo, X., Holdcroft, S., Mani, A. et al. (2011). *Physical Chemistry Chemical Physics* 13: 18055.
51 Kwon, S.H., Kang, H., Lee, J.H. et al. (2019). *Materials Today Communications* 21: 100625.
52 Inamuddin, A.M. and Asiri, A.M. (2017). *Organic-Inorganic Composite Polymer Electrolyte Membranes Preparation, Properties, and Fuel Cell Applications*, 1e. Springer.
53 Son, B., Oh, K., Park, S. et al. (2019). *International Journal of Energy Research* 43: 4157.
54 Vinothkannan, M., Kim, A.R., and Yoo, D.J. (2021). *RSC Advances* 11: 18351.
55 Vinothkannan, M., Kim, A.R., Ryu, S.K., and Yoo, D.J. (2022). *Journal of Membrane Science* 649: 120393.
56 Rao, A.S., Rashmi, K.R., Manjunatha, D.V. et al. (2021). *Materials Today: Proceedings* 35: 344.
57 Zhou, B., Pu, H., Pan, H., and Wan, D. (2011). *International Journal of Hydrogen Energy* 36: 6809.
58 Borup, R.L., Mukundan, R., and More, K.L. (2018). U.S. Department of Energy. https://www.osti.gov/servlets/purl/1440415 (accessed 26 November 2021).
59 Liu, L., Li, Y., Qiao, R. et al. (2021). *Energy & Fuels* 35: 12482.
60 Wang, S.H. and Lin, H.L. (2014). *Journal of Power Sources* 257: 254.
61 Mirfarsi, S.H., Parnian, M.J., Rowshanzamir, S., and Kjeang, E. (2022). *International Journal of Hydrogen Energy* 47: 13460.
62 Huang, Z., Lv, B., Zhou, L. et al. (2022). *Journal of Membrane Science* 644: 12099.
63 Seng, L.K., Masdar, M.S., and Shyuan, L.K. (2021). *Membranes* 11: 728.
64 Kannan, A., Aili, D., Cleemann, L.N. et al. (2020). *International Journal of Hydrogen Energy* 45: 1008.
65 Ozden, A., Shahgaldi, S., Li, X., and Hamdullahpur, F. (2019). *Progress in Energy and Combustion Science* 74: 50.

66 Park, S., Lee, J.W., and Popov, B.N. (2012). *International Journal of Hydrogen Energy* 37: 5850.
67 Yang, Y., Zhou, X., Li, B., and Zhang, C. (2021). *International Journal of Hydrogen Energy* 46: 4259.
68 Chen, W.H., Ko, T.H., Cheng, S.I. et al. (2014). *International Journal of Nanotechnology* 11: 993.
69 Yu, J., Yoshikawa, Y., Matsuura, T. et al. (2005). *Electrochemical and Solid-State Letters* 8: A152.
70 Waldrop, K., Wycisk, R., and Pintauro, P.N. (2020). *Current Opinion in Electrochemistry* 21: 257.
71 Tanaka, S. and Shudo, T. (2014). *Journal of Power Sources* 268: 183.
72 Tang, A., Crisci, L., Bonville, L., and Jankovic, J. (2021). *Journal of Renewable and Sustainable Energy* 13: 022701.
73 Wilkinson, D.P., Zhang, J.J., and Hui, R. (2009). *Proton Exchange Membrane Fuel Cells, Materials Properties and Performance*. ISBN 9781439806647. CRC Press, 460 P, 157 B/W Illustration, p. 313.
74 Shimpalee, S., Lilavivat, V., McCrabb, H. et al. (2016). *International Journal of Hydrogen Energy* 41: 13688.
75 Gerteisen, D. and Sadeler, C. (2010). *Journal of Power Sources* 195: 5252.
76 Pan, M., Zhang, H., and Li, K. (2015). *Rubber Chemistry and Technology* 88: 475.
77 Lin, C.W., Chien, C.H., Tan, J. et al. (2011). *Journal of Power Sources* 196: 1955.
78 Qiu, D., Liang, P., Peng, L. et al. (2020). *International Journal of Hydrogen Energy* 45: 5465.
79 Healy, J., Hayden, C., Xie, T. et al. (2005). *Fuel Cells* 5: 302.

8

Recent Advances in Microbial Fuel Cells for Sustainable Energy

Muhammad R. Sulaiman[1,2] *and Ram K. Gupta*[2]

[1] *Department of Plastic Engineering, Kansas Technology Center, Pittsburg State University, Pittsburg, KS, 66762, USA*
[2] *Department of Chemistry, Kansas Polymer Research Center, Pittsburg State University, Pittsburg, KS, 66762, USA*

8.1 Introduction

The growing global population gave rise to some serious challenges, including reliable energy supply, clean drinking water, fresh air to breathe, etc. These challenges need to be addressed as they would only get worse by delaying. The energy sources can be categorized into two main groups: renewable energy sources and nonrenewable energy sources. The nonrenewable sources include fossil-based fuels and nuclear energy. Fossil fuels are the conventional source of energy and have been utilized for decades. But, the adverse environmental effects and fast depleting oil reservoirs have started a new wave for exploring reliable energy sources. Similarly, harvesting energy via nuclear reactions was considered an excellent alternative to fossil fuels. But, substantial safety concerns related to the handling of spent fuel and radiations have been expressed. However, after the incident happened in a Japanese nuclear power plant in 2011, a plethora of questions have been raised regarding the reliability and safe implementation of nuclear technology for energy generation [1]. Therefore, a quest for finding a reliable and sustainable energy source was begun, which includes research about solar energy, wind energy, hydro energy, and various types of fuel cells, including microbial fuel cells (MFCs). These energy types are renewable and can be produced repeatedly over time.

Furthermore, in many world regions, the availability of clean water will be insufficient in the upcoming decades. Therefore, treating wastewater or wastewater recycling is among the methods that would help meet the demand. The wastewater treatment techniques that have been currently employed require considerable energy. The wastewater treatment system in the United States requires almost 2% of the total energy generated in the United States [2]. The required energy decreases the efficiency of the system, which makes the process expensive and infeasible. For this reason, various researches have been carried out to develop a process that is efficient enough to carry out on a large scale. At that time, the

Sustainable Energy Storage in the Scope of Circular Economy: Advanced Materials and Device Design, First Edition. Edited by Carlos Miguel Costa, Renato Gonçalves, and Senentxu Lanceros-Méndez.

discovery that microorganisms have the ability to generate electricity and treat wastewater simultaneously gave hope to the development of an efficient wastewater treatment system.

It all started when Potter, a British botanist, proposed a concept of generating electricity by employing microorganisms that help in oxidizing the organic molecules [3]. The proposal garnered tremendous attention as a potential technique for solving energy and wastewater-related problems. At the beginning of Potter's discovery, the technique was not entertained considerably. But, with the discovery that the electrons can be transferred directly to the electrodes by the microorganism in 1999, Potter's idea began to gain attention [4]. After that, the concept was employed practically for energy production and wastewater treatment. The last two decades witnessed immense attention of researchers, which resulted in the development of a sustainable energy source for electronic devices [5] and economic applications [6]. Potter's idea led to the development of a new type of fuel cell that can consume organic wastes present in the wastewater and transform them into electricity and also clean the water. This fuel cell was named MFCs.

In principle, various factors influence the productivity of the MFCs. These factors include the microorganism inoculum, the proton exchange membrane (PEM), the chemical materials present in the fuel, and the electrode material. The most important and most investigated factor is the electrode material since it has a direct impact on the productivity of the cell and commercializing process of MFCs. Till now, substantial research work has been done on electrode material to improve the MFC performance. But, before getting into the electrode materials, understanding the MFCs is significantly important. For this reason, architecture and the working principle of MFCs are discussed.

8.1.1 Introduction to Microbial Fuel Cells

The fuel cells are electrochemical devices that process the fuel and transform its chemical energy into electricity without undergoing combustion reaction. Different varieties of fuel cells have been researched, including proton-exchange membrane fuel cells (PEMFCs), direct methanol fuel cells (DMFCs), solid oxide fuel cells, alkaline fuel cells, and so on. MFC is another type of fuel cell that has been significantly studied. They have been produced by employing a PEM that separates cathodic and anodic compartments. The common PEMs used in MFCs are Nafion and poly(tetrafluoroethylene) (PTFE). The fuel for MFCs comprises organic material, usually some wastes, such as domestic wastewater or food sector industrial wastes. As the name suggests, MFCs employ microorganisms, which create a biofilm at the anode that works as the catalyst and oxidizes the organic matter of the fuel to produce electricity. The presence of O_2 in the anodic side resists the electron generation; therefore, an anaerobic environment needs to be provided. While at the cathode, an O_2 reduction reaction takes place, and water is produced. The schematic illustration of MFCs is shown in Figure 8.1.

8.1.2 Electron Transfer Mechanism

The power yield of MFCs is affected by numerous factors, such as membrane efficiency for H^+ ion transport, the organic materials contained by the wastewater, and electron transport (ET) rate (from microorganism to the anode) [8]. The bacterias employed for ET are known as exoelectrogens. These microorganisms can transport electrons from the oxidative

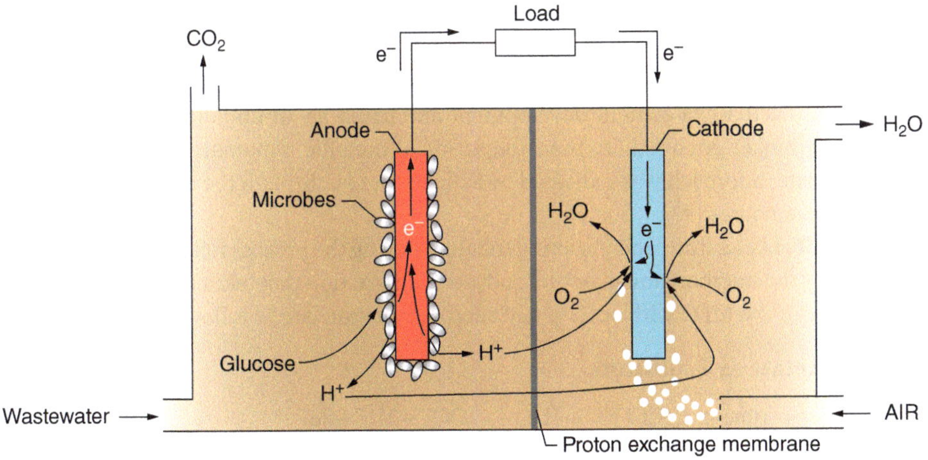

Figure 8.1 Schematic illustration of MFC [7]. *Source:* Ref. [7]/MDPI/CC BY-4.0.

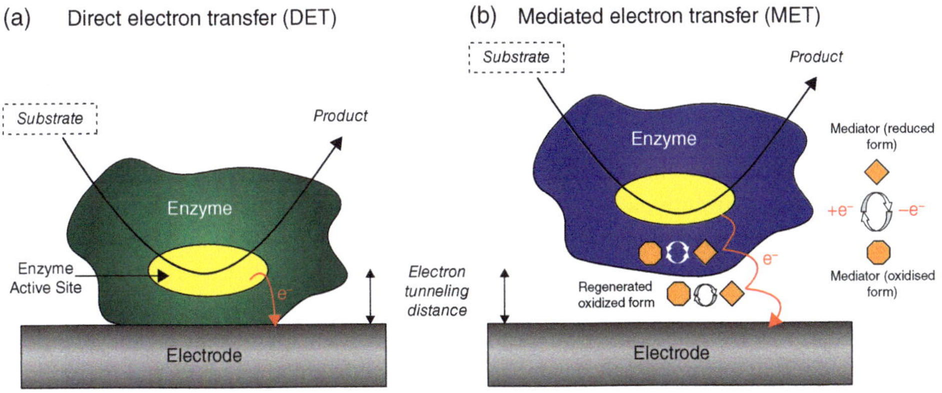

Figure 8.2 Diagrammatic illustration of an unsophisticated system of (a) direct electron transfer (DET) and (b) mediated electron transfer (MET) between the active spot of the microbe and the solid anode in the course of substrate oxidation. *Source:* Reproduced from Ref. [9]/With permission of Elsevier.

metabolic route to the surrounding. The well-known bacterias, which exhibit those characteristics, are called Geobacter and Shewanella. These bacterias can transfer electrons to their surroundings by three methods, such as direct ET, electron transfer with the help of the mediator, and ET via nanowires. The schematic illustration of direct electron transfer (DET) and mediated electron transfer (MET) is shown in Figure 8.2. The direct ET phenomenon is observed in Shewanella and Geobacter, which can transfer electrons to the anode without any help of the mediator. The cytochrome, which is the exterior layer on these bacterias, participates in transferring the electrons generated from NADH [10]. However, some bacterias classes, such as Pseudomonas and Shewanella, release particular chemicals; for example, flavins, which act as the shuttling molecule and carry out electron transportation from the exterior layer of the bacteria to the electrode [11].

Some bacterias also employ nanowires to directly transport electrons to the electrode, which includes some classes of Geobacter and Shewanella. These bacterias contain conductive appendages that help them conduct the electrons externally. These conductive nanowires are actually a cellular development and could be around 20 μm in length and report much higher conductivity than the artificial metallic nanomaterial [12]. Another electron transfer approach was indicated by Schaetzle et al. through the bacterial oxidation of expelled catabolites [13].

The quantity of electrons and protons generated during the process relies on the substrate consumed by the microorganism or biocatalysts. The reaction that takes place at the anodic and cathodic of the MFC with acetate and sucrose substrate are as follows:

a) **Using acetate as a substrate**

Anode compartment: $CH_3COO^- + H_2O \rightarrow 2CO_2 + 2H^+ + 8e^-$

Cathodic compartment: $O_2 + 4e^- + 4H^+ \rightarrow 2H_2O$

b) **Using sucrose substrate**

Anodic compartment: $C_{12}H_{22}O_{11} + 13H_2O \rightarrow 12CO_2 + 48H^+ + 48e^-$

Cathodic compartment: $O_2 + 4e^- + 4H^+ \rightarrow H_2O$

The produced H^+ ions are traveled from the anodic side to the cathodic side through the PEM. Meanwhile, the electrons get to the anode in two ways: with or without the help of a mediator. MFCs without the mediators are commercially promising and have significant advantages over conventional mediator-assisted MFCs since they do not need costly mediators to transfer electrons. Also, some of the mediators are toxic to the microorganism, which significantly lowers the cell performance.

Over the years, significant improvements in the MFCs technology have been witnessed. Despite that, various challenges still need to be addressed for increasing the practical application of that technology, for example, disturbance in both chambers and membrane resistivity during H^+ transfer. However, two main complications have been resolved regarding power production. (a) finding out that the power produced in MFC is directly related to the substrate concentration [14]. Therefore, during power production, higher than the specific value of the substrate concentration should be avoided [15]. (b) Since a significant portion of the total power generated in the MFCs is utilized to overcome the cell's internal resistance, the limited output is available in MFCs [15]. The main source of internal resistance (R_{int}) has been found to be the PEM that creates a partition between the anode and the cathode compartments of the MFCs. Also, the cathodes have been improved into biocathode, which can efficiently catalyze O_2 oxidation. Furthermore, various improvement factors have been investigated to strengthen the power production, such as designing MFCs with the elimination of PEM to reduce the R_{int} of the cell. Some PEM-free MFCs are single-chamber MFCs (SCMFCs), upflow MFC, and stacked MFCs [15]. Another type of MFCs is sediment MFCs. The soil and sediments are composed of wastes of living organisms, settled dead microorganisms and plankton, feces, and anthropogenic organic matters. In some places, these materials have enough carbon content to be utilized as a power production source. So, in MFCs, the anode is inserted into the anaerobic sediment, and the cathode remains in the overlaying water. Both the electrodes are connected to each other through an external circuit. The organic content in the sediment is eaten by the exoelectrogens and instantly transfer electrons outside the cell [16].

8.1.3 MFC Substrate

Substrates in MFCs are of great importance since they can influence the energy generation efficiency of the cell. In general, for energy production, the substrate can be simple to complex. The complex substrates contain complex organic contents that are good for the development of active microbes. In comparison, simple substrates are preferred for instant energy output. Glucose and acetate are the basic simple substrate for energy production. Other preferred substrates include lignocellulosic agricultural waste, brewery wastewater, starch processing water, landfill leachates, dye wastewater, and cellulose and chitin from industrial and municipal waste are some unusual substrates for energy generation from MFCs. The sanitary, food processing, and corn stovers wastewaters contain high organic content that feeds a great range of microorganisms in MFCs. Also, they contain specific microbes capable of separating sulfides, which are essential in treating wastewater [17]. Additionally, these substrates have a high quantity of growth facilitators that help grow electrochemically active microorganisms while treating the wastewater. This coexisting process minimizes the energy consumption for the treatment plant and also decreases the infeasible sludge generated by ongoing anaerobic production [18]. MFCs can efficiently treat leachate when attached in series along with producing electricity [19].

8.1.4 Electrode Materials

The efficient working of MFCs significantly relies on selecting an appropriate electrode material. The electrode materials have considerable influence on electrochemical efficiency, bacterial adhesion, and electron transfer. Various studies have been carried out to improve its power generation using various carbon materials, such as carbon fiber, carbon felt, carbon paper, and carbon nanotube-based composites. For practically implementing MFC, the price of the materials must be minimum, and the power output should be optimized. Excellent values of the properties, such as surface area, electrical conductivity, stability, and cost, are required while considering the material for anode and cathode.

In this chapter, the recent advances in electrode materials are discussed in detail. The effect of various electrode materials on power production and wastewater treatment is discussed.

8.2 Materials for Anode

In MFCs, the anode materials have significant importance due to their influence on the development of biofilms and ET from the microbe to the electron acceptors. Numerous carbon-based materials have been utilized as anode materials, including carbonaceous materials, metal and metal oxide-based, and conducting polymer. Recently, various enhanced materials, such as N_2-enriched materials, thermal-treated materials, chemical-treated materials, ammonia-modified materials, composite materials, have been investigated for the application of anode materials [20]. In this section, applications of these materials and their influence on MFC performance are discussed below in detail.

8.2.1 Conventional Carbonaceous Materials

Currently, the carbon-containing materials have the advantage of high surface area, less porosity, higher conductivity, excellent biocompatibility, high thermal stability, and superior ability of ET kinetics. Carbon-based materials, such as nanotubes, paper, cloth, meshes, brushes, felt, fiber, rod, reticulated vitreous carbon, and same graphitic structures, have already been employed [20, 21]. Among them, carbon cloth has extensively been utilized for the anode materials for MFCs. It exhibits a high surface area compared to the simple carbon paper, but it has superior porosity due to the presence of wide vacant cavities. Generally, carbon cloth exhibits superior electrical conductivity, exceptional flexibility, and improved mechanical durability compared to simple carbon paper. The unfavorable characteristics of the material are related to the elevated price and lack of stability against chemicals, which triggers fouling and decreases the prolonged durability of the anode [22].

Despite the superior porosity, carbon paper is not desired as the anode in MFCs because of its higher price. Carbon brushes are a very appealing material that exhibits excellent surface area and excellent area to volume ratio. They become conductive when associating with titanium and are primarily used as the anode in MFC. The ongoing research is being done to make it cost effective [23]. Similarly, carbon mesh is a low-priced material and is readily available commercially. It is less electrical conductive and has inadequate mechanical stability, which results in poor overall stability. It has the ability to be folded to produce three-dimensional (3D) high surface area electrodes. Carbon mesh was treated with heat, acetone, high temperature, and ammonia gas process by Wang et al. and employed as an anode in the MFC and found that it is the better alternative for high-priced carbon cloths and papers [24]. Another carbon-based material is carbon veils, which are cost effective and have moderate electrical conductivity and superior porosity, enhancing the MFC performance. The single carbon veil film has high brittleness and is usually folded to develop 3D anode material.

Moreover, carbon felts are another frequently utilized material for the MFC anode. It has superbly porous and has excellent electrical conductivity. Also, it is low cost and is mechanically stable based on material thickness [25]. Similarly, granular-activated carbon (GAC) having exceptional biocompatibility and low cost but have inadequate conductivity because of the superior porosity. The superior porosity restrains the electron transfer because of void spaces in the anode, which decreases the electrochemical activity. For addressing this issue, Borsje et al. investigated three GACs as MFC bioanodes for energy generation [26]. Two of them exhibit a high surface area and one has a low surface area. They noticed that the synthesized material produced energy 1.3 to 2 times more compared to the graphite one. Similarly, Li et al. utilized GAC as an MFC anode and obtained 2.5 times more power density compared to carbon cloths [27]. GAC was also installed as a double-chamber MFC and obtained high coulombic efficiency with superior energy output. During the study, it was found that the superior surface area of GAC resulted in the development of biofilm that surrounds the anode. Also, it was observed that the higher the oxidation rate of the substrate, the improved the whole energy production.

Graphite is another in the list of anode material for MFC. It has good biocompatibility, is mechanically strong, and has an adequate surface area that allows it to be employed as the anode in MFC. But, the drawbacks, such as higher price and lower conductivity, hinder it from meeting the commercial energy requirements. Commonly, graphite-based materials, such as plates, rods, and paper, have better efficiency as anode for MFC than plain carbon. In an experiment, Vijay et al. investigated the MFC performance with the anode made up of graphite felt [28]. For this experiment, he used the nuclear wastewater sludge as an inoculation source to eliminate the uranium. The biofilm development was observed on the anode that verifies that the bacterias are present on the anode, as shown in Figure 8.3. The uranium elimination rate was found to be 32% after 24 hours and 90% after 120 hours through this process. The obtained energy production was too little, such as 2.91 W/m^3, which is resulted from inadequately generated electrons. Moreover, bacterial development's durability on the graphite felt was not good since nuclear waste contains various unstable components. The optimum processing parameters for the pH and temperature and efficient anode materials are necessary for durable MFC operations industrially.

Other than graphite, an allotropic form of carbon called graphene has been extensively investigated as the anode in the MFCs. Graphene is a two-dimensional (2D) layered structure with excellent electrical conductivity, good mechanical characteristics, well thermally and chemically stability, and superior biocompatibility. Also, the diamagnetic

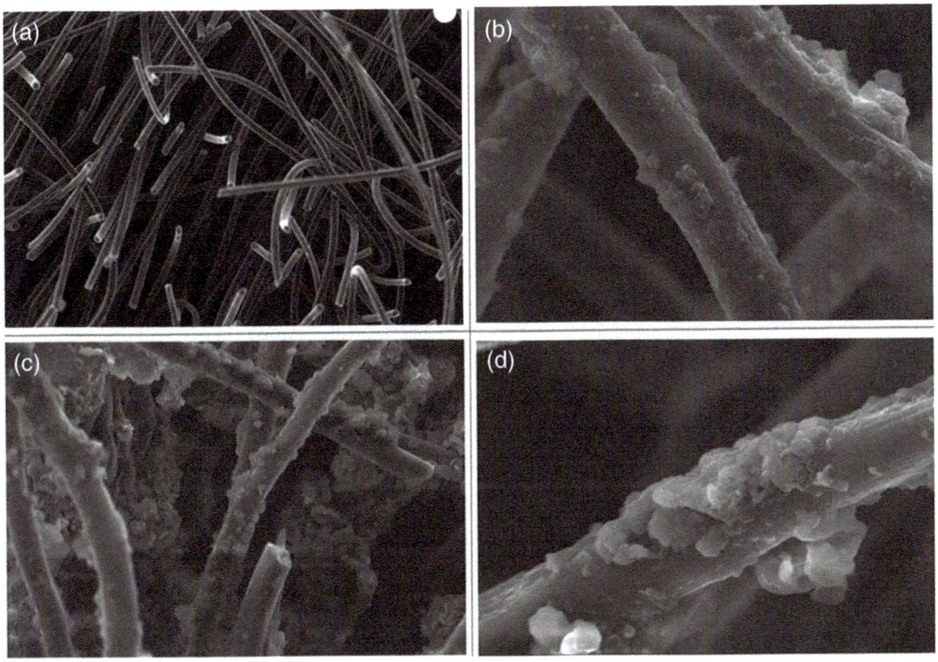

Figure 8.3 SEM images of graphite electrode-based biofilm (a) control MFC, (b)–(d) Test MFC at various magnification). *Source:* Vijay et al. [28]/With permission of Elsevier.

effect and nonlinear structure make it more effective than plain graphite [29]. Various researches have been done on employing graphene and its composites as MFC anode. For instance, a flexible 3D graphene sponge having micropores developed by an icy template has been utilized as an MFC anode [30]. The development of the 3D porous graphene structure via ice template at various ice crystal growth rates with the SEM images is presented in Figure 8.4. The micropores in the graphene sponge help attain superior power density than the carbon felt because of improved bacterial growth on the micropore-containing graphene sponge. But, the higher price of graphene sponge, such as \$2000/m^3, made it impracticable. Similarly, SnO_2 decorated reduced graphene oxide (rGO) was employed as the MFC anode and achieved five times better power density than the pure rGO [31]. The schematic illustration for the preparation of the SnO_2@rGO is shown in Figure 8.5. The high surface area and the electrical conductivity of SnO_2@rGO resulted in the excellent growth of bacteria and efficient electron transfer. Despite the exceptional energy production and wastewater treatment, graphene's higher price limits its wide-scale application impracticable. One technique to decrease the price of graphene-based materials is to prepare them from natural wastes. The materials obtained through this process were cost efficient and according to the primary requirement of the MFC anodes.

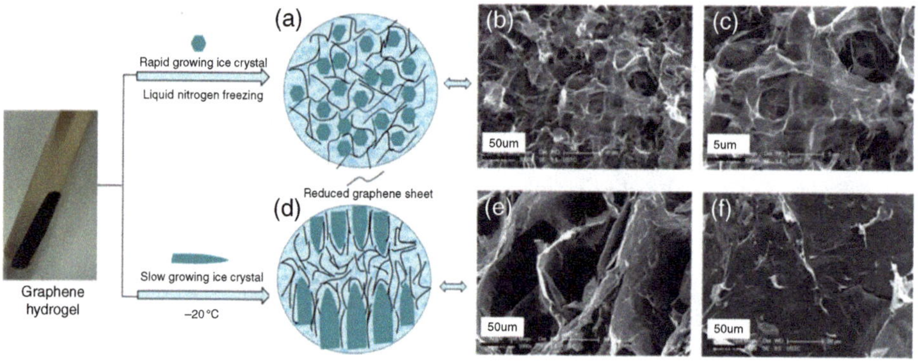

Figure 8.4 (a and d) Diagrammatic illustration of the preparation method of 3-dimensional porous graphene framework triggered by ice template at various development rates of ice crystals. SEM images of (b and c) graphene foam (e) graphene sponge (f) prepared by liquid nitrogen freezing (quick) and 10 C treatment in a refrigerator (slow), respectively. *Source:* Chen et al. [30]/With permission of Royal Society of Chemistry.

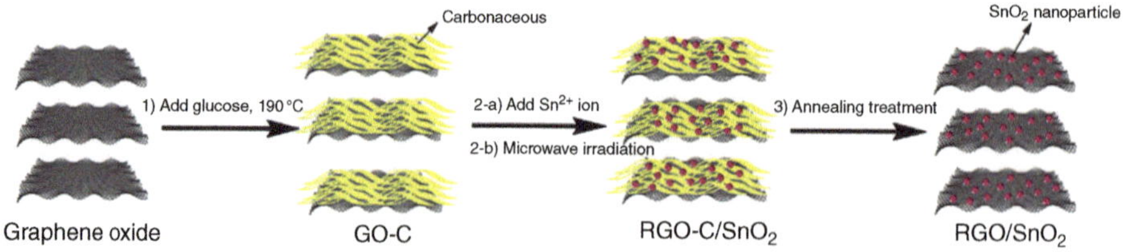

Figure 8.5 Diagrammatic representation of the RGO/SnO_2 synthesis method. *Source:* Reproduced from Ref. [31]/With permission Elsevier.

8.2.2 Metal and Metal Oxide-Based Anode for MFC

Metals and metal oxides have been known for better performances in fuel cells. In MFCs, most metals, such as Ag, Ni, Au, Al, Cu, Mo, Fe, Ti, and stainless steels (SS), have also displayed superior performance but also facing the durability issue because of their corrosiveness that restricts the microorganism adhesion and limits biocompatibility [32]. For instance, Mo-based anode material was prepared and utilized as the MFC anode by Yamashita et al. [33]. The power density of 1296 mW/m^2 was recorded, but due to its corrosive nature, the material displayed durability issues. Similarly, Yin et al. prepared 3D porous TiO_2 nanosized sheets. The material displayed superior active surface area that facilitated improving the electron transfer rate and diffuseness of molecules. However, when testing the pure Ti metal as an MFC anode, low adhesiveness of bacteria's and poor electron transportation via microorganism respiration were observed [34]. Different modification techniques have been developed to address the stability issue and utilize the superior conductivity of the metals. These techniques either employ carbon or conductive materials to modify the metal surface or undergo different treatments, such as ammonia, gas diffusion, and chemical oxidation. For this reason, the SS mesh treated with gas diffusion and utilized as the MFC anode by You et al., with an improved power density of 951.60 mW/m^2, was recorded [35]. Santoro et al. concluded their review with the remarks that the Au, Ag, Ti, and Pt metals can only be utilized as MFC anode for a short period because of their corrosive nature [22]. Furthermore, these metal are very expensive, which make them unfeasible for the practical application. For this reason, the transformation of the metals and metal oxide with the help of carbon-based material or conducting polymers can prove to be a big development regarding anode materials for MFCs. Until now, various metals, such as Zn, Cu, Ag, and metal oxides, like titanium oxide and zinc oxide, have been utilized to prepare the composite with conducting polymers like polycarbazole in order to improve MFC performance [36].

8.2.3 Natural Waste-Based Anode Material for MFC

Recently, MFCs anode materials have been prepared from natural biowastes, such as mushrooms and corn stems. The reason for extensively increasing interest in biowaste-derived anode materials is its various benefits, such as low cost, abundance, durability, and recyclability. Mostly, the biowastes are carbonized at various temperatures and environments that significantly improve the surface area and electrical conductivity of the material. The layered corrugated carbon (LCC) that are being utilized as the MFC anode offers an excellent layered surface area for bacterial growth. The improved biofilm development significantly improves the MFC performance in terms of current density. For instance, Chen et al. prepared low-cost LCC anode material and observed that the current density substantially increased from 200 A/m^2 to 390 A/m^2 with the increase of the number of layers from 3 to 6 [37]. The obtained current density from the synthesized LCC outperformed the regular graphite felt. The biowastes-derived anode electrodes for MFC are considered a proficient candidate due to their favorable characteristics and economic affordability. The anode materials with a 3D structure and meso and micropores were also produced via processing biowastes. The prepared material showed excellent electron transfer rate and electrochemical kinetics [38]. The 3D spongy materials are mostly prepared through the biowastes as the low cost and high activity materials.

Furthermore, the high surface area can significantly increase the anode kinetic, but the overall MFC activity is also affected by the internal resistance. So, maintaining low internal resistance is also a challenge for high-performing MFCs. For addressing this challenge, an anode material was synthesized by using Hibiscus Cannabinus Stem by Chen et al. [39]. The synthetic procedure involves the carbonization at the temperature of 1000°C in an N_2 atmosphere. The electrode was prepared by joining the SS wire with the prepared material. The Hibiscus Cannabinus maintained a 3D morphology with stainless steel wire and necessitated a mesopores-containing morphology for the better development of the bacterial film. The other variables, such as conductivity, resistance, durability, porosity, and biocompatibility, were found to be adequate for the MFC anode. The prepared material having the dimension of 10 nm in diameter and 15 nm in length only displayed the internal resistance of 10 Ω with the 3-times increment in the power density compared to the regular graphite anode. In a similar investigation, the pyrolyzed wild mushrooms, king mushrooms, and corn were employed as the anode in MFC and compared their results with the regular graphite anode [38]. The obtained results showed that the corn-based material had eight times better activity compared to the regular graphite anode. All the prepared material displayed highly porous structures with large surface area, as displayed in Figure 8.6.

Similarly, a tremendous amount of research has also been carried out on forest waste, papers, corn straw, various shells, such as almonds, and chestnuts, and silk cocoon. The biowastes used for the preparation of anode largely rely on the abundance of the plant in the particular region. For example, in Malaysia, palm oil is the abundant biowaste. Recently, bamboo charcoal was prepared by Jun et al. and employed as an MFC anode and resulted in 50% improved efficiency compared to the graphite [40]. Similarly, Hung et al. investigated the application of pyrolyzed coffee waste as the porous material for the MFC anode [41]. The schematic illustration for the preparation of the material is shown in Figure 8.7. Coffee has been largely utilized globally, and the produced waste is estimated to be six million tons each year. Therefore, synthesizing MFC anode material from coffee wastes can not only provide a renewable energy source but also help in dealing with coffee waste disposal. The team tested the material as the anode and obtained the power density of 3927 mW/m^2. There have been various biowaste materials that are not yet investigated for the MFC anode. Moreover, most of the investigated biowaste materials have not shown adequate electrical conduction efficiency to fulfill the current energy requirement for MFC.

Therefore, carbon and its allotropic forms, such as graphene, are considered the best material for the MFC anode because of their superior conductivity, high thermal and mechanical durability, biocompatibility, and abundance, etc. However, graphene can also be produced from waste materials by various techniques. Among them, the Hummer process is the general approach used for graphene preparation. The recent constraints of biowaste-based anode material, such as poor conductivity, can be enhanced by producing the composite using metal oxides with graphene-based materials. Thus, graphene-based composites, like GO/CuO, enhance the power density of MFC.

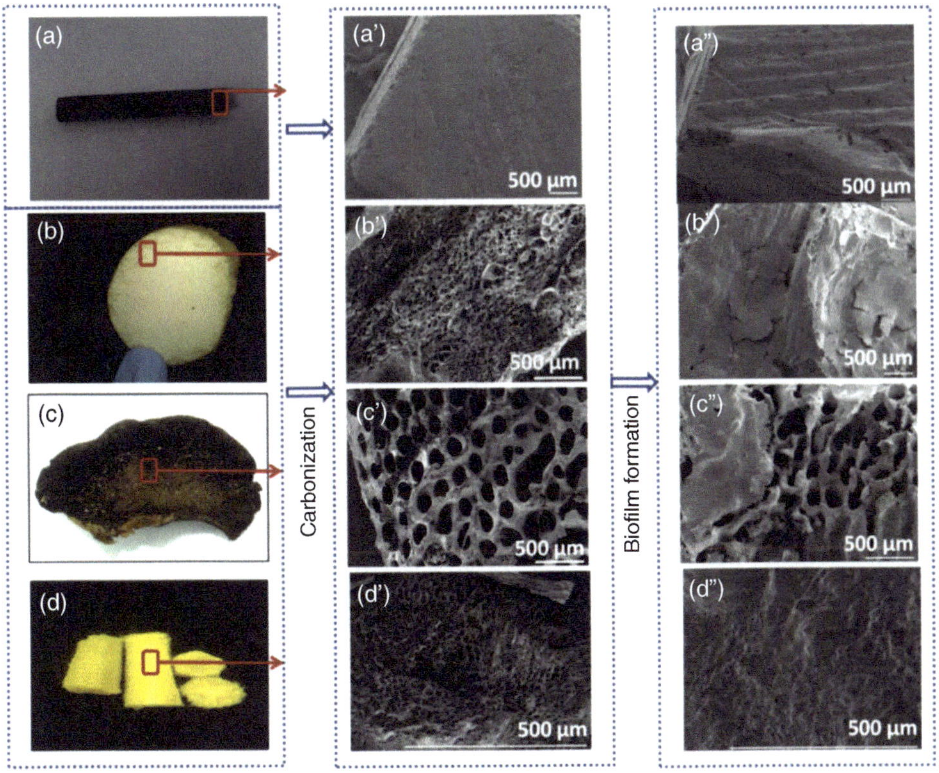

Figure 8.6 Electrode materials based on natural materials (a) graphite, (b) King mushroom, (c) wild mushroom, and (d) corn stem, with a marker that shows the analyzed portion, FE-SEM images of carbon electrodes after carbonization, and biofilm formation. *Source:* Karthikeyan et al. [38]/With permission of Elsevier.

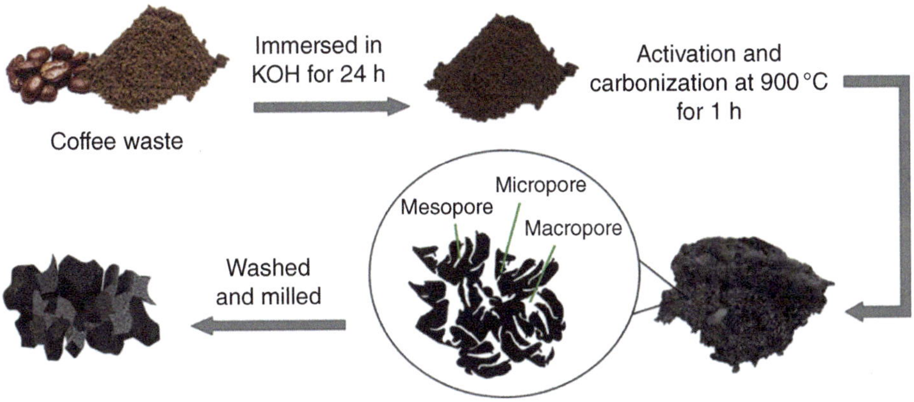

Figure 8.7 Schematic illustration of coffee waste-based activated carbon. *Source:* Hung [41]/With permission of American Chemical Society.

8.2.4 Modification Approaches for MFC Anode

The materials for the application of MFC need to meet various requirements, such as biocompatibility, superior conductivity, high surface area, and bacterial adhesion. The carbon, metals, and biowastes-derived material exhibit significantly improved MFC performance but lacked some of the necessary properties required for the long-run MFC process. Various modification approaches have been developed to improve their MFC performance, which helps achieve higher electricity production at higher stability. These modification approaches include treating the surface of the material, coating the anode with more stable materials, and making various composites, like graphene–TiO_2 and carbon felt-PANI. The surface modification can be done by using ammonia, thermal, acid, and electrochemical processes. However, different types of coating and composites are exceptional techniques to modify MFC productivity.

The ammonia-based surface treatment was carried out at 5% ammonia gas at 700 °C in a helium atmosphere. The power density of the synthesized material improved to 1970.0 mW/m^2 from 1640.0 mW/m^2 [42]. With the increased power density, the startup time also decreased significantly. The ammonia facilitates the development of the positive charge on the anode surface, which attracts the negatively charged bacterial cell wall. In this way, at the large surface area of the material, the ammonia treatment promotes adhesion and support fast electron transfer. Similarly, heat treatment is another surface modification method. However, this method only has a small contribution to the MFC performance because of the difficulty in adjusting the material structure only by thermal control at a specific temperature. In a study, Wang et al. observed only a 3% increase in the power density of the carbon mesh after heat treatment [43]. Moreover, treating the surface with the acid is considered a facile method, promoting efficient electron transfer by inserting the anode between the electrode surface and the bacteria. The surface treatment was done on the graphite felt via electrochemical oxidation. A 40% increase in power density was observed, which is attributed to high electron transfer from the bacteria to the anode.

Surface coating is another method for improving the surface properties of the MFC anode. The common coating materials are the conducting polymers, CNTs, metals, and various composites that facilitate enhance the MFC performance. Among others, CNTs-containing materials are considered efficient for coating and help in reducing the internal resistance and increasing the ET [44]. Similarly, coating a conducting polymer on the anode material significantly enhances the MFC performance, which is due to the increased surface area and the enhanced accumulated electrons on the anode. Moreover, the metal oxidation coating can also be done to achieve high MFC performance. In an investigation, the graphite rod was coated with the Au and Pd and received an exceptional 20 times higher current density with Au-coated electrons compared to plain graphite [45]. In the case of Pd coating, the material achieved 50% to 150% superior current density than the plain graphite rod.

Recently, conducting polymers have garnered tremendous attention for the modification of MFC anode material through coating or by making composites. The CPs facilitate in improving the electron transfer rate and surface area. For increasing the energy production efficiency, the conducting polymers are mixed with other nanosized materials, like metal and metal oxides. Among others, polyaniline, polypropylene, and PEDOT are extensively being used to serve the purpose. In an experiment, Hou et al. prepared an rGO and PANI composite and coated it on the carbon cloth [46]. The synthesized material displayed a superior power density of 1326 mW/m^2 compared to the pure rGO anode, attributed to the exceptional rGO conductivity and availability of high surface area to the PANI that further increase the MFC productivity. In another study, the SS-based MFC anode material was coated with PANI via electropolymerization technique [47]. The material exhibited excellent conductivity and offered superior current and power density of 1.4 A/m^2 and 780 mW/m^2, respectively.

Another approach for increasing the anode activity is by employing composite materials. The conventional MFC anode faced some serious challenges, such as expensiveness, lower electric conduction, and the abundance of materials. Recent researches indicated that composite materials exhibit excellent properties that can overcome challenges faced by the MFC anode. Much attention has been given to the metal–carbon composites, which include many polymeric composites, metal–graphite, and metal–carbon cloth composites. Various synthetic approaches for the preparation of composites materials have been employed. Among them, hydrothermal, solvothermal, and sol–gel methods are used commonly. Gong et al. indicated the excellence of the solvothermal method compared to the hydrothermal method, as displayed in Figure 8.8 [48]. The material $CuCo_2S_4$-rGO nanocomposite displayed exceptional electrochemical characteristics via both techniques.

Similarly, TiO_2 nanotubes/carbon paper-based composite material was synthesized via the sol–gel method. The synthesized MFC anode exhibited a superior surface area that facilitates higher bacterial adhesion and promotes efficient electron transfer [34]. Moreover, carbon brushes were altered with the help of Fe/Ni/AC as a modification agent by Liu et al. [49]. A superior energy density of 1092.0 mW/m^2 was achieved. Furthermore,

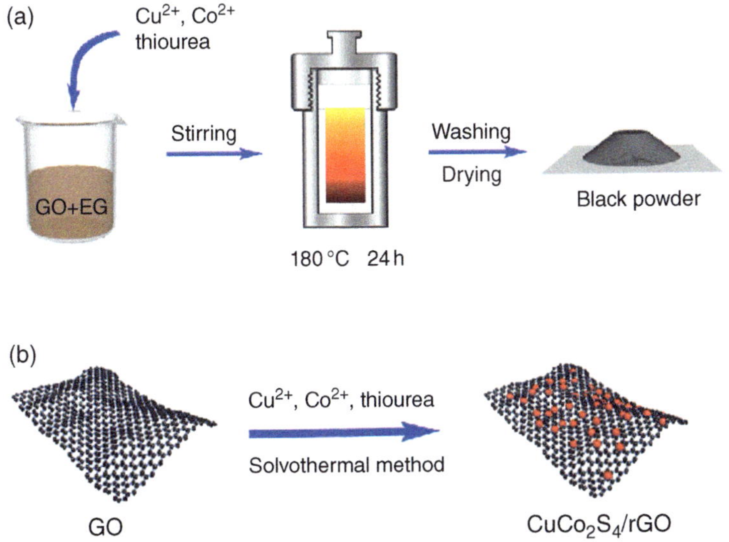

Figure 8.8 (a) Simple synthesis process and (b) the schematic illustration for the formation of CuCo2S4/rGO. *Source:* Reproduced from Ref. [48]/With permission of Elsevier.

composites based on conducting polymers were also widely researched. The MFC composite anode material based on PANI-natural loofah sponge was synthesized by Yuan et al. [50]. The prepared material displayed excellent power output of 1090.72 mW/m^2 and paved the way for natural material utilization to produce efficient MFC anodes.

8.3 Materials for Cathode

Cathode materials are also important in MFC because they catalyze O_2 reduction reactions (ORR) and influence overall cell efficiency. The O_2 reduction occurs at a three-phase interface, such as air, electrolyte, and electrode. The O_2 reduction can proceed via a four-electron transfer route, which produces H_2O, and a two-electron transfer route, which results in the production of H_2O_2. However, a four-electron transfer route is preferred since H_2O_2 compromises the electrode activity. Furthermore, MFC's cathode material should have superior mechanical properties, excellent catalytic activity, and higher ionic and electronic conductivity. The poor O_2 reduction reaction at neutral pH and lower temperature require higher overpotential, leading to a significant limitation on the MFC activity. Various materials have been used as the MFC cathode to enhance the performance of the cell. These materials are overviewed in Sections 8.3.1–8.3.4.

8.3.1 Pt-Based Cathode

Till now, Pt-based catalysts have been identified as the state of the art for the O_2 reduction reactions. But the higher price and scarcity of the material restrict its widespread application. Pt offers high surface area and low overpotential that make their application inevitable. Recently, researchers are working on either producing Pt-free cathodes or trying to reduce the quantity of Pt used in the catalyst by making composites or layering the different materials with Pt. For instance, Pt-deposited CNTs were prepared by Ghasemi et al. and obtained an increase in ORR activity [51]. Also, it was noted that the reduction in the Pt loading by 20 times did not significantly affect the ORR performance.

8.3.2 Nonprecious Metal Cathode

Various techniques and materials have been employed to fabricate nonprecious metal-based cathode for ORR. These materials usually contain transition metals, which are much cheaper than precious metals. For this reason, Yuan et al. took the benefit of π-π correlation of the metal and carbon and produced amino-functionalized CNTs with the Fe phthalocyanine support [52]. The material showed excellent electron transfer, and higher power output was achieved compared to the Pt–C cathodes.

Transition metal oxides are another important material that displayed tremendous potential as the cathode for MFC. For example, Morris et al. prepared a cathode material based on lead oxide [53]. They observed that the ORR performance of the material increased fourfolds to that of Pt/C catalyst, while the prepared lead oxide electrode had price eight times lower than that of Pt-catalysts.

8.3.3 Biocathodes

Due to the significant challenges faced by the conventional catalysts, such as high price, environmental friendliness, and complex fabrication process, the biocathodes have been produced. The biocathodes contain microbes that work as the catalyst and reduce O_2 by forming a bacterial film on the cathode surface. Typically, biocathodes can be classified into two categories, such as aerobic and anaerobic. In aerobic biocathodes, O_2 acts as an electron acceptor and produces H_2O_2 intermediate while utilizing transition metals, like Fe, Mn, as the electron transfer mediator. The electrons generated by the biofilm reduce the Fe(III) into Fe(II), which later got oxidized by O_2. However, in anaerobic biocathodes, nitrates and sulfates act as electron acceptors instead of O_2. In which the biofilm reduces the nitrates and sulfides. The potential for Mn, Fe, and nitrate is similar to that of O_2. However, the use of anaerobic biocathodes helps to avoid losing the electrons via O_2 that can penetrate into the anode via PEM.

8.3.4 Metal-Free Cathode

Recently, extensive researches have been carried out to find the best alternative to expensive metal-based cathodes. For this reason, various metal-free electrodes have been prepared that offer similar activity but at a lower price. The metal-free electrocatalysts are mostly doped with N_2 to increase catalytic activity and carry out electrons transfer via a four-electron transfer route. In a study, Gong et al. produced N_2-doped CNTs that showed an increased ORR activity, stability, and low overpotential as compared to the conventional Pt/C electrocatalyst [54]. In another study, N_2-doped graphene nanosheets achieved excellent electrocatalytic activity with a current output of about 8.75 mA/cm^2, superior to platinum/carbon catalysts.

8.4 Conclusion

The productivity of the MFC heavily relies on the electrode materials. The anode material for MFC should have unique properties, such as biocompatibility, excellent bacterial adhesion, superior activity, large surface area, and good conduction. For this reason, much importance has been given to the anode material and widely researched for better stability and cheap materials. Furthermore, materials that are toxic to the cells cannot be employed as an anode regardless of their other properties. Therefore, carbon-based materials are extensively used because they fulfill all the requirements and are cost effective. Surface treatment is another approach to improve the activity of the material. Similarly, efficient cathode materials are also required for improved ORR kinetics. Despite the higher efficiency of the Pt-based catalysts, the current research concentrates on finding cost effective and efficient alternatives. Finding efficient electrode materials not only increases the overall cell kinetics but also facilitates the widespread application of MFC.

References

1 Funabashi, Y. and Kitazawa, K. (2012). Fukushima in review: a complex disaster, a disastrous response. *The Bulletin of the Atomic Scientists* 68: 9–21.

2 United States Environmental Protection Agency. Energy efficiency for water utilities. https://www.epa.gov/sustainable-water-infrastructure/energy-efficiency-water-utilities (accessed 26 November 2021).

3 Potter, M.C. (1911). Electrical effects accompanying the decomposition of organic compounds. *Proceedings of the Royal Society of London. Series B, Containing Papers of a Biological Character* 84: 260–276.

4 Kim, B.H., Kim, H.J., Hyun, M.S., and Park, D.H. (1999). Direct electrode reaction of Fe(III)-reducing bacterium, Shewanella putrefaciens. *Journal of Microbiology and Biotechnology* 9: 127–131.

5 Shukla, A.K. and Suresh, P. (2004). Biological fuel cells and their applicaitons. *Current Science* 87: 14.

6 Lovley, D.R. (2006). Bug juice: harvesting electricity with microorganisms. *Nature Reviews. Microbiology* 4: 497–508.

7 Włodarczyk, P.P. and Włodarczyk, B. (2019). Wastewater treatment and electricity production in a microbial fuel cell with Cu–B alloy as the cathode catalyst. *Catalysts* 9: 12–15.

8 Liu, H., Cheng, S., and Logan, B.E. (2005). Power generation in fed-batch microbial fuel cells as a function of ionic strength, temperature, and reactor configuration. *Environmental Science & Technology* 39: 5488–5493.

9 Leech, D., Kavanagh, P., and Schuhmann, W. (2012). Enzymatic fuel cells: recent progress. *Electrochimica Acta* 84: 223–234.

10 Yang, Y., Xu, M., Guo, J., and Sun, G. (2012). Bacterial extracellular electron transfer in bioelectrochemical systems. *Process Biochemistry* 47: 1707–1714.

11 Yu, J., Cheng, G., and Luo, W. (2017). Ternary nickel-iron sulfide microflowers as a robust electrocatalyst for bifunctional water splitting. *Journal of Materials Chemistry A* 5: 15838–15844.

12 Malvankar, N.S., Vargas, M., Nevin, K.P. et al. (2011). Tunable metallic-like conductivity in microbial nanowire networks. *Nature Nanotechnology* 6: 573–579.

13 Schaetzle, O., Barrière, F., and Schröder, U. (2009). An improved microbial fuel cell with laccase as the oxygen reduction catalyst. *Energy & Environmental Science* 2: 96–99.

14 Chen, G.W., Choi, S.J., Lee, T.H. et al. (2008). Application of biocathode in microbial fuel cells: cell performance and microbial community. *Applied Microbiology and Biotechnology* 79: 379–388.

15 Sharma, Y. and Li, B. (2010). The variation of power generation with organic substrates in single-chamber microbial fuel cells (SCMFCs). *Bioresource Technology* 101: 1844–1850.

16 Kim, B.H., Doo Hyun Park, P.K.S., and, Kim, H.J. (1999). Mediator-less biofuel cell. US Patent 5,976,719.

17 Rabaey, K., Van De Sompel, K., Maignien, L. et al. (2006). Microbial fuel cells for sulfide removal. *Environmental Science & Technology* 40: 5218–5224.

18 Liu, H., Ramnarayanan, R., and Logan, B.E. (2004). Production of electricity during wastewater treatment using a single chamber microbial fuel cell. *Environmental Science & Technology* 38: 2281–2285.

19 Rodrigues, P.C., Lisboa-Filho, P.N., Mangrich, A.S., and Akcelrud, L. (2005). Polyaniline/polyurethane networks. II. A spectroscopic study. *Polymer* 46: 2285–2296.

20 Palanisamy, G., Jung, H.Y., Sadhasivam, T. et al. (2019). A comprehensive review on microbial fuel cell technologies: Processes, utilization, and advanced developments in electrodes and membranes. *Journal of Cleaner Production* 221: 598–621.

21 Kumar, G.G., Sarathi, V.G.S., and Nahm, K.S. (2013). Recent advances and challenges in the anode architecture and their modifications for the applications of microbial fuel cells. *Biosensors & Bioelectronics* 43: 461–475.

22 Santoro, C., Arbizzani, C., Erable, B., and Ieropoulos, I. (2017). Microbial fuel cells: From fundamentals to applications. A review. *Journal of Power Sources* 356: 225–244.

23 Cai, T., Meng, L., Chen, G. et al. (2020). Application of advanced anodes in microbial fuel cells for power generation: a review. *Chemosphere* 248: 125985.

24 Wang, K., Liu, Y., and Chen, S. (2011). Improved microbial electrocatalysis with neutral red immobilized electrode. *Journal of Power Sources* 196: 164–168.

25 Huong Le, T.X., Bechelany, M., and Cretin, M. (2017). Carbon felt based-electrodes for energy and environmental applications: a review. *Carbon (NY)* 122: 564–591.

26 Borsje, C., Liu, D., Sleutels, T.H.J.A. et al. (2016). Performance of single carbon granules as perspective for larger scale capacitive bioanodes. *Journal of Power Sources* 325: 690–696.

27 Li, F., Yarnykh, V.L., Hatsukami, T.S. et al. (2010). Scan-rescan reproducibility of carotid atherosclerotic plaque morphology and tissue composition measurements using multicontrast MRI at 3T. *Journal of Magnetic Resonance Imaging* 31: 168–176.

28 Vijay, A., Khandelwal, A., Chhabra, M., and Vincent, T. (2020). Microbial fuel cell for simultaneous removal of uranium(VI) and nitrate. *Chemical Engineering Journal* 388: 124157.

29 Haseen, U., Umar, K., Ahmad, H. et al. (2020). Graphene and its composites: applications in environmental remediation. In: *Modern Age Waste Water Problems*, 85–91. Springer.

30 Chen, W., Huang, Y.X., Li, D.B. et al. (2014). Preparation of a macroporous flexible three dimensional graphene sponge using an ice-template as the anode material for microbial fuel cells. *RSC Advances* 4: 21619–21624.

31 Mehdinia, A., Ziaei, E., and Jabbari, A. (2014). Facile microwave-assisted synthesized reduced graphene oxide/tin oxide nanocomposite and using as anode material of microbial fuel cell to improve power generation. *International Journal of Hydrogen Energy* 39: 10724–10730.

32 Rahimnejad, M., Adhami, A., Darvari, S. et al. (2015). Microbial fuel cell as new technol ogy for bioelectricity generation: a review. *Alexandria Engineering Journal* 54: 745–756.

33 Yamashita, T. and Yokoyama, H. (2018). Molybdenum anode: a novel electrode for enhanced power generation in microbial fuel cells, identified via extensive screening of metal electrodes. *Biotechnology for Biofuels* 11: 1–13.

34 Yin, T., Lin, Z., Su, L. et al. (2015). Preparation of vertically oriented TiO_2 nanosheets modified carbon paper electrode and its enhancement to the performance of MFCs. *ACS Applied Materials & Interfaces* 7: 400–408.

35 You, S.J., Wang, X.H., Zhang, J.N. et al. (2011). Fabrication of stainless steel mesh gas diffusion electrode for power generation in microbial fuel cell. *Biosensors & Bioelectronics* 26: 2142–2146.

36 Dominguez-Benetton, X., Srikanth, S., Satyawali, Y. et al. (2013). Enzymatic electrosynthesis: an overview on the progress in enzyme-electrodes for the production of electricity, fuels and chemicals. *Journal of Microbiology Biochemistry Technology* S6: 007.

37 Chen, S., He, G., Liu, Q. et al. (2012). Layered corrugated electrode macrostructures boost microbial bioelectrocatalysis. *Energy & Environmental Science* 5: 9769–9772.

38 Karthikeyan, R., Wang, B., Xuan, J. et al. (2015). Interfacial electron transfer and bioelectrocatalysis of carbonized plant material as effective anode of microbial fuel cell. *Electrochimica Acta* 157: 314–323.

39 Chen, S., Hou, H., Harnisch, F. et al. (2011). Electrospun and solution blown three-dimensional carbon fiber nonwovens for application as electrodes in microbial fuel cells. *Energy & Environmental Science* 4: 1417–1421.

40 Zhang, J., Li, J., Ye, D. et al. (2014). Tubular bamboo charcoal for anode in microbial fuel cells. *Journal of Power Sources* 272: 277–282.

41 Hung, Y.H., Liu, T.Y., and Chen, H.Y. (2019). Renewable coffee waste-derived porous carbons as anode materials for high-performance sustainable microbial fuel cells. *ACS Sustainable Chemistry & Engineering* 7: 16991–16999.

42 Cheng, S. and Logan, B.E. (2007). Ammonia treatment of carbon cloth anodes to enhance power generation of microbial fuel cells. *Electrochemistry Communications* 9: 492–496.

43 Wang, X., Cheng, S., Feng, Y. et al. (2009). Use of carbon mesh anodes and the effect of different pretreatment methods on power production in microbial fuel cells. *Environmental Science & Technology* 43: 6870–6874.

44 Sun, J.J., Zhao, H.Z., Yang, Q.Z. et al. (2010). A novel layer-by-layer self-assembled carbon nanotube-based anode: preparation, characterization, and application in microbial fuel cell. *Electrochimica Acta* 55: 3041–3047.

45 Fan, Y., Xu, S., Schaller, R. et al. (2011). Nanoparticle decorated anodes for enhanced current generation in microbial electrochemical cells. *Biosensors & Bioelectronics* 26: 1908–1912.

46 Hou, J., Liu, Z., and Zhang, P. (2013). A new method for fabrication of graphene/polyaniline nanocomplex modified microbial fuel cell anodes. *Journal of Power Sources* 224: 139–144.

47 Sonawane, J.M., Al-Saadi, S., Singh Raman, R.K. et al. (2018). Exploring the use of polyaniline-modified stainless steel plates as low-cost, high-performance anodes for microbial fuel cells. *Electrochimica Acta* 268: 484–493.

48 Gong, Y., Zhao, J., Wang, H., and Xu, J. (2018). $CuCo_2S_4$/reduced graphene oxide nanocomposites synthesized by one-step solvothermal method as anode materials for sodium ion batteries. *Electrochimica Acta* 292: 895–902.

49 Liu, Y., Fan, Y.S., and Liu, Z.M. (2019). Pyrolysis of iron phthalocyanine on activated carbon as highly efficient non-noble metal oxygen reduction catalyst in microbial fuel cells. *Chemical Engineering Journal* 361: 416–427.

50 Yuan, Y., Zhou, S., Liu, Y., and Tang, J. (2013). Nanostructured macroporous bioanode based on polyaniline-modified natural loofah sponge for high-performance microbial fuel cells. *Environmental Science & Technology* 47: 14525–14532.

51 Ghasemi, M., Shahgaldi, S., Ismail, M. et al. (2011). Activated carbon nanofibers as an alternative cathode catalyst to platinum in a two-chamber microbial fuel cell. *International Journal of Hydrogen Energy* 36: 13746–13752.

52 Yuan, Y., Zhao, B., Jeon, Y. et al. (2011). Iron phthalocyanine supported on amino-functionalized multi-walled carbon nanotube as an alternative cathodic oxygen catalyst in microbial fuel cells. *Bioresource Technology* 102: 5849–5854.

53 Morris, J.M., Jin, S., Wang, J. et al. (2007). Lead dioxide as an alternative catalyst to platinum in microbial fuel cells. *Electrochemistry Communications* 9: 1730–1734.

54 Gong, K., Du, F., Xia, Z. et al. (2009). Nitrogen-doped carbon nanotube arrays with high electrocatalytic activity for oxygen reduction. *Science (80-)* 323: 760–764.

Part IV

Sustainable Energy Storage Devices and Device Design

9

Multifunctional Sustainable Materials for Energy Storage

Michael Thielke and Ana J. Sobrido

School of Engineering and Materials Science, Chemical Engineering and Renewable Energy, Faculty of Science and Engineering, Queen Mary University of London, Mile End Road, London, E1 4NS, UK

9.1 Redox Flow Batteries as Alternative Energy Storage Technology for Grid-Scale and Off-Grid Applications

Sustainable energy storage plays a key role in the circular economy, underpinned by a transition to renewable energies and sustainable materials and devices. With an increased global focus on renewable energy generation through wind and solar, the need for low-cost sustainable energy storage is greater than ever. With the drastic fall in lithium-ion battery cost in recent years – 97% decrease in the last decades, i.e. 1 KWh cost US$ 7500 in 1991 to just US$ 181 in 2018 [1] lithium batteries have found a market in stationary storage applications, even though they are not particularly well suited to medium-to-large-scale grid storage (10 seconds of MWh), due to safety, lifetime, and cost issues. One of the most promising alternatives to medium-to-large energy storage technology is the redox flow battery (RFB). Compared to Li-ion batteries, RFBs exhibit high round trip efficiency, depth of discharge, fast responsiveness, reduced environmental impact, and the ability to decouple power and energy capacity [2, 3].

In a typical RFB, anolyte and catholyte are pumped through half-cells, separated from each other by an ion-exchange membrane. The electrodes, commonly carbon felts or papers, are crucial component of the RFBs, and their chemical and physical properties play a major role in both performance and durability of the battery (Figure 9.1). Among the most popular RFB chemistries is the all-vanadium redox flow battery (VRFB), pioneered by M. Skyllas-Kazacos et al. [4] This is due to their high reversibility, round-trip energy efficiency (EE) of about 85%, large power output, and reduction of crossover or cross-contamination, as it employs the same element in both anolyte and catholyte. A VRFB uses the four oxidation states of vanadium – V(II), V(III), V(IV), and V(V) and sulfuric acid as supporting electrolyte. Catholyte and anolyte solutions with the dissolved vanadium species are stored in two external tanks from which the redox active species flow to the core

Sustainable Energy Storage in the Scope of Circular Economy: Advanced Materials and Device Design, First Edition. Edited by Carlos Miguel Costa, Renato Gonçalves, and Senentxu Lanceros-Méndez.

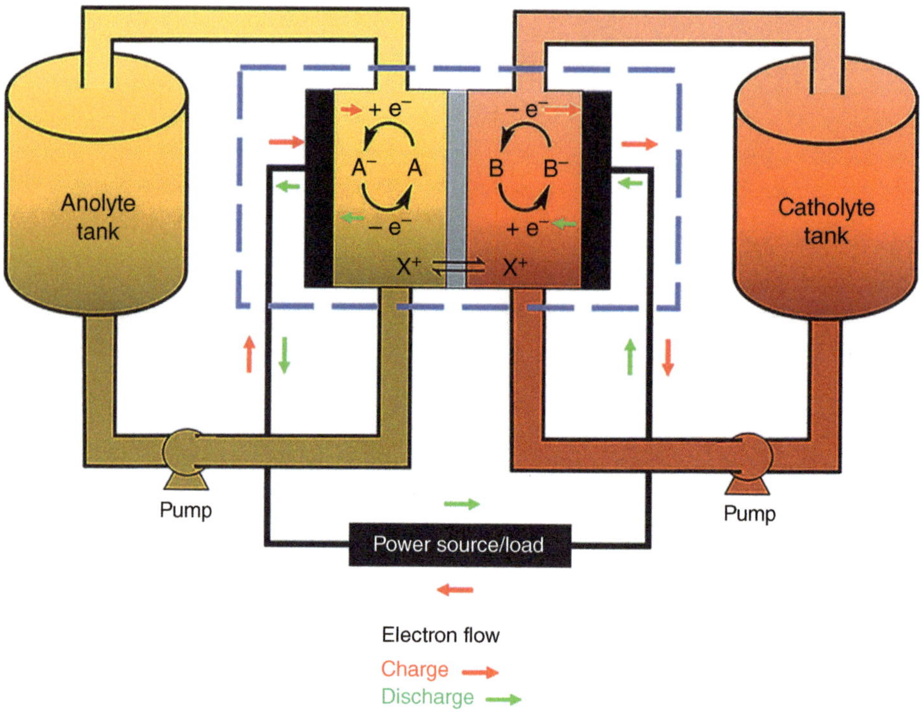

Figure 9.1 Schematic representation of a RFB. *Source:* Ref. [3]/With permission of Royal Society of Chemistry.

of the electrochemical cell, where the reactions take place, by using two peristaltic pumps. The reactions taking place in a VRFB during charge and discharge are detailed below:

Discharge

Positive electrode: $VO_2^+ + 2H^+ + e^- \rightarrow VO^{2+} + H_2O$

Negative electrode: $V^{2+} \rightarrow V^{3+} + e^-$

Charge

Positive electrode: $VO^{2+} + H_2O \rightarrow VO_2^+ + 2H^+ + e^-$

Negative electrode: $V^{3+} + e^- \rightarrow V^{2+}$

The standard open circuit voltage of a VRF is 1.26V at 25 °C and 50% state of charge (SoC), but due to ohmic losses and other side effects, the real voltage at 50% SoC is 1.4 and 1.6V at 100% SoC.

When the vanadium species are pumped through the electrochemical cell, they come into contact with the electrodes, where the redox reactions take place. In the typical configuration, an ionic conducting membrane (Nafion®) separates positive and negative electrode and enables the transport of H^+ during charge and discharge.

Although VRFBs continue to be the most extended chemistry for commercial rRFBs, alternative chemistries have been explored in the last decade, as vanadium is considered a strategic material, only found in a few places around the world (South Africa, Northwestern China, and Eastern Russia). Among the most developed is the zinc–bromine redox flow battery (ZBFB), a type of hybrid RFB, which uses a zinc metal negative electrode and bromine (Br_2) in the positive electrode (Figure 9.2) [5–8]. In a ZBFB, a solution of zinc bromide is stored in catholyte and anolyte is pumped through the reactor where the redox reactions take place and back into the tanks. During charge, Zn is plated from the electrolyte onto the negative electrode, whereas bromide is converted to bromine at the positive electrode and stored in a thick oil in a separate part of the tank. On discharge, the reverse process takes place and the metallic zinc dissolves in the electrolyte and bromine converts into bromide. The discharge process generates 1.8 V across a cell.

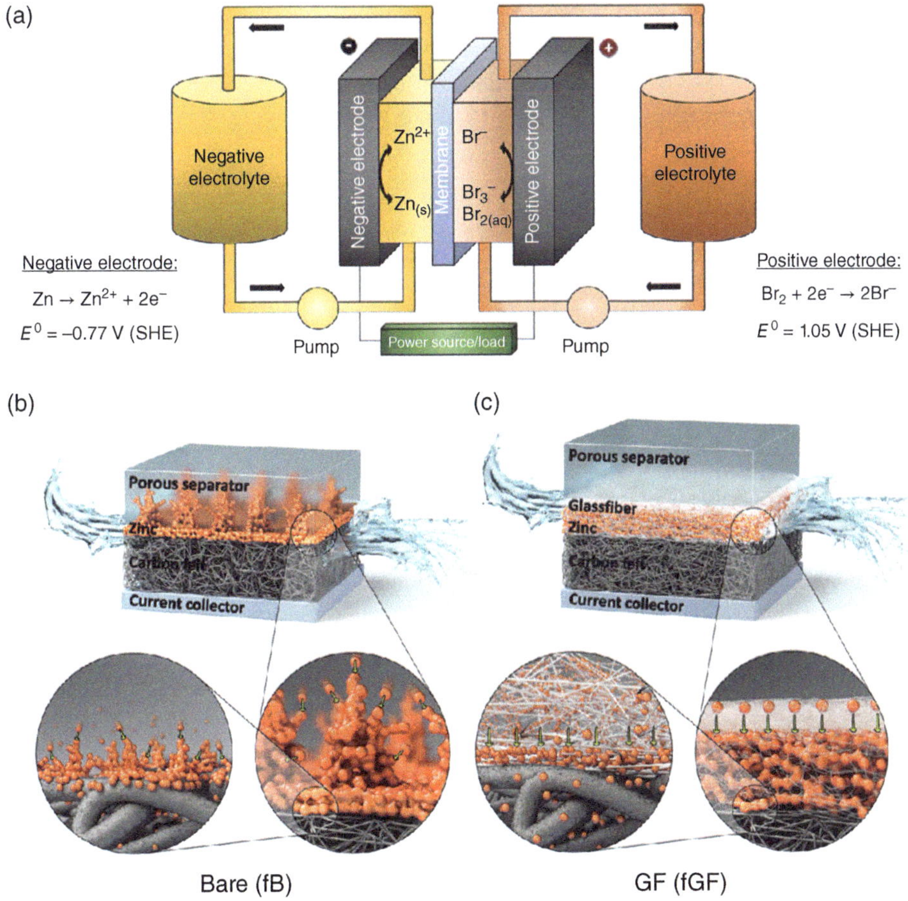

Figure 9.2 (a) The schematic representation of a ZBFB. (b) Conventional configuration without any interlayer (fB) shows the formation of a Zn dendrite, which pierces the separator. (c) Configuration using glass fiber to prevent damage of the separator via dendrite growth. *Source:* Kim [5]/With permission of American Chemical Society.

The major drawback in ZBFBs is the formation of metal dendrites as the battery is charged and discharged, which leads to a short life cycle and short circuits caused by dendrite perforation of the membrane separating both electrodes. There have been some strategies to prevent this [9], like the one presented in Figure 9.2c, consisting of a glass fiber (GF) interlayer able to induce a uniform deposition of Zn rather than the formation of dendrites, resulting in an enhancement of the reversibility and duration of the battery [5].

One of the limitations in the use of aqueous electrolytes in RFBs is that cell voltages are limited by the water electrolysis and redox potential of inorganic species are commonly low in nature. The use of organic solvents enables a wider potential window, of up to 6 V, as in the case of the acetonitrile [10]. Organometallic complexes constitute a main family of redox active species employed in nonaqueous RFBs that can present different metal centers (V, Mn, Ni, Co, and Fe) and ligands and exhibiting remarkable potential as redox active species in flow batteries [11–13].

Other strategies include the study of metal-free redox active organic molecules, consisting of abundant elements [14, 15]. Among them, it is worth mentioning quinone (Figure 9.3) [16–21] and TEMPO (2,2,6,6-tetramethylpiperidin-1-yl)oxyl) derivatives [22–29]. Dialkoxybenzene derivatives are another example of redox pairs showing promising performance in RFBs [30, 31].

9.1.1 Traditional Carbon Electrodes in Redox Flow Batteries

Electrodes are key components of any electrochemical device. Desirable properties include large specific surface area with plenty of active sites for catalytic reactions, excellent chemical stability, wide voltage window, high electrical conductivity, and efficient electron transfer that allows for fast kinetics and reversibility at the electrode/electrolyte interface that can then lead to optimal performance, power density, and avoidance of overpotentials. In the case of RFBs, the porous microstructure is also key to enable the efficient flow of the electrolyte carrying the redox active species through the electrode [32, 33]. The RFB electrodes must also be able to undergo compression, which is applied to reduce ohmic losses in the device.

Carbon is the preferred choice of electrode material in RFBs, and it has been employed in a variety of forms, including carbon felts, graphite felts, carbon papers, carbon cloths, graphite, and graphene oxide [34–39]. Figure 9.4 shows the most popular commercial carbon electrodes employed in RFBs [40]. All of them consist of entangled carbon fibers with high conductivity and surface area. In particular, commercial carbon (CH) and graphitic felts (GF) derived from polyacrylonitrile (PAN) have been widely studied due to their high conductivity, three-dimensional porous structure, and chemical inertness [41]. The fibrous structure allows the formation of high surface area based on the fiber diameter without compromising the mechanical stability. Especially in RFBs, these properties are important to maintain stability in the electrolyte flow and allowing an unhindered turbulent flow of electrolyte through the electrode with enough contact time for the reaction of the anolyte and catholyte.

Commercial carbon electrodes, however, suffer from reduced electrocatalytic activity and poor wettability due their high hydrophobic nature, limiting the overall performance

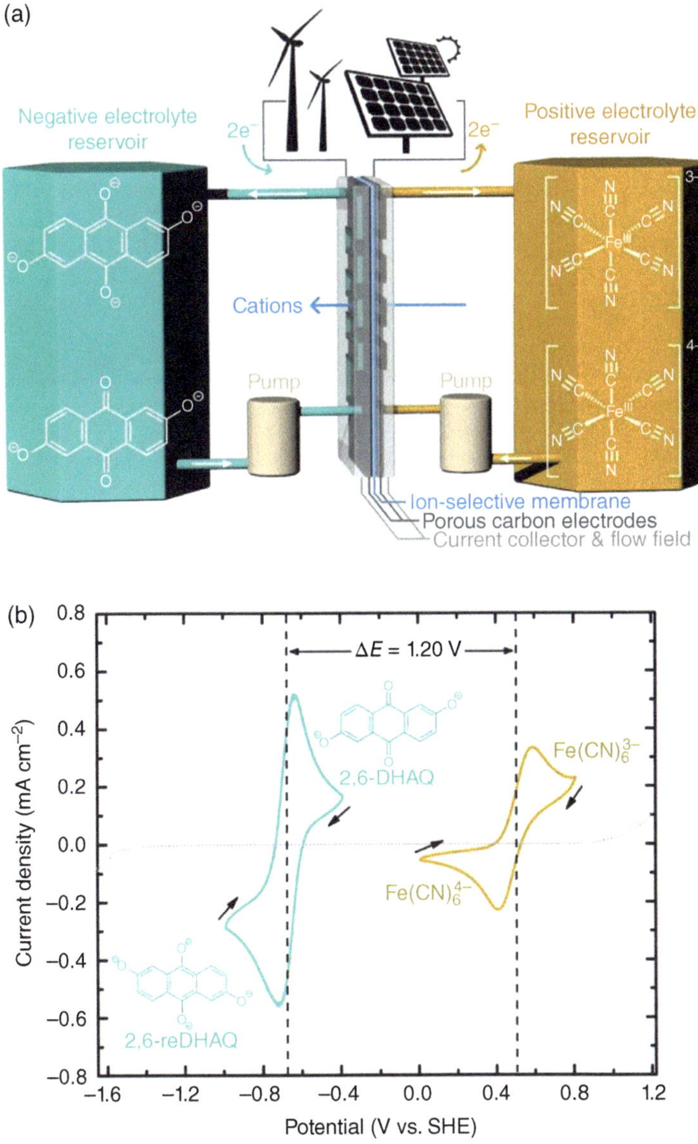

Figure 9.3 (a) Schematic of cell in charge mode. Essential components are labeled with color-coded lines and text. (b) Cyclic voltammogram of 2 mM 2,6-DHAQ and ferrocyanide scanned at $100\,mV\,s^{-1}$ on glassy carbon electrode. Dotted line represents CV of 1 M KOH background scanned at $100\,mV\,s^{-1}$ on graphite foil electrode. *Source:* [16]/With permission of American Association for the Advancement of Science.

of the battery [42, 43]. Two main approaches have been developed to enhance their electrocatalytic activity (Figure 9.5). The activation or modification of the surface, via thermal treatment or via functionalization of the carbon surface with heteroatoms, such as oxygen or nitrogen, constitutes one of the strategies to enhance the performance of the electrode [44–47].

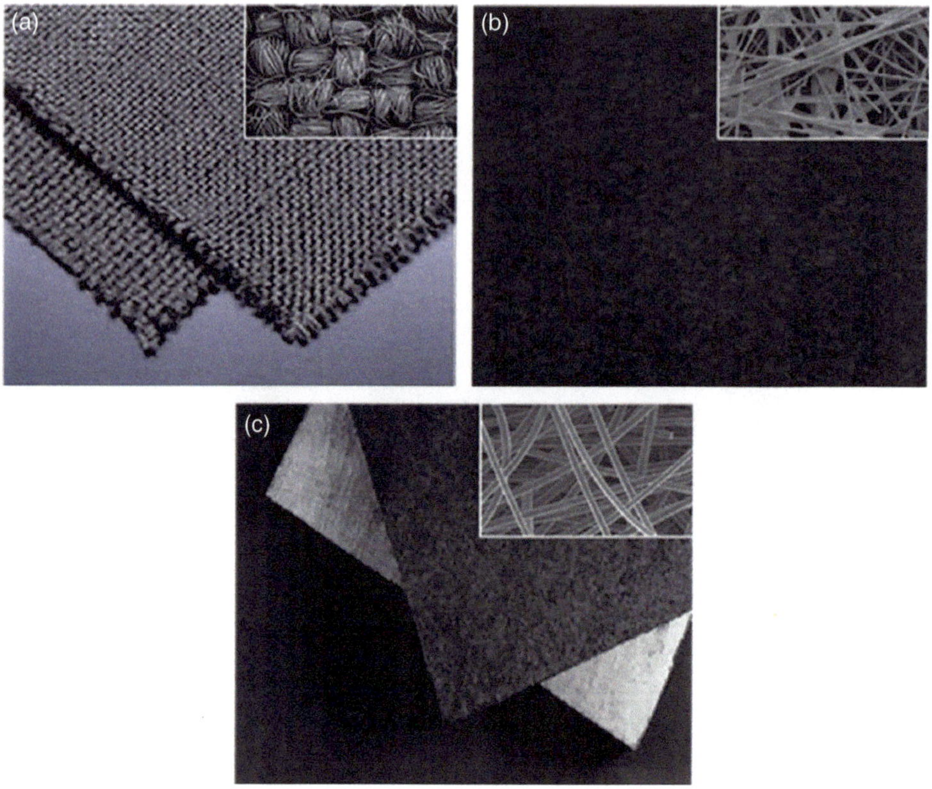

Figure 9.4 Comparison of current commercialized electrodes for RFB: (a) carbon cloth, (b) carbon paper, and (c) carbon felt. *Source:* Kim et al. [40]/With permission of Royal Society of Chemistry.

An increase in the number of surface heteroatoms transforms the carbon felt in hydrophilic, which ultimately leads to a better wettability with the electrolyte carrying the redox active species. The other strategy consists in using the carbon as a support for electrocatalyst nanoparticles that can contribute to enhance the kinetics of the redox reactions [48–51].

More radical approaches seek to control the porous microstructure of the carbon electrodes. In this category, electrospinning has emerged as a great choice to produce materials with tailored properties [52]. The first electrospun fibers used as electrodes in redox flow batteries were prepared using PAN as carbon source and obtaining fibers with diameters of 100–200 nm, an order of magnitude smaller than commercial carbon electrodes [53]. Although the formation of fibers of smaller diameter leads to high surface area for reactions, the mass transport properties can suffer, resulting in parasitic pumping power losses and inhomogeneity of reaction across the electrodes [54, 55]. Therefore, there have been considerable effort focused on the optimization of the microstructure of electrospun carbon electrodes, also including alignment of fibers [56], modification of the surface chemistry [57, 58], or decoration with nanoparticles [48]. A summary of the different approaches employed in recent years can be found in Figure 9.6.

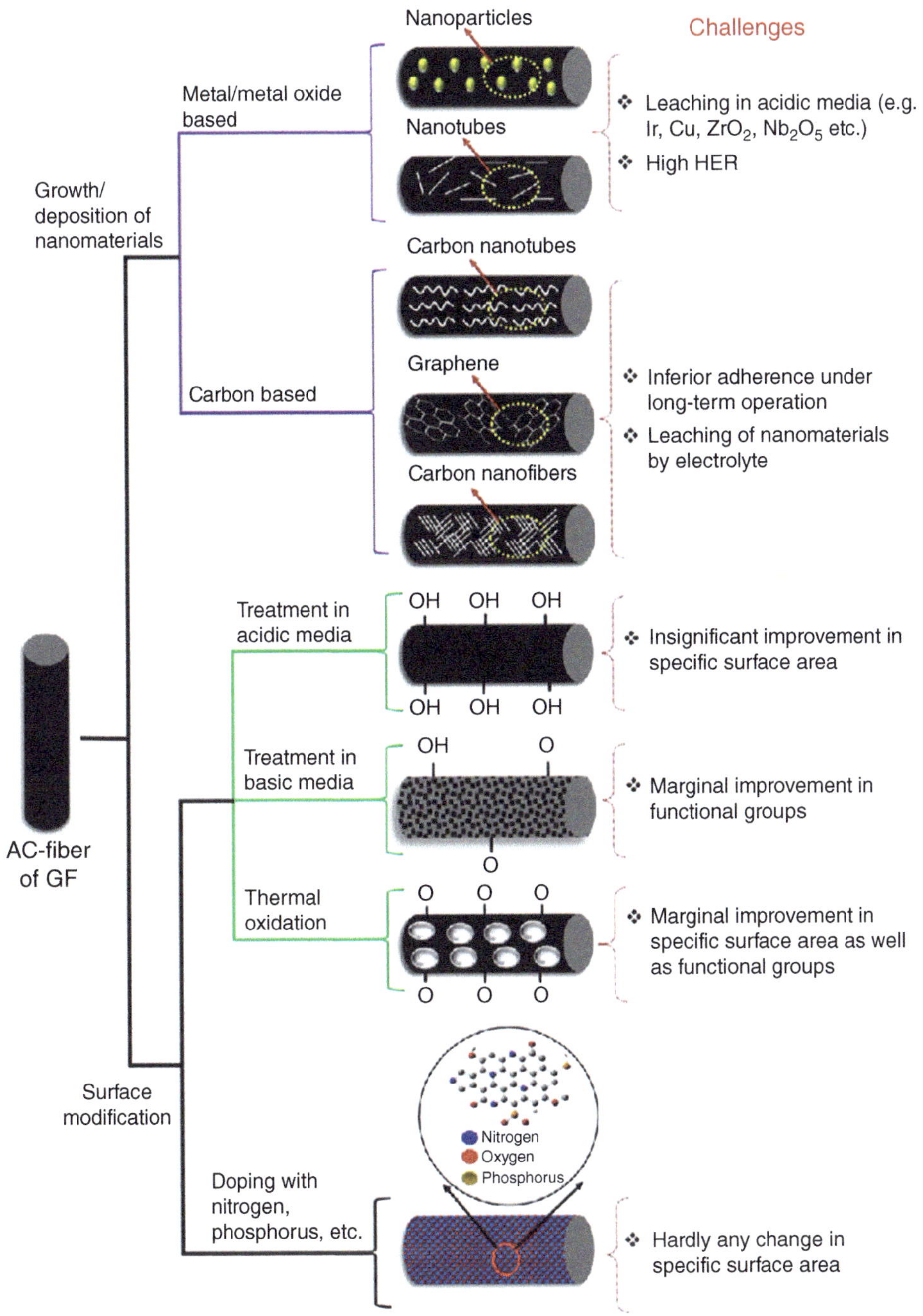

Figure 9.5 Overview of the methods adopted for GF modification for all-V RFB. *Source:* Ref. [44]/ With permission of American Chemical Society.

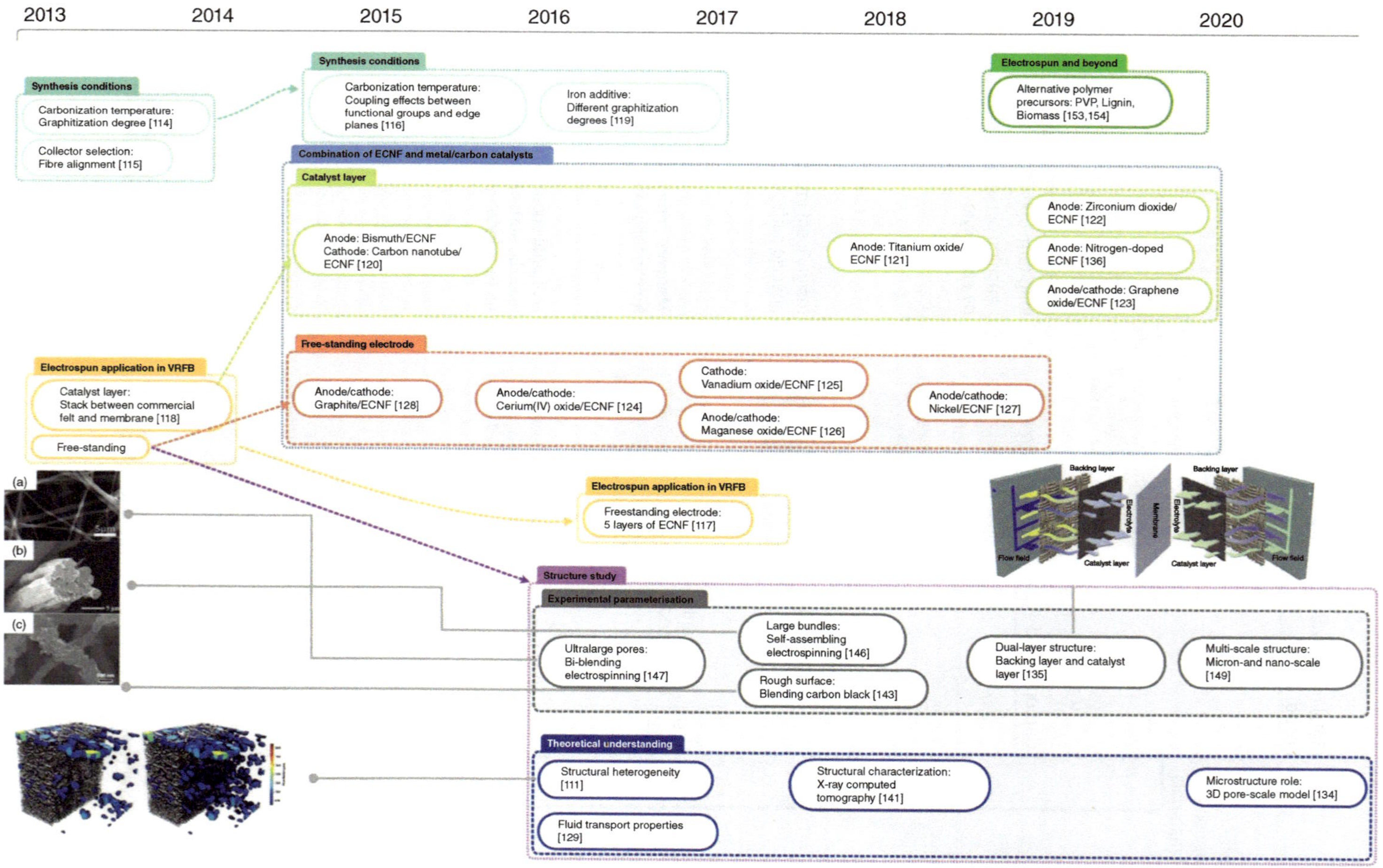

Figure 9.6 Historical development of electrospinning in RFBs since 2013, grouped into studies of similar aspects of electrospinning: effect of synthesis conditions (cyan), combination with catalysts (light blue), application in VRF (yellow), microstructural studies (pink), and the use of polymers other than PAN (green). *Source:* Weng et al. [52]/With permission of Elsevier.

9.1.2 Processing of Biomass Into Electroactive Materials

Despite the development of PAN-derived carbon electrodes, the future of renewable energies needs independency of finite resources, such as crude oil, therefore alternatives for the formation of electroactive materials from sustainable resources is an inevitable step toward a greener future. The use of biomass-derived carbon materials for energy storage devices has been of increasing interest for a few years now, particularly for supercapacitors [59–62] and metal-ion batteries [63–65]. However, only recently biomass materials for electrodes have been reported for RFBs, with an increasing amount of interest in alternative grid scale and stationary energy storage technologies.

Due to the fact that virtually all biomass material consists of a significant percentage of carbon, the source formation of electroactive carbon for energy applications can be taken from a large variety of materials. The properties of the resulting carbon will depend on the biomass source. The most common approach for the fabrication of electroactive carbon from biomass is to convert the biomass fabrication of activated carbon (AC). Hydrothermal carbonization (HTC) is the most widely used method for the conversion of organic compounds to submicron and nanostructured ACs through high pressure and temperature in aqueous media. A further carbonization step at higher temperature under protective gas atmosphere leads to ACs with increased graphitization. Electrospinning can also be used to process biomass materials into electroactive carbons [66], enabling the formation of biomass-derived carbon in the form of fibers.

Heteroatoms present in the biomass and its intrinsic structure can often be found in the resulting carbon and have great influence on the electrocatalytic properties. Heteroatom doping on the biomass-derived carbons can be particularly important in the case of RFB electrodes and can either be obtained by using heteroatom-containing biomass [67] materials or heteroatom-donating additives, such as urea [68], ammonium oxalate [69], or ferric ammonium citrate [70]. A similar but more dominant effect toward the etching can be seen when KOH is used as an additive in the carbonization to achieve high porosity [71, 72].

9.1.3 Examples of Biomass-Derived Electrodes for Redox Flow Batteries

Electroactive carbon materials that are suitable for electrodes in energy materials can be made from a variety of biomass sources. The challenge in RFBs is to produce a robust three-dimensional conductive structure that has enough stability to withstand the flow of electrolyte. This must also happen without building up pressure through clogging and remain on a high enough contact time to initiate the redox reaction of the active species contained in anolyte and catholyte that store the electrochemical energy. Although fibrous carbon structures, such as graphitic felt, have proven to be suitable porous electrodes with high conductivity and low cost, these show poor kinetics toward the desired electrochemical processes and have limited porous microstructures. Moreover, they are produced from petrol-derived compounds.

Biomass-derived electrodes for RFBs can be made from either a biomass-derived AC that is used to decorate the surface of existing carbon felts to improve the electrocatalytic activity and increase the electrochemical active surface area (ECSA) or from full-biomass derived freestanding carbon fibers via electrospinning.

Examples for the biomass-derived ACs are reported in numerous studies. Most of them involve an activation step, such as thermal or hydrothermal treatment, and a higher temperature carbonization step under protective gas atmosphere to achieve high surface area and improved electrocatalytic properties [73]. The source of the initial biomaterial allows to control which heteroatom compositions, porous structures, and other morphologies of the resulting carbon can be produced. AC has been widely researched as electrode because of its high porosity, electrical conductivity, and stability in acidic media, enabling it to provide more active sites for electrocatalytic reactions [74, 75].

Heteroatoms lead to an introduction of active sites onto the carbon surface to boost the performance of the biomass-derived carbon electrodes, increase the charge accumulation, and promote the adsorption of positive and negative charges of the electrode. Experimental studies show, doping heteroatoms on the carbon surface is a very promising method to increase their electrochemical activity of the electrodes in RFBs [76–78]. Increasing the surface-active sites by doping accelerates the adsorption of ions [79], and differences in electronegativity between carbon and doping atoms (such as N, P, O, and S) lead to changes in the electrochemical properties of the carbon [80, 81]. The source of the used biomass has an impact of the resulting carbon properties that can be seen in a direct comparison of using nitrogen-containing chitin from shrimp shell or pinewood to form AC under identical condition, by performing a HTC at 300 °C, followed by carbonization in nitrogen at 850 °C (Figure 9.7) [67]. While the BET surface of the pine-derived AC was around four times higher ($436\,m^2\,g^{-1}$ vs. $102\,m^2\,g^{-1}$), the determined ECSA was determined more than two and a half times larger for the chitin-derived AC, resulting in an overall higher performance that can directly be attributed to the higher resulting ECSA. Another chitin-derived AC reported was made from a lower temperature HTC at 200 °C, resulting in a six times higher BET surface area ($616\,m^2\,g^{-1}$) and a more than double ECSA than that of the 300°, leading to a much higher overall activity as shown in CV, underlining the important effect of the hydrothermal temperature for the BET surface area and the nonproportional increase of the ECSA.

A common approach to drastically increase the BET surface area to values higher than $1000\,m^2\,g^{-1}$ is the use of additional KOH in the activation step followed by carbonization, as performed with orange peel [71] ($1901\,m^2\,g^{-1}$), used tea bags [72] ($2085\,m^2\,g^{-1}$), or sugarcane bagasse [82] ($1255\,m^2\,g^{-1}$) or in addition with a HTC on fish scales [83] ($1492\,m^2\,g^{-1}$).

An alternative way of reaching a comparable high surface area was reported by using coffee beans to form AC at a temperature of 850 °C followed by a steam-based activation [84]. Starting from a biochar with a BET surface from less than $1\,m^2\,g^{-1}$, AC samples with increased BET surface areas of $541\,m^2\,g^{-1}$ could be achieved by performing the steam activating process for 1 hour, $817\,m^2\,g^{-1}$ for 2 hours, and $1113\,m^2\,g^{-1}$ for 3 hours. Yet, in the electrochemical characterization, a direct comparison of a biochar-modified graphite plate electrode showed a higher anodic and cathodic peak currents and narrower potential peak separation for both redox couples than the AC after three hours of activation. This could be linked to the decreased nitrogen in the sample (2.1% vs. 0.7% in ultimate analysis), showing that a large surface area alone cannot necessarily leads to a better electrochemical performance.

Next to the given heteroatoms and functionalities that are introduced through the chemical composition of the biomass, additives in the activation and carbonization process can also be used for functionalities that improve electrocatalytic properties for the redox

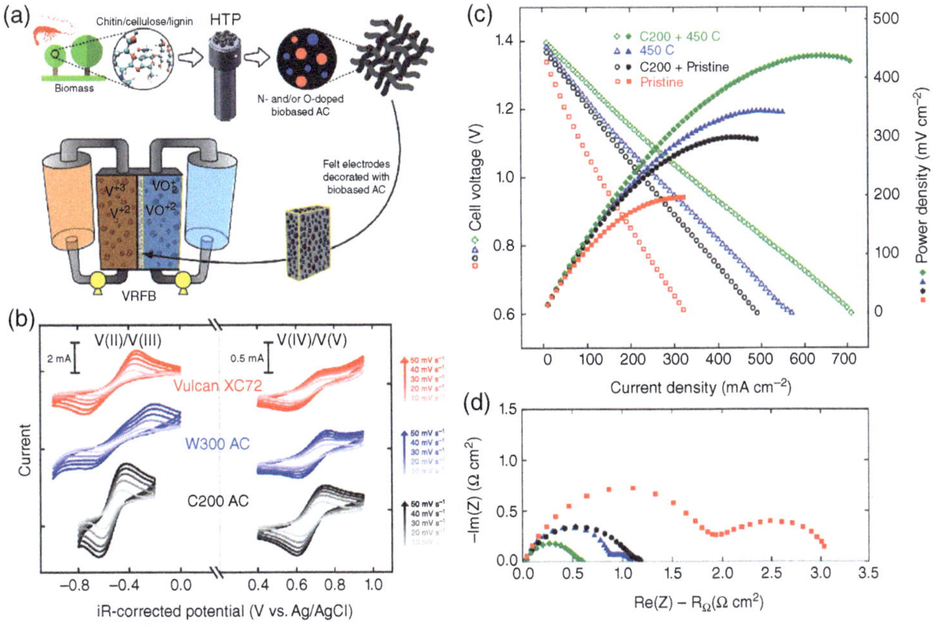

Figure 9.7 (a) Scheme showing biomass from either shrimp shell-derived chitin or from pine wood subjected to hydrothermal process to produce ACs that are then incorporated into felt electrodes. The electrodes are then implemented into a VRFB to evaluate their performance. (b) Cyclic voltammograms of Vulcan XC72, W300 AC (pine wood), and C200 AC (Chitin) with 1 M V(II) in 2 M H_2SO_4 and 0.1 M V(IV) in 0.1 M H_2SO_4 at scan rates of 10, 20, 30, 40, and 50 mV s^{-1}. (c) Discharge polarization and power density curves and (d) Nyquist plots from EIS at OCV for pristine felts, C200 AC + pristine felt, felt treated at 450 °C, C200 AC + felt treated at 450 °C. *Source:* Wan [67] With permission of American Chemical Society.

couples. Alternative additives with metal-containing components can be used to improve the electrocatalytical properties of the resulting AC. Iron-containing additives, for example, were used to form AC from a tropical nut-bearing tree named scaphium scaphigerum, which was activated by soaking in deionized water and HTC. Mixing the hydrothermal product with iron(III) chloride and subsequent carbonization at 800 °C in argon, followed by HCl and water washing (Figure 9.8) [85]. Iron(III) chloride as additive led to an increase in the surface area for the resulting porous carbon (615.6 m^2 g^{-1} vs. 259.8 m^2 g^{-1}), a higher degree of graphitization shown by Raman spectroscopy and XPS, leading to a higher discharge capacity and EE than the pristine carbon felt in a VRFB.

An alternative iron additive that was reported on scaphium scaphigerum in an identical procedure was potassium ferrate (K_2FeO_4) [86]. During the carbonization, the effect of potassium ferrate on the carbon is similar in increasing the graphitization, as well as depositing oxygen and etching the surface through a forming alkaline, resulting in a large specific surface area of m^2 g^{-1} (262.3 m^2 g^{-1} without potassium ferrate). The Fe-doped AC on carbon fibers exhibited an improved electrocatalytic performance, lower charge transfer resistance, and improved charge transfer compared to a pristine GF. Another iron-based active agent for improving the resulting electrochemical properties that has been reported is ferric ammonium citrate. In a study on obtaining AC from kiwi, ferric ammonium citrate

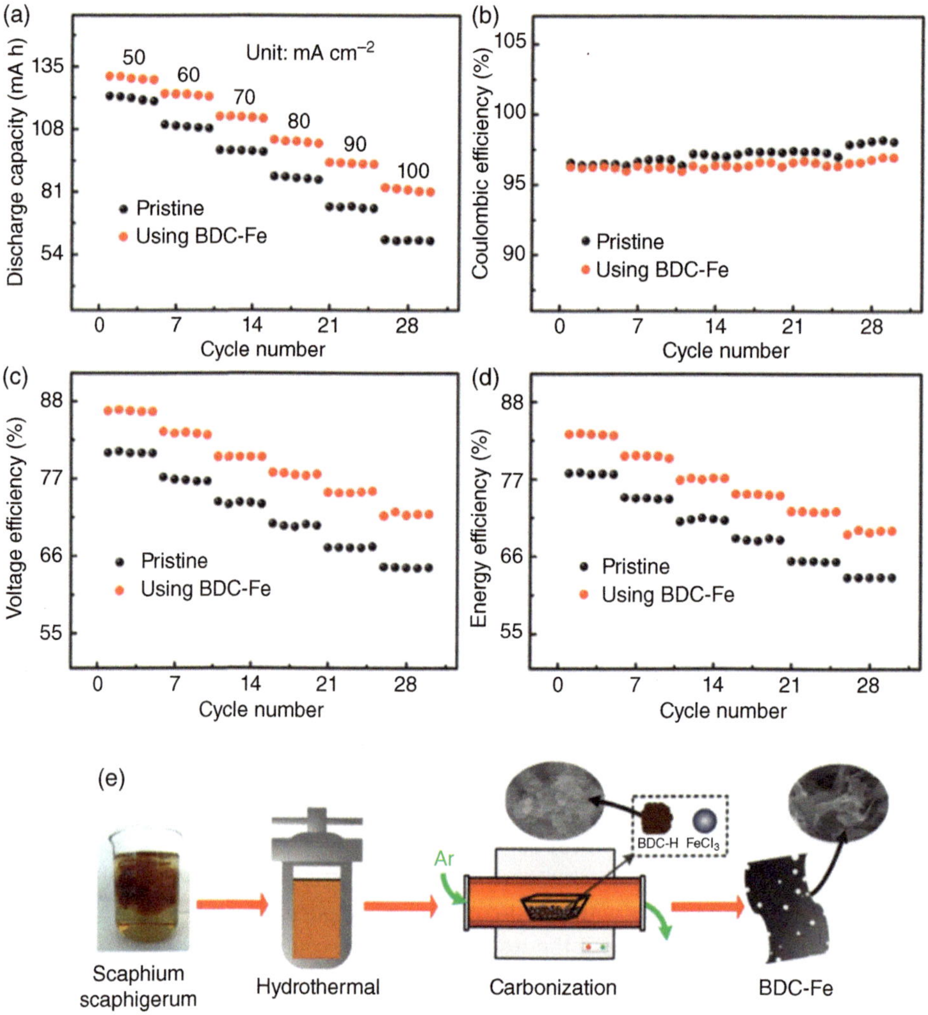

Figure 9.8 (a) Discharge capacity, (b) CE, (c) VE, (d) EE for pristine cell and biomass BDC-Fe modified cell at current density of 50–100 mA cm^{-2}. (e) Schematic representation of the synthesis of the BDC-Fe material. *Source:* Jiang et al. [85], IOP Publishing.

was added during the carbonization process resulting in highly graphitized N-doped porous carbon materials for RFBs [70]. At high temperatures, ferric ammonium citrate decomposes into iron- and ammonia-based compounds, leading to an introduction of N-containing functional groups and contributing to the formation of a porous structure. The electrocatalytic activity of both redox pairs shows a highly increased performance using ferric ammonium citrate. EIS measurements suggested a control by charge transfer and diffusion processes, with a significantly decreased ohmic resistance with ferric ammonium citrate in the process, caused by increased conductivity, wettability, and the decreased charge transfer resistance, leading to faster electrochemical kinetics.

A different salt that has been reported to improve the resulting AC is zinc chloride, added in the HTC with fresh sal wood sawdust to promote the formation of oxygen functionalities and high porosity on the surface [87]. After being carbonized at 850 °C under nitrogen, the sawdust-derived AC exhibited a BET-surface area of 1857 $m^2 g^{-1}$ and significantly higher anodic and cathodic peaks in the CV of the decorated carbon felt, as well as a lower ohmic and charge transfer resistances, which was attributed to the additional oxygen functionalities at the surface.

Metal-free alternative for introducing a doping can be found based on nitrogen-containing molecules, such as urea. The previously mentioned seeds of scaphium scaphigerum can also be carbonized in the presence of urea to introduce nitrogen in the resulting AC. After HTC and subsequent the freeze-drying, a mixture with urea (4 : 1 by mass) was carbonized in argon at 850 °C to obtain N-doped AC with a content of 8.2% by XPS [88]. It was shown that electrochemical activity improved toward the V^{2+}/V^{3+} redox pair, with a ΔE_p of 141 mV (vs. 451 mV of undoped AC), which lead to charge–discharge experiments of a single VRFB cell with AC modified negative electrode 81.5% larger current density than that of pristine cell (1680.9 mA h vs. 926.3 mA h) and an improved EE by 9.8% compared with the pristine cell (58.9%) at 150 mA cm^{-2}. A similar metal-free approach for additive N-doping was reported by using ammonium oxalate as a nitrogen source and persimmon as biomass [68]. After HTC, the carbonization was performed on a homogenous mix with ammonium oxalate in different weight ratios (1 : 5, 1 : 10, 1 : 15) and carbonized at 750 °C. The CV curves indicated that all samples have improved catalytic performance toward the VO^{2+}/VO_2^+ reaction and the highest current density on the 1 : 10 sample. A starting discharge capacity of 108.9 mA h (23.9 mA h higher than pristine electrode) in charge–discharge tests in a VRFB configuration was reported at 50 mA cm^{-2}, and a capacity retention of 89.9% after 50 cycles with an EE of 79.9% (vs. 75.9%) was reported, proving the improved electrocatalytic activity of carbon electrodes through the ammonium oxalate additive.

The decoration with biomass derived is a first step toward the formation of sustainable electrodes. Fully sustainable electrodes need to be independent from fossil-based GF. Since the fibrous structure has proven to best suitable for the formation of the electrodes in redox batteries, the carbonization of naturally occurring fibrous materials should be a good starting point toward the formation of freestanding full biomass electrodes. Based on the naturally given fibrous structure of cotton, carbonized cotton pads can be carbonized at 1000 °C in argon, resulting in carbon fibers with a BET surface area of 16.6 $m^2 g^{-1}$ (Figure 9.9) [69]. Due to the high oxygen content, the resulting carbon fibers show a hydrophilic behavior, which also led to improved electrocatalytic behavior compared to a commercial thermally treated carbon paper, and show in the lower potential separation and higher anodic and cathodic peak potential. In an RFB, the carbonized cotton electrode exhibited lower overpotential and improved charge/discharge performance and a higher overall EE than the commercial thermally treated carbon paper, proving the need of defects and functional groups to improve the electrocatalytic activity.

Silk is another naturally occurring fibrous material that can be carbonized to produce conducting carbon materials. Their naturally high nitrogen content of up to 15% allows the formation of heteroatom-rich carbon electrodes. The silk fibers of the bombyx mori cocoon have been employed as a source for carbon fibers undergoing a HTC process in 0.06 M Na_2CO_3 followed by a carbonization step at 1000 °C under argon atmosphere. The resulting

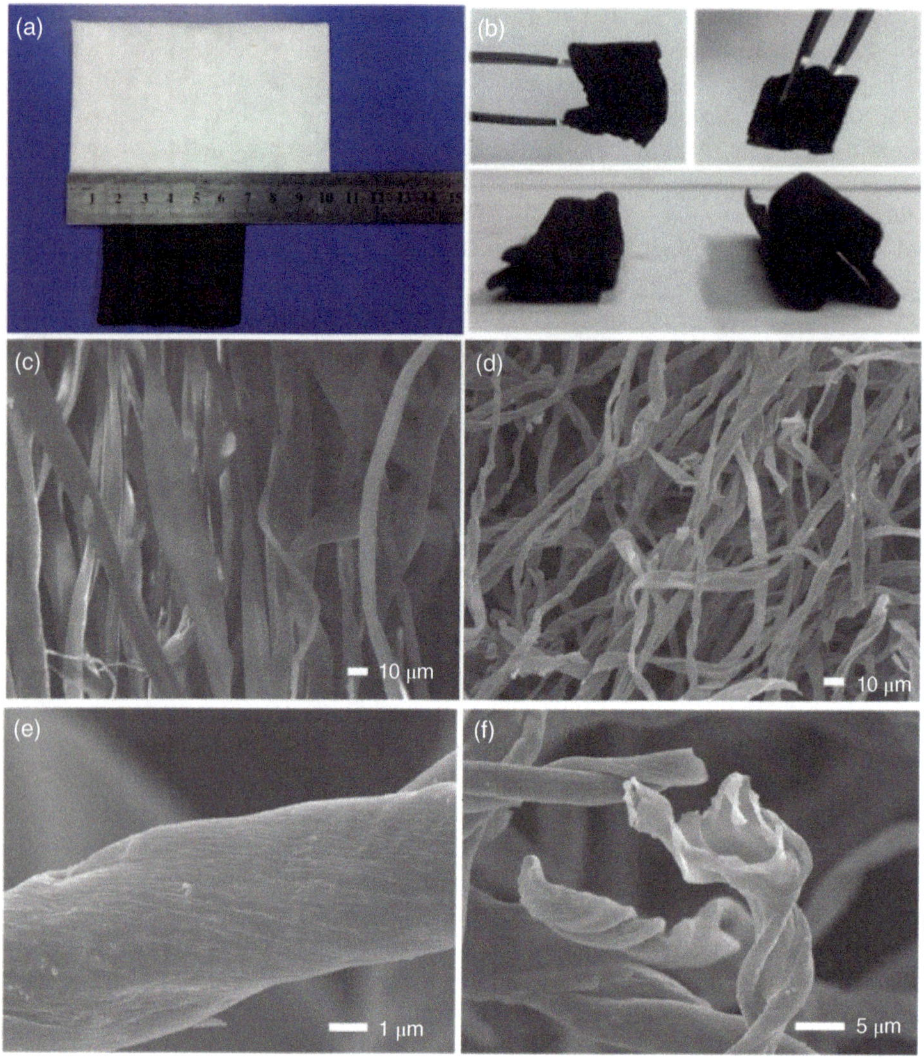

Figure 9.9 (a) The photograph of the pristine cotton pad and carbonized cotton (CC); (b) excellent mechanical flexibility of CC electrode in different folded states; SEM images of (c) pristine cotton and (d) CC; SEM images of (e) the surface of individual carbonized cellulose fiber, and (f) cross profile of carbonized fibers. *Source:* Zhang et al. [69]/With permission of Elsevier.

fibers contained 31.05% oxygen and 5.81% nitrogen [89]. The effect of the high heteroatom functionalities on the carbon surface led to a hydrophilic soaking effect on water, unlike the commercial carbon paper (CP) and oxidized CP, which were used for comparison and were both hydrophobic. The oxygen functional groups and nitrogen led to remarkable improvement CV measurements for the electrochemical performance for the V^{2+}/V^{3+} and VO^{2+}/VO_2^+ reaction compared to the commercial carbon papers and a lower charge transfer resistance and the faster mass transfer process in the EIS measurement. These superior electrochemical properties could be observed when used in as an electrode for the VRFB. The silk-based carbon fibers exhibited a lower charge/discharge overpotential, and higher capacities in comparison with CP and oxidized CP electrodes as well as better electrochemical reversibility and activity MC. Furthermore, the average discharge capacity was 83%, a 20% higher EE than that with CP electrode at the current density of 100 mA cm^{-2} with a higher a higher stability.

Alternatively to silk, silk protein-derived carbon fabrics have also been reported to be isolated in order to obtain heteroatom-rich carbon fibers after carbonization at 800 °C in argon, followed by an additional thermal treatment between 1200 and 2800 °C under argon atmosphere to increase the degree of carbonization [90]. As expected for heteroatom-rich carbon fiber, a considerably lower peak separation for both redox pairs in the VRFB was observed in the CV when compared with commercial carbon felts. This was attributed to the enhancing effect of the high electron density of pyridinic-N structure in the electron transfer kinetics for vanadium redox reactions, as well as a lower charge transfer. The silk protein-derived carbon fibers exhibited higher specific capacity of 32.5 Ah l^{-1} (vs. 22.2 Ah l^{-1}, measured between 0.8 and 1.6 V at 40 mA cm^{-2}) and a capacity retention after 100 cycles of 91%, compared to 80%, and EE value of 86.8% at 40 mA cm^{-2} (compared to 76.5), stating the potential of heteroatom-rich biomass-derived electrodes.

Electrospinning is an established method for the formation of submicron and nanosized fibers. This method is extremely versatile, enabling the fine control of the porous microstructure and surface chemistry of the resulting fibers by just adjusting the processing parameters. While most of the research is based on the electrospinning and carbonization of PAN and related polymers, recent works have shown that lignin is a suitable material for obtaining high-performing electrodes [66, 91]. Since pure lignin has a relatively low and broad molecular weight distribution, which usually does not allow the formation a stable fiber through entanglement of the polymeric chains in the electrospinning process, it is usually electrospun in combination of with a plasticizer to improve the entanglement and allow for the formation of a continuous mat. A comparison of lignin from three different isolation processes (ethanol organosolv, Kraft lignin [KL], and phosphoric acid lignin [PL]) used in combination with polyethylene oxide (PEO) as plasticizer showed that the PL-derived carbon fibers exhibited more intense peaks for oxidation and reduction of V species [91]. Adding conductive carbon during the electrospinning process of KL increased the electrocatalytic activity and results in a higher intensity of the signals related to reduction and oxidation and lowering the potential differences to 0.4 V for V^{5+}/V^{4+} and 0.2 V for V^{3+}/V^{2+}.

Another example of the use of lignin in electrospinning was performed using high molecular lignin fraction, resulting in fibers with diameters of 750–1000 nm, that were then carbonized by preheating at 250 °C for 30 minutes, followed by thermal treatment at 800 °C for 2 hours [66]. The electrocatalytic activity toward the redox reactions of vanadium (V^{2+}/V^{3+} and VO^{2+}/VO_2^+) in CV measurements was significantly better than the commercial fiber mats, which is a very promising result toward the implementation of lignin-derived carbons as electrodes in RFBs.

The addition of P favors the VO^{2+}/VO_2^+ reaction in the case of the commercial electrode. However, the introduction of P in the electrospun carbon fibers was counterproductive, resulting in a decrease in electrochemical performance, which was attributed to a reduction of active sites due to steric hindrance of the porous structure through phosphate group (Figure 9.10). Nevertheless, these results are encouraging and suggest that by tuning the microstructure and surface chemistry of the fibers could lead to materials that out-perform current electrodes.

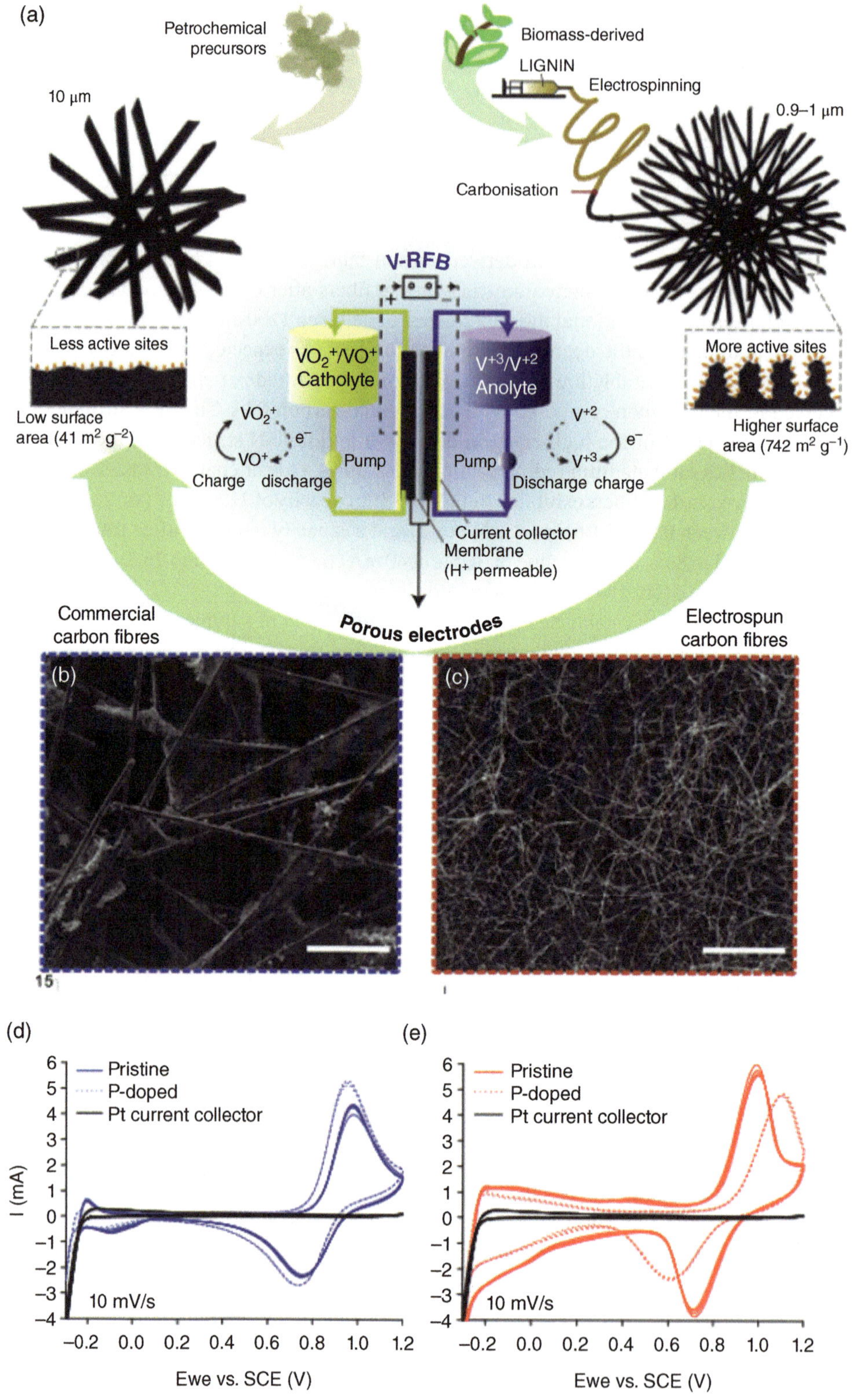

Figure 9.10 (a) Scheme comparing a high surface area, sustainable electrospun carbon electrode material with a commercial fibrous nonwoven carbon paper for their application in VRFB. SEM images of both studied materials: (b) Commercial Sigracet® GDL29AA and (c) CNF electrospun carbon nanofiber mat; scale bars correspond to 100 mm; cyclic voltammograms for samples with exposed areas of 0.79 mm^2 for: (d) GDL29AA (solid lines: blue: pristine, black: Pt current collector, dashed lines: P-doped) and (e) CNF (solid lines: red: pristine, black: Pt current collector, dashed lines: P-doped) 66. *Source:* (b) and (c): Ribadeneyra et al. [66]/With permission of Elsevier.

References

1 Ziegler, M.S. and Trancik, J.E. (2021). *Energy & Environmental Science* 14: 1635.
2 Sanchez-Diaz, E., Ventosa, E., Guarnieri, M. et al. (2021). *Journal of Power Sources* 481: 228804.
3 Arevalo-Cid, P., Diaz, P., Mendes, A., and Azevedo, J. (2021). *Sustainable Energy & Fuels* 5: 5366.
4 Skyllas-Kazacos,M., Rychcik, M., and Robins, R. (1986). All vanadium redox battery. AU Patent 575247.
5 Kim, R., Jung, J., Lee, J.-H. et al. (2021). *ACS Sustainable Chemistry & Engineering* 9: 12242–12251.
6 Butler, P.C., Eidler, P.A., Grimes, P.G. et al. (2002). Zinc/bromine batteries. In: *Handbook of Batteries*, 3e (ed. D. Linden and T.B. Reddy), 39.1–39.20. New York: McGraw-Hill.
7 Yin, Y., Yuan, Z., and Li, X. (2021). *Physical Chemistry Chemical Physics* 23: 26070.
8 Suresh, S., Ulaganathan, M., and Pitchai, R. (2019). *Journal of Power Sources* 438: 226998.
9 Leibin, G., Hui, G., Haili, H. et al. (2020). *Frontiers in Chemistry* 8: https://doi.org/10.3389/fchem.2020.00557.
10 Chiba, K., Ueda, T., Yamaguchi, Y. et al. (2011). *Journal of the Electrochemical Society* 158: A1320–A1327.
11 Noack, J., Roznyatovskaya, N., Herr, T., and Fischer, P. (2015). *Angewandte Chemie International Edition* 54: 9775–9808.
12 Gong, K., Fang, Q., Gu, S. et al. (2015). *Energy and Environmental Science* 8: 3515–3530.
13 Hogue, R.W., Armstrong, C.G., and Toghill, K.E. (2019). *ChemSusChem* 12: 4506–4515.
14 Leung, P., Shah, A.A., Sanz, L. et al. (2017). *Journal of Power Sources* 360: 243–283.
15 Wang, W. and Sprenkle, V. (2016). *Nature Chemistry* 8: v204–v206.
16 Lin, K., Gerhardt, M.R., Tong, L. et al. (2015). *Science* 349: 1529–1532.
17 Er, S., Suh, C., Marshak, M.P., and Aspuru-Guzik (2015). *Chemical Science* 6: 885–893.
18 Liu, Y., Lu, S., Chen, S. et al. (2019). *ACS Applied Energy Materials* 2: 2469–2474.
19 Huskinson, B., Marshak, M.P., Suh, C. et al. (2015). *Nature* 505: 195–198.
20 Tong, L., Jing, Y., Gordon, R.G., and Aziz, M.J. (2019). *ACS Applied Energy Materials* 2: 4016–4021.
21 Yang, B., Hoober-Burkhardt, L., Krishnamoorthy, S. et al. (2016). *Journal of the Electrochemical Society* 163: A1442–A1449.
22 Liu, T., Wei, X., Nie, Z. et al. (2016). *Advanced Energy Materials* 6: 1501449.
23 Liu, Y., Goulet, M.A., Tong, L. et al. (2019). *Chemistry* 5: 1861–1870. https://doi.org/10.1016/j.chempr.2019.04.021.
24 Orita, A., Verde, M.G., Sakai, M., and Meng, Y.S. (2016). *Journal of Power Sources* 321: 126–134.
25 Janoschka, T. et al. (2015). *Nature* 527: 78–81.
26 Winsberg, J., Janoschka, T., Morgenstern, S. et al. (2016). *Advanced Materials* 28: 2238–2243.
27 Winsberg, J., Hagemann, T., Janoschka, T. et al. (2016). *Angewandte Chemie International Edition* 56: 686–711.
28 Duan, W., Vemuri, R.S., Milshtein, J.D. et al. (2016). *Journal of Materials Chemistry A* 4: 5448–5456.

29 Wang, W., Xu, W., Cosimbescu, L. et al. (2012). *Chemical Communications* 48: 6669–6671.

30 Huang, J., Cheng, L., Assary, R.S. et al. (2014). *Advanced Energy Materials* (6): 5, 1401782.

31 Huang, J., Su, L., Kowalski, J.A. et al. (2015). *Journal of Materials Chemistry A* 3: 14971–14976.

32 Simon, B.A., Gayon-Lombardo, A., Pino-Munoz, C.A. et al. (2022). *Applied Energy* 306: 117678.

33 Zhang, D., Forner-Cuenca, A., Taiwo, O.O. et al. (2020). *Journal of Power Sources* 447: 227249.

34 Kim, J., Lim, H., Jyoung, J.-Y. et al. (2017). *Carbon* 111: 592–601.

35 Kaneko, H., Nozaki, K., Wada, Y. et al. (1991). *Electrochimica Acta* 36: 1191–1196.

36 Chakrabarti, M.H., Brandon, N.P., Hajimolana, S.A. et al. (2014). *Journal of Power Sources* 253: 150–166.

37 Nibel, O., Taylor, S.M., Pătru, A. et al. (2017). *Journal of the Electrochemical Society* 164: A1608.

38 Deng, Q., HuangYang, X.-Y., Zhang, X. et al. (2022). *Advanced Energy Materials* 12: 2103186.

39 He, Z., Lv, Y., Zhang, T. et al. (2022). *Chemical Engineering Journal* 427: 131680.

40 Kim, K.J., Park, M.-S., Kim, Y.-J. et al. (2015). *Journal of Materials Chemistry A* 3: 16913–16933.

41 Kazacos, M. and Skyllas-Kazacos, M. (1989). *Journal of the Electrochemical Society* 136: 2759–2760.

42 Derr, I., Bruns, M., Langner, J. et al. (2016). *Journal of Power Sources* 325: 351–359.

43 Trogadas, P., Taiwo, O.O., Tjaden, B. et al. (2014). *Electrochemical Communications* 48: 155–159.

44 Gautam, R.K., Kapoor, M., and Verma, A. (2020). *Energy & Fuels* 34: 5060–5071.

45 Sun, B. and Skyllas-Kazacos, M. (1992). *Electrochimica Acta* 37: 1253–1260.

46 Kim, K.J., Kim, Y.J., Kim, J.H., and Park, M.S. (2011). *Materials Chemistry and Physics* 131: 547–553.

47 Flox, C., Rubio-Garcia, J., Skoumal, M. et al. (2013). *Carbon* 60: 280–288.

48 Schnucklake, M., Cheng, M., Maleki, M., and Roth, C. (2021). *Journal of Physics: Materials* 4: 032007.

49 Jiang, Y., Cheng, G., Li, Y. et al. (2021). *Chemical Engineering Journal* 415: 129014.

50 Taş, M. and Elden, G. (2022). *Energy Storage* 4 (2): e265.

51 Mariyappan, K., Ragupathy, P., and Ulaganathan, M. (2021). *Journal of the Electrochemical Society* 168: 090566.

52 Weng, Y., Kok, M.D.R., Victoria Tafoya, J.P. et al. (2021). *Journal of Energy Chemistry* 59: 492–529.

53 Wei, G., Liu, J., Zhao, H., and Yan, C. (2013). *Journal of Power Sources* 241: 709–717.

54 Zeng, L., Sun, J., Zhao, T.S. et al. (2020). *International Journal of Hydrogen Energy* 45: 12565–12576.

55 Xu, Z., Xiao, W., Zhang, K. et al. (2020). *Journal of Power Sources* 450: 227686.

56 Flox, C., Fabrega, C., Andreu, T. et al. (2013). *RSC Advances* 3: 12056–12059.

57 Wei, G., Gao, Z., Wei, Z. et al. (2015). *Physical Chemistry Chemical Physics* 17: 20368–20375.

58 Wei, G., Su, W., Wei, Z. et al. (2016). *Electrochimica Acta* 199: 147–153.

59 Yang, H., Ye, S., Zhou, J., and Liang, T. (2019). *Frontiers in Chemistry* 7: 274. https://doi.org/10.3389/fchem.2019.00274.

60 Shaker, M., Ghazvini, A.A.S., Cao, W. et al. (2021). *New Carbon Materials* 36: 546–572. https://doi.org/10.1016/s1872-5805(21)60038-0.

61 Bi, Z., Kong, Q., Cao, Y. et al. (2019). *Journal of Materials Chemistry A* 7: 16028–16045. https://doi.org/10.1039/c9ta04436a.

62 Gopalakrishnan, A. & Badhulika, S. *Journal of Power Sources* 2020, 480, https://doi.org/10.1016/j.jpowsour.2020.228830

63 Yuan, H., Liu, T., Liu, Y. et al. (2019). *Chemical Science* 10: 7484–7495. https://doi.org/10.1039/c9sc02743b.

64 Liedel, C. (2020). *ChemSusChem* 13: 2110–2141. https://doi.org/10.1002/cssc.201903577.

65 Liu, J., Yuan, H., Tao, X. et al. (2020). *EcoMat* 2: https://doi.org/10.1002/eom2.12019.

66 Ribadeneyra, M.C., Grogan, L., Au, H. et al. (2020). *Carbon* 157: 847–856. https://doi.org/10.1016/j.carbon.2019.11.015.

67 Wan, C.T.-C., Barreiro, D.L., Forner-Cuenca, A. et al. (2020). *ACS Sustainable Chemistry & Engineering* 8: 9472–9482. https://doi.org/10.1021/acssuschemeng.0c02427.

68 Zhang, T., Zhu, Y., Lv, Y. et al. (2021). *Ionics* 27: 4771–4781. https://doi.org/10.1007/s11581-021-04236-1.

69 Zhang, Z.H., Zhao, T.S., Bai, B.F. et al. (2017). *Electrochimica Acta* 248: 197–205. https://doi.org/10.1016/j.electacta.2017.07.129.

70 Cheng, D., Tian, M., Wang, B. et al. (2020). *Journal of Colloid and Interface Science* 572: 216–226. https://doi.org/10.1016/j.jcis.2020.03.069.

71 Maharjan, M., Bhattarai, A., Ulaganathan, M. et al. (2017). *Journal of Power Sources* 362: 50–56. https://doi.org/10.1016/j.jpowsour.2017.07.020.

72 Abbas, A., Abbas, S., Bhattarai, A. et al. (2021). *Journal of Power Sources* 488: https://doi.org/10.1016/j.jpowsour.2020.229411.

73 Culebras, M., Geaney, H., Beaucamp, A. et al. (2019). *ChemSusChem* 12: 4516–4521. https://doi.org/10.1002/cssc.201901562.

74 Radford, G.J.W., Cox, J., Wills, R.G.A., and Walsh, F.C. (2008). *Journal of Power Sources* 185: 1499–1504. https://doi.org/10.1016/j.jpowsour.2008.08.020.

75 Zhou, H., Zhang, H., Zhao, P., and Yi, B. (2006). *Electrochimica Acta* 51: 6304–6312. https://doi.org/10.1016/j.electacta.2006.03.106.

76 Fu, S., Zhu, C., Song, J. et al. (2017). *Electroanalysis* 29: 1469–1473. https://doi.org/10.1002/elan.201600787.

77 Huang, Y., Deng, Q., Wu, X., and Wang, S. (2017). *International Journal of Hydrogen Energy* 42: 7177–7185. https://doi.org/10.1016/j.ijhydene.2016.04.004.

78 Yang, D.-S., Lee, J.Y., Jo, S.-W. et al. (2018). *International Journal of Hydrogen Energy* 43: 1516–1522. https://doi.org/10.1016/j.ijhydene.2017.11.145.

79 Wang, S., Zhao, X., Cochell, T., and Manthiram (2012). *Journal of Physical Chemistry Letters* 3: 2164–2167. https://doi.org/10.1021/jz3008744.

80 Jiang, H.R., Shyy, W., Zeng, L. et al. (2018). *Journal of Materials Chemistry A* 6: 13244–13253. https://doi.org/10.1039/c8ta03388a.

81 Schnucklake, M., Eifert, L., Schneider, J. et al. (2019). *Beilstein Journal of Nanotechnology* 10: 1131–1139. https://doi.org/10.3762/bjnano.10.113.

82 Mahanta, V., Raja, M., and Kothandaraman, R. (2019). *Materials Letters* 247: 63–66. https://doi.org/10.1016/j.matlet.2019.03.045.

83 He, Z., Cheng, G., Jiang, Y. et al. (2020). *International Journal of Hydrogen Energy* 45: 3959–3970. https://doi.org/10.1016/j.ijhydene.2019.12.045.

84 Krikstolaityte, V., O. E. Y. Joshua, A. Veksha *et al. Batteries* 2018, 4, https://doi.org/10.3390/batteries4040056

85 Jiang, Y., Cheng, G., He, Z. et al. (2019). *Journal of the Electrochemical Society* 166: A3918–A3926. https://doi.org/10.1149/2.0941915jes.

86 Lv, Y., Li, Y., Han, C. et al. (2020). *Journal of Colloid and Interface Science* 566: 434–443. https://doi.org/10.1016/j.jcis.2020.01.118.

87 Maharjan, M., Wai, N., Veksha, A. et al. (2019). *Journal of Electroanalytical Chemistry* 834: 94–100. https://doi.org/10.1016/j.jelechem.2018.12.049.

88 Jiang, Y., Du, M., Cheng, G. et al. (2021). *Journal of Energy Chemistry* 59: 706–714. https://doi.org/10.1016/j.jechem.2020.12.013.

89 Wang, R. and Li, Y. (2019). *Journal of Power Sources* 421: 139–146. https://doi.org/10.1016/j.jpowsour.2019.03.023.

90 Lee, M.E., Jang, D., Lee, S. et al. (2021). *Applied Surface Science* 567: https://doi.org/10.1016/j.apsusc.2021.150810.

91 Vivo-Vilches, J. F., A. Celzard, V. Fierro *et al. Nanomaterials (Basel)* 2019, 9, https://doi.org/10.3390/nano9010106

10

Sustainable Energy Storage Devices and Device Design for Sensors and Actuators Applications

Reeya Agarwal[1], *Sangeeta Singh*[1], *and Ahmed E. Shalan*[2,3]

[1] *Department of Electronics and Communication Engineering, Microelectronics & VLSI Lab, National Institute of Technology, Patna, 800005, India*
[2] *BCMaterials, Basque Center for Materials, Applications and Nanostructures, Martina Casiano, UPV/EHU Science Park, Barrio Sarriena s/n, Leioa, 48940, Spain*
[3] *Central Metallurgical Research and Development Institute (CMRDI), Helwan, Cairo, 11421, Egypt*

10.1 Introduction of Sustainable Energy Storage Devices

Although there are numerous applications for wearable sensors and therapeutic devices, they must be continuously powered to maintain their dependability. The development of wearable electrochemical energy storage devices using supercapacitors (SCs) and batteries has helped meet this demand [1]. Wearable energy storage device research has focused on characteristics like stretchability, self-healing, and transparency because of the human body's curvature, safety concerns, and the wide range of possible applications. However, today's wearable energy storage systems have lower energy densities than previous generations because of the limited thickness and surface area of flexible electronics [2]. Electrochemical energy storage, in part because of SCs' self-discharging feature, eventually runs out of storage capacity. Regular recharging from an immobile electrical grid is required to extend the working time, which causes issues for users [3]. Wearable energy harvesting devices can be used in conjunction with energy storage devices to restore SCs/batteries or even to enable a fully self-sustaining wearable electronics system at a reasonable cost. Biofuel cells and thermoelectric generators have become increasingly popular in recent years as a way to harvest energy from the human body [4]. Figure 10.1 describes advanced energy storage devices.

As a result, this is because most wearable devices are made to be worn comfortably on the user's body. Body energy depends more on human movements or continuous metabolic activity than wearable solar cells, so it is a more reliable power source for wearable devices [6]. In contrast, wearable energy harvesters cannot power a self-powered sensing patch, but they can work with a wide range of sensors. There are two versions of this proof-of-concept model. Energy harvesters could be a useful tool for determining a person's level of intensity [7]. For human body energy harvesting devices, the output power is inversely

Sustainable Energy Storage in the Scope of Circular Economy: Advanced Materials and Device Design, First Edition. Edited by Carlos Miguel Costa, Renato Gonçalves, and Senentxu Lanceros-Méndez.

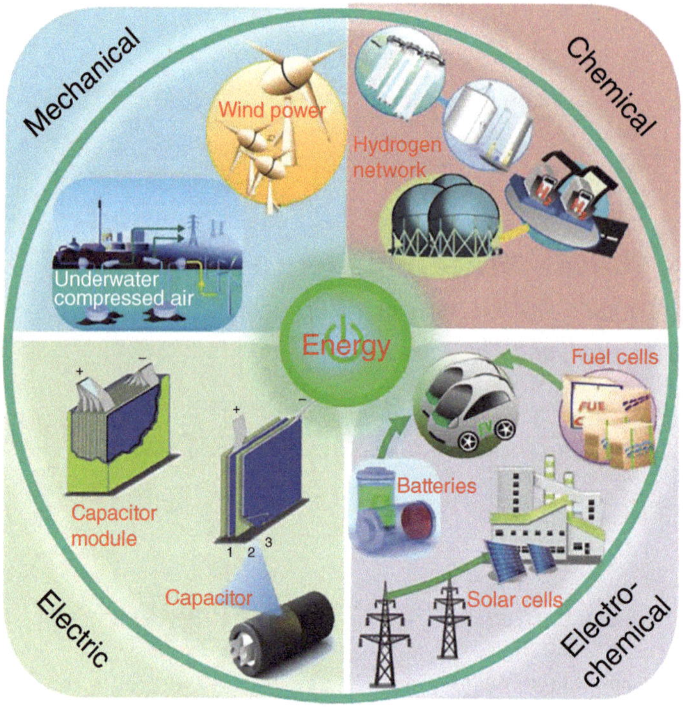

Figure 10.1 Advanced Energy Storage Devices. *Source:* Liu et al. [5]/With permission of John Wiley & Sons.

proportional to the intensity of human physiological signals such as movement frequency, sweat/blood lactate/glucose concentration, or body temperature. Using human body energy to power extra-biological signals or external stimuli is a fascinating idea [8]. A temperature sensor or a chemical sensor, for example, could be powered by triboelectric nanogenerators (TENGs). Instead of a charging procedure and additional surgery to change the energy source, further development of the technology could lead to an implanted self-sustaining system that uses human-body energy harvesters [9]. There are implantable pacemakers, like the TENG and BFC, that can power the devices. Microelectronics, meanwhile, is finding new uses for human-body energy as the technology shrinks, becomes more flexible, and uses less energy. Scientists are working to develop rechargeable flexible SCs/batteries that harness the energy generated by body motion, heat, and biofuels to provide long-term power for worn or implanted devices to benefit from these advantages [10]. Figure 10.2 describes sustainable and versatile energy storage devices.

The widespread use of wearable SCs and batteries will render stationary power plants obsolete. Combining TENG/PENGs with SC/battery may greatly expand the practical use of human-body energy harvesters because these devices can now produce stable output for the performance of TENG/PENGs, BFCs, and TEGs, which are affected by movement status and the surrounding environment [12]. When it comes to human-body energy harvesters, conformability and mechanical compliance are often used because they must come

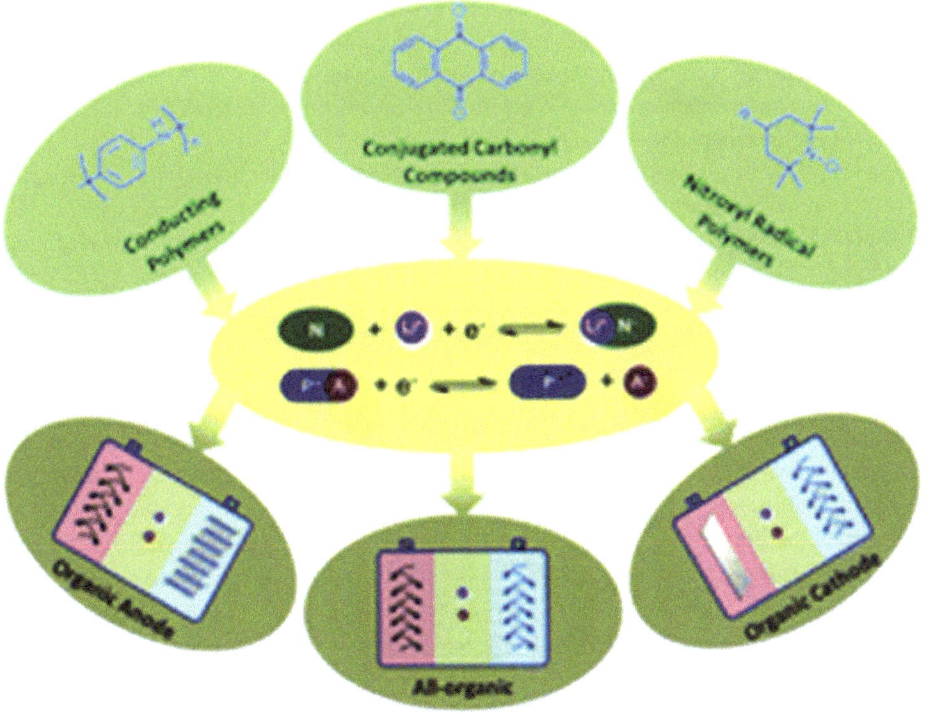

Figure 10.2 Sustainable and Versatile Energy Storage Devices . *Source:* Reproduced from Zhou et al. [11]/Royal Society of Chemistry/CC BY 4.0creativecommons.org/licenses/by/4.0/.

into contact with soft curvilinear tissues like the skin, heart, or eyes. There have been a variety of platforms used to develop wearable energy harvesters with specific characteristics such as comfort, biocompatibility, flexibility, stretchability, washability, or ease of fabrication such as thin polymer films and tattoos [13]. To be fully integrated with current energy harvesting technologies, energy storage devices must be fully functional and produced in the same way as energy harvesters. Using wireless sensor networks to collect environmental data is common when environmental data is needed for decision-making purposes [14]. Keeping tabs on natural and manmade environmental changes is referred to as "environmental monitoring." There are a variety of methods for monitoring the environment in real-time including aerial and satellite remote sensing, sensor networks, and laboratory analysis of field samples. It is their job to watch for changes in the environment's chemistry, biology, or demographics. There is something for everyone in these ecosystems: Deserts have everything, from frigid nights to tropical rain forests to Arctic wastelands. The development of environmental nodes is fraught with scientific challenges [15]. It is difficult for them to access the public power system because they are often located in remote areas. Understanding which architecture and operating methods will save the most energy for the nodes in question can be difficult. An energy-efficient design combined with a data-for-energy trade-off can help a node last longer. As a result, data collection will move at a slower pace [16]. Choosing an appropriate power source for the network's operation is the first step in embedded wireless systems and networks (EWSN) design. EWSN devices deployed in the field are typically powered by main (non-rechargeable) batteries. Energy harvesting techniques frequently make use of primary batteries and other forms of energy storage [17]. The number of portable electronic devices will rise steadily in the coming decades as the internet of things (IoT) expands.

When artificial intelligence (AI) and highly integrated microelectronics are combined to create energy storage frameworks, such as electronic textiles, intelligent sensors, and medical implants with high-tech capabilities, a wide range of useful applications can be created (see Figure 10.3). Miniaturized components and microfabrication technologies, which are becoming increasingly popular, may help improve the energy density of practical micro-energy storage devices [19]. When energy is scarce, MESDs must be adaptable and reliable. MESDs, on the other hand, tend to fracture under chemical or mechanical stress, resulting in the loss of electrical contact. As a result, a micropower source's mechanical performance must match its electrochemical performance. For thousands of years, textiles have been an indispensable part of daily life, and that is still the case today [20]. They have recently gotten a lot of attention because of their unusual combination of porosity and flexibility.

The rest of the chapter is divided into 8 sub-sections: where Section 1.2 describes the literature review, Section 1.3 describe the need for sustainable energy storage devices and Energy storage approaches whereas sustainable and ecofriendly energy storage is described in Sections 1.4 and 1.5 describes different energy storage mechanisms, Section 1.6 describes different novel 2D materials for energy storage, 2D materials for energy storage devices and challenges facing 2D energy technology. Section 1.7 describes nature-inspired materials for sensing and energy storage applications and Section 1.8 describes the summary of the chapter.

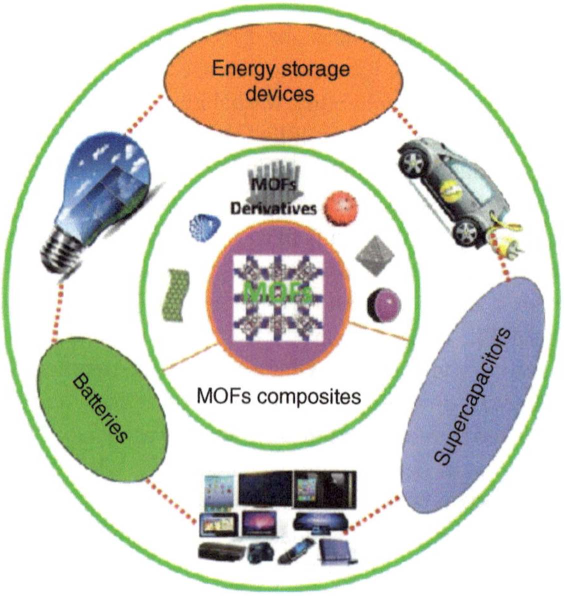

Figure 10.3 Metal-Organic Frameworks for Energy Storage Devices. *Source:* Tahira et al. [18]/With permission of Elsevier.

10.2 Literature Survey

The term "energy storage" refers to a technique that transforms electricity into a form that can be stored due to the difficulty in storing it (e.g. electrochemical). Later, the saved energy may be put to use [21]. Each type of energy storage system has its own set of benefits and drawbacks, so shopping around is a good idea. The application dictates the technology to be used. When using energy storage units for environmental monitoring, many factors must be taken into account. The most important are the compact size, adequate capacity, and minimal environmental impact of the unit [22]. An energy storage subsystem's overall performance has a large impact on a sensor node's overall efficiency The type of energy storage used has an impact on the node's size, cost, and lifespan. Figure 10.4 shows some common storage devices for powering sensor nodes for environmental monitoring [23].

Muhammad Rakibul Islam et al. [2021] E-waste is generated when electrical goods are used, and this non-biodegradable e-waste is hazardous to both human health and the environment in large quantities. With bio-polymer-based nanoparticles, environmentally friendly and long-lasting devices may be possible. Gelatin/single-walled carbon nanotube (Gel/SWCNT) nanocomposites were created using water casting in this study, and they were made in an easy and low-cost way. It was discovered that SWCNT influenced the nanocomposite material's structure, surface morphology, and electrical and electrochemical properties in a study that was conducted. Researchers found that by examining the FTIR and field emission scanning electron microscopy (FESEM) data on the interaction between SWCNTs and the Gel matrix, the interaction was noticeably improved. The hydrophobic SWCNTs in the nanocomposites became hydrophobic as a result of the addition of hydrophilic SWCNTs to the Gel matrix [24].

Narendra Pal Singh et al. [2021] Polymer-based batteries have become increasingly popular and widely used in recent years. A wide range of redox-active polymers with a variety of structural topologies and colors are available to researchers for use in creating active materials. Redox polymers can be oxidized or reduced reversibly, and this property makes them redox polymers. Polymer-based batteries can be used to create devices with

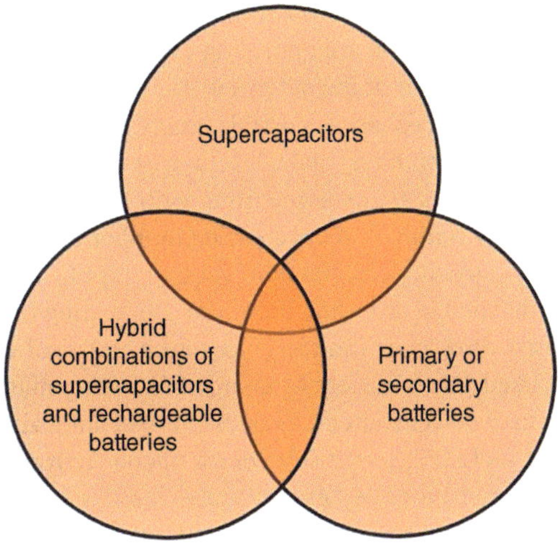

Figure 10.4 These devices are often used to power environmental sensor nodes.

flexible or even bending electrodes and to achieve high power densities. The battery can be produced in an environmentally friendly manner because no (heavy) metals are used in the materials. Conjugated polymers with distinct isolated redox centers, as well as non-conjugated polymers with clearly defined redox moieties, are the subjects of this investigation. In the future, polymer-based batteries will enable a slew of innovative new applications for energy storage technologies [25].

Libu et al. [2021] Today's wearable gadgets run on a combination of Li-ion batteries and various energy harvesters, which means they need to be powered by a reliable energy source like solar panels. Despite the dangers, they pose and the long-term harm they will do to the environment, numerous treatments are in use today. Health monitoring, prosthetics, and sports all use wearables powered by dependable, long-term energy sources. Wearables that produce power in the W to mW range, which is sufficient for the operation of sensing and communication components, could be developed using devices that combine energy production with energy storage. Components that store and produce sweat, as well as sensors that use sweat as a signal source [26].

R. Shruthi et al. [2021] Data collection, processing, and analysis are all part of precision agriculture, which aims to increase agricultural output while using fewer resources. One of the most widely used methods of modern agriculture is precision agriculture. Agriculture contributes significantly to the US economy. Precision agriculture uses the IoT and wireless sensor networks (WSN) in conjunction with other technologies to provide an intelligent farm management system. Agronomists can use IoT to help them make better decisions about their crops because it provides them with real-time updates on their progress. Agricultural applications become game changers as the IoT monitors and transmit information without the need for human intervention. It is possible to integrate IoT systems with agricultural operations by using sensors and actuators that can respond to a wide range of inputs and provide live data. Precision agriculture's primary sensors, which vary in size depending on the operation's scale, can benefit both small and large farms. It is the goal of this research to gain a better understanding of how sensors can be used in agriculture, to characterize the different adoption levels of the IoT in agriculture, and to investigate various sensing methods for effectively monitoring agricultural metrics [27].

Khairunisak Abdul Razak et al. [2021] When it comes to harvesting energy, piezoelectric materials are appealing due to the possibility of converting mechanical energy to electric power or the other way around. Since scientists first started looking into piezoelectric materials as a possible energy collection system, several decades have passed. Researchers have looked into a wide range of materials and architectural designs. Researchers are experimenting with harnessing mechanical energy to collect power from the environment. An increasing number of studies using piezoelectric materials, based on fluid/structure interaction, human input, and vibration energy harvesting, have been published in scientific literature. Depending on the situation, the absorbed energy can be used directly or indirectly to power electrical equipment. For example, researchers have looked into a variety of energy-harvesting materials such as concrete energy harvesters, sensor power sources, microelectromechanical systems, and implantable medical electronics, including power sources for microelectromechanical systems. The use of a physiological activity to collect energy *in vivo* and charge the device from within has been demonstrated in more recent studies, and this represents a significant advance. Materials like lead zirconate titanate, barium titanate, zinc oxide, potassium sodium niobate, and poly(ethylene glycol) have had their piezoelectric energy harvesting properties studied. Vinylidene fluoride is a chemical compound. Piezolectrochemistry, which stands for piezoelectricity-chemistry, has recently been developed by researchers as a new technique for collecting and storing energy. As a result of this innovation, future energy collection systems will be even more environmentally friendly than current systems [28].

Ben Richard Hughes et al. [2021] Research on self-charging power systems has largely focused on the low energy collecting component because it is so important. Low-power devices like wearables, wireless sensor networks, portable electronics, and LED lights have a high demand for energy storage. This research provides a comprehensive assessment of various types of energy storage that can meet that demand. According to this article, there has been progressing in the integration of low-energy harvesting and energy storage and management systems research, and this includes current scientific advancements and relevant studies. When developing the self-sustaining power unit, scientists took several factors into account, including the design, distribution networks, efficiency, and compatibility with other components. Before settling on a self-charging power unit for a specific application, it is critical to weigh the technical, economic, and environmental implications thoroughly. In-depth research now shows the challenges and new research topics that need to be explored to develop a more sustainable and reliable integrated system that can be used in real-world situations rather than just in the lab, instead [29].

Jia Guangwen et al. [2021] Water evaporates in every part of the world, even in Antarctica. To keep up with this trend, manufacturers are turning to smart materials that convert electrical impulses from water stimulation directly into mechanical motion. The development of materials with fast response times, broad-scale deformation, and long-term stability is a difficult issue in the development of novel functional materials for moisture-gradient actuators. Oxygen groups and d-spacing can be used to adjust the MXene/GO composition ratio, which can be done by adjusting oxygen groups and d-spacing. MGO_3 bending speed is astoundingly quick. The activation of MGO_3 remained constant throughout the experiment due to the presence of GO, which inhibited the oxidation of MXene. It was also extremely stable across a wide range of cycles. MGO_3 actuators can be used to create devices that mimic flower blooming, transport objects, and act as a non-contact control switch. Because of its linear response to humidity, MGO_3 is a good option for respiratory monitoring in terms of biocompatibility. It was found that MGO films have a wide range of potential applications in smart actuators and sensing devices as well as biological and health care research [30].

H. Zhao et al. [2021] The public's interest in pavement monitoring, communication, and energy distribution has grown as the electric car era has progressed. Pavement engineering technologies such as thermal, piezoelectric, and photovoltaic are all discussed in depth in this paper. For maximum efficiency, solar panels should be used to replace the surface layer and installed on top of the thermoelectric module and piezoelectric transducer, by the ambient energy distribution principle. Power density and Levelized Cost of Energy (LCOE) were used to compare three different technologies' performance in terms of cost-effectiveness. The ownership costs of piezoelectric transducers were higher than those of thermoelectric modules. When compared to other technologies, photovoltaics had the highest power density. These new technologies had a wide range of uses because of.

Electrical devices such as wireless remote sensors for pavement disease and traffic conditions can be powered by piezoelectric technology, which may also be used to power other low-power electronic devices. Because of its high power density, solar pavement has found use in a wide range of fields. The urban heat island effect and pavement rutting were both reduced by utilizing thermoelectric technology. The long-term durability, safety, and overall life-cycle cost of sustainable energy technologies should be thoroughly investigated. Alternative energy technologies like photovoltaic thermal (PV/T) systems and others like them, such as hybrid energy production technologies, should also be considered [31].

Martin Kaltenbrunner et al. [2021] A significant and far-reaching impact of technology has been felt by society in recent years. Digital and robotic devices are becoming increasingly important to everyone, from newborns to the elderly. Even though progress is heartening, it comes at a price for our ecosystem, with rising energy consumption, greenhouse gas emissions, and environmental degradation due to everything from production to inappropriate disposal. Materials research is working hard to find solutions to one of society's most pressing problems: climate change. New soft, biologically inspired electronics and robotics have the unique capacity to become like their natural antitypes in terms of environmental effect and ecological impact when it comes to performance and performance. These are still under construction. A growing number of soft and bioinspired robotic components, from actuators to energy storage and electrical components, make use of sustainable materials. Everything from in vivo microbots to biohybrid machines to completely biodegradable but highly effective actuators is possible with biobased robotics, and the possibilities are nearly limitless. Recent advances in robotics point to a more sustainable future that is both environmentally friendly and long-lasting [32].

José María Portalo et al. [2021] Examples of the digital revolution in traditional energy systems include smart grids and microgrids, which are intelligent energy facilities. As a result of these new paradigms, the authors claim, renewable energy use, sustainable development, and grid resilience will all rise. Smart grid and microgrid experimental designs were found to have a problem with heterogeneity in hardware and software requirements and the use of commercially available equipment during testing. Experimental validation, specification of accompanying equipment of industrial/proprietary or open-source nature, and concretization are often insufficient in their respective fields, despite previous research providing a diverse range of designs. To address these problems, the research presented in this paper proposes a novel multi-layered architecture for the deployment and monitoring of heterogeneous microgrids. To better organize the hardware and software components, the architecture is divided into six functional levels [33].

Aguilar J. et al. [2021] Buildings in urban areas consume a large amount of energy, so this is a hot topic for R&D. System integrators are expected to grow in importance over the next few years, as more buildings implement BEMS AI techniques have been and will continue to be critical to technological advancement. Using AI methods to improve smart building energy management systems, this study conducts a comprehensive literature review. When it comes to environmental goals like increasing the environment's energy efficiency, an autonomous management system needs to do a lot of specialized things like data processing and decision-making. As a result, the author discovered new areas of research, such as the development of self-driving data analysis cycles and systems with a large number of agents [34].

Azam Ali et al. [2021] Recent advances in mechanical energy harvesting (MEH) technology have resulted in a flurry of discoveries. Using Cite Space's bibliometric analysis, MEH's growth patterns and current research status were visualized. Web of Science Core Collection (WOSCC) performed an advanced keyword search on MEH between the years 2000 and 2020 and found the original publications. It was then decided which institutions and countries had the most prolific writers by visualizing author and institution collaboration networks. It was possible to refine our initial findings by using co-citation networks to identify those that had the greatest impact on them. Using keyword analysis, the cluster analysis revealed the structure of knowledge and the primary subject groupings, revealing current hot research areas and new frontiers for future research. Researchers set out to look at the key MEH research institutes on a micro, meso, and macro scale to see how they collaborated and where they got their knowledge on MEH issues. Moving forward, MEH research will place an increased emphasis on the current state of knowledge and hotspots. According to grants and collaborative countries studies, policy support for MEH research from China, the United States, the United Kingdom, and South Korea is highly advantageous. Before focusing on the practical applications of energy harvesting technology in real-world settings, the research team compared and contrasted different energy harvesting systems [35].

Peng Shuhua et al. [2021] Wearable medical equipment, electronic skin, and soft robots will all benefit from the development of elastomeric and flexible electronics that can expand and flex. Stretchable electronic devices need to be able to withstand mechanical strain without failing and maintain the materials' electrical conduction capabilities despite their stretchability. Conventional metals and silicon-based semiconductors can only dream of such a feat. Shortly, polymer nanocomposites with excellent stretching and electrical capabilities have the potential to revolutionize a wide range of applications, such as stretchable conductors and electrodes, as well as deformable energy harvesting and storage devices. Other relevant topics, such as wearable sensors for biophysical signals, are also discussed in this article. Defiance of restrictions calls for careful consideration of the selection and hybridization of nanomaterials as well as new microstructure designs and the application of simple production procedures [36].

Aliakbari Raouf et al. [2021] The need for better energy storage systems and materials has grown as people's understanding of and reliance on electronics has grown. It is the goal of this study to show that poly ortho-aminophenol and its composites can be used as a form of energy storage in the long term. Capacitive qualities (specific capacitances) for each composite are well-documented, and comparisons to other electrode materials are also available. In the following section, all of these composites are compared to a wide range of different electrode materials. In-depth research has been done on a few of these materials' electrochemical test graphs (cyclic voltammetry [CV], GCD, and electrochemical impedance spectroscopy [EIS]). POAP composites outperform pure POAP and other composites electrochemically, according to the study's findings. They have also been tested on real smartphones and are the best of the bunch. These electrodes' morphological and structural properties have also been examined (scanning electron microscopy [SEM], transmission electron microscopy [TEM], Brunauer–Emmett–Teller [BET], Fractal dimension, and components). A synergistic relationship has been found between the various parts. By using this polymer, electrodes with improved performance and customizable surfaces can be created. The study concludes with a few final thoughts and predictions for the future. Due to their electrochemical and morphological properties, these composites could be used as supercapacitor electrodes in batteries [37].

Chimay J. Anumba et al. [2021] New and better ways to complete projects on time, safely, and within budget are constantly being sought out by the construction industry. Building processes have been automated, but with limited success in reducing long-standing risks that have bedeviled the business. Industry 4.0 is expected to reduce waste, improve project efficiency, and boost productivity. However, the successful adoption of many new technologies, such as virtual design modeling, sensing, data analysis, storage and communication, human–computer interfaces, and robotics, will be crucial to the success of the shift. Automating and real-time control, both of which are currently lacking, will be made possible with the development of digital twins and cyber-physical systems. An electronic copy of an existing physical system can be created by using sensors and actuators in a digital twin. However, in the case of a cyber-physical system, electronic components can be integrated into a physical system. Despite their success in reducing non-fatal injuries and improving safety management, the report says, digital twins and cyber-physical systems are only just now finding use in the construction industry. Cyber-physical systems and digital twins are already being used in other industries, and this paper argues that the construction industry should do the same. The next generation of digital twin and cyber-physical systems is defined in this study as the integration of developing technology with physical construction or constructed facilities. According to this research paper, using digital twins and next-generation cyber-physical system scenarios can help companies increase employee productivity, improve worker health and safety, and better manage building system lifecycles while also improving workforce competence [38].

Dipankar Bandyopadhyay et al. [2021] To create a more environmentally friendly and environmentally sustainable planet, the use of natural resources for energy collection must be re-examined. Traditional energy collection methods emphasize centralized production rather than decentralized delivery as opposed to a complicated distribution network. Modern portable devices, however, are concerned about the economic viability of incorporating such energy. Miniaturized power harvesters that can extract high-density electricity from their surroundings are needed for all of these devices, including smartphones, laptops, iPads, and even Google Glasses. Unique micropower generators have been developed to capture high-density energy to supplement lithium-ion or lithium-polymer battery technology, and some of these have been invented. Over the last few decades, advanced research in nanoscale technologies, such as triboelectric nanogenerators and thermoelectric devices, has progressed at an alarming pace. This classification also includes piezoelectric devices. Many different applications with various power consumption ranges can be implemented using technological advancements spanning from the micro to the nanoscale, as shown in prototypes [39].

Jain Rachna et al. [2021] Cyber-physical systems, which are becoming increasingly popular, make extensive use of actuators, sensors, and networking technology. Cyber-physical social systems have emerged as a result of significant advances in the field. Among the technologies that are part of Industry 4.0 are the IoT, Big Data, Cloud Computing, and Intelligent Systems (IS). Cyber-physical systems connect devices on the physical network to the internet. In addition to transportation and weather, other fields such as robotics and smart buildings can all benefit from the technology. For a cyber-physical systems (CPS), it is all about how cyber and physical components interact with each other as well as the IoT. It is also about security via firewalls and cryptosystems as well as robustness in the system as well as a human intervention both inside and outside the loop [40].

Wang et al. [2021] A flaw detector and an energy source go hand in hand in dynamic processing. Electricity, gas, and heat are among the many energy sources that make up the integrated energy system (IES). Numerous significant challenges are addressed and examined, including collaborative planning, optimization, threat assessment, status assessment, and general situational awareness. Machine learning has significant strategic and practical relevance for relevant studies and practices in China's energy system development, as well as for relevant studies and practices in the United States. The integrated energy system uses machine learning technology. Fault detection in China's energy-based district heating system is the primary goal of regional internet research (RIR) and energy development in the country. The integrated energy system utilizing machine learning technology helps to stimulate the economy and create jobs more effectively than the traditional electricity distribution system [41].

Gao Shan et al. [2021] When it comes to sports training and competition, motion monitoring of the lower limbs is essential. No cost-effective and energy-saving option currently exists that reduces computational complexity without compromising on cost. This motion-capture and energy-harvesting lower-limb gizmo was 3D printed and put to the test. Low-frequency biomechanical energy harvesting and lower-limb motion sensing are made possible, as are other applications, using sliding block-rail piezoelectric generators (S-PEG) and ratchet-based triboelectric nanogenerators (R-TENG). Mechanical features have been designed into a new S-PEG to convert 3D motion in the lower limbs to 1D linear sliding on the rail during the manufacturing process. When used in conjunction with the IoT architecture, an S-PEG operating at a very low frequency can deliver high output power to self-sustaining wireless sensor devices. It is possible to convert rotation angles to pulse counts by using digitalized triboelectric output devices such as R-TENGs, and vice versa. Additional physical factors can be estimated to enhance sensory perception. The Consumer Electronics Show will feature physical rehabilitation demonstrations, virtual reality human–machine interfaces, and sports monitoring (CES 2017). Lower-limb motion monitoring is in high demand, and this hybrid system was created to meet that demand [42].

Guo Xinge et al. [2021] Cities and rural areas are using more and more wireless sensor nodes, and energy-collection technologies are becoming more popular as dependable power sources. Wireless sensor nodes. Energy harvesters using electromagnetic induction, electrostatic effect, and piezoelectric effect have been around microelectromechanical systems for a long time since the concept was first floated (MEMS). Due to wearable technology's increasing energy consumption, new nanogenerators, such as the piezoelectric and triboelectric, have been developed as viable solutions for harvesting various types of mechanical power. Battery storage, wireless data transmission, and other useful components have been integrated to produce systems that are both efficient and environmentally friendly. AI is advancing at such a rapid pace that authors expect the development of AI-assisted self-sustaining systems, which will be critical in ushering in the next 5G era [43].

10.3 Need for the Sustainable Energy Storage Devices

The use of clean energy can help protect the environment by reducing pollution in the air and water, as well as damage caused by coal mining and natural gas extraction [44]. Creating electricity is the only activity that contributes to greenhouse gas emissions. While we wait for new fossil-fuel infrastructure to be built, continue to invest in policies at the state and federal levels that promote increased production of renewable energy and increased demand for clean energy from consumers and businesses [45].

10.3.1 Reduce First

Americans waste a lot of electricity because it is used inefficiently. Everyone can save money and help the environment at the same time by using less energy. This does not have to come at the expense of one's quality of life [46, 47]. Even the smallest alterations can result in significant financial savings. Here are three energy-saving tips consumers can do right away as shown in Figure 10.5.

Electricity generation contributes to environmental pollution at a rate second only to that of transportation in the United States. Coal and natural gas are the primary sources of electricity generated in the United States. Fossil fuel energy production hurts the environment. Reduce pollution by generating electricity from renewable sources [48]. Renewable energy

Figure 10.5 Three Energy-Saving Tips.

sources (such as the sun and wind) do not contribute to climate change like fossil fuels do (CO_2). This has a beneficial effect on combating global warming.

10.3.2 Electricity Generation and Health

- Sulfur dioxide (SO_2) emissions, which contribute to acid rain, account for 66% of the country's total sulfur dioxide (SO_2) emissions. According to the American Lung Association, sulfur dioxide can cause asthma attacks as well as fine particle generation, both of which are detrimental to respiratory well-being [49].
- Nitrogen oxides (NO_x), which interact with sunlight to form ground-level ozone and smog, are produced by electricity generation, which accounts for 29% of total nitrogen oxides (NO_x). According to the American Lung Association, excessive levels of NOx in children and adolescents increase the risk of developing respiratory disorders [50].
- Ozone (O_3), a beneficial gas in the upper atmosphere, is found in the stratosphere. Lower-atmosphere ozone, on the other hand, is responsible for smog. Transportation and energy generation are the primary contributors to ground-level ozone formation. Asthma attacks and a greater risk of lifelong lung disease for children who grow up in areas with high levels of ozone pollution are all linked to inhaling ozone, according to the American Lung Association [51].
- In the context of air pollution, particulate matter (PM) is more generally known as soot. Those who have lung or heart illness are at greater risk of injury from exposure to particulate matter (e.g. asthma, bronchitis, and emphysema).
- Global climate change is fueled in part by greenhouse gases like carbon dioxide (CO_2). Fossil fuel burning's long-term impacts could be considerably more dangerous than air pollution's current toll on human life. As the climate warms, tropical diseases may become more prevalent, and deaths from severe weather may rise [52].

An extremely poisonous element called mercury is emitted into the environment by coal-fired power plants. Fish and other animals' fat cells contain large amounts of mercury. Humans are exposed to mercury when they consume seafood [53]. Permanent liver and central nervous system damage caused by mercury exposure can result in cognitive and motor impairment as well as hearing loss as well as slurred speech. Mercury can cause birth problems and developmental disorders in children if consumed by pregnant or nursing mothers. Because mercury builds up in living species, it is constantly being recycled in the environment as it goes up the food chain [54].

10.3.2.1 The Economic Benefits of Using Renewable Energy Sources are Numerous

Fuel diversity and increased energy security are provided by renewable energy, as well as the decreased danger of fuel spills and reduced reliance on imported fuels. As an added benefit, the use of renewable energy helps to protect the nation's limited natural resources [55].

10.3.2.2 Protection of the Energy Supply

Fuel diversity and increased energy security are provided by renewable energy, as well as decreased fuel spill risks and less reliance on imported fuels. As an added benefit, the use of renewable energy helps to protect the nation's limited natural resources [56].

10.3.2.3 Increasing the Economy

More jobs are being created on average because the renewable energy industry employs more people than the fossil fuel industry. More money for families is generated by this sector. This benefits the renewable energy supply chain as well as other businesses that are not related [57].

10.3.2.4 Stability of the Currency

Renewable energy sources such as wind, solar, hydro, and geothermal do not require fuel or transportation, so their costs are more predictable. Incentives for consumers to use renewable energy sources are available from some power companies [58].

10.3.2.5 Electricity and the Environment

Traditional power generating emits a wide range of harmful substances into the environment as a result of the combustion of fossil fuels. Many of the same chemicals that are bad for our health also have negative effects on the ecosystem [59]. Fossil fuel-based electricity generation is to blame for:

- About 37% of the country's emissions are from carbon dioxide (CO_2), potent greenhouse gas and a major contributor to climate change. Carbon dioxide is released into the atmosphere when fossil fuels are burned. With the increased frequency and intensity of heat waves, more severe droughts, an increase in severe storms, and an increase in infectious disease spread due to climate change, it's no secret that climate change is a major environmental threat [60].
- When sulfur dioxide (SO2) and rainwater combine, it causes acid rain, which affects 66% of the country. Forests, crops, and other vegetation suffer damage from acid rain, and the plants can eventually die as a result. Aside from making lakes and rivers ecologically "dead," acidification is also a problem. Additionally, acidification affects the chemistry of the soil, allowing dangerous heavy metals to leach into nearby bodies of water and the earth. Sulfur dioxide also has the effect of speeding up the deterioration of stone and paint, destroying several historical structures [61].
- As much as a year's worth of mercury can be circulated in the air and transferred thousands of kilometers from its point of origin. Methylmercury builds up in fish's fatty tissue and is regularly regenerated as it climbs further and further up the food chain. Mercury is toxic to the liver and central nervous system, and it has been linked to birth abnormalities in animals and humans alike [62].
- Smog and ground-level ozone are created when nitrogen oxides (NOx) combine with sunlight. In lakes and streams, nitrogen oxide deposition promotes algal blooms. Because of oxygen depletion, fish and other aquatic life become extinct. There is evidence that nitrogen dioxide causes lung illness in animals [63].
- Reduced vision (haze) is mostly caused in the United States by particulate matter. These soot particles made of ash (which may contain heavy metals, radioactive isotope atoms, or isotope decay products), can transport and deposit trace metals like mercury hundreds of miles away from their source. Because soot discolors and corrodes stone and other construction materials, many of our nation's landmarks are harmed [64]. Particles

that have traveled a great distance settle on the ground or in water, generating the effects depicted in Figure 10.6.

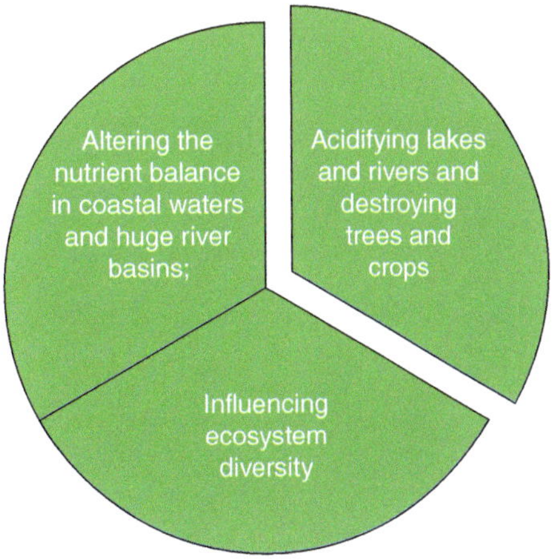

Figure 10.6 Effects by Particles that have traveled a great Distance on the Ground/Water.

10.3.3 Energy Storing Approaches

To achieve a flexible renewable energy output and grid integration, efficient energy storage is a critical pillar of the energy transition [65]. To make the necessary transition to a carbon-free electrical grid, determine which storage solutions are the most effective. Energy can be produced, transported, and changed with relative ease in all three of these three ways. However, there has not been a practical, simple, or cost-effective solution to store it until now [66]. As a result, electricity must be produced continually to meet demand, and this necessitates the use of storage devices to help integrate renewable energy sources and prevent clean energy supply from dropping during times of low demand while also improving grid efficiency and security. Wind and solar energy, which are both becoming increasingly popular as fossil fuels are phased out, require improved electrical energy storage (EES) systems so that the electrical grids remain stable and every green megawatt generated is put to good use [67]. Figure 10.7 shows the energy storage systems overview.

10.3.4 Storage Systems for Large Amounts of Energy

Because electricity cannot be kept in its pure state, it must be converted into mechanical or chemical energy [69]. Supply chains benefit from storage systems at every point. There are three types of energy storage systems: large-scale storage, which is used in places where GW scale is needed; grid storage, which is used in the form of power generation assets, and end-user storage (which is used at the residential level and works with kW), which is storage [70] that is integrated into the power grid. Figure 10.8 describes the energy storage systems classification.

10.3.4.1 Electrochemical Storage

Electrochemical energy storage refers to secondary batteries that store energy electrochemically [72]. Battery oxidation–reduction reverse reactions electrochemically convert chemical energy into electric energy stored in their active components. In the present day and age, batteries are available in a wide range of sizes to meet a variety of requirements. Commercially available secondary batteries can be divided into the following basic categories based on their electrochemical system as shown in Figure 10.9.

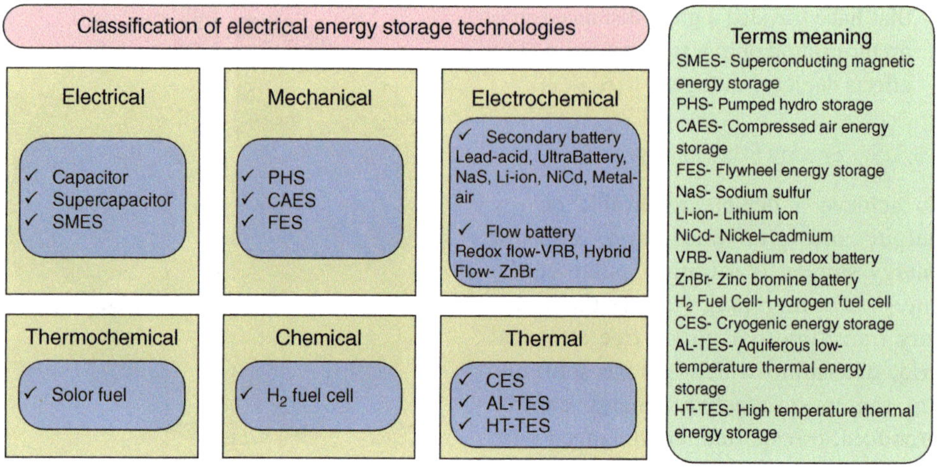

Figure 10.7 Energy Storage Systems Overview. *Source:* Reproduced from Choton et al. [68]/With permission from Elsevier.

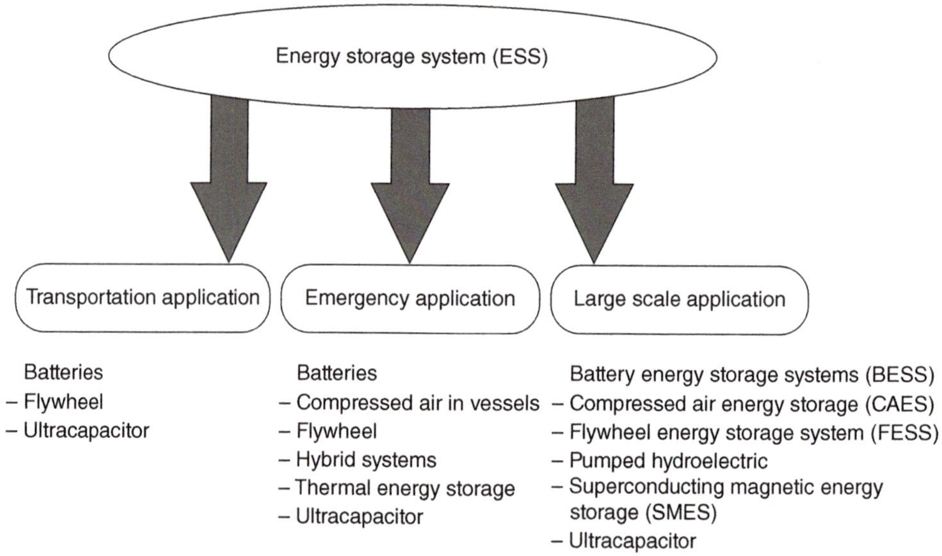

Figure 10.8 Energy Storage Systems Classification. *Source:* Reproduced from Hamdi et al. [71]/With permission from Elsevier.

The development of sustainable energy technologies relies heavily on EES devices. Solar or wind-generated electricity can meet our energy needs while also being environmentally friendly [74]. These renewable energy solutions, however, only provide electricity on demand, necessitating the development of efficient and dependable techniques for storing it. It has been estimated that second-to-second fluctuations cost tens of billions of dollars per year [75]. To exploit large-scale solar or wind-based electricity generation, new EES technologies must be developed. Improved EES systems are also necessary to allow the

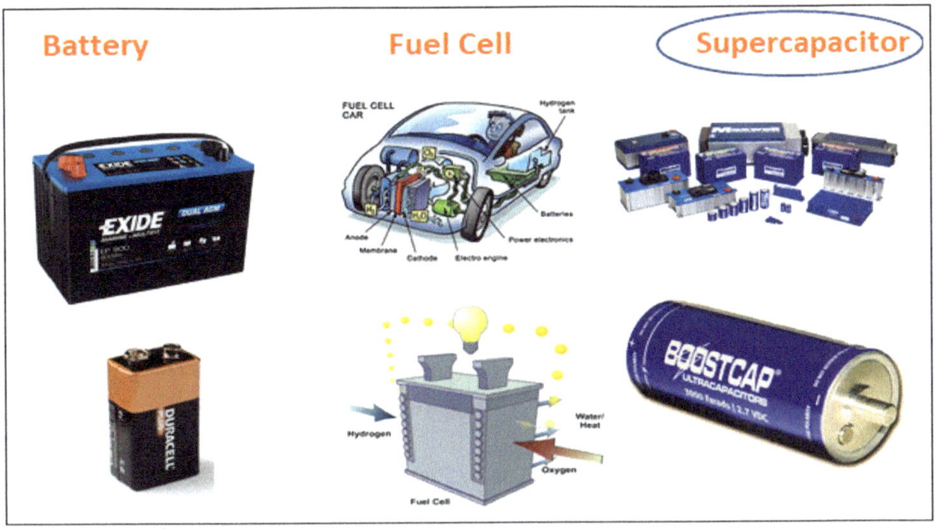

Figure 10.9 Types of Electrochemical Energy Storage Applications. *Source:* Reproduced from Samarjeet et al. [73]/MDPI/CC BY 4.0.

widespread usage of HEVs, PHEVs, and all-electric vehicles practically [76]. Portable electronic devices like laptops and smartphones need better electrical energy storage (ESS) performance, reliability, and efficiency as shown in Figure 10.10.

10.3.4.2 Thermochemical Storage

With a contemporary energy system based on a high proportion of renewable sources and high energy efficiency, thermal energy storage is becoming increasingly important in our society [78]. Due to the integration of solar and wind power into highly efficient technologies like combined heat and power generation, new storage technologies are required, which puts additional pressure on the industry's heat storage professionals. Future energy systems will need more compact and long-term storage than current ones [79]. High-specific storage capacities and low heat losses will be advantageous in future energy storage advancements. Thermal energy storage has been shown in numerous studies over the last few years to be a fundamental technology for producing highly effective short- and long-term thermochemical energy storage for various purposes, such as solar thermal systems or cogeneration systems. Chemical bonds between suitable materials can store energy for long periods with almost no energy loss [80]. At the same time, high levels of energy storage density can be achieved. Both of these characteristics will be required for future energy storage applications. TCES research has expanded rapidly in recent years, particularly in the areas of material creation and material optimization. The term TCES refers to a system for storing heat using thermochemical reactions [81]. This activity focuses primarily on chemical and thermal properties such as increasing energy storage density, improving thermal conductivity, or enhancing cycle stability. Storage and its components, such as the reactor, have also received some consideration. These areas must be covered if the methods are to be commercially available. High-temperature heat storage devices increase the

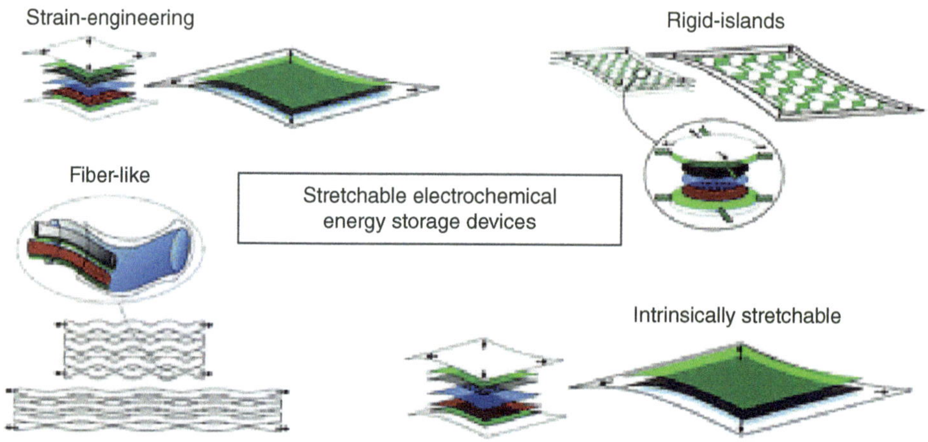

Figure 10.10 Stretchable Electrochemical Energy Storage Devices. *Source:* Reproduced from David et al. [77]/Royal Society of Chemistry/CC BY 4.0.

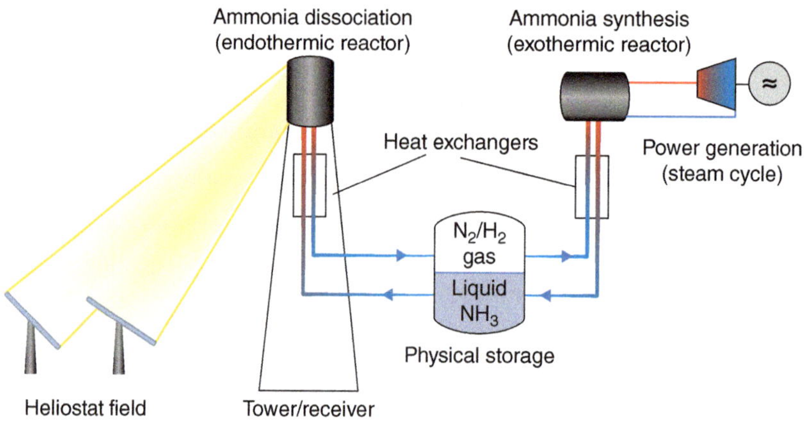

Figure 10.11 Thermochemical Energy Storage System. *Source:* Reproduced from Henner et al. [83]/With permission of AIP Publishing.

efficiency of power plants and the recovery of process heat [82]. Thermochemical processes, even at high temperatures, offer a significant storage capacity option. As depicted in Figure 10.11, the thermochemical energy storage system's mechanism is explained.

10.3.4.3 Thermochemical Energy Storage (TCES): Physical Fundamentals

More than 90% of thermal energy storage processes are based on sensible heat storage technologies [84]. High-temperature applications can make use of solid storage materials like ceramics or liquids like molten salts. Sensible heat storage has a long history and a solid technical foundation. There are only a few examples of TCES systems in use. TCES systems are in a completely different scenario than the rest of the industry. While traditional thermal energy storage methods use reversible chemical or physical interactions, TCES systems

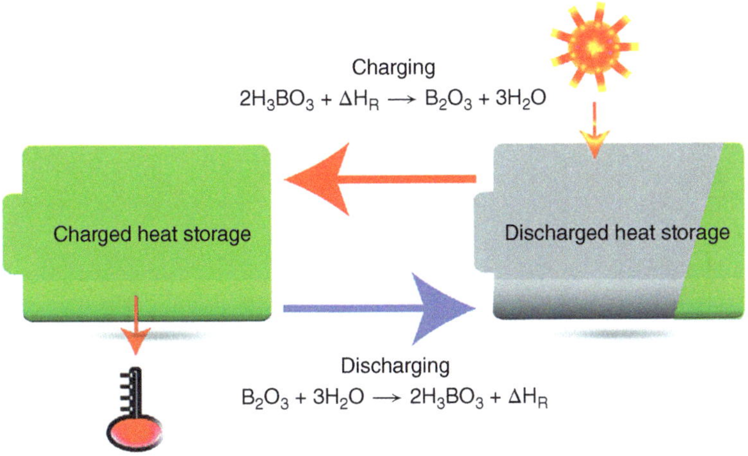

Figure 10.12 A high potential candidate for thermochemical energy storage. *Source:* Reproduced from Clemens et al. [87]/MDPI/CC BY 4.0creativecommons.org/licenses/by/4.0/.

store and release energy through non-thermal means [85]. Adsorption and absorption are thermochemical sorption processes; reversible chemical reactions are thermochemical reversible chemical reactions [86]. Solid–gas and solid-liquid thermochemical reactions are further subdivided. Figure 10.12 depicts a promising thermochemical energy storage candidate.

10.3.4.4 Thermal Energy Storage

Thermal energy storage can be achieved in a variety of ways (TES). Thermal waste can be stored and used indefinitely depending on the exact technology used [88]. This is true at all scales: from a single process to a whole city or region. To demonstrate how to utilize this technology, anyone can store summer heat for use in the winter or winter cold for use in the summer (Seasonal thermal energy storage). As well as ice-slush, large masses of native earth or bedrock can be stored in tanks, as can deep aquifers located between impermeable strata, which can be reached with heat exchangers through boreholes [89]. A thermal energy storage system is shown in Figure 10.13, which explains how it works.

Grit and water can be stored in pits that are lined with these materials, and the tops of the pits can be insulated for added warmth. Another technique for storing thermal energy is peak shaving, which involves generating heat and electricity during off-peak hours while using cheaper off-peak power [91]. Energy systems powered almost entirely by renewable sources of energy require short-term heat storage to balance the high proportions of variable renewable electricity output and integrate the electricity and heating sectors [92]. Explain thermal energy storage systems in Figure 10.14.

10.3.4.5 Chemical and Hydrogen Energy Storage

Chemical energy storage is well-suited to the great transition challenge from fossil fuels to renewable energy production forms like wind and solar because it matches the intermittent, variable, and dispersed generation of various energy sources with geographic consumption

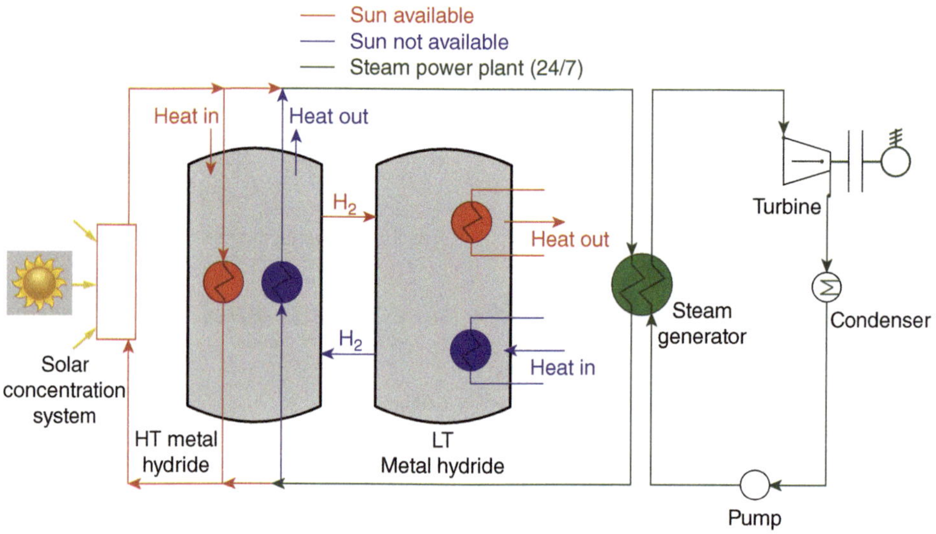

Figure 10.13 Thermal Energy Storage System. *Source:* Reproduced from Bo et al. [90]/MDPI/CC BY 4.0creativecommons.org/licenses/by/4.0/.

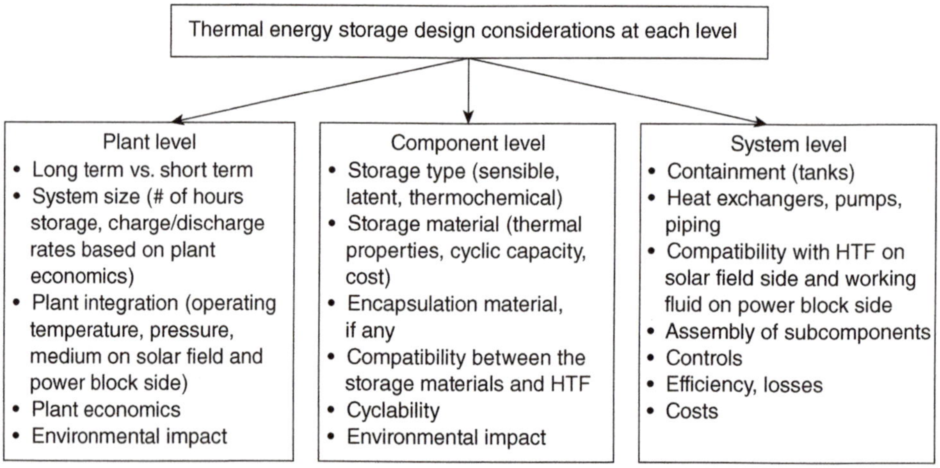

Figure 10.14 Thermal Energy Storage Technologies and Systems. *Source:* Reproduced from Kuravi et al. [93] with permission from Elsevier.

demands [94]. Renewable energy-producing regions are frequently found to be located far from areas where there is a high demand for the energy they produce. Currently, chemical energy storage methods that use liquid hydrocarbons and electrochemical such as reversible batteries, biomass, or gas are the most common. When compared to liquid fuels like diesel, gasoline, or liquid methane, even the most advanced electrochemical or charge storage devices like lithium-ion batteries or ultracapacitors have low volumetric energy densities [95]. Figure 10.15 illustrates the mechanism of a chemical energy storage system.

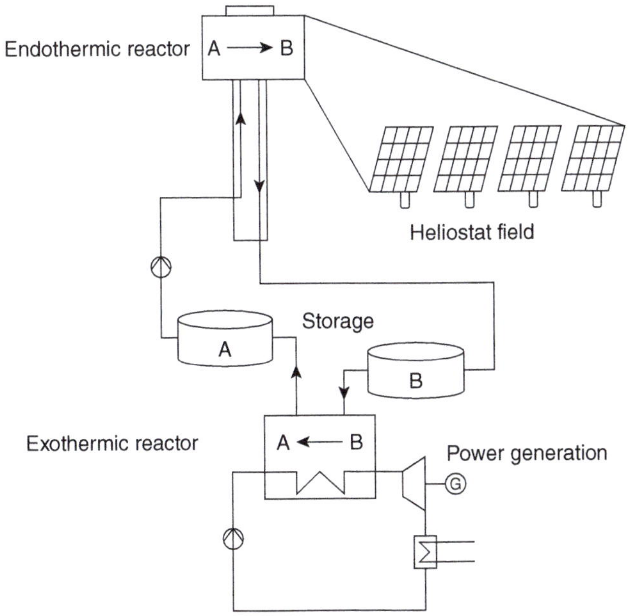

Figure 10.15 Chemical energy storage System Mechanism. *Source:* Adapted from Georg et al. [96].

Storage of hydrogen energy in batteries or capacitors from wind and solar is currently difficult at scale (see Figure 10.16). Chemical energy storage and energy carrier must be distinguished. Only intermittent renewable energy sources necessitate the use of energy storage [97]. An EES system is a common application for the energy storage systems discussed here. Hydrogen energy storage has the distinct advantage of being a general-purpose energy storage solution that can be used in a wide range of applications. Natural gas is enticing as a low-carbon energy source because it produces no carbon dioxide when used (although its production might release carbon dioxide). This means that hydrogen energy storage has the potential to be a game-changer in the energy storage industry [98]. However, it will only be feasible if the hydrogen economy is fully developed. This fuel economy relies on hydrogen rather than fossil fuels. Hydrogen does not naturally occur in large quantities on Earth, despite its widespread use in the chemical industry Instead, a variety of raw materials can be combined to create it. While natural gas still provides the vast majority of today's hydrogen, the element can also be synthesized from water and other compounds like alcohol. The latter is an organic compound [99]. Either method necessitates the use of energy, which can be stored in the hydrogen and released at a later date. Instead of being limited to a few applications like the original energy sources, hydrogen has numerous possibilities. There is no hydrogen economy as of yet, and creating one will require a lot of new infrastructure to get started. To meet the demands of various users, hydrogen will have to be produced on a large scale and distributed in various ways [100]. Some products may be transported via pipeline, while others will be transported to distribution centers via tanker, depending on where they are manufactured. Because hydrogen is a gas even at extremely low temperatures, devising a way to store it in a concentrated form will almost

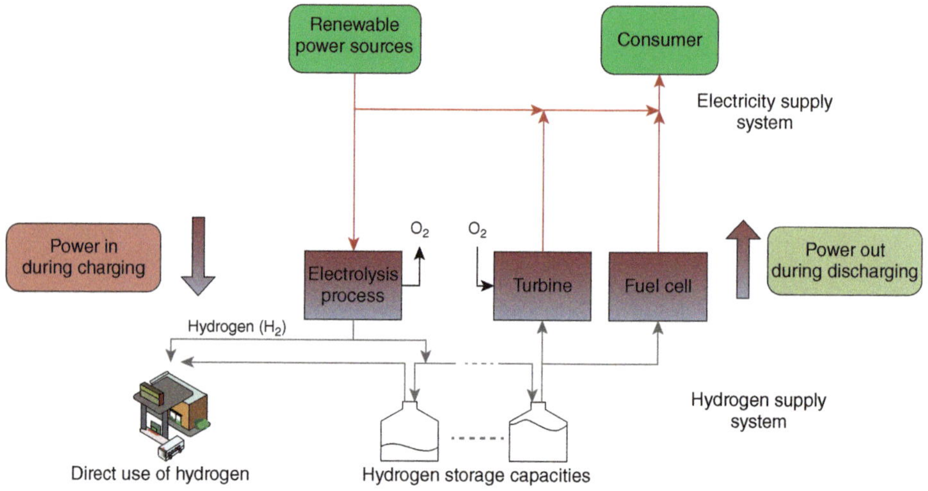

Figure 10.16 Hydrogen Energy Storage. *Source:* Georg et al. [96]/With permission of Smart Energy for Europe Platform GmbH.

certainly necessitate utilizing highly complex technology. For the time being, steel cylinders under high pressure are the only option. As an alternative, hydrogen has a higher energy density, which is advantageous [101]. Hydrogen would likely have to be largely produced electrolytically from water rather than chemically from fossil fuels today to develop a carbon-free economy and deliver the enormous amounts required. Hydrogen can be made from water, but it is currently inefficient due to the need for energy. Water has a lot of uses in the universe [102]. There needs to be a way to generate hydrogen for all other uses from hydrogen storage, so it has a future. It will be necessary to use low-cost renewable electricity sources. Whether or not it is more cost-effective to produce hydrogen in this manner than to use electrical power directly will be determined by numerous criteria; however, the answer is still unclear today [103]. However, the adaptability of hydrogen as a form of energy storage makes it very appealing. It is easy to make with water, but it is pricey right now. Gas turbines, gas-fired steam turbines, reciprocating engines, and fuel cells are just a few of the combustion plants that can make use of natural gas once it is been produced. Open-air burning yields only water as a byproduct [104]. Using hydrogen to power fuel-cell cars is a viable option; however, this technology is still prohibitively expensive when compared to other zero-emissions options. It is a great alternative to natural gas because it can be used to heat and cook in the home. The hydrogen economy is typically envisioned as having a few large manufacturing facilities, but there may be advantages to having a large number of smaller, more widely dispersed ones [105].

10.4 Sustainable and Ecofriendly Energy Storage

Using energy sustainably means not compromising future generations' ability to meet their own needs to meet current demands. There are numerous ways to define sustainable energy. Greenhouse gas emissions and economic inequity are among the most commonly cited causes of global warming [106]. Renewable energy sources such as wind, hydropower, solar electricity, and geothermal energy are all better for the environment than traditional energy sources like coal and petroleum. Some renewable energy projects, such as harvesting forests for biofuels, may have significant negative effects on the environment [107]. The role of nonrenewable energy sources in the energy mix is debatable. Because of concerns about radioactive waste, nuclear proliferation, and accidents, nuclear power, despite being a low-carbon source with mortality rates comparable to those of wind and solar. One advantage of switching from coal to natural gas is that it has a lower climate impact, but this could delay the adoption of more environmentally friendly solutions [108]. Carbon capture and storage (CCS) can be integrated into power plants to reduce CO_2 emissions. The downside is that this is a pricey solution that has only been partially implemented. Off-grid battery applications using sustainable materials are shown in Figure 10.17.

Reducing greenhouse gas emissions to levels consistent with the 2015 Paris Agreement will necessitate a system-wide revolution in energy production, distribution, storage, and consumption [110]. Combustion of fossil fuels and biofuels is a major contributor to air pollution, which is blamed for 7 million deaths annually. With this, moving to a low-carbon energy system is beneficial to human health in numerous ways. While meeting climate change targets and benefiting the health and economies of developing countries, it is possible to make electricity and clean cooking accessible to everyone [111]. Mitigation scenarios for climate change offer ways to transition away from fossil fuels and toward low-emission electricity generation. An eco-friendly energy storage system's secondary battery is defined in Figure 10.18 as a biodegradable battery and its biodegradation mechanism.

As energy-intensive technologies and processes become more difficult to electrify, many scenarios predict that hydrogen fuel derived from low-emission sources will become increasingly important [113]. To accommodate higher proportions of fluctuating renewable energy, electrical networks require flexibility through infrastructure such as energy storage. It will be necessary to shift to clean energy sources and conserve energy in infrastructure and technology to reduce carbon emissions (such as buildings and transportation systems). To reduce energy-related greenhouse gas emissions, emerging technologies that

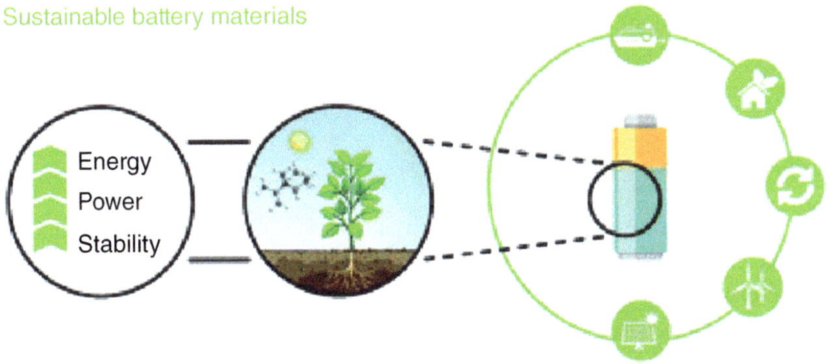

Figure 10.17 Sustainable materials for off-grid battery applications. *Source:* Elizabeth et al. [109]/ With permission of Royal Society of Chemistry.

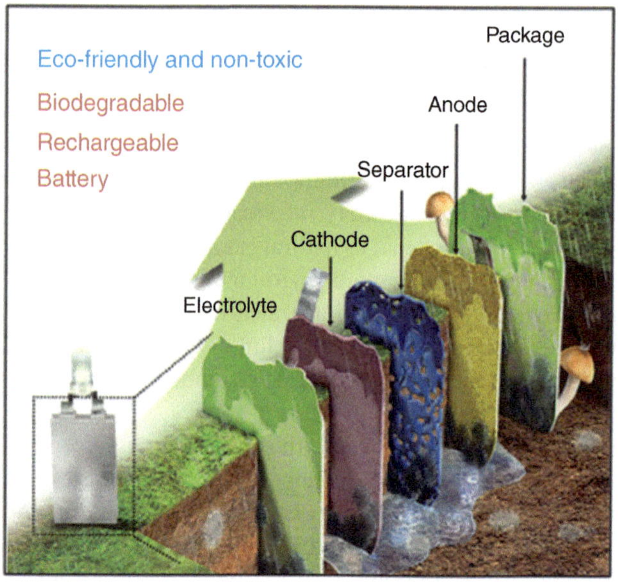

Figure 10.18 biodegradable battery and its biodegradation mechanism. *Source:* Lee et al. [112]/ With permission of John Wiley & Sons.

have not yet matured can be used [114]. Due to falling costs, this segment of the market has grown significantly and is expected to continue to do so. Well-thought-out government policies that promote energy system reform can reduce greenhouse gas emissions while also improving air quality. This often has the additional benefit of increasing energy security. Carbon pricing, renewable portfolio standards, the elimination of fossil fuel subsidies, and the development of electrification and sustainable transportation-supporting infrastructures are some of the policymaking approaches currently under consideration [115]. Figure 10.19 depicts sensing and energy storage hierarchical materials inspired by nature.

Recent technological advancements have reduced the cost of wind and solar power generation significantly. As a result, it is expected that renewable energy will grow at its fastest rate in the United States through 2050. However, the cost of energy storage is still prohibitive. Production of renewable energy is sporadic; it is dependent on the weather [117]. To prevent this energy from being wasted.

10.4.1 Longer Charges

There are many uses for batteries now, from small electronics like alkaline batteries to larger ones like the lithium-ion batteries found in cars and laptops. Despite this, much work remains to be done [118]. The ability to store solar electricity for use at night or extend the range of electric vehicles with large-capacity batteries with long discharge times could be extremely useful. There are not many of these batteries on the market right now. Despite this, new estimates suggest that by 2050, up to 100 GW worth of these batteries will have

Figure 10.19 Nature-inspired hierarchical materials for sensing and energy storage. *Source:* Chungping et al. [116]/With permission of Royal Society of Chemistry.

been constructed When compared to Hoover Dam, it is a 50-fold increase in power [119]. This has the potential to have a significant impact on renewable energy's long-term viability. About 70% of the world's cobalt reserves are found in the Democratic Republic of the Congo, where workers are subjected to appalling conditions during the extraction process. Scientists have developed new recycling procedures for lithium and cobalt batteries. New material batteries are also a goal for the team. Lithium has many of the same properties as sodium, but it is much more readily available [120].

10.4.2 Safer Batteries

It is also important to make batteries safer, which is currently a top priority. A battery's electrolyte particles move around when it is being used to balance the electricity that is flowing from its positive terminal [121]. The components in liquid electrolytes can catch fire if they are not handled carefully. The battery can catch fire or melt if there is a leak in its construction. Scientists are working to improve battery life by developing solid electrolytes. These batteries may be ready for use in electric vehicles as early as 2026, based on promising results from lab tests [122]. Particles have a much harder time moving around in solids than they do in liquids. Although solid-state batteries are better suited for consumer electronics and electric vehicles, scientists are investigating all-liquid battery designs known as flow batteries for large-scale energy storage. In these devices, the electrolyte and electrodes are both liquids. Battery charging speeds are now lightning fast, and large batteries are no longer a logistical challenge. However, research is working to bring the price of these systems closer to the average American family's budget [123].

10.4.3 Storing Sunlight as Heat

Batteries are the most common form of renewable energy storage, but there are less expensive alternatives. Using solar concentrators, for instance, mirrors are used to focus sunlight, which heats the salt to a liquefying temperature and then evaporates it [124]. Heat the molten salt and a generator with steam generated from coal or nuclear power, as is done in traditional power plants. They can be used to generate electricity when it is cloudy or dark outside. Another power-generating method could be used in conjunction with this idea. Wind turbines generate electricity even when it is not windy, which can be used to heat salt later. It is still expensive to use concentrated solar power (CSP) because of the equipment involved [125]. If energy storage and generation are to remain competitive, they must become more efficient. One way to do this is to raise the temperature of the salt, which allows for more efficient power generation. Figure 10.20a–e depicts materials and devices for electrochemical energy storage that are inspired by elements found in nature [127].

10.4.4 Advanced Renewable Fuels

CSP plants have the potential to assist stabilize the electric grid in the short term [128]. Wind turbines and solar panels produce more electricity than the utilities' customers need, therefore utilities can store the extra in these fuels until they are needed. Because hydrogen and ammonia have a higher energy density per pound than batteries, they can be used in applications where batteries are inefficient. Natural gas or other nonrenewable fossil fuels are used to make these fuels today, but the processes are incredibly inefficient [129]. Hydrogen and other fuels are currently being produced using renewable electricity by scientists. Hydrogen fuel, for example, can be made by electrically splitting water molecules. The most difficult part is finding ways to make the process more effective while also reducing costs. A great amount of money may be made by using renewable, never-ending energy [130].

10.5 Different Energy Storage Mechanisms

Storage of mechanical energy is possible in water pushed to higher altitudes, or by transferring solid materials to higher altitudes via pumped-storage methods [131] (gravity batteries). When electrical demand is high, other commercial mechanical methods include compressing air and using flywheels to transform electric energy into internal energy (also known as kinetic energy) and then back again [132]. The following is a comprehensive overview of available energy storage systems as shown in Figure 10.21.

10.5.1 Hydroelectricity

Hydroelectric dams with reservoirs can generate electricity when needed most. When water demand is low, a reservoir holds it back until it is needed again [133]. The overall effect is very similar to pump storage because it does not have a pumping loss. When other power sources are overproducing, a hydroelectric dam reduces its output, which is similar to storing energy in other power plants. Dams are one of the most effective methods of storing energy because only the timing of their generation changes. A hydroelectric turbine can be up and running in a matter of minutes [134].

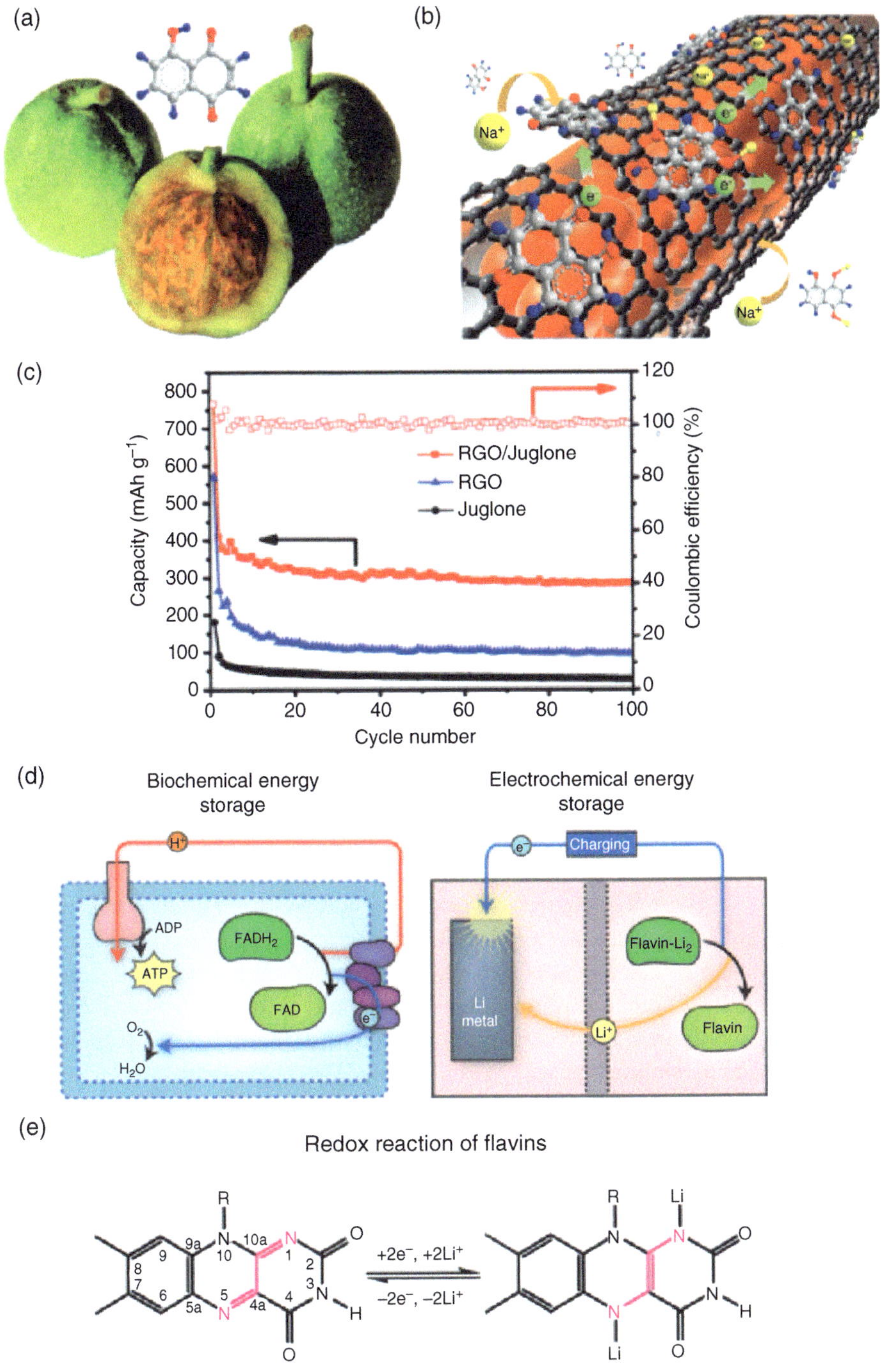

Figure 10.20 Nature-inspired electrochemical energy-storage materials and devices. a) Chemical structure of the Juglone molecule that can be derived from waste walnut epicarp. b) Schematic illustration showing the π−πconjugation of the Juglone molecule with rGO nanosheets and the reversible sodium-ion insertion/extraction mechanism of the hybridized electrodes. c) Capacity retention of Juglone/rGO electrode compared with rGO and Juglone electrodes at a current density of 0.1 A g−1. d) In the left, flavins work as a redox cofactor to facilitate energy storage from nutrients into high-energy ATP in organisms. In the right panel, inspired by biological energy transduction, flavins are proposed to serve as an active material for electrochemical lithium ions storage. e) The electrochemical redox reaction of flavin molecule with reversible lithium ioninsertion/extraction. *Source:* Lin Guo et al. [126]/With permission of John Wiley & Sons.

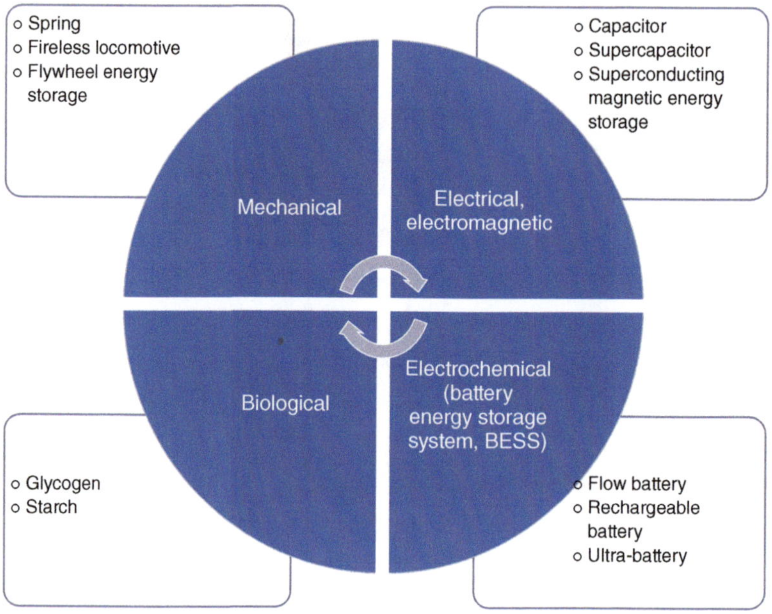

Figure 10.21 Available Energy Storage Systems.

10.5.2 Hydroelectric Power Was Generated and Then Transferred

During times of low electricity demand, water can be pumped from a lower source to a higher reservoir [135]. When the need for electricity rises, a turbine converts water into electricity. Reversible turbine-generator systems are capable of performing pump and turbine functions (usually a Francis turbine design). The height difference between two bodies of water is commonly used by facilities. In contrast to the "pump-back" system, which uses pumped storage in conjunction with traditional hydroelectric plants that use natural stream flow to move water between reservoirs, pure pumped-storage facilities only use pumped storage [136].

10.5.3 A Compressor That Produces Compressed Air

Overage energy is used to compress air for later power generation using compressed air energy storage (CAES). In such applications as my locomotives' propulsion, small-scale systems have been employed for a very long time [137]. A salt dome, for example, is used to store compressed air below. There is a gap between the unpredictability of output and the load that compressed-air energy storage (CAES) can fill. CAES storage meets consumers' energy needs by storing energy that is readily available when it is needed. Wind and solar power are two examples of renewable energy [138]. As a result, when they do not provide enough electricity, other sources of energy are needed to keep up with demand. When renewable energy sources produce too much energy, compressed-air energy storage units can be used to store the excess. When the demand for electricity rises or energy resources become scarcer, the stored energy can be used [139]. Air compression generates heat, hence after compression, the air is warmer. Heat is required for expansion.

10.5.4 Flywheel

When a flywheel is cranked up to an extremely high speed, it stores energy as rotational motion in the form of flywheel energy storage (FES). Energy conservation means that as more energy is added to the flywheel, its rotational speed increases, and as more energy is taken away, it decreases [140]. Flywheel acceleration and deceleration are typically accomplished with the help of electric motors in most FES systems. There is some consideration given to mechanical energy-based devices. These flywheels can reach their top speed in a matter of minutes ("charge").

10.5.5 Gravitational Pull of a Massive Object

A system that raises and lowers solid masses using an electric motor/generator can store and release energy. As much as 85% of the stored energy can be recovered with this technology. Using ancient vertical mine shafts or specifically designed towers, heavy weights can be winched up to store energy and let to be released slowly through a controlled descent [141]. It is planned to open in Edinburgh, Scotland, in 2020, as a prototype for vertical retail. In 2013, the California Independent System Operator worked with potential energy storage, such as gravity energy storage. It looked at how earth-filled hopper train cars were transported from lower to higher altitudes using electric locomotives [142]. The following are some further ideas as shown in Figure 10.22.

10.5.6 Thermal

It is possible to temporarily store or remove heat using thermal energy storage (TES).

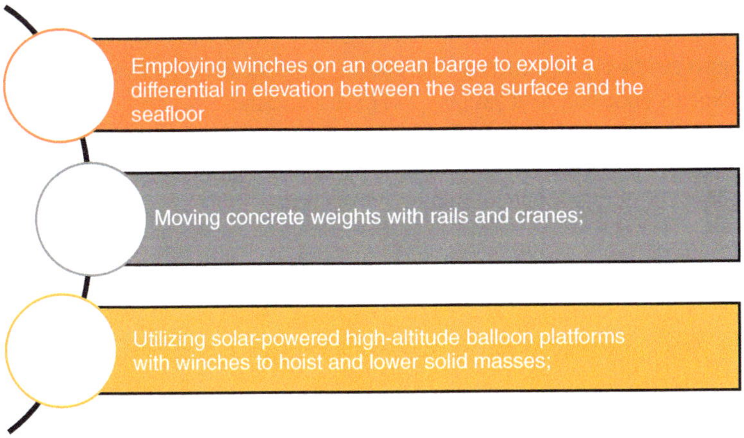

Figure 10.22 Gravitational Pull of a Massive Object.

10.5.7 Thermal Heat Sensitiveness

The internal warmth of the stored energy can be used as a method of sensible heat storage. Heat and cold are stored in winter thermal energy storage (WTES) to prevent wastage months after they are generated [143]. Aquifers, borehole networks, or a water-filled mine can all be used to store material, depending on the geological substrate. Seasonal thermal energy storage (STES)-based projects typically pay for themselves within 4–6 years of going live.

10.5.8 Latent Heat Thermal (LHTES)

Thermal energy storage devices that utilize latent heat transfer heat to or from a material to change its phase [144]. Phase changes include processes such as melting, solidification, vaporization, and liquification. One such substance is a phase change material (PCM). Materials used in LHTESs often have high latent heat, so when they reach a certain temperature, the phase shift absorbs a large amount of energy, far more than sensible heat. Water's latent heat of vaporization is used in a steam accumulator, an LHTES that converts from a liquid to a gas [145]. Ice storage air conditioning systems use off-peak electricity to store cold by freezing water. Ice releases its cold during the melting process, which can be used to keep things cool when demand is high.

10.5.9 Charging System for the Carnot Battery

When using resistive heating or heat pumps, Rankine or Brayton cycles can be used to convert thermal energy back to electrical energy. This technique has been studied for retrofitting existing coal-fired power plants to fossil-fuel-free generation systems. Energy surpluses from intermittent renewable energy sources fuel high-temperature heat storage, which replaces coal-fired boilers [146].

10.5.10 Lithium-Ion Battery

When using a flow battery, the cell is charged or discharged by passing a solution through a membrane, where ions are swapped out. Practical applications for chemically calculating cell voltage use the Nernst equation and ranges [147]. The volume of a container determines how many solutions can be stored in it. Fuel cells and electrochemical accumulators are complementary technologies, and flow batteries are a hybrid of the two. Commercial uses exist for long-half-cycle storage, such as backup power grids.

10.5.11 Supercapacitor

Electrochemical capacitors without typical solid dielectrics are known as electric double-layer capacitors (EDLCs) or ultracapacitors. They are also referred to as EDLCs or supercapacitors. Capacitance is influenced by two storage techniques: double-layer capacitance and pseudocapacitance [148]. Between traditional capacitors and rechargeable batteries, there is a technological chasm that is filled by supercapacitors. A high energy density is achieved by using ultra-high-density polymer capacitors. Supercapacitors have power densities that are 10 to 100 times greater than batteries, despite their lower specific energy and energy density. As a result, the charge/discharge cycles are drastically reduced. Their charge-discharge cycles are far greater than those of batteries [149]. There are numerous uses for supercapacitors, including the following:

- Static random-access memory's low supply current makes it ideal for memory backup (SRAM)
- A variety of power sources can be used to provide power to vehicles such as buses and trains, as well as cranes and elevators.

10.5.12 Chemical

The process of converting electricity into liquid fuel, such as propane or butane, is known as gas power. Using electricity, water is electrolyzed to produce hydrogen and oxygen. In the first instance, hydrogen is either injected into the natural gas distribution system or used as a transportation fuel [150]. As soon as that is done, the methane can be released into the natural gas pipeline. The third method involves using a wood gas generator or a biogas plant to produce hydrogen, which is then added to the biogas upgrader's output gas.

10.5.13 Hydrogen

The chemical element hydrogen serves as a form of energy storage. Hydrogen fuel cells are devices that use hydrogen to produce electricity [151]. When renewables account for less than 20% of total grid demand, their economic impact on the grid is minimal. As a result, as total grid demand rises, so does the demand for additional external storage It is possible to create and expand ionic hydrogen from any of these sources. The naturally occurring aluminum oxide barrier on the metal can be removed and water can be added to produce hydrogen [152]. Commercially viable systems to generate hydrogen from recycled aluminum cans are still a long way off and far more sophisticated than electrolysis systems. It's been no problem for Imperial Chemical Industries to store large amounts of gaseous hydrogen in their underground storage facilities for many years. "Power paste," a magnesium and hydrogen-based fluid gel, releases hydrogen when it comes into contact with water [153]. Magnesium powder and hydrogen are combined to form magnesium hydride, which is heated and subjected to five to six times the pressure that it will be subjected to in a rocket engine to create a power paste. An ester and a metal salt are added at the end to complete the process.

10.5.14 Electrochemical

An electrochemical cell is a component of a rechargeable battery. "Secondary cell" refers to an electrochemical reaction that can be reversed by an electrical current. Some types of rechargeable batteries are also available in disposable battery form factors [154]. For a higher initial cost, rechargeable batteries are preferable because they may be recharged for pennies on the dollar and re-used numerous times. Rechargeable battery chemistries that are common include the following as shown in Figure 10.23.

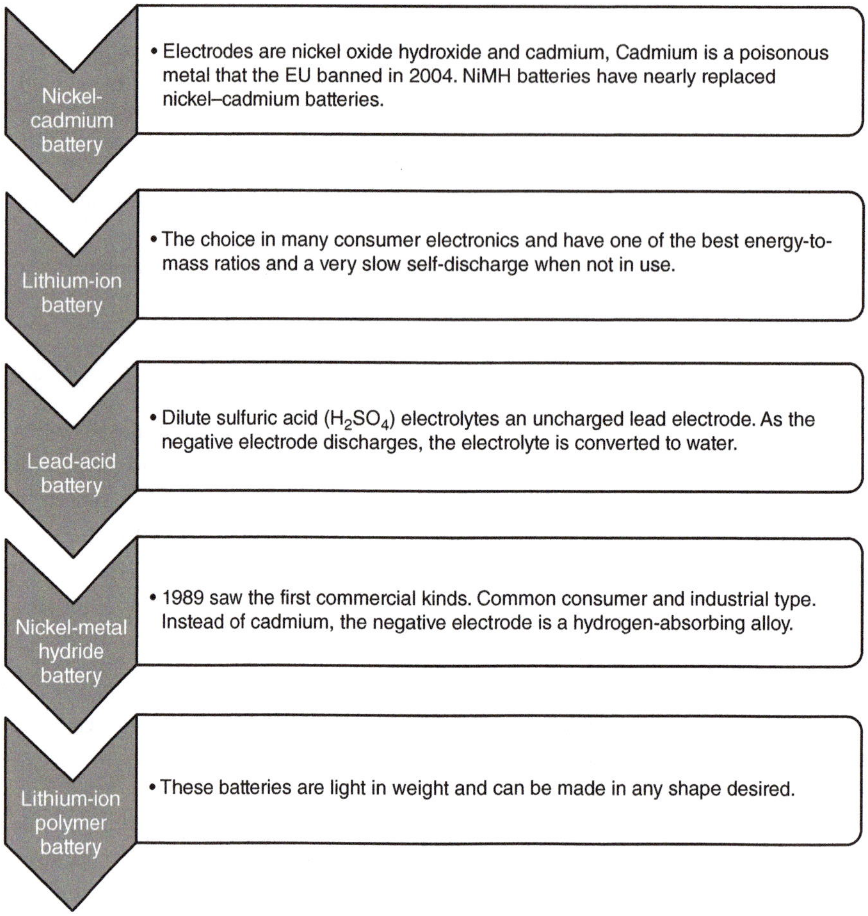

Figure 10.23 Rechargeable Battery Chemistries.

10.5.15 Methane

Methane has the chemical formula CH_4 and is the most basic hydrocarbon. Hydrogen is more difficult to store and move than methane. Infrastructure for storage and combustion (such as pipelines, gasometers, and power plants) is in place and fully functional [155]. Synthetic natural gas (also known as SNG) is made by combining hydrogen and oxygen in a series of reactions. Hydrogen reacts with carbon dioxide in the Sabatier process to produce methane and water. Methane can be stored and used to generate electricity in the future. Because the wastewater is recycled, this method uses a lot less water [156]. A nearby power station uses pure oxygen to burn methane, which eliminates nitrogen oxides during the electrolysis process. As a result of the combustion of methane, CO_2 and water are produced. Water recycling for further electrolysis and CO_2 recycling for Sabatier process improvement are also options. Reuse of reaction waste products occurs during methane combustion and storage. Instead of being a cost, CO_2 has economic value when used as part of an energy storage vector rather than as part of CCS [157].

10.5.16 Biofuels

Biodiesel, vegetable oil, alcohol fuels, and biomass are just a few of the biofuels that can be used in place of fossil fuels. A variety of chemical processes can convert coal, natural gas, plant, and animal biomass, as well as organic waste into short hydrocarbons that can be used to replace current hydrocarbon fuels [158]. Chemical compounds include Fischer–Tropsch diesel, methanol, dimethyl ether, and syngas. During World War II, Germany was heavily reliant on this diesel source because of the country's limited access to crude oil sources. Similarly, coal is used to make the vast majority of South Africa's diesel fuel.

10.5.17 Aluminum

Several academics have suggested aluminum as a potential energy storage material. When compared to lithium, its equivalency is roughly four times as large [159]. Aluminum may be used to generate hydrogen and thus energy by reacting it with water. The first step is removing the natural oxide layer, which can be accomplished through pulverization, chemical interactions with caustic chemicals, or alloying. Using a method known as Hall–Héroult to recycle the aluminum oxide produced as a waste product makes the reaction theoretically sustainable [160]. Aluminum could be utilized to store energy produced by the Hall-Heroult Process more efficiently than direct solar electrolysis if the process is powered by solar or wind energy.

10.5.18 Ways Utilizing Electricity

A capacitor is a name given to a passive two-terminal electrical component used to store energy electrostatically [161]. There are numerous shapes and sizes of practical capacitors, each comprising two or more metal plates separated by a dielectric (i.e. insulator). A capacitor can be used as a temporary battery or rechargeable energy storage while it is not connected to its charging circuit. Capacitors are commonly used to keep a gadget powered when the batteries need to be changed. A capacitor's electrostatic field stores energy between the plates [162]. In the presence of a potential difference between the conductors, a capacitor connected across a battery creates an electric field, causing a positive charge (+Q) to build on one plate and a negative charge (−Q) to accumulate on the other. A long enough connection between a battery and a capacitor prevents current from flowing through the capacitor [163]. In contrast, an accelerating or alternating voltage can supply the capacitor's leads with a displacement current. Capacitor plates and dielectric layers both store charge. Capacitance increases when the distance between conductors shrinks and the surface area of the conductors expands. Using high voltage breakdown to regenerate the dielectric after a capacitor failure may one day make self-healing capacitors a reality [164]. Wires and leads introduce inductance and resistance to the circuit. Researchers are looking into the quantum effects of nanoscale capacitors used in digital quantum batteries.

10.5.19 Magnetic Materials with Superconductivity

Direct current passing through a cooled superconducting coil can store energy in a magnetic field created by the current. In a SMES system, find a superconducting coil, a power conditioning system, and a refrigerator [165]. Once the superconducting coil is charged, the magnetic energy may be stored indefinitely because the current does not degrade once it is charged. When the coil is discharged, the energy that was stored in it is released back into the network. The inverter/rectifier wastes about 2–3% of the available energy in each direction. SMEs use the least amount of electricity for storing energy compared to other technologies [166]. The round-trip efficiency of SMES systems is at least 95%. High energy requirements of refrigeration and the price of superconducting wire make short-term storage like power quality improvement impractical, hence SMES is used. It also aids in power system balancing.

10.6 Different Novel 2D Materials for Energy Storage

Energy conversion and storage have become major global challenges in recent years, making it incredibly challenging to develop an energy or power source technology that is both efficient and reliable at the same time [167]. The reason for this is that scientists are striving to develop a form of energy that is both environmentally friendly and long-lived. Based on these characteristics, it has been determined that electrochemical energy conversion and storage is one of the most viable solutions because its energy consumption is low, reliable, and ecologically friendly. Fuel cells, primary/secondary batteries, and supercapacitors are all electrochemical energy storage and conversion devices. The focus of our review is on energy storage; thus, researchers discuss a wide range of elements and characteristics of energy storage technologies [168]. Researchers employ rechargeable batteries (secondary) and electrochemical capacitors (ECs) as energy storage technologies that convert chemical energy into electrical energy. The hybrid supercapacitor battery is a recent innovation in this industry. Batteries store electrical energy by using electrodes with faradic redox reactions (cathode and anode). Since the reactions occur at a higher voltage at the cathode, both positive and negative electrodes are interchangeable when referring to cathodes [169]. Because it creates electrical double layers (EDL) inside an EC, it is referred to as a supercapacitor. Redox processes at electrode materials can, under some circumstances, cause charge storage (pseudo capacitor). A hybrid supercapacitor battery's complete cell must have both a redox-type battery electrode and an EDL-type capacitor electrode. Secondary batteries (specifically Li-ion batteries) are frequently employed in practical applications due to their high specific energy and specific capacity [170]. Figure 10.24 shows 2D nanomaterials for electrochemical energy storage.

Rechargeable batteries, on the other hand, have some limitations, including limited power output, a slow charge–discharge rate, a low-capacity rate, and poor cyclic stability. These negatives must be taken into account when making any decisions [172]. As a result of these drawbacks, the supercapacitor is the best battery replacement available today. Supercapacitor technology has many advantages, including high power density, quick charge-discharge cycles, high-rate capability, and enormous cyclic stability. These are just a few. However, there is a problem with the supercapacitor's specific energy. EDCs have a low storing capacity since they use electrostatic ion adsorption as their only energy storage method [173]. Unlike an EDLC, which only stores a little amount of charge on an electrode material's surface, a pseudocapacitor holds a big amount of charge on the electrode material's surface and also produces higher specific energy. Pseudocapacitors control how well

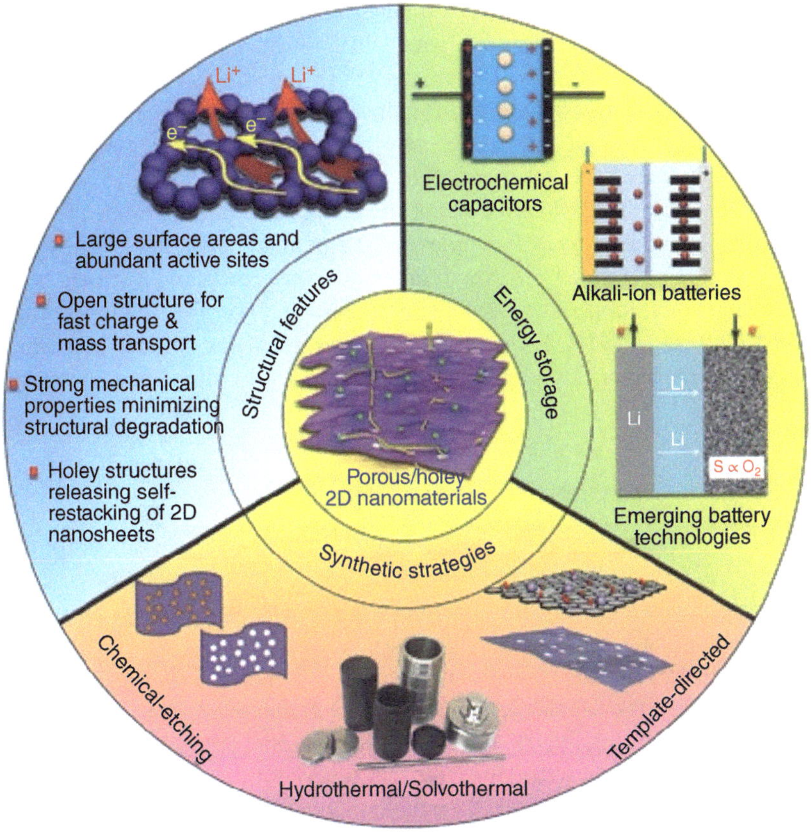

Figure 10.24 2D Nanomaterials for Electrochemical Energy Storage. *Source:* Peng et al. [171]/With permission of John Wiley & Sons.

a device can deal with issues like high energy and power density. Using a pseudocapacitor, on the other hand, has significant drawbacks [174]. Scientists are actively working on developing a new type of storage material known as a hybrid supercapacitor (HSC) or hybrid supercapacitor battery. This technology makes use of an asymmetric supercapacitor with electrodes that are both battery and capacitor types. When using battery-type materials as the anode and electrical double-layer materials as the cathode, organic Li salt compounds function as the electrolyte [175]. This novel sort of storage material may take advantage of the benefits of both batteries and supercapacitors, such as their high specific capacities and energies. While HSCs have high specific energy at low current densities, the problem is that as the current density rises, so does the value of specific energy. Due to battery-type materials' poor kinetics, HSC is constrained. ion adsorption–desorption in the electrolyte slows Li intercalation and deintercalation in battery-type anode materials compared to that of capacitors in the cathode materials [176]. An electrode in a charge storage device determines a substance's kinetics by the mobility and diffusion of electrolyte ions in the solution. This means that the material's morphology is crucial in making it better at storing charge in every way. Nanoparticles are frequently used as electrode materials in charge storage devices because of their small size and low cost. Nanomaterials can be divided into four broad categories: (zero-dimensional) 0D, (one-dimensional) 1D, (two-dimensional) 2D, and (three-dimensional) 3D. 0D nanomaterials include the likes of quantum dots, yolk-shells, and core-shells [177]. These strongly resemble spherical particles. An example of a one-dimensional (or flat-sheet) Nanomaterial is a nanoparticle. When it comes to 2D materials, their dimensions are not in the nanometric range, and their thickness is made up of a few atomic layers. Examples of 2D nanomaterials include graphene, metal-layered double hydroxides (LDH), several transition metal oxides and hydroxides, as well as transition metal dichalcogenides. 3D nanomaterials are bigger than nanoparticles, but they still offer advantages [178]. Porous carbon spheres, graphene aerogel, and other 3D nanomaterials are common. To store electrical energy, all of these nanomaterials were used.

10.6.1 2D Materials for Energy Storage Devices

Meeting the world's growing energy demands will be humanity's most difficult task in the twenty-first century. All three aspects of energy – production, transit, and storage – are the subject of active research today [179]. Nanomaterials are intriguing prospects for use in next-generation energy-related technologies due to the emergence of new electrical, mechanical, and optical capabilities at the nanoscale, as well as increased surface area, fascinating dynamics, and controllable nanostructures. 2D materials with atomically thin layers that stack via van der Waals (VDW) forces are a well-studied class of nanomaterials for electrical, optical, and energy applications [109]. 2D materials have expanded considerably since graphene's discovery in the 1970s. Graphene, silicon, phosphorene, and hexagonal boron nitride (h-BN) are examples of single-atom-thick 2D materials that can be used in many applications. bulk layered materials can be thinned to a few atoms thick layers such as transition metal dichalcogenides (TMDs), transition metal carbides, nitrides, and carbonitrides (Mxenes), and transition metal oxides (TMOs). Solar energy storage systems, photovoltaic devices, piezoelectric devices, as well as thermoelectric devices, have all been widely studied. The focus of this viewpoint will be on the use of 2D materials as catalysts and electrodes in metal ion batteries (MIBs) and supercapacitors [180]. As an alternative to noble-metal catalysts, 2D materials have been studied as catalysts for the synthesis of chemical fuels like H2 or methanol. With its transition from semiconducting (2H) to metallic (1T) phases and the exposure of its edge sites, the TMD MoS_2 has shown remarkable catalytic efficiency in hydrogen evolution processes (HER). Heteroatom-doped graphene has shown remarkable catalytic efficiency in oxygen evolution reactions (OER), oxygen reduction reactions (ORR), and HER. 2D materials have also been investigated as electrodes in intercalation-based energy storage systems, according to researchers [181]. While carrier density can be utilized to produce superconductivity or promote semiconducting-to-metallic phase transitions, intercalation inserts ions or molecules reversibly into the VDW gaps between 2D materials to fine-tune their properties. Intercalants have an impact on optical properties, interlayer distance, and mechanical strain, as well as 2D films [182]. Researchers are aiming to improve the safety, energy density, and affordability of MIBs by using intercalation chemistry. A promising family of 2D electrodes or electrode support

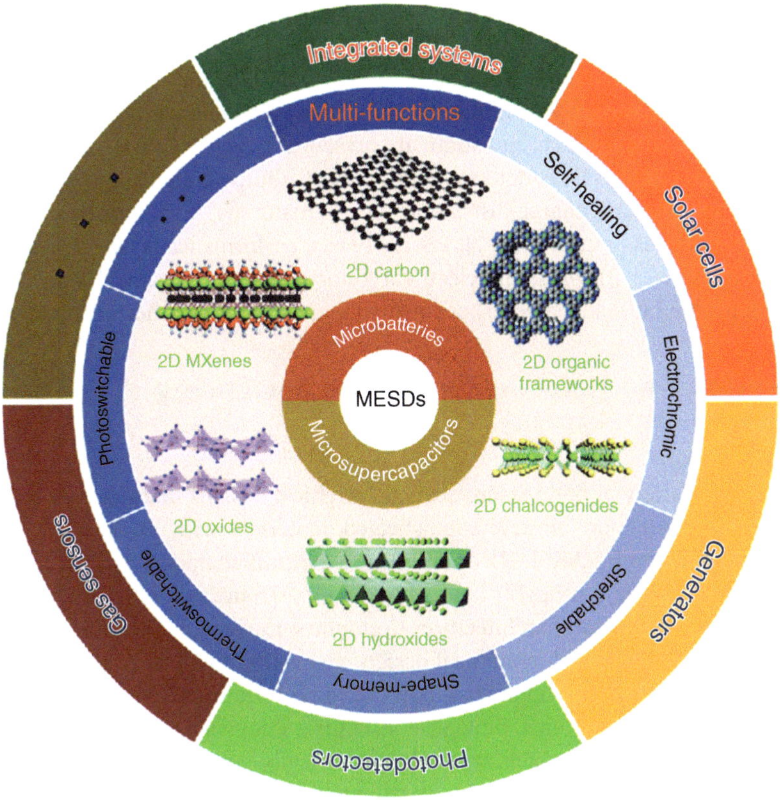

Figure 10.25 2D Materials for miniaturized energy storage devices. *Source:* Panpan et al. [183]/ With permission of Royal Society of Chemistry.

materials for MIBs includes TMDs, MXenes, and graphene or reduced-graphene oxide (rGO). Figure 10.25 describes 2D materials for miniaturized energy storage devices.

10.6.2 Challenges Facing 2D Energy Technology

In the field of two-dimensional materials for energy applications, research focuses on the evaluation of prototype device performance by metric comparison with industry standards. The metric-driven analysis is critical for discovering new 2D materials [184]. In terms of understanding how these materials work fundamentally, however, this approach falls short. Industrial integration needs mechanical knowledge due to the additional challenges presented by mass production, which are not present in lab-scale prototypes. It is not uncommon for lab-made 2D materials to have flaws that prevent them from having the desired qualities [185]. Mechanistic knowledge of how these faults affect material properties and how to selectively synthesize them is required to repeat lab-scale results using an industrial synthesis approach. Because of the structural similarities between 2D systems, a mechanistic understanding of one system can be used to improve existing technologies in other systems. The atomic structure has a key role in 2D materials' catalytic activity and ability to assimilate intercalants. Due to this, the structure of certain materials can drastically modify their conductivity [186]. Many studies have focused on the 2H and 1T′ phases of 2D TMDs because of their semiconducting and semimetallic properties. Structure-electronic phase transitions can occur simultaneously with intercalant injection. Intercalants' ability to permeate into 2D structures when defects and strain are present can be impacted by this. The electrochemistry of these materials is determined by this attribute, so understanding the nature of chemical bonding and charge distribution in 2D materials, as well as how they change throughout reactions, is crucial [187]. Energy device performance is heavily influenced by interactions between 2D materials, electrical connections, and structural supports, and these interactions are intimately tied to both structure and chemical bonding [188].

10.7 Nature-Inspired Materials for Sensing and Energy Storage Applications

Nature contains a vast variety of nano/micro-scale hierarchical designs that serve as endless inspiration for the creation of new sophisticated materials with distinctive features [189]. For artificial hierarchical 3D architectures, natural materials have a wide variety of scaffolds that can be thoroughly studied to obtain extremely significant knowledge. The desirable ability of natural architectures to combine the attributes of their separate components into an intelligent hierarchical design means that the total of their parts is usually more efficient and energy-optimized than the sum of their parts alone [190]. In terms of well-managed energy systems, biological structures conversion and storage are shining examples. Pili nanowires, which are highly electrically conductive routes, have been found to facilitate long-range extracellular electron transfer in various types of microbial systems. These ideas have stimulated a lot of research into the development of bioelectronic interfaces, which might be utilized to manufacture microbial electrolysis cells or fuel cells by using the electrical connections between cells as green electronic materials [191]. A new generation of high-performance rechargeable energy storage devices has been developed thanks to the use of redox biomolecules in a series of life-sustaining chemical reactions by Nature. This is due to their impressive electrochemical properties, high structural flexibility, chemical diversity, and ability to be integrated into electronic systems [192]. Figure 10.26 shows nature-inspired materials for sensing and energy storage applications.

Platforms for smart energy storage devices with advanced electrochemical properties that mimic incomparable features of natural sources such as skin's self-healing properties, the olfactory sensory neuron systems of mosquitoes, or even the combination of naturally occurring cellular scaffolds with electrically conductible nanomaterials have been developed [194]. Because of their superior electrochemical properties, such as reversible capacity and energy density, nature-inspired devices are an attractive alternative for meeting future energy society demands while also serving as excellent candidates for cutting-edge sensing and energy storage technology development [195].

10.7.1 Sensing and Energy Storage Artificial Nano and Microstructures

The hierarchical topologies and adjustable properties of sophisticated nanomaterials, such as nacre, bone, and bamboo, have inspired the intelligent design of these materials [196]. It is possible that replicating the structural arrangements and interactions found in biological

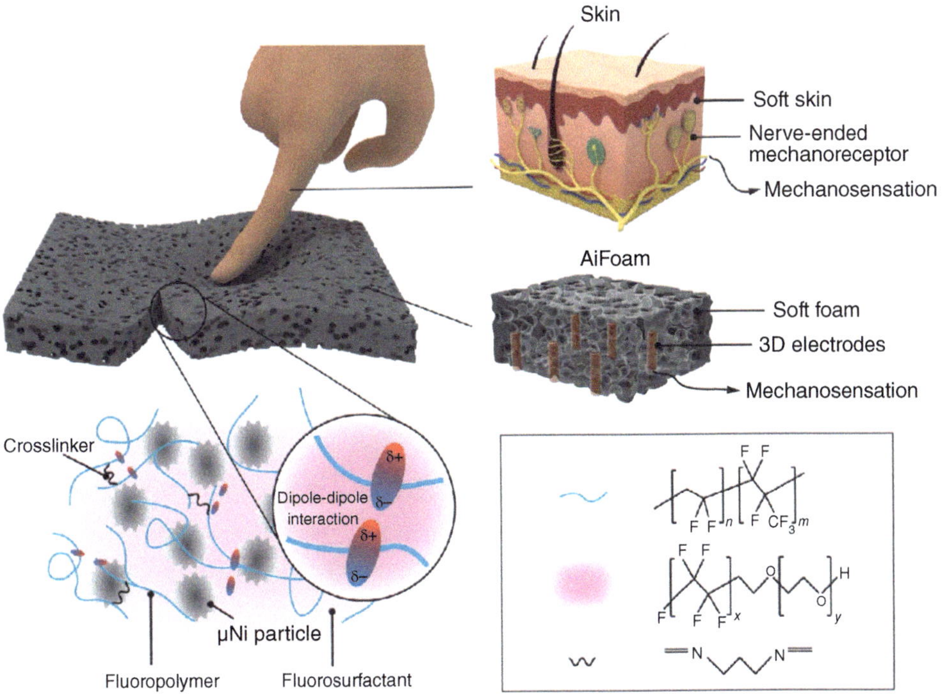

Figure 10.26 Nature-inspired materials for sensing and energy storage applications. *Source:* Hongchen et al. [193]/Springer Nature/CC BY 4.0.

and natural systems may lead to high-performance materials with a wide range of application options and superior behavior compared to manufactured products. The complex interconnections formed in biological systems, which allow for unique and complicated capabilities, have encouraged scientists to mimic neurological systems or skin designs for the preparation of nano and microstructures [197]. Figure 10.27 describes flexible pressure sensors with microstructures.

10.7.2 Bioinspired Hierarchical Nanofibrous Materials

There are several features inherent in structural proteins that enable them to serve certain activities in biological systems, including providing structural rigidity and organization. Examples of hierarchical natural materials with sizes ranging from nano to macro include self-assembling biomolecular complexes [199]. Nanofibrous materials with interconnected and heterogeneous pores of various length scales can play a variety of roles, including asymmetric diffusion and signal transmission. They can also be used to direct cells. While inspired by natural systems, synthetic hierarchical nanofibrous architectures have presented a challenge to the scientific community, eager to reproduce the form and function of natural objects for usage across a wide range of industrial and healthcare domains Bioinspired hierarchical nanofibrous materials have been synthesized using a variety of processes, including electrospinning, three-dimensional printing, and lithography [200].

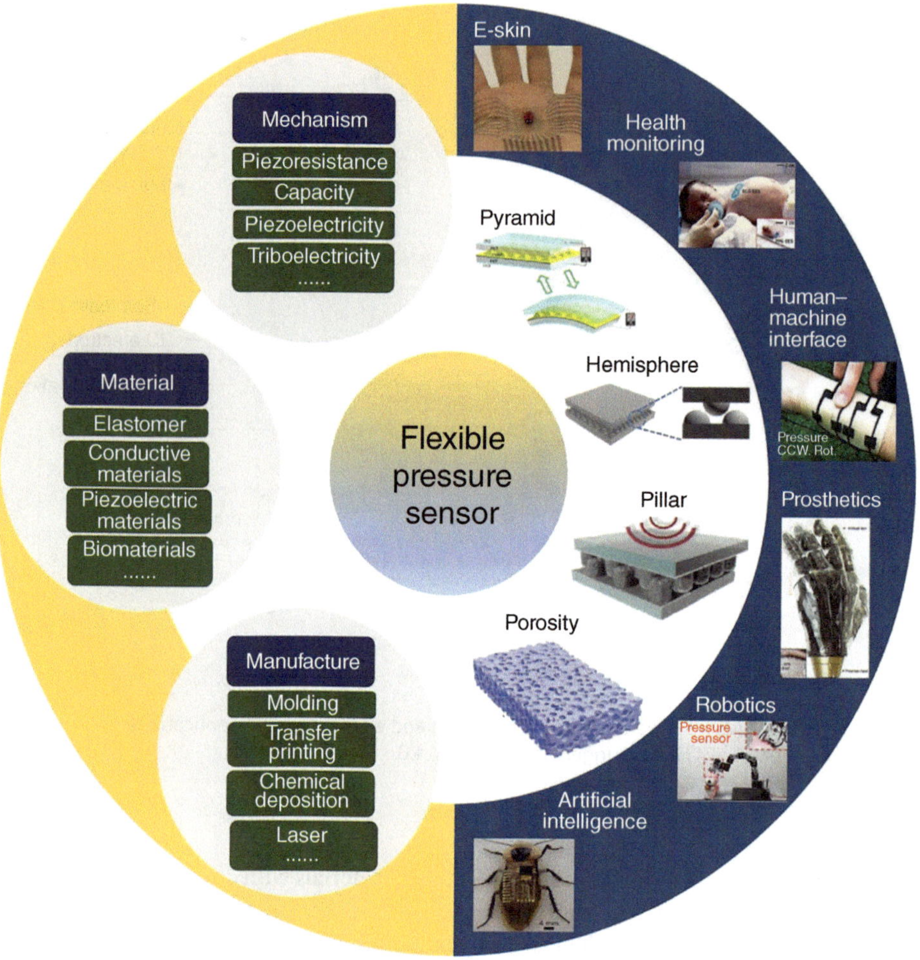

Figure 10.27 Flexible pressure sensors with microstructures. *Source:* Ruitao et al. [198]/John Wiley & Sons/CC BY 4.0.

The challenge remains, however, in altering structures concurrently at the molecular and macro sizes because molecular factors may interfere with management at larger scales, which may have an impact on molecule arrangements. Their materials use self-assembly techniques to construct a controlled macrostructure geometry with micromechanical limitations and a directed nanofibril network generated by solvent removal dynamics [201]. Figure 10.28 biological and bioinspired materials: energy storage system.

10.7.3 Nature-Inspired Polymer Nanocomposites

High mechanical properties, such as strength and toughness, as well as low weight, can be achieved by hierarchical patterns in natural polymeric materials [203]. One endeavor has been to simulate natural processes to create natural polymeric materials. For example, synthetic nacre was created by ice-templating and mineralization, which is the same

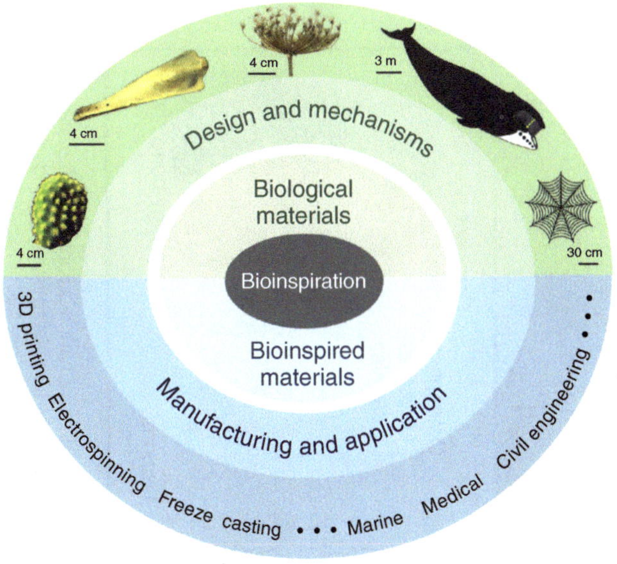

Figure 10.28 Biological and bioinspired materials: Energy Storage System. *Source:* Bin et al. [202]/ With permission of Elsevier.

bioprocess utilized to make natural nacre. Because of the biopolymer's high mechanical strength, little nacre grains can be joined together to form larger ones, which can be used to make platelets and organic layers [204]. Figure 10.29a–c shows high-performance nanocomposites inspired by nature.

These materials are also referred to as "bricks and mortar". Another natural polymer composite material is made of calcium carbonate and chitin–protein fibers organized in a planar twisting plywood pattern on the lobster's cuticle [206]. Bone structure and function have also been replicated in a variety of experiments. It has been found that bone microstructures with collagen and hydroxyapatite nanocrystals have a hierarchical structure, with fibers that are parallelly organized to create lamellae and structured in concentric twisting sheets. Three significant qualities of a steel-based material, analogous to the hierarchical structure of bone's presence in various phases, nanolamination, and metastability have been established [207]. Graphene and graphene oxide-based polymer nanocomposites, for example, have undergone bioinspired design in various research in the literature. To duplicate the features of polymer-based architectures found in nature and understand their relationships, scientists are still having difficulty simulating them. As a result, the ultimate properties of synthetic materials inspired by nature are influenced by how building components interact at various scales in hierarchical frameworks [208].

10.7.4 Skin-Inspired Hierarchical Polymer Materials

The skin has numerous functions such as protection, temperature sensing and regulation, and aiding in waste disposal and absorption. It is the most important part of the integumentary system [209]. Hierarchical laminated materials were inspired by such an organ

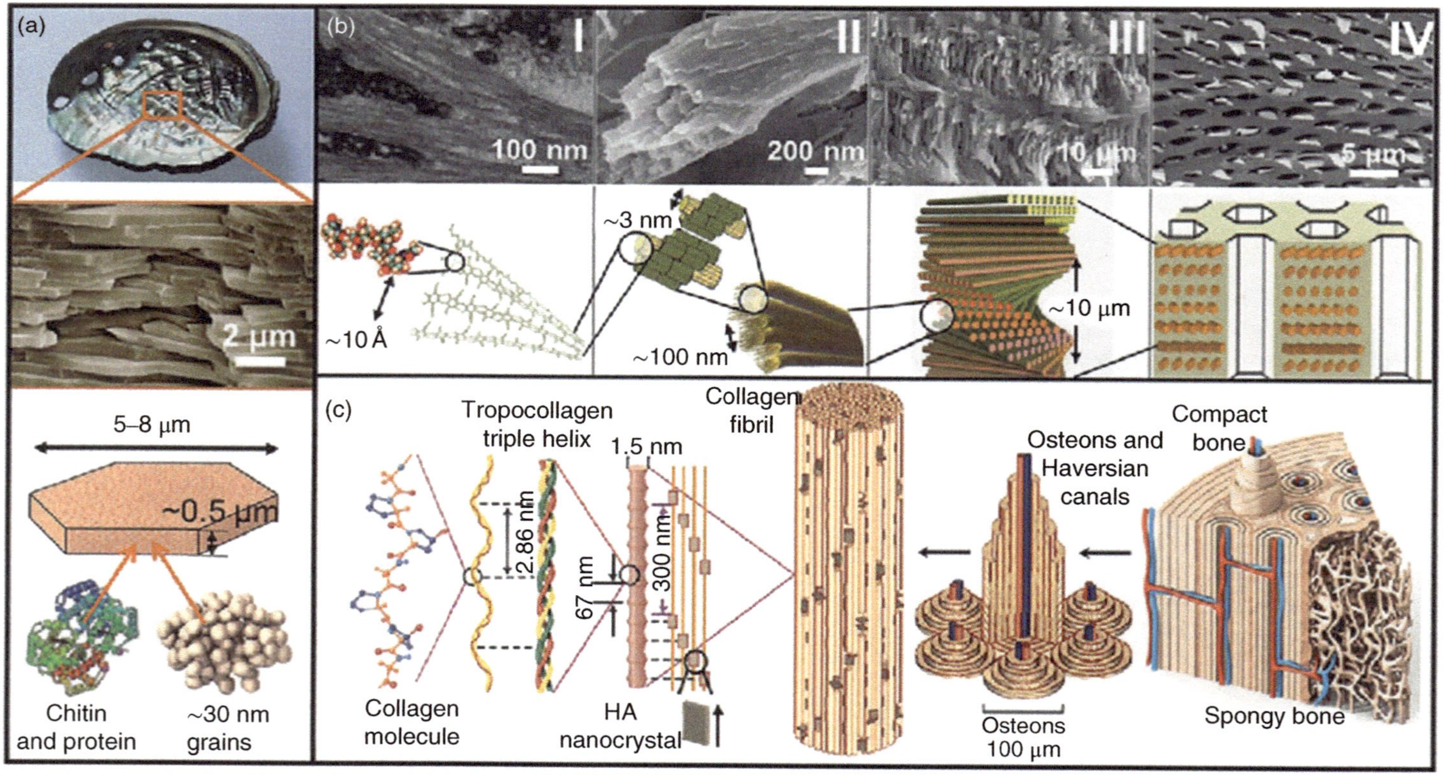

Figure 10.29 High-performance nanocomposites inspired by nature. (a) Layered hierarchical architecture of nacre with inorganic CaCO3 platelets and organic layers. (b) Hierarchical architecture of the lobster cuticle: i) α-chitin, ii) mineralized chitin–protein fibers, iii) twisted plywood, and iv) cuticle with pore canals. (c) Hierarchical architecture of bone. The orderly architecture of bone is assembled from two major nanophases: collagen fibrils made from collagen molecules and HA nanocrystals distributed along the collagen. *Source:* Qunfeng et al. [205]/With permission of John Wiley & Sons.

because of its great sensitivity and flexibility to various stimuli such as pressure, pain, temperature, and so on. Human finger motions have been used to construct self-powered triboelectric flex sensors that can transfer audio and text from human finger gestures. It was a hierarchical network that could be mechanically adjusted, stretched to accommodate changes in the environment, and yet function properly [210]. Figure 10.30a–h shows skin-inspired hierarchical polymer materials.

10.7.5 Neuron-Inspired Network Materials

Synapses, or cell-to-cell interactions, are how neurons communicate with one another. These three bio elements – soma, dendrites, and axons – are what make up neurological systems [212]. Dendrites and axons are responsible for transmitting electrical and chemical signals between cells. The signal transfer mechanism of such highly efficient systems has recently been emphasized as highly conductive materials based on neuron-inspired networks. The electronic conductivity and mechanical durability of multi-walled CNTs made them ideal for representing axons and dendrites in a neural network mimicking interconnectivity [213]. In addition, the flower-like shape of MoS_2 nanosheets, which in this case acted as the soma, enhanced diffusion processes and dramatically boosted active surface areas. The created nervous-system-like conductive network displayed electrochemically outstanding performance, demonstrating the huge potential of Nature-inspired structures as engineering structures, which was impressive. MoS_2 and reduced graphene oxide were combined with vacuum-aided filtering and high-temperature thermal reduction to create conductive multilayer materials [214]. Electronic skin materials inspired by neurons' signal transmission style have recently been found to respond not only to physical impulses but also to chemical changes linked with acid-based and oxidative/reductive alterations, according to recent findings. This neuron-inspired network was constructed using polypyrrole, which has reversible chemical characteristics and high conductivity, as well as CNTs, which make use of its outstanding electrical and mechanical capabilities [215]. Cross-linking of polypyrrole chains to the CNT surface was achieved by hydrogen bond interactions. Researchers discovered that polypyrrole was star-shaped, simulating dendritic and axon activity. Figure 10.31 describes emerging materials for neuromorphic devices and systems.

10.7.6 Tunable Energy Storage Materials

For today's civilization, it is critical to create energy storage materials that are both wearable and foldable, have high energy densities, and have extended cycle lives. Scientists have already made significant progress toward this aim by imitating the unique features of naturally ordered systems [217]. Recently, the material science community has become increasingly interested in the adaptability of interfacial interactions in natural materials for the fabrication of bioinspired films for supercapacitor applications. Interfacial synergy can be perfectly duplicated in the lab for natural materials such as p–p conjugated bonds and hydrogen bonding and electrostatic interactions, opening a flood of possibilities for the development of high-energy density and high-power density supercapacitor materials [218]. These interactions have led to the creation of new supercapacitor biocomposites with exceptional tensile strength, excellent electrical conductivity, and ultralong

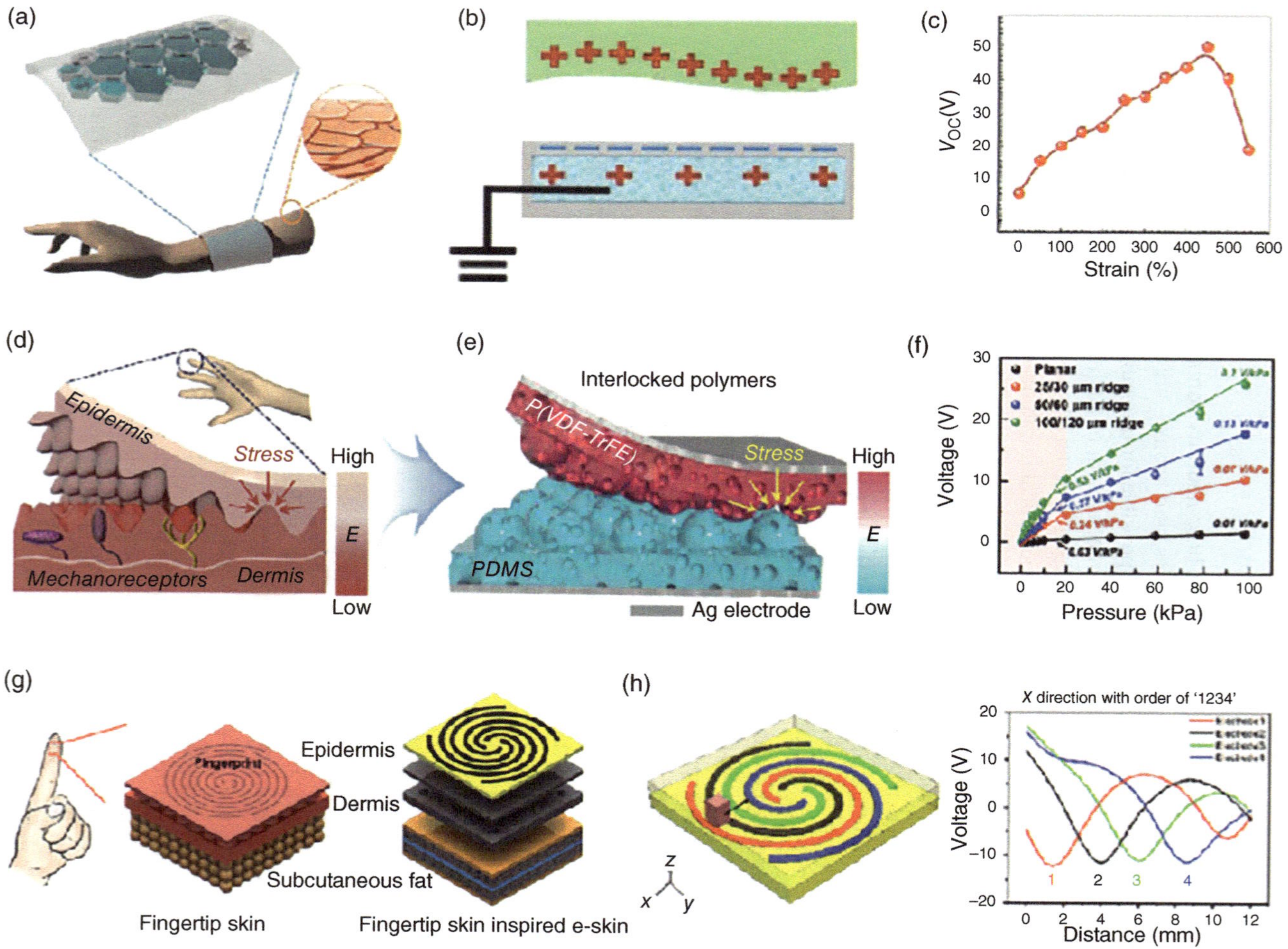

Figure 10.30 Skin-inspired hierarchical polymer materials. TENGs with bioinspired structures from human skin. (a) Schematic illustration of the biological cells bioinspired TENG. (b) Schematic illustration of the working mechanism. (c) Output voltage under strains. (d) Schematic illustration of the epidermis–dermis structure and hierarchical nanoporous. (e) Structure of the interlocked microridge structure bioinspired triboelectric sensor. (f) Pressure sensitivity of the sensor. (g) Schematic illustration of fingertip skin inspired integrated e-skin system. (h) Fingerprint inspired triboelectric sliding sensor and its simulated response when sliding along the X-direction. *Source:* Reproduced from Wenjian et al. [211]/Elsevier/CC BY 4.0creativecommons.org/licenses/by/4.0/.

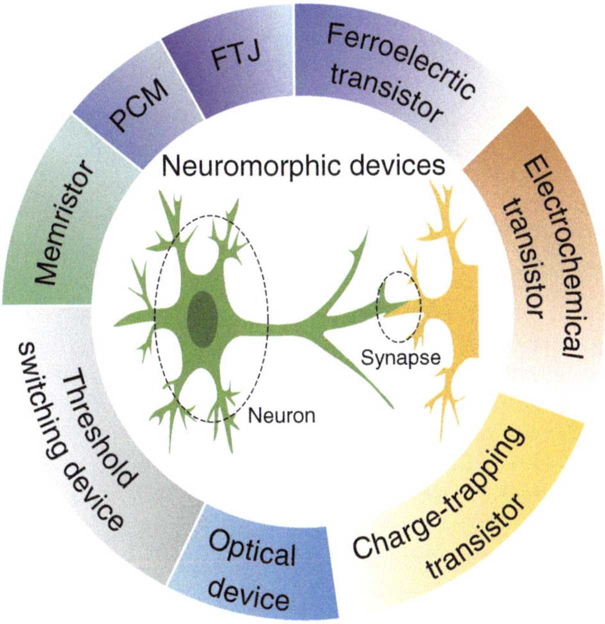

Figure 10.31 Emerging Materials for Neuromorphic Devices and Systems. *Source:* Min-Kyu et al. [216]/With permission of Elsevier.

cyclability. Graphene supercapacitors inspired by nacre, for example, have shown remarkable electrical and mechanical capabilities by constructing supercapacitor electrodes using interfacial interaction techniques. Because of noncovalent bonding interactions, graphene and polypyrrole (PPy) layers make flexible supercapacitors [219]. 2D–2D nanohybrid interfacial connections were employed to construct extraordinarily flexible domains and boost the electrical conductivity of the resulting biofilms. The p–p stacking, hydrogen bonding, and electrostatic interactions between the polymer polyaniline (PANI) fibers and graphene oxide (GO) nanosheets have also been used to produce flexible energy storage devices. Wu and colleagues increased the mechanical performance and specific capacitance of the prior supercapacitor-based natural nacre electrodes using an effective technique for generating alloy-site-polyaniline–graphene oxide (HPA–GO) nanohybrid films [220]. It was discovered that when PANI was grafted onto hallosytes and rGO NSs, the high tensile strength was attributable to interactions between the hydrogen bonding systems of the two molecules themselves. This groundbreaking breakthrough paves the way for new developments, such as the development of hierarchically structured supercapacitor materials with extremely high tensile strengths modeled after nacre. The nano/microfluidic frameworks of this film are linked in three dimensions and are extremely conductive. Both the well-distributed ion channels found in cell membrane proteins and the hierarchical structure of natural nacre served as inspiration for their creation [221]. Non-covalent p–p stacking interactions between rGO and CNT dramatically affected the fluidic nano/microarchitecture properties. One of the goals of developing this new energy storage technology was to increase the power density and specific energy density of the final product. In terms of

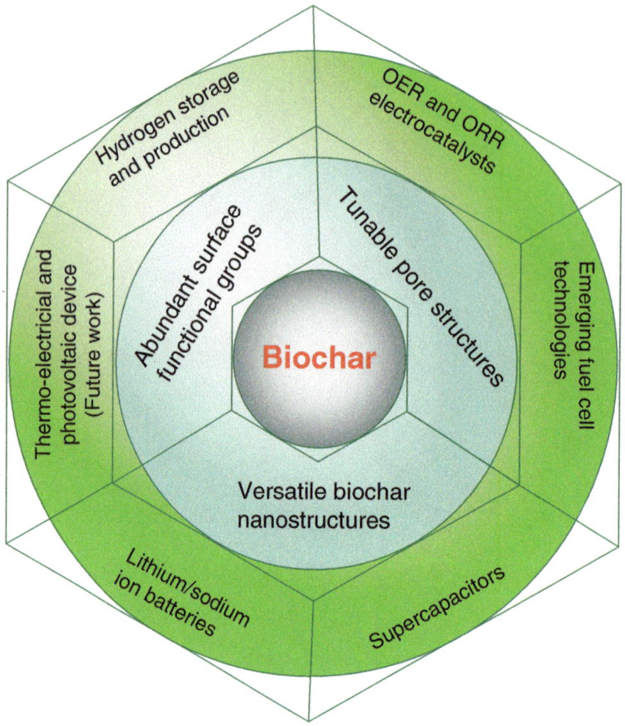

Figure 10.32 Emerging applications of biochar-based materials for energy storage. *Source:* Han et al. [223]/With permission of Royal Society of Chemistry.

power density and specific energy density, this technique beat prior studies on organic electrode materials. High-performance organic electrodes for diverse rechargeable batteries were constructed for the first time using the fundamental understanding of the synergistic interfacial interactions between their components [222]. Figure 10.32 describes emerging applications of biochar-based materials for ES and conversion.

10.7.7 Tunable Sensing Materials

The creation of new functional materials that may have both strength and toughness, two desirable mechanical qualities that are sometimes mutually exclusive, is now a major structural issue in materials science [224]. Many artificial materials have a high strength-to-toughness ratio, but those that are reciprocally tough have low strength properties, limiting their technical utility. With the combination of synergistic interfacial interactions arising at the hierarchical framework, Nature has successfully dealt with this issue. As a result, in nature, multilayer hierarchical structures built with "bricks and mortar" construction methods combine strength and durability [225]. Natural materials abound, but nacre stands out due to its extraordinary mechanical capabilities, which are the result of synergistic interactions between the interfaces of 2D aragonite calcium carbonate platelets and 1D nanofibrillar proteins in the layered structure of the material's nacre. Graphene-based synthetic nacre materials with exceptional strength and toughness have been

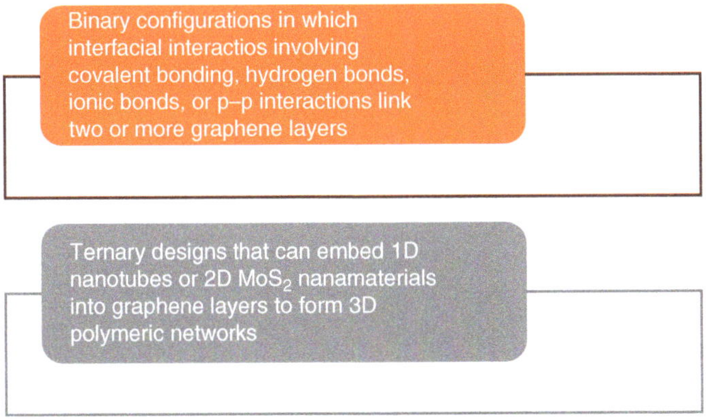

Figure 10.33 Classification of Artificial nacre materials based on graphene.

developed by researchers. These materials were inspired by clues found in nature [226]. Artificial nacre materials based on graphene are classed as shown in Figure 10.33.

When combined with the aforementioned nanoarchitectures, different types of interfacial interactions can be employed to create highly customizable bioinspired materials with superior mechanical properties. In terms of tensile strength and young's modulus, the generated nanoparticles outperformed numerous other nanomaterials, such as graphdiyne nanoplatelets, CNTs, and nanoclays [227]. The capacity of nature-inspired materials to combine numerous desired qualities through synergistic interfacial interactions has lately benefited smart technological gadgets. The development of high-performance sensing materials has been driven by the synergistic interfacial properties of these materials, which have helped resolve the sensitivity/stretchability contradiction. This technology uses a combination of inorganic "brick" and organic "mortar" building blocks to control crack development and disperse loading energy through interfacial interactions to create nano heterostructures with high sensitivity and great stretchability [228]. The dynamic designs that result from finely tweaking the covalent and non-covalent interactions to maximize the overall properties are revolutionizing electronic sensing. An electrically assisted 3D printing method utilizes graphene nanoplatelets (GNPs) in a photocurable resin to simulate the characteristic nacre properties of hierarchically ordered artificial materials. The unexpected hardness of these 3D artificial nacre materials is on par with actual nacre. Individual GNPs operating together in a hierarchical network have a synergistic effect, which is most likely the case [229]. Nacre structures printed using additive manufacturing have good electron transfer (ET) characteristics running perpendicular to the GNP orientation, but considerable electrical resistance running perpendicular. Synergistic interfacial interactions have been proven to be significant in the anisotropic properties of these three-dimensional designs because of the covalent bonds, p–p interactions, and hydrogen bonding that coexist at graphene nanoplatelet surfaces [230]. Through these past interactions, the nacre-inspired nano-laminated framework was able to detect crack propagation and depletion processes independently, which opened up new possibilities for future research into wearable sensors that are self-aware by nature [231]. Figure 10.34 describes hierarchically structured self-healing sensors with tunable positive/negative piezoresistivity.

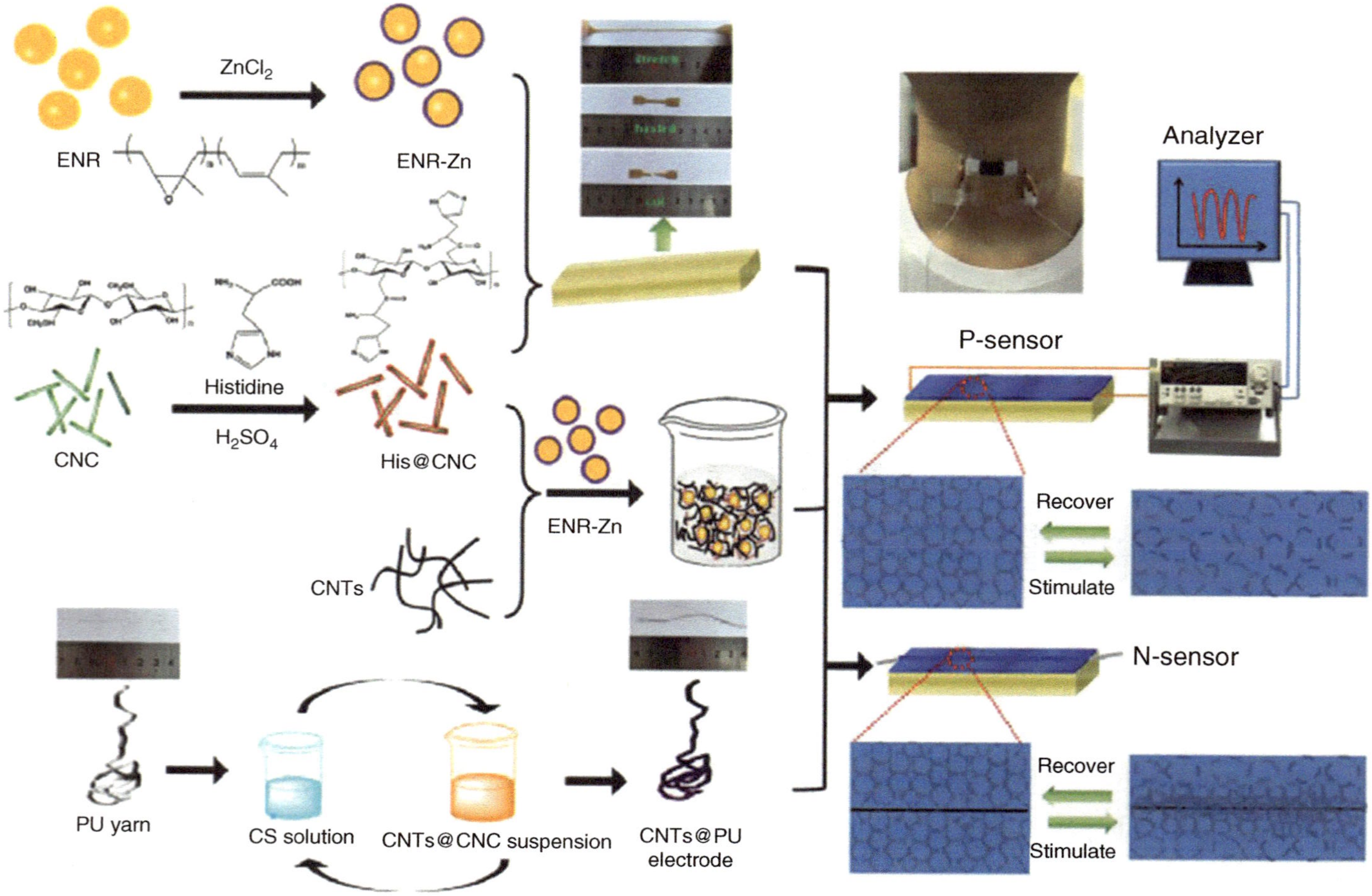

Figure 10.34 Hierarchically structured self-healing sensors with tunable positive/negative piezo resistivity. *Source:* Xuehui et al. [232]/With permission of John Wiley & Sons.

Human skin's potential to boost its elastic moduli quality through the synergistic interactions of its building blocks has recently inspired researchers. As a result, 3D hierarchical nanostructures have given rise to a new class of ultrasensitive and ultra-flexible wearable sensors [233]. According to the findings, the tests' interlocked structured biofilms were considerably more sensitive to pressure than planar films that lacked the network. These findings offer a novel approach to the integration of hierarchical biological materials into flexible and wearable sensing devices for a wide range of practical applications, such as display panels, automotive electronics, artificial e-skins, and smart controllers and switches [234]. Several skin-inspired sensors worked well because of the interfacial interactions in 3D hierarchically structured materials, which were partly responsible. The dynamic combination of building blocks into hierarchical nanostructures has, for example, improved the capabilities of various flexible sensors by combining conductive nanomaterials, such as CNT-based materials, metallic nanowires, or metal nanoparticles with supramolecular networks. This has opened up new avenues for the development of smart skin-inspired wearable electronic sensors with multifunctional capabilities [235].

10.7.8 Bioinspired Batteries

Nature-inspired supercapacitor materials have piqued the interest of researchers because they could be utilized to make energy storage devices with high energy density, high coulombic efficiency, and extraordinarily long cycling stability that are low-cost, renewable, and environmentally friendly [236]. They are more helpful as electrochemical pseudocapacitive devices thanks to the ion-transport channels supplied by nature-inspired hierarchical structures, which speed up charge transport in the material network. Ren and his colleagues were inspired by the integrated structure and synergistic function of a Texas cactus species when constructing binder-free high-performance supercapacitors. Researchers employed a new design technique inspired by plant growth processes to create low-cost, ultra-efficient supercapacitor materials. New genetically engineered resilin-like proteins (RLPs) have been developed by researchers from the Universities of Helsinki and Turku, which have different recognition sites for rGO and NFC molecular signals [237]. This research was inspired by natural nacre hierarchies as well as matrix-mediated proteins in biological nanosystems. rGO and NFC surface-attached RLP proteins were used to assemble supramolecular nacre-like biocomposite materials. The pH-tunable rGO/NFC interlayer space has a major impact on the biomaterial's ability to store electrochemical charges. The mechanical and energy storage capacities of novel nano biomaterials were altered as a result of pH-dependent structural alterations. rGO and NFC layers cannot separate when RLP protein connectors are stretched because of the tight twist in the connectors. At neutral pH, proteins assume their most extended shape, while fusion proteins undergo greater amounts of conformational change, adopting their most extended configuration and resulting in the breaking of linkages [238]. Fusion proteins the development of stretchable nano-bio supercapacitors may be aided by this dynamic supramolecular composite in the future. A hierarchically structured hydrogel polyelectrolyte network intercalated between two electrodes of nanowire-PANI electrodeposited on carbon fiber was used to construct a free-additive, superconducting, self-healing supercapacitor material [239]. Figure 10.35 describes the bioinspired molecular redesign of a multi-redox catholyte for high-energy non-aqueous organic redox flow batteries.

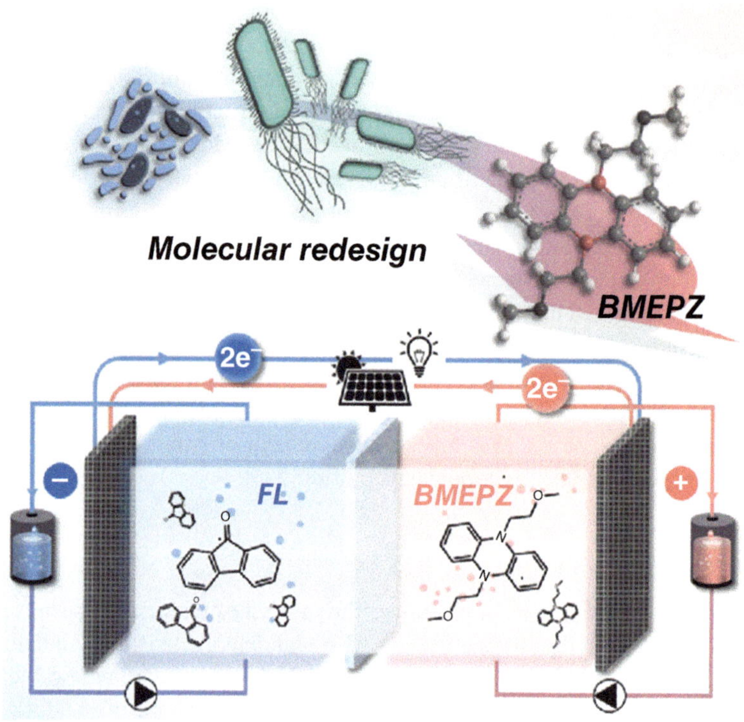

Figure 10.35 Bioinspired molecular redesign of a multi-redox catholyte for high-energy non-aqueous organic redox flow batteries. *Source:* Kwon et al. [240]/With permission from Elsevier.

10.7.9 Bioinspired Energy Storage Devices

The development of bioinspired and hierarchically organized battery electrode materials is a new way of energy storage. As several recent and exciting examples have revealed, natural systems have the immense potential to inspire human ingenuity and lead to the development of the next generation of superior energy storage nanosystems [241]. As a result of their many unique advantages, wood-inspired materials have piqued researchers' interest in energy storage applications. These advantages include hierarchically porous interconnected networks, excellent mechanical properties, excellent electronic conductivity, and tunable multifunctional structures, which promote fast electron-and-ion transport and thus improve electrochemical performances Successfully manufactured rechargeable batteries based on wood-inspired materials exhibit high efficiency [242]. The cathode is built of CNT/Ru–coated F-Wood, the separator membrane is made of glass fiber, and the anode is constructed of lithium foil in a high-end wood-based Li–O_2 battery. The CNT/Ru-coated F-Wood was created as a nod to the multiphase transport process seen in trees. While traveling through the CNT framework, electrons, oxygen gas, and ions adopt a tri-pathway hierarchical structure. Membrane-free structures or those with excessive amounts of electrolyte were compared to the capacitance behavior of structures using the tri-pathway material to determine how structural variables affected electrochemical performance [243]. There are several advantages to using a wood-based cathode over a previous Ru/N doped CNT material, including the smaller overpotentials, greater area capacity, and increased cycle and mechanical stability. As proven here, a Nature-inspired method can spark wearable electrochemical energy storage systems with high sustainability and outstanding performance. Natural materials have been used as templates to manufacture lithium-ion battery electrode materials. To construct hierarchical popcorn-like structures, Nb_2O_5 bioinspired materials as puffed rice carbon were employed as templates as (PL-Nb_2O_5). Because the created hierarchical porosity network significantly limited the ion migration pathways, the reaction kinetic rates at high currents were enhanced, and the solid-state energy storage device's overall performance was benefited [244]. It is just one example of how the variety of natural materials and structural arrangements make them a valuable resource for the development of high-power electrodes for use in solid-state energy storage devices. Three-dimensional (3D) structures can be formed by several types of microorganisms in nature that can assemble biominerals [245]. Figure 10.36 describes bioinspired energy storage and harvesting devices.

Biomimetic templating methods, which examine how molecular structures self-assemble in nature, have been created to synthesize nanomaterials at lower temperatures [247]. Due to the inspiration provided by this natural omen, researchers created hierarchical porous graphitic carbon spheres from a-cyclodextrin self-assembly on a template (pluronic F127 micelle cluster). Nanostructured carbons with remarkable electrochemical properties [248] for LIBs can be made using this technique because it incorporates large surface areas, hierarchically porous structures, a high level of graphitization, and a significant amount of heteroatom motif content into the carbon structure. A new generation of ultra-efficient bioinspired batteries may be created by using the self-assembly capabilities of natural systems as a design guide [249].

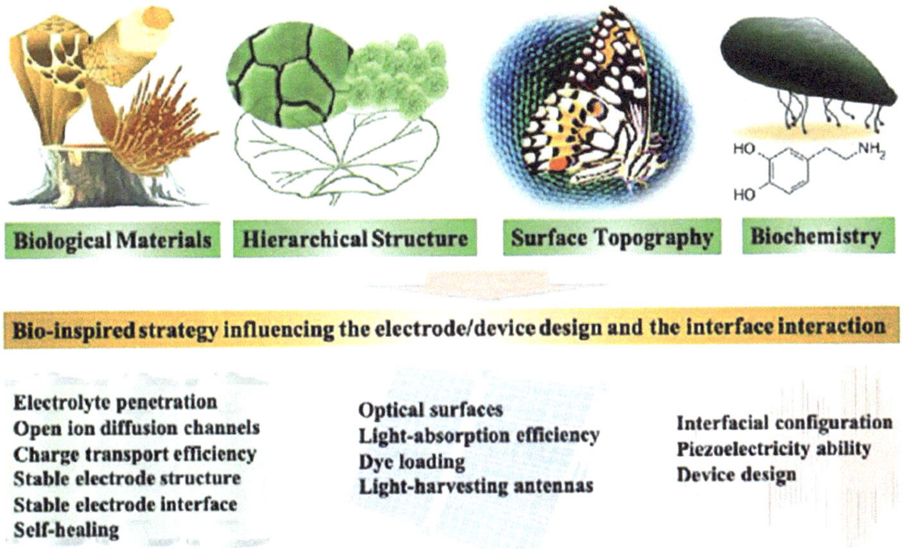

Figure 10.36 Bioinspired Energy Storage and Harvesting Devices. *Source:* Jing et al. [246]/With permission of John Wiley & Sons.

10.8 Conclusions

Sustainable energy storage devices are discussed extensively in this chapter. The use of sustainable energy storage devices, among other issues, is also discussed in detail. Sustainable energy storage devices are discussed in detail in the following chapters. An examination of the literature regarding substantial energy storage systems is presented in the next chapter, which was done by several different authors. The need for sustainable energy storage devices has also been discussed in this chapter, which includes sections on how to reduce first, electricity generation and health, economic benefits of using renewable energy sources, protection of the energy supply, increasing gross domestic product (GDP), maintaining currency stability, and electricity and the environment, amongst other subjects. First and foremost, reduce power generation and health care costs. This chapter also covers energy-storing techniques such as storage systems for massive amounts of energy, electrochemical energy, thermochemical storage, and its physical principles, thermal energy storage, chemical, and hydrogen energy storage, and other subjects selected by the author. The chapter also addresses sustainable and ecologically friendly energy storage, including subjects such as longer charge periods, better batteries, storing sunlight as heat, enhanced renewable fuels, and a slew of other issues. Besides that, the chapter discusses various energy storage mechanisms such as hydroelectricity, the mechanism by which hydroelectric power is generated and transferred, a compressor that produces compressed air, a flywheel, gravitational pull of a massive object, thermal storage, thermal heat sensitiveness, latent heat thermal, charging system for the Carnot battery, lithium-ion battery, supercapacitor, chemical storage such as hydrogen and electrochemical storage such as methane, the ability to turn liquid into solids such as water, In the following chapter, researchers will learn about novel 2D materials for energy storage, 2D materials for energy storage devices, challenges facing 2D materials in energy technology, nature-inspired materials for sensing and energy storage applications, and artificial nano and microstructures for sensing and energy storage. Researchers will also learn about bioinspired hierarchical nanofibrous materials, nature-inspired polymer nanocomposites, skin-inspired hierarchical polymer materials, neuron-inspired hierarchical polymer materials, and neuron-inspired hierarchical polymer materials.

References

1 Huang, J., Xia, S., Li, Z. et al. (2021). Applications of four-dimensional printing in emerging directions: review and prospects. *Journal of Materials Science and Technology* 91: 105–120.

2 Renugadevi, N., Saravanan, S., and Sudha, C.M.N. (2021). IoT based smart energy grid for sustainable cities. *Materials Today: Proceedings*.

3 Gao, M., Wang, P., Jiang, L. et al. (2021). Power generation for wearable systems. *Energy & Environmental Science* 14 (4): 2114–2157.

4 Jasim, N.A., Haider, T.H., and Rikabi, S.A.L. (2021). Design and implementation of smart city applications based on the internet of things. *International Journal of Interactive Mobile Technologies* 15: 13.

5 Liu, J., Wang, J., Xu, C. et al. (2018). Advanced energy storage devices: basic principles, analytical methods, and rational materials design. *Advanced Science* 5 (1): 1700322.

6 Li, X., Xiaojuan, C., Zhaoyu, J. et al. (2021). Recent progress in conductive polymers for advanced fiber-shaped electrochemical energy storage devices. *Materials Chemistry Frontiers* 5 (3): 1140–1163.

7 Bu, L., Zhang, Y., Liu, H. et al. (2021). An IIoT-driven and AI-enabled framework for smart manufacturing system based on a three-terminal collaborative platform. *Advanced Engineering Informatics* 50: 101370.

8 Samal, N. and Shiney, O.J. (2021). Energy harvesting using piezoelectric transducers: a review. *Journal of Scientific Research* 65 (3): 163–176.

9 Song, J.-H., Min, S.-H., Kim, S.-G. et al. (2022). Multi-functionalization strategies using nanomaterials: a review and case study in sensing applications. *International Journal of Precision Engineering and Manufacturing-Green Technology* 9 (1): 323–347.

10 Berry, C., Chan, L., Duan, M. et al. (2021). *Aquaponics Monitoring and Control System: Piponics: An Open-Source, Internet of Things Device and Mobile Application*. University of British Columbia Library.

11 Song, Z. and Zhou, H. (2013). Towards sustainable and versatile energy storage devices: an overview of organic electrode materials. *Energy & Environmental Science* 6 (8): 2280–2301.

12 Engler, N. and Krarti, M. (2021). Review of energy efficiency in controlled environment agriculture. *Renewable and Sustainable Energy Reviews* 141: 110786.

13 Eapen, D.E., Suresh, R., Patil, S. et al. (2021). A system engineering perspective on electrochemical energy technologies and a framework for application-driven choice of technology. *Renewable and Sustainable Energy Reviews* 147: 111165.

14 Bansal, M., Goyal, A., and Choudhary, A. (2021). Industrial Internet of Things (IIoT): a vivid perspective. In: *Inventive Systems and Control*, 939–949. Singapore: Springer.

15 Cao, H., Zeng, X., Wu, L. et al. (2021). A hybrid wind and rainwater energy harvesting system for applications in sea-crossing bridges. *Ocean Engineering* 234: 109267.

16 Sharma, A. and Sharma, P. (2021). Chapter 6, Energy harvesting technology for IoT edge applications. *Smart Manufacturing-When Artificial Intelligence Meets the Internet of Things*, 1–14.

17 Almalki, F., Alsamhi, S.H., Sahal, R. et al. (2021). Green IoT for eco-friendly and sustainable smart cities: future directions and opportunities. *Mobile Networks and Applications* 1–25.

18 Mehtab, T., Yasin, G., Arif, M. et al. (2019). Metal-organic frameworks for energy storage devices: batteries and supercapacitors. *Journal of Energy Storage* 21: 632–646.

19 Samal, R., Bhat, M., Kapse, S. et al. (2021). Enhanced energy storage performance and theoretical studies of 3D cuboidal manganese diselenides embedded with multiwalled carbon nanotubes. *Journal of Colloid and Interface Science* 598: 500–510.

20 Latino, M.C. and Neri, G. (2021). Chemoresistive metal oxide gas sensor: working principles and applications. *Atti della Accademia Peloritana dei Pericolanti-Classe di Scienze Fisiche, Matematiche e Naturali* 99 (S1): 41.

21 Liu, L. and Choi, S. (2021). Miniature microbial solar cells to power wireless sensor networks. *Biosensors and Bioelectronics* 177: 112970.

22 Mohsen, S., Zekry, A., Youssef, K. et al. (2021). A self-powered wearable wireless sensor system powered by a hybrid energy harvester for healthcare applications. *Wireless Personal Communications* 116 (4): 3143–3164.

23 Chen, J. and Lee, P.S. (2021). Electrochemical supercapacitors: from mechanism understanding to multifunctional applications. *Advanced Energy Materials* 11 (6): 2003311.

24 Alam, R.B., Ahmad, M.H., and Islam, M.R. (2021). Bio-inspired gelatin/single-walled carbon nanotube nanocomposite for transient electrochemical energy storage: an approach towards eco-friendly and sustainable energy system. *Heliyon* 7 (7): e07468.

25 Chauhan, N.P.S., Jadoun, S., Rathore, B.S. et al. (2021). Redox polymers for capacitive energy storage applications. *Journal of Energy Storage* 43: 103218.

26 Manjakkal, L., Yin, L., Nathan, A. et al. (2021). Energy autonomous sweat-based wearable systems. *Advanced Materials* 33: 2100899.

27 Madhumathi, R., Arumuganathan, T., and Shruthi, R. (2022). Internet of things in precision agriculture: a survey on sensing mechanisms, potential applications, and challenges. In: *Intelligent Sustainable Systems*, 539–553. Singapore: Springer.

28 Nor, N.M., Hamzah, H.H., and Razak, K.A. (2021). Recent advancement in sustainable energy harvesting using piezoelectric materials. In: *Sustainable Materials for Next Generation Energy Devices*, 221–248. Elsevier.

29 Calautit, K., Nasir, D.S.N.M., and Hughes, B.R. (2021). Low power energy harvesting systems: state of the art and future challenges. *Renewable and Sustainable Energy Reviews* 147: 111230.

30 Jia, G., Zheng, A., Wang, X. et al. (2021). Flexible, biocompatible and highly conductive MXene-graphene oxide film for smart actuator and humidity sensor. *Sensors and Actuators B: Chemical* 346: 130507.

31 Wang, J., Xiao, F., and Zhao, H. (2021). Thermoelectric, piezoelectric and photovoltaic harvesting technologies for pavement engineering. *Renewable and Sustainable Energy Reviews* 151: 111522.

32 Hartmann, F., Baumgartner, M., and Kaltenbrunner, M. (2021). Becoming sustainable, the new frontier in soft robotics. *Advanced Materials* 33 (19): 2004413.

33 González, I., Calderón, A.J., and Portalo, J.M. (2021). Innovative multi-layered architecture for heterogeneous automation and monitoring systems: application case of a photovoltaic smart microgrid. *Sustainability* 13 (4): 2234.

34 Aguilar, J., Garces-Jimenez, A., R-Moreno, M.D. et al. (2021). A systematic literature review on the use of artificial intelligence in energy self-management in smart buildings. *Renewable and Sustainable Energy Reviews* 151: 111530.

35 Azam, A., Ahmed, A., Kamran, M.S. et al. (2021). Knowledge structuring for enhancing mechanical energy harvesting (MEH): an in-depth review from 2000 to 2020 using CiteSpace. *Renewable and Sustainable Energy Reviews* 150: 111460.

36 Peng, S., Yu, Y., Wu, S. et al. (2021). Conductive polymer nanocomposites for stretchable electronics: material selection, design, and applications. *ACS Applied Materials & Interfaces* 13: 43831–43854.

37 Aliakbari, R., Kowsari, E., Marfavi, Y. et al. (2021). Comprehensive study on poly ortho-aminophenol composite electrodes and their utilization for supercapacitor applications and green energy storage: a review. *Journal of Energy Storage* 44: 103365.

38 Akanmu, A.A., Anumba, C.J., and Ogunseiju, O.O. (2021). Towards next-generation cyber-physical systems and digital twins for construction. *Journal of Information Technology in Construction (ITcon)* 26 (27): 505–525.

39 Thakur, S., Dasmahapatra, A.K., and Bandyopadhyay, D. (2021). Multifaceted usage of miniaturized energy technologies for sustainable energy harvesting. In: *Sustainable Fuel Technologies Handbook*, 369–405. Academic Press.

40 Jain, R. (2021). Challenges and applications of cyber-physical systems. In: *Design, Applications, and Maintenance of Cyber-Physical Systems*, 1–17. IGI Global.

41 Wang, P., Poovendran, P., and Manokaran, K.B. (2021). Fault detection and control in integrated energy system using machine learning. *Sustainable Energy Technologies and Assessments* 47: 101366.

42 Gao, S., He, T., Zhang, Z. et al. (2021). A motion capturing and energy harvesting hybridized lower-limb system for rehabilitation and sports applications. *Advanced Science* 8: 2101834.

43 Guo, X., Liu, L., Zhang, Z. et al. (2021). Technology evolution from micro-scale energy harvesters to nanogenerators. *Journal of Micromechanics and Microengineering* 31: 093002.

44 Pandya, S., Mistry, M., Parikh, P. et al. (2021). Precision agriculture: methodologies, practices, and applications. In: *Proceedings of Second International Conference on Computing, Communications, and Cyber-Security*. Singapore: Springer.

45 Rotib, H.W., Nappu, M.B., Tahir, Z. et al. (2021). Electric load forecasting for Internet of Things smart home using hybrid PCA and ARIMA algorithm. *International Journal of Electrical and Electronic Engineering & Telecommunications* 10 (6).

46 Cioara, T., Anghel, Ionut, Antal, Marcel, et al. (2021). An Overview of Digital Twins Application Domains in Smart Energy Grid. arXiv:2104.07904.

47 Hassan, R.J., Zeebaree, S.R.M., Ameen, S.Y. et al. (2021). State of art survey for iot effects on smart city technology: challenges, opportunities, and solutions. *Asian Journal of Research in Computer Science* 32–48.

48 Leng, J., Zhou, M., Xiao, Y. et al. (2021). Digital twins-based remote semi-physical commissioning of flow-type smart manufacturing systems. *Journal of Cleaner Production* 306: 127278.

49 Abeykoon, C., McMillan, A., and Nguyen, B.K. (2021). Energy efficiency in extrusion-related polymer processing: a review of state of the art and potential efficiency improvements. *Renewable and Sustainable Energy Reviews* 147: 111219.

50 Mohol, S.S. and Sharma, V. (2021). Functional applications of 4D printing: a review. *Rapid Prototyping Journal* 27: 1501–1522.

51 Nurelmadina, N., Hasan, M.K., Memon, I. et al. (2021). A systematic review on cognitive radio in low power wide area network for industrial IoT applications. *Sustainability* 13 (1): 338.

52 Martins, P., Lopes, S.I., Cruz, A.M.R. et al. (2021). Towards a smart & sustainable campus: an application-oriented architecture to streamline digitization and strengthen sustainability in academia. *Sustainability* 13 (6): 3189.

53 Kermani, M., Adelmanesh, B., Shirdare, E. et al. (2021). Intelligent energy management based on SCADA system in a real Microgrid for smart building applications. *Renewable Energy* 171: 1115–1127.

54 Kundaria, N., Mohanty, S.S., Varjani, S. et al. (2021). A review on integrated approaches for municipal solid waste for environmental and economical relevance: monitoring tools, technologies, and strategic innovations. *Bioresource Technology* 342: 125982.

55 Mishu, M.K., Rokonuzzaman, M., Pasupuleti, J. et al. (2021). An adaptive TE-PV hybrid energy harvesting system for self-powered iot sensor applications. *Sensors* 21 (8): 2604.

56 Achour, Y., Ouammi, A., and Zejli, D. (2021). Technological progress in modern sustainable greenhouses cultivation as the path towards precision agriculture. *Renewable and Sustainable Energy Reviews* 147: 111251.

57 Do, T.C., Dang, T.D., Dinh, T.Q. et al. (2021). Developments in energy regeneration technologies for hydraulic excavators: a review. *Renewable and Sustainable Energy Reviews* 145: 111076.

58 Hesham, R., Soltan, A., and Madian, A. (2021). Energy harvesting schemes for wearable devices. *AEU International Journal of Electronics and Communications* 145: 153888.

59 Deb, A., Goswami, D., Bhowmik, B. et al. (2022). Analysis of piezoelectric energy harvesting schemes and a proposition for state-of-the-art approach in hydro-electric power generation. In: *Renewable Energy Optimization, Planning, and Control*, 173–183. Singapore: Springer.

60 Goswami, Y. and Sharma, S. (2021). Graphene reinforced biopolymer nanocomposites in energy storage applications. In: *Graphene-Based Biopolymer Nanocomposites*, 233–250. Singapore: Springer.

61 Otoni, C.G., Azeredo, H.M.C., Mattos, B.D. et al. (2021). The food–materials nexus: next generation bioplastics and advanced materials from agri-food residues. *Advanced Materials* 33: 2102520.

62 Andronie, M., Lăzăroiu, G., Iatagan, M. et al. (2021). ustainable cyber-physical production systems in big data-driven smart urban economy: a systematic literature review. *Sustainability* 13 (2): 751.

63 Sathiyanathan, P., Wu, L., Schmidt, T.W. et al. (2021). Piezoelectric-piezocapacitive hybrid sensor based on electrospun Poly (vinylidene fluoride)-Poly (octafluoropentyl acrylate)-sulphonated Poly (phenylene sulfide) blend nanofiber. *Sensors and Actuators A: Physical* 331: 112993.

64 Carlucci, F. (2021). A review of smart and responsive building technologies and their classifications. *Future Cities and Environment* 7 (1): 10.

65 Nasab Azimi, M., Zand, M., and Eskandari, M. (2021). Optimal planning of electrical appliance of residential units in a smart home network using cloud services. *Smart Cities* 4 (3): 1173–1195.

66 Jaibalaganesh, T., Jayaraman, K., Raman, K. et al. (2021). Fault identification and isolation for components of photovoltaic energy storage system based on IEEE21451 standard. *International Journal of Numerical Modelling: Electronic Networks, Devices and Fields* 34 (2): e2819.

67 Abass, O. and Dawodu, A.A. (2021). Towards SDG-9: an IoT architecture for converging OT-IT in industry 4.0 and digital economy era. *University of Ibadan Journal of Science and Logics in ICT Research* 6 (1, 2): 57–67.

68 Das, C.K., Bass, O., Kothapalli, G. et al. (2018). Overview of energy storage systems in distribution networks: placement, sizing, operation, and power quality. *Renewable and Sustainable Energy Reviews* 91: 1205–1230.

69 Varshney, A., Garg, N., Nagla, K.S. et al. (2021). Challenges in sensors technology for industry 4.0 for futuristic metrological applications. *Mapan* 36: 1–12.

70 Campolo, C., Genovese, G., Iera, A. et al. (2021). Virtualizing AI at the distributed edge towards intelligent IoT applications. *Journal of Sensor and Actuator Networks* 10 (1): 13.

71 Abdi, H., Ivatloo, B.M., Javadi, S. et al. (2017). Energy storage systems. In: *Distributed Generation Systems*, vol. 7, 333–368. Elsevier Inc.

72 Elmouatamid, A., Naitmalek, Y., Ouladsine, R. et al. (2021). A MicroGrid system infrastructure implementing IoT/big-data technologies for efficient energy management in buildings. In: *Advanced Technologies for Solar Photovoltaics Energy Systems*, 571–600. Cham: Springer.

73 Siwal, S.S., Zhang, Q., Devi, N. et al. (2020). Carbon-based polymer nanocomposite for high-performance energy storage applications. *Polymers* 12 (3): 505.

74 Zhao, L., Liu, Z., Chen, D. et al. (2021). Laser synthesis and microfabrication of micro/ nanostructured materials toward energy conversion and storage. *Nano-Micro Letters* 13 (1): 1–48.

75 Zhou, Z., Zheng, C., and Liao, H. (2021). Software-defined machine-to-machine communication for smart energy management in power grids. In: *Green Internet of Things (IoT): Energy Efficiency Perspective*, 43–51. Cham: Springer.

76 Ma, X., Gao, W., Du, F. et al. (2021). Rational design of electrochemiluminescent devices. *Accounts of Chemical Research* 54 (14): 2936–2945.

77 Mackanic, D.G., Chang, T.-H., Huang, Z. et al. (2020). Stretchable electrochemical energy storage devices. *Chemical Society Reviews* 49 (13): 4466–4495.

78 Chan, C.Y., Wang, Z., Jia, H. et al. (2021). Recent advances of hydrogel electrolytes in flexible energy storage devices. *Journal of Materials Chemistry A* 9 (4): 2043–2069.

79 Blum, S.A., Zahrebelnei, F., Nagata, N. et al. (2021). Experimental design to enhance dopamine electrochemical detection using carbon paste electrodes. *Brazilian Journal of Analytical Chemistry* 8 (32): 178–197.

80 Suryadevara, N.K. (2021). Energy and latency reductions at the fog gateway using a machine learning classifier. *Sustainable Computing: Informatics and Systems* 31: 100582.

81 Mohanraj, T. and Jegadeeshwaran, R. (2021). Introduction to Industry 4.0. In: *Tribology of Polymer and Polymer Composites for Industry 4.0*, 113–127. Singapore: Springer.

82 Bag, S., Wood, L.C., Telukdarie, A. et al. (2021). Application of Industry 4.0 tools to empower the circular economy and achieving sustainability in supply chain operations. *Production Planning and Control* 1–23.

83 Kerskes, H. (2016). Chapter 17 - Thermochemical energy storage. In: *Storing Energy* (ed. T.M. Letcher), 345–372. Oxford: Elsevier.

84 Yang, J., Wang, Z., Yao, J. et al. (2021). Effect and mechanism of various pre-freezing temperatures on the deflection response of a highly biocompatible hydrogel artificial muscle. *Sensors and Actuators A: Physical* 332: 113126.

85 Xu, C., Song, Y., Han, M. et al. (2021). Portable and wearable self-powered systems based on emerging energy harvesting technology. *Microsystems & Nanoengineering* 7 (1): 1–14.

86 Gharat, P.V., Bhalekar, S.S., Dalvi, V.H. et al. (2021). Chronological development of innovations in reflector systems of parabolic trough solar collector (PTC)-A review. *Renewable and Sustainable Energy Reviews* 145: 111002.

87 Huber, C., Jahromy, S.S., Jordan, C. et al. (2019). Boric acid: a high potential candidate for thermochemical energy storage. *Energies* 12 (6): 1086.

88 Nizam, P.A., Gopakumar, D.A., Pottathara, Y.B. et al. (2021). Nanocellulose-based composites: fundamentals and applications in electronics. In: *Nanocellulose Based Composites for Electronics*, 15–29. Elsevier.

89 Sajjad, M. and Wen, L. (2021). Covalent organic frameworks-based nanomaterials: design, synthesis, and current status for supercapacitor applications: a review. *Journal of Energy Storage* 39: 102618.

90 Li, B., Li, J., Shao, H. et al. (2018). Mg-based hydrogen absorbing materials for thermal energy storage—a review. *Applied Sciences* 8 (8): 1375.

91 Shen, F., Li, Z., Guo, H. et al. (2021). Recent advances towards ocean energy harvesting and self-powered applications based on triboelectric nanogenerators. *Advanced Electronic Materials* 7: 2100277.

92 Gong, X., Yang, Q., Zhi, C. et al. (2021). Stretchable energy storage devices: from materials and structural design to device assembly. *Advanced Energy Materials* 11 (15): 2003308.

93 Kuravi, S., Trahan, J., Goswami, D.Y. et al. (2013). Thermal energy storage technologies and systems for concentrating solar power plants. *Progress in Energy and Combustion Science* 39 (4): 285–319.

94 Rokonuzzaman, M., Mishu, M.K., Amin, N. et al. (2021). Self-sustained autonomous wireless sensor network with integrated solar photovoltaic system for internet of smart home-building (IoSHB) applications. *Micromachines* 12 (6): 653.

95 Wang, L., Wu, X., Su, B.S.Q.-w. et al. (2021). Enzymatic biofuel cell: opportunities and intrinsic challenges in futuristic applications. *Advanced Energy and Sustainability Research* 2 (8): 2100031.

96 Fuchs, G., Lunz, B., Leuthold, M. et al. (2012). *Technology Overview on Electricity Storage*, 26. Aachen, Juni: *ISEA*.

97 Karad, S. and Thakur, R. (2021). Efficient monitoring and control of wind energy conversion systems using Internet of things (IoT): a comprehensive review. *Environment, Development and Sustainability* 23: 1–18.

98 Parandeh, S., Etemadi, N., Kharaziha, M. et al. (2021). Advances in triboelectric nanogenerators for self-powered regenerative medicine. *Advanced Functional Materials* 31: 2105169.

99 Gao, X., Pishdad-Bozorgi, P., Shelden, D.R. et al. (2021). Internet of things enabled data acquisition framework for smart building applications. *Journal of Construction Engineering and Management* 147 (2): 04020169.

100 Steinbuch, M., Oomen, T., and Vermeulen, H. (2021). Motion control, mechatronics design, and Moore's law. *IEEJ Journal of Industry Applications* 11: 21006010.

101 Chatterjee, D.P. and Nandi, A.K. (2021). A review on the recent advances in hybrid supercapacitors. *Journal of Materials Chemistry A* 9 (29): 15880–15918.

102 Zhu, Y., Gonzalez-Rosillo, J.C., Balaish, M. et al. (2021). Lithium-film ceramics for solid-state with ionic devices. *Nature Reviews Materials* 6 (4): 313–331.

103 Sharma, A., Sharma, S., and Gupta, D. (2021). A review of sensors and their application in the internet of things (IOT). *International Journal of Computer Applications* 975: 8887.

104 Mohsen, S. (2021). Hybrid energy harvester for medical sensor node toward real-time healthcare monitoring. *Proceedings of Engineering and Technology Innovation* 18: 43.

105 Guo, X., He, T., Zhang, Z. et al. (2021). Artificial intelligence-enabled caregiving walking stick powered by ultra-low-frequency human motion. *ACS Nano* 15: 19054–19069.

106 Williams, A.J., Torquato, M.F., Cameron, I.M. et al. (2021). Survey of energy harvesting technologies for wireless sensor networks. *IEEE Access* 9: 77493–77510.

107 Jamroen, C., Chanon, F., Wipa, K. et al. (2021). A novel UV sensor-based dual-axis solar tracking system: implementation and performance analysis. *Applied Energy* 299: 117295.

108 Shobole, A.A. and Wadi, M. (2021). Multiagent systems application for the smart grid protection. *Renewable and Sustainable Energy Reviews* 149: 111352.

109 Navalpotro, P., Castillo-Martínez, E., and Carretero-González, J. (2021). Sustainable materials for off-grid battery applications: advances, challenges, and prospects. *Sustainable Energy & Fuels* 5 (2): 310–331.

110 Lee, H., Jiang, Z., Yokota, T. et al. (2021). Stretchable organic optoelectronic devices: design of materials, structures, and applications. *Materials Science & Engineering R: Reports* 146: 100631.

111 Zalke, J.B., Pandey, S.R., Nandanwar, R.V. et al. (2021). An inductor less piezoelectric energy harvesting interface circuit using gyrator induced voltage flip technique. *Circuit World*.

112 Lee, M.H., Lee, J., Jung, S.-K. et al. (2021). A biodegradable secondary battery and its biodegradation mechanism for eco-friendly energy-storage systems. *Advanced Materials* 33 (10): 2004902.

113 Wang, L., He, T., Zhang, Z. et al. (2021). Self-sustained autonomous wireless sensing based on a hybridized TENG and PEG vibration mechanism. *Nano Energy* 80: 105555.

114 Javaid, M., Haleem, A., Singh, R.P. et al. (2021). Significance of sensors for Industry 4.0: roles, capabilities, and applications. *Sensors International* 2: 100110.

115 Feng, X., Zhang, Y., Kang, L. et al. (2021). Integrated energy storage system based on triboelectric nanogenerator in electronic devices. *Frontiers of Chemical Science and Engineering* 15 (2): 238–250.

116 Xu, C., Puente-Santiago, A.R., Rodríguez-Padrón, D. et al. (2021). Nature-inspired hierarchical materials for sensing and energy storage applications. *Chemical Society Reviews* 50: 4856–4871.

117 Ren, Z., Xu, J., Le, X. et al. (2021). Heterogeneous wafer bonding technology and thin-film transfer technology-enabling platform for the next generation applications beyond 5G. *Micromachines* 12 (8): 946.

118 Ibaseta, D., García, A., Álvarez, M. et al. (2021). Monitoring and control of energy consumption in buildings using WoT: a novel approach for smart retrofit. *Sustainable Cities and Society* 65: 102637.

119 Aabid, A., Raheman, M.A., Ibrahim, Y.E. et al. (2021). A systematic review of piezoelectric materials and energy harvesters for industrial applications. Sensors 21 (12): 4145

120 Zhuo, F., Qiao, H., Zhu, J. et al. (2021). Perspective on antiferroelectrics for energy storage and conversion applications. *Chinese Chemical Letters* 32 (7): 2097–2107.

121 Algamili, A.S., Khir, M.H.M., Dennis, J.O. et al. (2021). A review of actuation and sensing mechanisms in MEMS-based sensor devices. *Nanoscale Research Letters* 16 (1): 1–21.

122 Devi, N., Sahoo, S., Kumar, R. et al. (2021). A review of the microwave-assisted synthesis of carbon nanomaterials, metal oxides/hydroxides and their composites for energy storage applications. *Nanoscale* 13: 11679–11711.

123 Behara, Naga Sudha Rani, and Putti Srinivasa Rao. "Application of energy harvesting in the field of mechanical vibrations." *IOP Conference Series: Materials Science and Engineering*. Vol. 1057. No. 1-9. IOP Publishing, 2021 012024.

124 Alsamhi, S.H., Afghah, F., Sahal, R. et al. (2021). Green internet of things using UAVs in B5G networks: a review of applications and strategies. *Ad Hoc Networks* 117: 102505.

125 Hasan, M.A.M., Wu, H., and Yang, Y. (2021). Redox-induced electricity for energy scavenging and self-powered sensors. *Journal of Materials Chemistry A* 9: 19116–19148.

126 Wang, H., Yang, Y., and Lin, G. (2017). Nature-inspired electrochemical energy-storage materials and devices. *Advanced Energy Materials* 7 (5): 1601709.

127 Zhao, T., Xu, M., Xiao, X. et al. (2021). Recent progress in blue energy harvesting for powering distributed sensors in the ocean. *Nano Energy* 88: 106199.

128 Amrillah, T., Hermawan, A., Wulandari, C. et al. (2021). Crafting the multiferroic $BiFeO_3$-$CoFe_2O_4$ nanocomposite for next-generation devices: a review. *Materials and Manufacturing Processes* 36 (14): 1579–1596.

129 Hao, D., Zhang, T., Guo, L. et al. (2021). A high-efficiency, portable solar energy-harvesting system based on a foldable-wings mechanism for self-powered applications in railways. *Energy Technology* 9 (4): 2000794.

130 Roumie, J. (2021). Analytical and comparative study on the design of kinetic facades and their daylight performance in buildings. *Al-Manara University Journal* 1 (3).

131 Abavisani, I., Rezaifar, O., and Kheyroddin, A. (2021). Multifunctional properties of shape memory materials in civil engineering applications: a state-of-the-art review. *Journal of Building Engineering* 44: 102657.

132 Saleem, A. and Jazani, D. (2021). Presence aware power saving mode (PA-PSM) development for IoT devices for energy conservation. *Open Journal of Energy Efficiency* 10 (3): 81–96.

133 Park, S., Rosca, E., and Agarwal, N. (2021). Driving social impact at the bottom of the Pyramid through the internet-of-things enabled frugal innovations. *Technovation* 118: 102381.

134 Pivoto, D.G.S., Almeida, L.F.F., Righi, R. et al. (2021). Cyber-physical systems architectures for the industrial internet of things applications in Industry 4.0: a literature review. *Journal of Manufacturing Systems* 58: 176–192.

135 Pal, N.R., Bello Pérez, R.E., Corchado, E. et al. (2023). Advances in intelligent systems and computing. 1435.

136 Zahoor, S. and Mir, R.N. (2021). An IoT enabled smart city: assessing the applications, resource constraints, existing solutions, and research directions. *International Journal of Computing and Digital System* 12: 269–283.

137 Alsufyani, A., Alotaibi, Y., Almagrabi, A.O. et al. (2021). Optimized intelligent data management framework for a cyber-physical system for computational applications. *Complex & Intelligent Systems* 1–13.

138 Frate, G.F., Ferrari, L., and Desideri, U. (2021). Energy storage for grid-scale applications: technology review and economic feasibility analysis. *Renewable Energy* 163: 1754–1772.

139 Sharma, P., Jain, S., Gupta, S. et al. (2021). Role of machine learning and deep learning in securing 5G-driven industrial IoT applications. *Ad Hoc Networks* 123: 102685.

140 Kaswan, Kuldeep Singh, and Jagjit Singh Dhatterwal. "The Use of Machine Learning for Sustainable and Resilient Buildings." *Digital Cities Roadmap: IoT-Based Architecture and Sustainable Buildings* (2021): 1–62.

141 Song, L., Huang, Z., Guo, S. et al. (2021). Hierarchically architected polyvinylidene fluoride piezoelectric foam for boosted mechanical energy harvesting and self-powered sensor. *ACS Applied Materials & Interfaces* 13 (31): 37252–37261.

142 Riccio, A., Sellitto, A., Ameduri, S. et al. (2021). Shape memory alloys (SMA) for automotive applications and challenges. In: *Shape Memory Alloy Engineering*, 785–808. Butterworth-Heinemann.

143 Shah, K.A., Meng, F., Li, Y. et al. (2021). A synthesis of feasible control methods for floating offshore wind turbine system dynamics. *Renewable and Sustainable Energy Reviews* 151: 111525.

144 Dahiya, N., Dalal, S., and Jaglan, V. (2021). 7 Efficient green solution. *Green Internet of Things for Smart Cities: Concepts, Implications, and Challenges* 113.

145 Mehmood, M.Y., Oad, A., Muhammad, A. et al. (2021). Edge computing for IoT-enabled smart grid. *Security and Communication Networks* 2021: 16.

146 Kulkarni, V., Sahoo, S.K., and Mathew, R. (2021). Applications of internet of things for microgrid. *Microgrid Technologies* 405–428.

147 Sun, Z., Zhu, M., and Lee, C. (2021). Progress in the triboelectric human–machine interfaces (HMIs)-moving from smart gloves to AI/haptic enabled HMI in the 5G/IoT era. *Nano Energy Advances* 1 (1): 81–120.

148 Xu, C., Li, G., Ning, Y. et al. (2021). Banana peel derived nitrogen-doped porous carbon with enhanced electrocatalytic activity for complete oxidation of methanol under room temperature. *Sensors and Actuators B: Chemical* 344: 130112.

149 Vitali, G., Francia, M., Golfarelli, M. et al. (2021). Crop management with the IoT: an interdisciplinary survey. *Agronomy* 11 (1): 181.

150 Wang, Y., Dai, M., Wu, H. et al. (2021). Moisture induced electricity for self-powered microrobots. *Nano Energy* 90: 106499.

151 Sreenilayam, S.P., Ahad, I.U., Nicolosi, V. et al. (2021). Mxene materials based printed flexible devices for healthcare, biomedical and energy storage applications. *Materials Today* 43: 99–131.

152 Merazzo, K.J., Lima, A.C., Rincón-Iglesias, M. et al. (2021). Magnetic materials: a journey from finding north to an exciting printed future. *Materials Horizons* 8: 2654–2684.

153 Wemyss, A.M., Ellingford, C., Morishita, Y. et al. (2021). Dynamic polymer networks: a new avenue towards sustainable and advanced soft machines. *Angewandte Chemie* 133 (25): 13841–13852.

154 Liu, H., Yu, C., and Wu, H. (2021). Potential applications of smart device recognition in industry. In: *Smart Device Recognition*, 259–294. Singapore: Springer.

155 Javaid, M., Haleem, A., Rab, S. et al. (2021). Sensors for daily life: a review. *Sensors International* 2: 100121.

156 Zhang, S., Malič, B., Li, J.-F. et al. (2021). Lead-free ferroelectric materials: prospective applications. *Journal of Materials Research* 36: 985–995. https://doi.org/10.1557/s43578-021-00180-y.

157 Elvitigala, C.S. (2021). Low-cost hardware-in-the-loop (hil) simulator for simulation and analysis of embedded systems with non-real-time applications. LSU Master's Theses. 5262. https://digitalcommons.lsu.edu/gradschool_theses/5262. 10.31390/gradschool_theses.5262.

158 Kim, H., Choi, H., Kang, H. et al. (2021). A systematic review of the smart energy conservation system: from smart homes to sustainable smart cities. *Renewable and Sustainable Energy Reviews* 140: 110755.

159 Ryan, K.R., Down, M.P., and Banks, C.E. (2021). Future of additive manufacturing: overview of 4D and 3D printed smart and advanced materials and their applications. *Chemical Engineering Journal* 403: 126162.

160 Zhang, Z., Shi, Q., He, T. et al. (2021). Artificial intelligence of toilet (AI-Toilet) for an integrated health monitoring system (IHMS) using smart triboelectric pressure sensors and image sensor. *Nano Energy* 90: 106517.

161 Manikandan, M. and Rajkumar, S.P. (2021). Multidisciplinary design approach for solar-powered tri-lobed HALESA. *Energy Conversion and Management* 245: 114616.

162 Farooq, A., Seyedmahmoudian, M., Horan, B. et al. (2021). Overview and exploitation of haptic tele-weight device in virtual shopping stores. *Sustainability* 13 (13): 7253.

163 Cao, X., Xiong, Y., Sun, J. et al. (2021). Piezoelectric nanogenerators derived self-powered sensors for multifunctional applications and artificial intelligence. *Advanced Functional Materials* 31: 2102983.

164 Mehrpouya, M., Vahabi, H., Janbaz, S. et al. (2021). 4D printing of shape memory polylactic acid (PLA) components: a scientific study from the fabrication techniques to the applications. *Polymer* 230: 124080.

165 Xu, Z., Congran, J., Andrew, C. et al. (2021). Implantable cardiac Kirigami-inspired lead-based energy harvester fabricated by enhanced piezoelectric composite film. *Advanced Healthcare Materials* 10 (8): 2002100.

166 Fu, H., Mei, X., Yurchenko, D. et al. (2021). Rotational energy harvesting for self-powered sensing. *Joule* 5: 1074–1118.

167 Sharma, S., Sudhakara, P., Borhana Omran, A.A. et al. (2021). Recent trends and developments in conducting polymer nanocomposites for multifunctional applications. *Polymers* 13 (17): 2898.

168 Javaid, M., Haleem, A., Singh, R.P. et al. (2021). Significance of Quality 4.0 towards comprehensive enhancement in the manufacturing sector. *Sensors International* 2: 100109.

169 Wu, Y., Zhao, X., Shang, Y. et al. (2021). Application-driven carbon nanotube functional materials. *ACS Nano* 15: 7946–7974.

170 Chen, G., Zhao, X., Andalib, S. et al. (2021). Discovering giant magnetoelasticity in a soft matter for electronic textiles. *Matter* 4: 3725–3740.

171 Peng, L., Fang, Z., Zhu, Y. et al. (2018). Holey 2D nanomaterials for electrochemical energy storage. *Advanced Energy Materials* 8 (9): 1702179.

172 Benavides, D.J., Arevalo, W., Gonzalez, L.G. et al. (2022). Method of monitoring and detection of failures in PV system based on machine learning. *Revista Facultad de Ingeniería Universidad de Antioquia* 102: 26–43.

173 Parisi, S., Holzbach, M., and Rognoli, V. (2021). Paving the way to post-digital smart materials. Experiments on human perceptions of a bio-inspired cellulose-based responsive interface. In: *Design Culture (S)*, 1308–1324. Cumulus the Global Association of Art and Design Education and Research.

174 Rahman, M.H., Werth, H., Goldman, A. et al. (2021). Recent progress on electroactive polymers: synthesis, properties, and applications. *Ceramics* 4 (3): 516–541.

175 Haroun, A.F., Le, X., Gao, S. et al. (2021). Progress in micro/nanosensors and nano energy for future AIoT-based smart home applications. *Nano Express* 2: 022005.

176 Huang, J., Zhao, M., Hao, Y. et al. (2021). Recent advances in functional bacterial cellulose for wearable physical sensing applications. *Advanced Materials Technologies* 7: 2100617.

177 Maimani, F., Calderón, A.A., Yang, X. et al. (2021). A 7-milligram miniature catalytic-combustion engine for millimeter-scale robotic actuation. *Sensors and Actuators A: Physical* 341: 112818.

178 Shaikh, E., Bashar, A., and Mohammad, N. (2021). Recent applications of computing and mobility technologies to modern manufacturing. In: *2020 International Conference on Communications, Signal Processing, and their Applications (ICCSPA)*, 1–6. IEEE.

179 Liu, W. (2021). Smart sensors, sensing mechanisms and platforms of sustainable smart agriculture realized through the big data analysis. *Cluster Computing* 1–15.

180 Dalaeen, A., Hosny, J., Khan, Y., and Ahmad, A. (2021). Synthesis and application of nanocomposite reinforced with decorated multi-walled carbon nanotube with luminescence quantum dots. *Advances in Nanoparticles* 10 (2): 75–93.

181 Amato, A., Quarto, A., and Di Lecce, V. (2021). An application of the cyber-physical system and multi-agent technology to demand-side management systems. *Pattern Recognition Letters* 141: 23–31.

182 Cattaruzzi, J. and Gaspari, J. (2021). Taxonomical investigation of self-sufficient kinetic building envelopes. *Journal of Architectural Engineering* 27 (4): 03121001.

183 Zhang, P., Wang, F., Yu, M. et al. (2018). Two-dimensional materials for miniaturized energy storage devices: from individual devices to smart integrated systems. *Chemical Society Reviews* 47 (19): 7426–7451.

184 Wiklund, J., Karakoç, A., Palko, T. et al. (2021). A review on printed electronics: fabrication methods, inks, substrates, applications, and environmental impacts. *Journal of Manufacturing and Materials Processing* 5 (3): 89.

185 Yasin, H.M., Zeebaree, S.R.M., Sadeeq, M.A.M. et al. (2021). IoT and ICT based smart water management, monitoring and controlling system: a review. *Asian Journal of Research in Computer Science* 42–56.

186 Gulati, K., Boddu, R.S.K., Kapila, D. et al. (2021). A review paper on wireless sensor network techniques in Internet of Things (IoT). *Materials Today: Proceedings* 51.

187 Jiang, C., Li, X., Lian, S.W.M. et al. (2021). Wireless technologies for energy harvesting and transmission for ambient self-powered systems. *ACS Nano* 15: 9328–9354.

188 Liu, Y., Shang, S., Mo, S. et al. (2021). Soft actuators built from cellulose paper: a review on actuation, material, fabrication, and applications. *Journal of Science: Advanced Materials and Devices* 6: 321–337.

189 Anshari, M. and Almunawar, M.N. (2021). Adopting open innovation for SMEs and industrial revolution 4.0. *Journal of Science and Technology Policy Management* 13: 405–442.

190 Zhang, Y., Zhu, Y., Zheng, S. et al. (2021). Ink formulation, scalable applications and challenging perspectives of screen printing for emerging printed microelectronics. *Journal of Energy Chemistry* 63: 498–513.

191 Panda, D.K. and Das, S. (2021). Smart grid architecture model for control, optimization and data analytics of future power networks with more renewable energy. *Journal of Cleaner Production* 301: 126877.

192 Xiong, S., Lin, M., Wang, L. et al. (2021). Defects-type three-dimensional Co_3O_4 nanomaterials for energy conversion and low-temperature energy storage. *Applied Surface Science* 546: 149064.

193 Guo, H., Tan, Y.J., Chen, G. et al. (2020). Artificially innervated self-healing foams as synthetic piezo-impedance sensor skins. *Nature Communications* 11 (1): 1–10.

194 Yang, J., Jihua, H., Tong, W. et al. (2021). Sizing and optimization of on-grid hybrid renewable energy systems considering hydroelectric energy storage. *Journal of Intelligent Fuzzy Systems* 1–16.

195 Gupta, R., Sossan, F., and Paolone, M. (2021). Countrywide PV hosting capacity and energy storage requirements for distribution networks: the case of Switzerland. *Applied Energy* 281: 116010.

196 Uchman, W., Kotowicz, J., and Li, K.F. (2021). Evaluation of a micro-cogeneration unit with integrated electrical energy storage for residential application. *Applied Energy* 282: 116196.

197 Sepulveda, N.A., Jenkins, J.D., Edington, A. et al. (2021). The design space for long-duration energy storage in decarbonized power systems. *Nature Energy* 6 (5): 506–516.

198 Tang, R., Lu, F., Liu, L. et al. (2021). Flexible pressure sensors with microstructures. *Nano Select* 2 (10): 1874–1901.

199 Wegener, M., Schneider, J.V., Malmquist, A. et al. (2021). Techno-economic optimization model for polygeneration hybrid energy storage systems using biogas and batteries. *Energy* 218: 119544.

200 Kim, M., Lim, H., Xu, X. et al. (2021). Sorghum biomass-derived porous carbon electrodes for capacitive deionization and energy storage. *Microporous and Mesoporous Materials* 312: 110757.

201 Tarashandeh, N. and Karimi, A. (2021). Utilization of energy storage systems in congestion management of transmission networks with an incentive-based approach for investors. *Journal of Energy Storage* 33: 102034.

202 Wang, Y., Naleway, S.E., and Wang, B. (2020). Biological and bioinspired materials: structure leading to functional and mechanical performance. *Bioactive Materials* 5 (4): 745–757.

203 Shao, X., Zhang, X., Liu, Y. et al. (2021). Metal chalcogenide-associated catalysts enabling CO_2 electroreduction to produce low-carbon fuels for energy storage and emission reduction: the catalyst structure, morphology, performance, and mechanism. *Journal of Materials Chemistry A* 9 (5): 2526–2559.

204 Temiz, M. and Dincer, I. (2021). Concentrated solar-driven thermochemical hydrogen production plant with thermal energy storage and geothermal systems. *Energy* 219: 119554.

205 Peng, J. and Cheng, Q. (2017). High-performance nanocomposites inspired by nature. *Advanced Materials* 29 (45): 1702959.

206 Benalcazar, P. (2021). Optimal sizing of thermal energy storage systems for CHP plants considering specific investment costs: a case study. *Energy* 234: 121323.

207 Giarola, S., Molar-Cruz, A., Vaillancourt, K. et al. (2021). The role of energy storage in the uptake of renewable energy: a model comparison approach. *Energy Policy* 151: 112159.

208 Gage, S.H., Bailey, J.J., Finegan, D.P. et al. (2021). Internal insulation and corrosion control of molten chloride thermal energy storage tanks. *Solar Energy Materials and Solar Cells* 225: 111048.

209 Kumar, P., Singh, P.K., Nagar, S. et al. (2021). Effect of different concentrations of functionalized graphene on charging time reduction in a thermal energy storage system. *Materials Today: Proceedings* 44: 146–152.

210 Zhu, H. and Ouahada, K. (2021). A distributed real-time control algorithm for energy storage sharing. *Energy and Buildings* 230: 110478.

211 Li, W., Pei, Y., Zhang, C. et al. (2021). Bioinspired designs and biomimetic applications of triboelectric nanogenerators. *Nano Energy* 105865.

212 Pu, X. and Wang, Z.L. (2021). Self-charging power system for distributed energy: beyond the energy storage unit. *Chemical Science* 12 (1): 34–49.

213 Díez, N., Fuertes, A.B., and Sevilla, M. (2021). Molten salt strategies towards carbon materials for energy storage and conversion. *Energy Storage Materials*.

214 Udawalpola, R., Masuta, T., Yoshioka, T. et al. (2021). Reduction of power imbalances using battery energy storage system in a bulk power system with extremely large photovoltaics interactions. *Energies* 14 (3): 522.

215 Xia, Y., Xu, Q., Qian, H. et al. (2021). Bilevel optimal configuration of generalized energy storage considering power consumption right transaction. *International Journal of Electrical Power & Energy Systems* 128: 106750.

216 Kim, M.-K., Park, Y., Kim, I.-J. et al. (2020). Emerging materials for neuromorphic devices and systems. *Iscience* 101846.

217 Stengler, J., Bürger, I., and Linder, M. (2021). Performance analysis of a gas-solid thermochemical energy storage using numerical and experimental methods. *International Journal of Heat and Mass Transfer* 167: 120797.

218 Preethika, M., Nagarajan, R.D., Shetty, B.H. et al. (2021). Recent trends in the applications of thermally expanded graphite for energy storage and sensors–a review. *Nanoscale Advances.*

219 Igbinovia, M.O. (2021). Internet of things in libraries and focus on its adoption in developing countries. *Library Hi Tech News* 38: 13–17.

220 Zhao, P., Zhang, S., Gou, F. et al. (2021). The feasibility survey of an autonomous renewable seawater reverse osmosis system with underwater compressed air energy storage. *Desalination* 505: 114981.

221 Chaudhary, M., Tyagi, S., Gupta, R.K. et al. (2021). Surface modification of cathode materials for energy storage devices: a review. *Surface and Coatings Technology* 127009.

222 Ortiz, C., Valverde, J.M., Chacartegui, R. et al. (2021). Scaling-up the calcium-looping process for CO_2 capture and energy storage. *Kona Powder and Particle Journal* 38: 189–208.

223 Liu, W.-J., Jiang, H., and Han-Qing, Y. (2019). Emerging applications of biochar-based materials for energy storage and conversion. *Energy & Environmental Science* 12 (6): 1751–1779.

224 Silva, F. and O'Regan, B. (2021). An innovative smart grid framework for integration and trading. In: *E3S Web of Conferences*, vol. 294, 1–37. EDP Sciences.

225 Pritoni, M., Paine, D., Fierro, G. et al. (2021). Metadata schemas and ontologies for building energy applications: a critical review and use case analysis. *Energies* 14 (7): 2024.

226 Fu, W., Turcheniuk, K., Naumov, O. et al. (2021). Materials and technologies for multifunctional, flexible or integrated supercapacitors and batteries. *Materials Today.*

227 Mück, T., Donyanavard, Bryan, Maity, Biswadip et al. (2021). MARS: middleware for adaptive reflective computer systems. arXiv:2107.11417.

228 Pyo, S., Lee, J., Bae, K. et al. (2021). Recent progress in flexible tactile sensors for human-interactive systems: from sensors to advanced applications. *Advanced Materials* 2005902.

229 Zhang, J. (2021). Distributed network security framework of energy internet based on internet of things. *Sustainable Energy Technologies and Assessments* 44: 101051.

230 Yaïci, W., Krishnamurthy, K., Entchev, E. et al. (2021). Recent advances in the Internet of Things (IoT) infrastructures for building energy systems: a review. *Sensors* 21 (6): 2152.

231 Sun, M., Wang, W., Zheng, P. et al. (2021). A novel road energy harvesting system based on a spatial double V-shaped mechanism for near-zero-energy toll stations on expressways. *Sensors and Actuators A: Physical* 323: 112648.

232 Liu, X., Su, G., Guo, Q. et al. (2018). Hierarchically structured self-healing sensors with tunable positive/negative piezoresistivity. *Advanced Functional Materials* 28 (15): 1706658.

233 Tetik, H., Wang, Y., Sun, X. et al. (2021). Additive manufacturing of 3D aerogels and porous scaffolds: a review. *Advanced Functional Materials* 2103410.

234 Li, Z., Wang, J., Xu, Y. et al. (2021). Green and sustainable cellulose-derived humidity sensors: a review. *Carbohydrate Polymers* 270: 118385.

235 Li, B., Hou, B., and Amaratunga, G.A.J. (2021). Indoor photovoltaics, The Next Big Trend in solution-processed solar cells. *InfoMat* 3 (5): 445–459.

236 Ziar, H., Manganiello, P., Isabella, O. et al. (2021). Photovoltatronics: intelligent PV-based devices for energy and information applications. *Energy & Environmental Science* 14 (1): 106–126.

237 Promphet, N., Ummartyotin, S., Ngeontae, W. et al. (2021). Non-invasive wearable chemical sensors in real-life applications. *Analytica Chimica Acta* 338643.

238 Riaz, A., Sarker, M.R., Saad, M.H.M. et al. (2021). Review on comparison of different energy storage technologies used in micro-energy harvesting, WSNs, low-cost microelectronic devices: challenges and recommendations. *Sensors* 21 (15): 5041.

239 Mao, Z., Li, F., Su, Q. et al. (2021). Application of energy-saving control strategy in air conditioning terminal equipment based on the constant temperature difference of chilled water. *Case Studies in Thermal Engineering* 28: 101409.

240 Kwon, G., Giyun, K., Lee, M.H. et al. (2019). Bio-inspired molecular redesign of a multi-redox catholyte for high-energy non-aqueous organic redox flow batteries. *Chemistry* 5 (10): 2642–2656.

241 Sadeeq, M.A.M. and Zeebaree, S. (2021). Energy management for internet of things via distributed systems. *Journal of Applied Science and Technology Trends* 2 (02): 59–71.

242 Pan, H., Qi, L., Zhang, Z. et al. (2021). Kinetic energy harvesting technologies for applications in land transportation: a comprehensive review. *Applied Energy* 286: 116518.

243 Sreekantha, D.K., Koujalagi, A., Girish, T.M. et al. (2021). Internet of things (IoT) enabling technologies and applications—a study. In: *Advances in Artificial Intelligence and Data Engineering*, 1425–1442. Singapore: Springer.

244 Shukla, P., Saxena, P., Madhwal, D. et al. (2021). Battery-operated resistive sensor based on electrochemically exfoliated pencil graphite core for room temperature detection of LPG. *Sensors and Actuators B: Chemical* 343: 130133.

245 Jiang, M., Lu, Y., Zhu, Z. et al. (2021). Advances in smart sensing and medical electronics by self-powered sensors based on triboelectric nanogenerators. *Micromachines* 12 (6): 698.

246 Ren, J., Liu, Q., Pei, Y. et al. (2021). Bioinspired energy storage, and harvesting devices. *Advanced Materials Technologies* 2001301.

247 Rennane, A., Benmahmoud, F., Cherif, A.T. et al. (2021). Design of autonomous multi-sensing passive UHF RFID tag for greenhouse monitoring. *Sensors and Actuators A: Physical* 331: 112922.

248 Rahman, M.T., Rana, S.M.S., Maharjan, P. et al. (2021). Ultra-robust and broadband rotary hybridized nanogenerator for self-sustained smart-farming applications. *Nano Energy* 85: 105974.

249 Jayachandran, M., Reddy, C.R., Padmanaban, S. et al. (2021). Operational planning steps in smart electric power delivery system. *Scientific Reports* 11 (1): 1–21.

11

Sustainable Energy Storage Devices and Device Design for in the Scope of Internet of Things

Vitor Correia[1], *Carlos Miguel Costa*[2,3,4], *and Senentxu Lanceros-Méndez*[5,6]

[1] *Algoritmi Research Center, University of Minho, Campus de Azurém, Guimarães, 4800-058, Portugal*
[2] *Physics Centre of Minho and Porto Universities (CF-UM-UP), University of Minho, 4710-057, Braga, Portugal*
[3] *Laboratory of Physics for Materials and Emergent Technologies, LapMET, University of Minho, 4710-057, Braga, Portugal*
[4] *Institute of Science and Innovation for Bio-Sustainability (IB-S), University of Minho, 4710-057, Braga, Portugal*
[5] *BCMaterials, Basque Center for Materials, Applications and Nanostructures, UPV/EHU Science Park, 48940, Leioa, Spain*
[6] *IKERBASQUE, Basque Foundation for Science, 48009, Bilbao, Spain*

11.1 Introduction

Internet of things (IoT) devices are increasingly an integral and necessary part of our society, taking into account the technological advances and the benefits that these devices offer in the way we interact among us and with the world around us ubiquitous in our daily activities, offering a clear improvement in our quality of life.

The IoT concept can be described as the intelligent connections between different people and devices, tasks are based on the information generated by sensors, associated with a large amount of information at the cloud computing level, all fed by thousands of millions of devices [1, 2] (Figure 11.1).

In general, it is possible to define four representative pillars of IoT systems:

- **Things and sensorization**: Objects/devices are equipped with sensors or sensor systems, which allow the collection of information from their surroundings, considering that information according to the functions assigned to them, and based on the stimuli they receive.
- **Data**: The data collected is stored internally and transmitted among its peers, as well as with other components present in the cloud. The origin of the information may or may not be identified as a way of guaranteeing privacy issues.
- **Connectivity**: Numerous information platforms and means of communication can be associated as a way of guaranteeing the interconnection of devices and their connection to the cloud.
- **Analytics**: The data collected is analyzed at various stages of the network: at the local level (device), in the mesh, and/or in the cloud, as a way of optimizing the response time

Sustainable Energy Storage in the Scope of Circular Economy: Advanced Materials and Device Design, First Edition. Edited by Carlos Miguel Costa, Renato Gonçalves, and Senentxu Lanceros-Méndez.

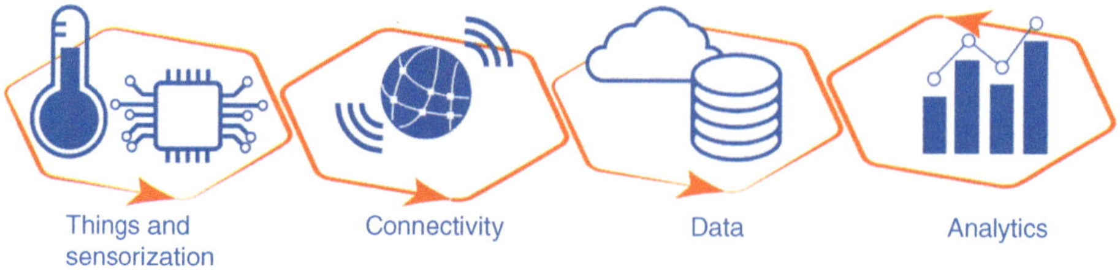

Figure 11.1 Schematic representation of the building blocks of the IoT technology.

of the devices and energy performance, while guaranteeing security and privacy. As processing moves away from the data source, the weight and volume of the data increase, opening the possibility for the implementation of cloud computing, artificial intelligence, big data, machine learning, and other future technological solutions, that may arise.

Over the last 20 years (the earliest literature on the IoT was published in 2002; [3]), there have arisen an increasing number of possibilities and fields of application of this new paradigm, including smart mobility [4], health [5], agriculture [6], industry 4.0 [7], and the consumer market where it is already found in many objects of our daily lives, such as household appliances, clothes, or toys, among others. In fact, it has been a strong and rapid growth in the number of these interconnected devices and according to IHS Markit, 125 billion IoT devices are expected to be connected to each other by 2030 [8], this number is highly influenced by the emergence of new communication networks, as it is the case of recent 5G networks and next generations with the exploration of new frequencies with larger bandwidth [9]. The establishment of specific protocols and standards for IoT devices will translate into an even larger acceleration in the emergence of new devices by reducing development and standardization time, in order to increase the efficiency and capabilities of the systems.

Physically, these devices generally include four main blocks: Sensors/signal conditioning, power/energy management, data processing, and communication, as represented in Figure 11.2.

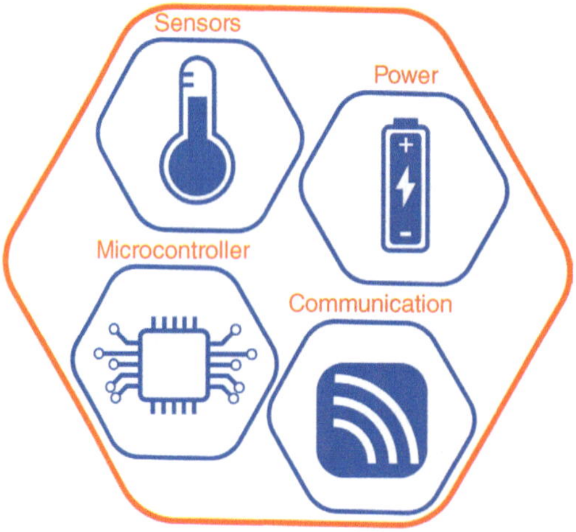

Figure 11.2 Schematic representation of the type blocks of IoT systems.

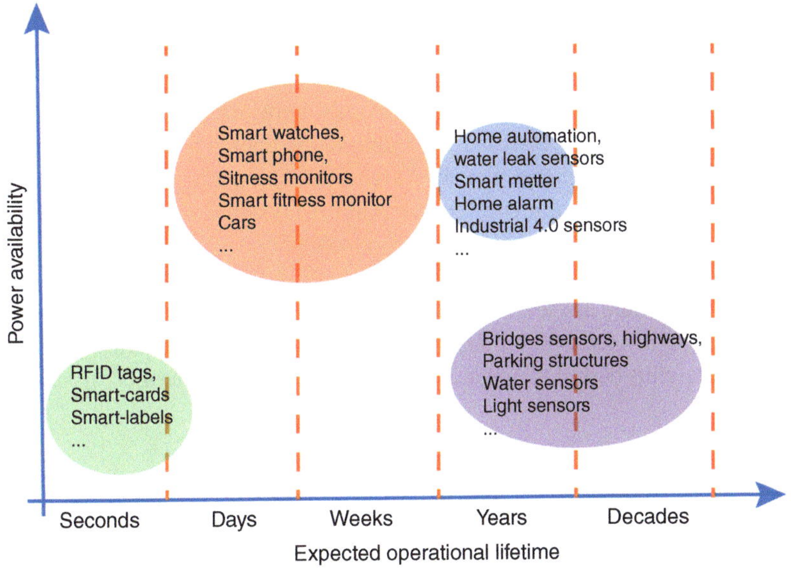

Figure 11.3 Scheduling the application of IoT systems and their corresponding lifetime. *Source:* Adapted from Jayakumar et al. [10].

Despite the strong evolution in terms of consumption and energy generation, the problem of powering these billions of devices represents one of the great challenges to implement the IoT concept, given the strict restrictions in terms of environments (temperature, humidity, corrosive, explosive, or volume, among others) in which they are placed, access limitation, geometry and energy/power needs, which sometimes represent difficult challenges to overcome. On the other hand, most of these devices may have a useful life from a few days to several years, which again corresponds to a strong constraint for the massive use of these technologies, Figure 11.3.

Due to cost, inconvenience, or infeasibility of being plugged, most of these IoT devices are expected to be unplugged and powered by batteries and/or energy harvesting. In addition, strict restrictions on the format and size of the devices also imply restrictions at the level of energy storage capacity on site, increasing the problem, as most of these devices require a long operational lifetime, which can vary from a few days to several days, weeks or even years.

The growing development and implementation of electronics in general, accompanied by the development and implementation of energy storage systems, does not only present serious challenges in terms of manufacturing methods [11] but also increasing concerns in terms of sustainability, ranging from materials and processes to waste disposal and end-of-life recyclability of the product [12], given the increasing number of devices and their lifespan [10].

Considering the diversity of applications in which IoT systems take part, as well as the possible applications, the methods of powering these systems are also very varied, depending on their application. The main power sources used in IoT devices include batteries, capacitors, fuel cells, solar cells, electric grids, or energy harvesting systems.

Regardless of the energy source used, or even in the case of systems that generate their own energy, energy storage systems are a key element to ensure the correct functioning of the devices, and in many cases, they operate as a primary energy source.

Regarding energy storage, the main systems currently being implemented include batteries, supercapacitors (SCs), and capacitors, according to the energy required, charging/discharging time, life cycle, and energy density (Figure 11.4). In this area, the factor of resource sustainability and environmental impact has also become a focus of concern in the process of developing and implementing new solutions, considering the large numbers involved.

With respect to batteries, there are several technologies currently used. The battery market for applications in IoT has focused on chemical batteries, thin film batteries, printed batteries, and solid-state chip batteries. Regarding chemical batteries, the most used are

Figure 11.4 Energy performance factors according to the type of energy storage system [13]. *Sources:* Nippon Chemi-Con/RS Components Ltd.; Focus Technology Co., Ltd.; TemplateMonster/Lowes.com; KEMET Corporation; Bematik.com/cablematic.com.

lithium batteries and alkaline batteries, among others [14], applied in the most different areas of activity and environment, including consumer market: smart locks, security cameras, smart meters, smart watches, smart glasses, body cameras, and hearing aids; health market: heart rate monitors, blood pressure monitors, blood glucose meters, and fall detectors, among others; retail: contactless points of sale, and smart tags; smart agriculture: agricultural drones, agricultural robots, soil moisture sensors and radio frequency identification (RFID) tags for livestock, among others [15].

Due to their high energy density, lower memory effect, and low discharge rate, lithium-ion batteries (LIBs), have been widely used in portable electronic devices and, increasingly in electric vehicle mobility [16]. LIBs are composed of a cathode, an anode, and a separator, most often immersed in a liquid electrolyte and sealed in a stainless steel, aluminum, or plastic bag. The anode is typically composed of graphite as active material, coated by a Cu sheet current collector. On the other hand, the cathode can comprise different types of active materials, including lithium metal oxides or phosphides, such as $LiCoO_2$ [17], $LiMn_2O_4$ [18], $LiNi_xCo_yMn_zO_2$ [19], $LiNi_xCo_yAl_zO_2$ [20], and $LiFePO_4$ (LFP) [21], among others.

With respect to capacitors, the demand is also growing following the technological evolution, presenting fast charge/discharge and excellent cycle stability. They are highly beneficial in the storage of energy obtained from energy harvesting systems, allied to their great safety of use [22]. Significant advances have been made in the field of ceramic and tantalum SCs and capacitors. Tantalum is one of the rarest and most valuable metals that can be found in nowadays electronic devices and is characterized by superior electronic and refractory characteristics, including high capacitance/unit volume, high dielectric constant, thermal stability over a wide temperature range, high resistance to oxidation, and excellent frequency are the main features, allowing, therefore, high-tech and high-integration applications. However, although tantalum corresponds to one of the large parts of the composition of capacitors, other materials are also part of their composition, such as Ta 36%, SiO_2 45%, Ni 6%, Fe 1.5%, Ag 0.5%, and organics 11%.

Due to the great demand for energy storage systems, the pressure on the extraction of rare earths is also increasing. Further, it is important to mention that a large part of their extraction occurs in underdeveloped countries, often with soft health, security, and environmental criteria.

Taking into account the expected evolution of materials used for the coming decades, it becomes imperative to search for alternative paths that guarantee the sustainability of the used resources, which represents a relevant challenge.

The way to ensure the sustainability of energy storage systems together with electronic devices involves three areas of activity along the life cycle of storage systems, as shown in Figure 11.5:

- New materials and manufacturing methods for energy storage systems allow for increased performance, or, on the other hand, lead to reducing the need to exploit resources. Another path is to use more sustainable materials, including biomaterials, with reduced environmental impact in the production and disposal phases.
- Optimization of design and electronic architectures, to reduce energy needs, represents a parallel path to the previous one, thus enabling some solutions for new materials, or optimizing the use of traditional raw materials.

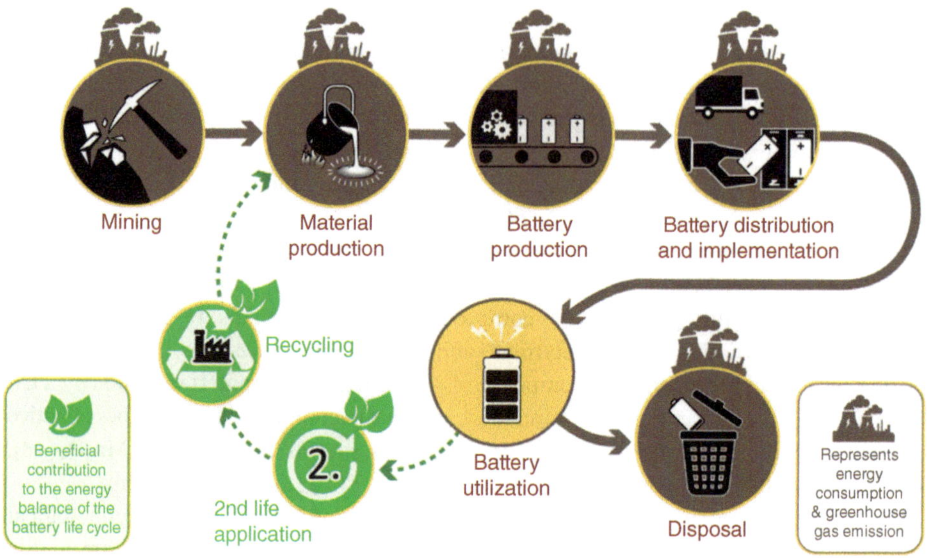

Figure 11.5 Production life cycles of energy storage systems.

- Recyclability of the systems will be one of the most critical for the coming decades, given the expected high volume of waste expected, based on current technologies of energy storage system and their use.

11.2 New Materials and Manufacturing Methods for Batteries

The battery technologies that are nowadays partially used in electronic devices, lithium-ion (Li-ion) batteries, bring a series of performance, environmental, and safety concerns. LIBs, despite being the most used technology (representing about 37% of the total) (Figure 11.6) [24], are prone to thermal runaway that can result in fires and explosions, they have a limited number of charge/discharge cycles (Table 11.1), and they use scarce materials, where their mining has negative impacts on the environment. Further, they are also difficult to separate and recycle, and a high impact on the manufacturing process also exists, as shown by the study by Dühnen et al. [25], which presents various life cycle assessment datasets on LIB cells and shows that producing 1 Wh of storage capacity requires a cumulative energy demand of 328 Wh and causes greenhouse gas emissions of 110 g CO_2 equivalent on average.

Thus, it will be necessary to find advanced solutions for battery technology, the search for alternative materials for the various constituents of batteries being the bases of most of the current approaches. Studies have been presented, which make use of anodic materials from biomass from the most diverse sources [26], however, so far, the results are still far from being industrially applicable. With respect to cathodic materials, there are also numerous attempts to escape critical raw materials, including Co and Nigiven their toxic character, combined with their high extraction costs. Phosphatizing-type cathodic

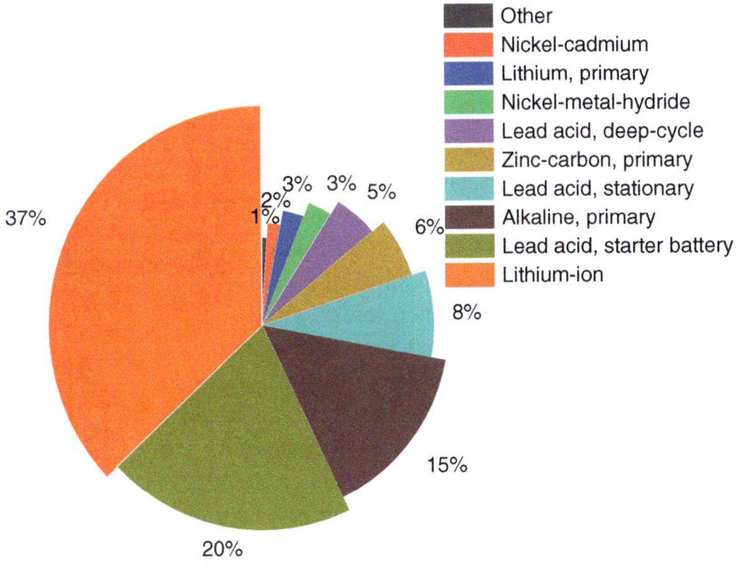

Figure 11.6 Markets for rechargeable batteries according to technology. *Source:* Adapted from Zhao et al. [23].

materials such as LFP or $LiFe_{1-x}Mn_xPO_4$ with $0 \leq x \leq 1$ are considered ecologically more suitable [26]. The high structural stability of the polyanionic phosphate network and the relatively low operating voltage window, which prevents unwanted parasitic reactions from the battery electrolyte, allow for a long cycle and lifetime for LFP-based LIBs. They do not meet the requirements of the automotive industry in terms of energy content at the material level (min. 680 Wh kg^{-1} and 3600 Wh l^{-1}) but they could be a solution for IoT devices.

Another possibility includes borates, sulfates, and silicates and involves a variety of central redox-active atoms (e.g. V, Mn, Fe, Co, Ni, and Mo) [27]. In comparison to layered transition metal oxides ($LiMO_2$), the formula mass of polyanionic host structures is relatively high, which results in low theoretical capacities for compounds with only one Li^+ ion. This disadvantage can be compensated with compounds with multiple Li^+ ions per formula unit. Li metal silicates (Li_2MSiO_4, with M = Mn, Fe) [28] show, for example, favorable characteristics: they contain only low-cost and abundant elements and contain two Li ions per formula unit, which in turn results in high theoretical capacities of ≈330 mAh g^{-1} [27], of which more than half could already be realized experimentally [28]. However, Li metal silicates exhibit very low electronic conductivities, which are commonly through the use of nanometric active particles coated with relatively high amounts of carbon, which translates into reduced energy density achievable at the electrode level.

The electrolyte plays a key role in battery performance. It must present high ionic and low electronic conductivity, a wide electrochemical window associated with high chemical and thermal stability. Room temperature ionic liquids represent an alternative to electrolytes based on organic solvents, are classified as "green solvents" [29], and show high thermal stability, low flammability, low volatility, and a wide electrochemical stability window (ESW) [30], allowing improved recyclability compared to carbonates (solution present in current Li batteries). Despite the good indicators, their biodegradability is still being studied, being very dependent on the chemical nature of the cations and anions [28].

In turn, as an alternative to traditional separators, built on polymeric microporous membranes, the search for more sustainable materials has been the focus of investigation.

New biocompatible and/or biodegradable materials obtained from renewable sources have been presented as viable solutions to replace synthetic polymeric membranes. Nano cellulose fibers (Figure 11.7) have shown to be a promising candidate for the next generation of "green" and renewable batteries, due to their low cost and ecological nature [32]. Due to their hierarchical porous structure and rough yet absorbent surface properties, various active electrochemical materials (e.g. battery electrode and SC electrode materials) and electrolytes can be easily incorporated into the paper structure, demonstrating excellent capacity per area (8.6 $mAh\,cm^{-2}$), with energy capacities compatible with conventional cells ~60 vs ~30 $mAh\,g_{cell}^{-1}$) [33].

Despite being based on non-environmental friendly manufacturing, silicon *wafer batteries* present numerous advantages for IoT devices, since they solve several of the current

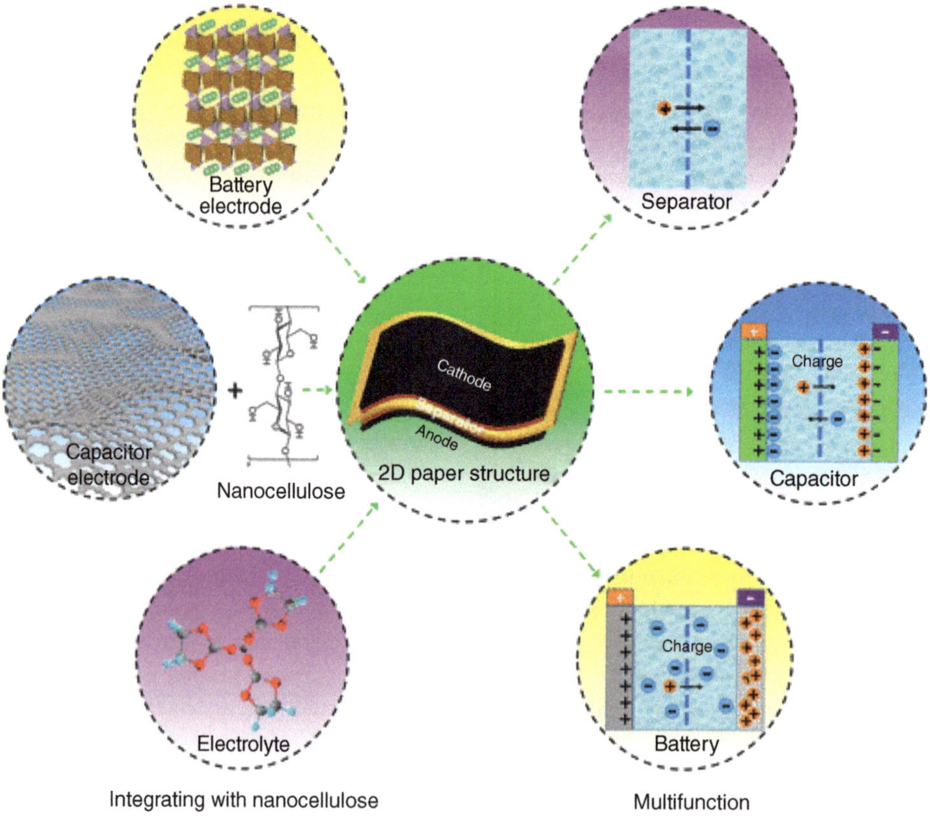

Figure 11.7 Examples of application of cellulose in the field of energy storage. *Source:* Chen and Hu [31]/With permission of American Chemical Society.

challenges of lithium batteries, being able to reach capacities up to 10× (680 Wh kg^{-1}) the current capacities [34]. Their main advantage relies on the fact that they use the same manufacturing processes used in the production of integrated circuits (IC), ICs that represent the majority of IoT devices. Each 12-in. wafer allows the implementation of about 36 million on-chip micro batteries, the small active batteries carrying a small enough charge and having enough space between cells to prevent dendrites from growing, which is among the main drawbacks of traditional batteries, leading to lose capacity over time.

As they share the same manufacturing process with the IC, the combination of these two technologies makes it possible to integrate the various blocks that make up an IoT system on a single chip, reaching high levels of integration and performance.

11.3 New Materials and Manufacturing Methods for Supercapacitors

Similarly, to batteries, capacitors in general have also been the target of technological evolution in terms of manufacturing processes, with a special focus on miniaturization by increased energy density per unit volume, based on advanced materials and designs.

In particular, advanced materials have allowed significant advances in next-generation SCs [13].

SCs can store charge electrostatically based on three principles and operations: charge separation at the electrode/electrolyte interface [35], fast redox reactions near the electrode surface (pseudocapacitance) [36], and hybrid mechanisms [36], where rapid storage of charges in SCs being due to surface or near-surface charge storage mechanisms at the electrode surface. Hybrid capacitance broadly means the combination of pseudo-capacitive materials and a double layer to nurture the advantages of both storage mechanisms to improve overall capacitance [37]. One of the main features of the SC mechanism is the linear dependence of capacitance on potential.

The latest progress in new materials and device sets focus on five major architectures for SCs: three-electrode "half-battery" architecture, two-electrode SC, flexible solid-state SC, fiber SC, and planar SC (micro-SCs).

With respect to materials, the materials that are generally employed as electric double-layer capacitor (EDLC) electrode materials and pseudocapacitive electrodes include transition metal oxide, sulfides, and conductive polymers, but may also include various carbon-doped materials [38].

The performance of SCs is greatly influenced by the electrolyte and electrode material, as well as by the operating window, related to the electrode and electrolyte compatibility and their conjugate operating range. Thus, the advancement of new SC technologies requires rational electrode design along with the proper use of electrolytes.

The materials used in the manufacture of electrodes for SC are characterized by having an effective potential, demanding high active electrochemical sites, porosity, high theoretical capacitance, and good electrochemical conductivity [39]. Much research has focused on adapting the morphology and structure of various electrode materials to the targeted storage mechanism [40].

Adaptation of architectures will also have a significant impact on the performance of the materials used as electrodes as well as on the performance of the device as a whole [41].

It is noteworthy to mention that there has been increasing work on developing new energy storage devices based on next-generation smart materials/technologies based on earth-abundant and economical materials. However, it is difficult to determine whether these new architectures can outperform their commercial counterparts, given the ambiguity in the comparison, based on the ideal conditions at which these tests are performed.

Thus, it is advisable to adopt the known parameters of commercial SCs as a standard for comparison, with one of the highest rated voltage of 2.85 V made available by Maxwell technologies [42]. The energy density varies widely from a maximum of 7.5 Wh kg^{-1} for the LS Mtron cell [43] with the Yunasko symmetrical hybrid cell offering high powers of 41 kW kg^{-1} [44].

There are several studies reporting solutions that point to the development of optimized electrodes in terms of real applicability, however, standard performance evaluation has not been reported for the developed materials used as electrode.

Thus, nickel–cobalt sulfide-graphene hybrid (NCS-GH) electrodes rich in defects in honeycomb-like structures have been reported. The incorporation of graphene not only served as a template for the synthesis but also helped to improve the overall conductivity and electrochemical active sites [45].

The NCS-GH electrodes are characterized by a specific capacitance of 1186.7 F g^{-1} at 1 A g^{-1} which when introduced in an NCS/GH/activated carbon device provides 46.4 Wh kg^{-1}, corresponding to a power density of 400 Wh kg^{-1} [45]. Ligand-free 1D single crystalline Cu_2S nanopillars on Cu substrate led to a specific capacity of 750 mF cm^{-2} at a current density of 2 mA cm^{-2}. Stable electrodes maintained a capacitance of 82.3% at a high current density of 40 mA cm^{-2} after 20 000 cycles. An asymmetric device based on activated carbon, AC/SC-Cu_2S, provides an energy density of 2.69 mW h cm^{-3} and a power density ranging from 10.9 to 218.4 mW cm^{-3} for a maximum potential difference of 1.6 V [46]. Ligand-free electrodes are recognized to provide high electrochemical activity due to their improved surface area, electrode wettability, and electrical conductivity, and therefore are being extensively investigated [47].

Microsupercapacitors allow ultra-high discharge rates, which is a relevant issue for numerous high-performance applications. Further, the miniaturization of SCs allows to open new application areas [48]. Further, studies have been presented on a solid-state SC using sulfur-doped graphene and ionic gel electrolyte, achieving a volume capacitance of ~582 F cm^{-3} at 10 mV s^{-1} and 8.1 F cm^{-3}, even at high discharge rates of 2000 V s^{-1}, and with a power density approaching 1191 W cm^{-3} [49]. Orangi et al. presented printed 2D Ti_3C_2 T_x Mxenes electrodes with interdigital architectures and obtained a capacitance per unit area of 1035 mF cm^{-2} at 2 mV s^{-1} while achieving an energy density of 51.7 uW h cm^{-2} [50].

In a greener approach, a new concept of manufacturing SCs based on cellulose fibers was presented [51]. Industrial flax waste was converted into fiber form SCs through wet spinning and in situ deposition of polypyrrole (PPy). The regenerated cellulose fibers collected from the ethyl alcohol coagulation bath showed excellent mechanical characteristics. After PPy deposition, the fiber electrodes showed high specific capacitance (510.7 F g^{-1} at 0.5 A g^{-1}) and excellent coulombic efficiency (99.3% at 10 A g^{-1}). Paper-based capacitors have been also presented, the best results being obtained for PPy-coated composite paper capacitors, with normalized gravimetric ($127\,F\,g^{-1}$) and volumetric ($122\,F\,cm^{-3}$) capacitances at $300\,mA\,cm^{-2}$ [31].

Despite the good results achieved in these new generations of environmental friendlier SCs, studies of temporal stability and degradation are still insufficient for these solutions to be used in commercial applications.

On the other hand, it is also important to point out that cellulose is considered a sustainable solution, but the various materials used in the construction process of the capacitors do not have this character.

11.4 New Designs to Optimize the Management and Energy Needs of the Devices

Another aspect, which makes it possible to improve the sustainability of energy storage systems, involves reducing the consumption of electronic devices, leading to an increase in the useful life of the systems and, therefore, a reduction of the environmental impact of waste. In this scope, numerous advances have been achieved, based on the improved integration capacity of microfabrication systems of digital circuits, with the technology currently being in 3 nm, and 2 nm technology is already developed. In comparative terms, and according to Samsung, "Optimized 3 nm process achieves 45% reduced power usage, 23% improved performance and 16% smaller surface area compared to 5 nm process" [52]. Based on this consumption reduction factor combined with the increase in efficiency, this is one of the factors that most influence energy consumption.

It is therefore expected that, with the emergence of even smaller (more efficient) manufacturing technologies, a greater reduction in energy consumption will occur.

This evolution is also felt in the remaining components used for the manufacture of electronic circuits as a result of the dimensional reduction and optimization of the circuits with a view to reducing energy losses due to heating. Data communication circuits are one of the pillars of IoT systems, being also a significant part of the energy consumption of the system. These circuits have also been the subject of study and optimization with a view to energy reduction, having in recent decades witnessed significant reductions.

Another of the major factors that significantly influence the energy consumption of devices are the algorithms and their optimization. A large percentage of IoT systems operate in a transient state, so consumption is directly related to the ratio between the time in sleep mode and the active state (sensor readings, data processing, and communication). Thus, the optimization of the processing algorithms and the ability to manage the partial activations of the electronic circuit, by regulating the loading and unloading rates of the storage devices, allows significant energy savings during the lifetime of the devices.

The combination of these three factors and their optimization when designing and developing IoT systems, are essential for the sustainability of the resources used in the devices, since a correct design of the devices combined with optimization of their use can easily extend the longevity of energy storage systems to more than double, which implies a reduction of more than 100% of the raw materials needed at the time of their production, with the same percentage reduction at the time of their disposal/recycling.

Temperatures and humility are other extremely relevant factors in energy storage elements, conditioning both their durability and the safety of the devices. These factors must also be considered at the time of design, with the correct selection of the technology to be used, as well as at the time of installation/use in an attempt to mitigate these conditions.

11.5 Recycling Solutions for Energy Storage Systems

In addition to reducing the consumption of natural resources, the ability to reuse and recycle the resources used in IoT devices, in particular in their energy storage systems, strongly influences the sustainable character of the device.

The recycling of used batteries, SCs, and capacitors has also been a main focus of research, as a way of alleviating the urgent demand for raw materials while avoiding the inadequate disposal of electrolytes, toxic solvents, or heavy metals, among others, causing environmental pollution and raising environmental and safety concerns [53]. Further, taking into account the current and rising prices of rare earths, in addition to being increasingly necessary, recycling processes are economically advantageous with respect to mining, as it is the case, for example, with the cost of recycling lithium, compared to its extraction [24, 53].

On the other hand, in the field of commercial capacitors, there are also numerous efforts and technologies dedicated to the recyclability of these elements, with utilization rates of around 93.5% for Ta, one of the most expensive elements in the manufacture of capacitors [54]. Electrolytic capacitors are also subject to a recycling process, being one of the types of capacitors used in current electronic circuits, with recovery rates of elements in the order of 87.9% to 98%, depending on the type of capacitor [55].

As for the new generations of batteries, based on so-called sustainable materials, recyclability is also a fact to be explored given the large numbers involved, the impact related to the disposal of these materials being also potentially significant. Nevertheless, there are currently no studies on the recyclability of these materials, with only material degradation studies being reported.

Similar to the new generations of environmental friendly [25] and biodegradable [56] batteries, capacitor and SC technology are also following the same path, suffering from similar concerns and challenges.

11.6 Conclusions

There are a variety of energy storage system solutions capable of being used in IoT devices, satisfying most of the requirements and application conditions, in terms of available energy and power. However, with respect to sustainability, there are already proposals for potential solutions, but most of them are still in the laboratory development stage and are expected to reach the market during the next decade.

Thus, in the immediate future, much of the work on sustainability will rely on the reduction and optimization of energy consumption of electronic systems, through the miniaturization of the systems and their complete integration.

Recyclability will also play an extremely important role over the next decade as a way of reducing the environmental impact of the current generations of energy storage systems (batteries and capacitors/SCs), composed of rare and expensive materials with major environmental impacts at the time of extraction and disposal. It is thus expected that future generations of energy storage systems will be designed considering, at the level of materials, processes, and designs, that recycling will maintain this central role in what corresponds to the sustainability of technological development.

Acknowledgments

The authors thank the Fundação para a Ciência e Tecnologia (FCT) for financial support under the framework of Strategic Funding grants UIDB/04650/2020, UID/FIS/04650/2020, and UID/EEA/04436/2020, and under projects MIT-EXPL/TDI/0033/2021, POCI-01-0247-FEDER-046985, 2022.03931.PTDC and PTDC/FIS-MAC/ 28157/2017. Carlos M. Costa thank the FCT for financial support under contract 2020.04028.CEECIND. Vitor Correia thanks FCT for the junior researcher contract (DL57/2016). Financial support from the BIDEKO Project, funded by MCIN/AEI, NextGenerationEU, PRTR, and from the Basque Government Industry Department under the ELKARTEK program is acknowledged.

References

1 Tu, M., Lim, M.K., and Yang, M.F. (2018). IoT-based production logistics and supply chain system – Part 2. *Industrial Management & Data Systems* 118: 96–125.
2 Dang, L.M., Piran, M.J., Han, D. et al. (2019). A survey on internet of things and cloud computing for healthcare. *Electronics* 8: 768.
3 Wang, J., Lim, M.K., Wang, C., and Tseng, M.-L. (2021). The evolution of the Internet of Things (IoT) over the past 20 years. *Computers & Industrial Engineering* 155: 107174.
4 El-Din, D.M., Hassanien, A.E., and Hassanien, E.E. (2020). Information integrity for multi-sensors data fusion in smart mobility. In: *Toward Social Internet of Things (SIoT): Enabling Technologies, Architectures and Applications* (ed. A.E. Hassanien, R. Bhatnagar, N.E.M. Khalifa and M.H.N. Taha), 99–121. Springer.
5 Leong, Y.-Y. and Chen, Y.-C. (2020). Cyber risk cost and management in IoT devices-linked health insurance. *The Geneva Papers on Risk and Insurance - Issues and Practice* 45: 737–759.
6 Baranwal, T.N. and Pateriya, P.K. (2016). Development of IoT based smart security and monitoring devices for agriculture. In: *2016 6th International Conference - Cloud System and Big Data Engineering (Confluence)*, 597–602. IEEE.
7 Khan, I.H. and Javaid, M. (2021). Role of internet of things (IoT) in adoption of industry 4.0. *Journal of Industrial Integration and Management* 0: 2150006.
8 Markit, I. (2017). https://www.eenewseurope.com/en/connected-iot-devices-to-reach-125-billion-by-2030-says-ihs-markit/ (accessed 26 November 2021).
9 Oughton, E.J., Lehr, W., Katsaros, K. et al. (2021). Revisiting wireless internet connectivity: 5G vs Wi-Fi 6. *Telecommunications Policy* 45: 102127.

10 Jayakumar, H., Raha, A., Kim, Y., et al. (2016). Energy-efficient system design for IoT devices. 2016 21st Asia and South Pacific Design Automation Conference (ASP-DAC), pp. 298–301.

11 Wang, C., Xia, K., Wang, H. et al. (2019). Advanced carbon for flexible and wearable electronics. *Advanced Materials* 31: 1801072.

12 Jung, Y.H., Chang, T.-H., Zhang, H. et al. (2015). High-performance green flexible electronics based on biodegradable cellulose nanofibril paper. *Nature Communications* 6: 7170.

13 Ji, C., Mi, H., and Yang, S. (2019). Latest advances in supercapacitors: from new electrode materials to novel device designs. *Chinese Science Bulletin* 64: 9.

14 Wójcicki, K., Biegańska, M., Paliwoda, B., and Górna, J. (2022). Internet of things in industry: research profiling, application, challenges and opportunities — a review. *Energies* 15: 1806.

15 Boursianis, A.D., Papadopoulou, M.S., Diamantoulakis, P. et al. (2022). Internet of things (IoT) and agricultural unmanned aerial vehicles (UAVs) in smart farming: a comprehensive review. *Internet of Things* 18: 100187.

16 Yun, L., Linh, D., Shui, L. et al. (2018). Metallurgical and mechanical methods for recycling of lithium-ion battery pack for electric vehicles. *Resources, Conservation and Recycling* 136: 198–208.

17 Liang, Y., Zhao, C.-Z., Yuan, H. et al. (2019). A review of rechargeable batteries for portable electronic devices. *InfoMat* 1: 6–32.

18 Luo, F., Wei, C., Zhang, C. et al. (2020). Operando X-ray diffraction analysis of the degradation mechanisms of a spinel $LiMn_2O_4$ cathode in different voltage windows. *Journal of Energy Chemistry* 44: 138–146.

19 Yang, L., Xi, G., and Xi, Y. (2015). Recovery of Co, Mn, Ni, and Li from spent lithium ion batteries for the preparation of $LiNixCoyMnzO_2$ cathode materials. *Ceramics International* 41: 11498–11503.

20 Xia, S., Huang, W., Shen, X. et al. (2020). Rearrangement on surface structures by boride to enhanced cycle stability for $LiNi_{0.80}Co_{0.15}Al_{0.05}O_2$ cathode in lithium ion batteries. *Journal of Energy Chemistry* 45: 110–118.

21 Kretschmer, K., Sun, B., Su, D. et al. (2015). Scalable preparation of $LiFePO_4$/C Nanocomposites with sp^2-coordinated carbon coating as high-performance cathode materials for lithium-ion batteries. *ChemElectroChem* 2: 2096–2103.

22 Yang, Q., Liu, Y., Xiao, L. et al. (2018). Self-templated transformation of MOFs into layered double hydroxide nanoarrays with selectively formed Co_9S_8 for high-performance asymmetric supercapacitors. *Chemical Engineering Journal* 354: 716–726.

23 Zhao, Y., Pohl, O., Bhatt, A.I. et al. (2021). A review on battery market trends, second-life reuse, and recycling. *Sustainable Chemistry* 2: 167–205.

24 Costa, C.M., Barbosa, J.C., Gonçalves, R. et al. (2021). Recycling and environmental issues of lithium-ion batteries: advances, challenges and opportunities. *Energy Storage Materials* 37: 433–465.

25 Dühnen, S., Betz, J., Kolek, M. et al. (2020). Toward green battery cells: perspective on materials and technologies. *Small Methods* 4: 2000039.

26 Liu, C., Li, F., Ma, L.-P., and Cheng, H.-M. (2010). Advanced materials for energy storage. *Advanced Materials* 22: E28–E62.

27 Andre, D., Kim, S.-J., Lamp, P. et al. (2015). Future generations of cathode materials: an automotive industry perspective. *Journal of Materials Chemistry A* 3: 6709–6732.

28 Islam, M.S., Dominko, R., Masquelier, C. et al. (2011). Silicate cathodes for lithium batteries: alternatives to phosphates? *Journal of Materials Chemistry* 21: 9811–9818.

29 Earle, M.J. and Seddon, K.R. (2000). Ionic liquids. Green solvents for the future. *Pure and Applied Chemistry* 72: 1391–1398.

30 Lewandowski, A. and Świderska-Mocek, A. (2009). Ionic liquids as electrolytes for Li-ion batteries—an overview of electrochemical studies. *Journal of Power Sources* 194: 601–609.

31 Chen, C. and Hu, L. (2018). Nanocellulose toward advanced energy storage devices: structure and electrochemistry. *Accounts of Chemical Research* 51: 3154–3165.

32 Jo, J.H., Jo, C.-H., Qiu, Z. et al. (2020 153). Nature-derived cellulose-based composite separator for sodium-ion batteries. *Frontiers in Chemistry* 8: 1–12.

33 Choi, K.-H., Cho, S.-J., Chun, S.-J. et al. (2014). Heterolayered, one-dimensional nanobuilding block mat batteries. *Nano Letters* 14: 5677–5686.

34 Eshraghi, N., Berardo, L., Schrijnemakers, A. et al. (2020). Recovery of nano-structured silicon from end-of-life photovoltaic wafers with value-added applications in lithium-ion battery. *ACS Sustainable Chemistry & Engineering* 8: 5868–5879.

35 Antonietti, M., Chen, X., Yan, R., and Oschatz, M. (2018). Storing electricity as chemical energy: beyond traditional electrochemistry and double-layer compression. *Energy & Environmental Science* 11: 3069–3074.

36 Zhou, L., Li, C., Liu, X. et al. (2018). Chapter 7 – Metal oxides in supercapacitors. In: *Metal Oxides in Energy Technologies* (ed. Y. Wu), 169–203. Elsevier.

37 Zhu, G., He, Z., Chen, J. et al. (2014). Highly conductive three-dimensional MnO_2–carbon nanotube–graphene–Ni hybrid foam as a binder-free supercapacitor electrode. *Nanoscale* 6: 1079–1085.

38 Devi, B., Jain, A., Roy, B. et al. (2020). Cobalt-embedded N-doped carbon nanostructures for oxygen reduction and supercapacitor applications. *ACS Applied Nano Materials* 3: 6354–6366.

39 Lokhande, A.C., Teotia, S., Shelke, A.R. et al. (2020). Chalcopyrite based carbon composite electrodes for high performance symmetric supercapacitor. *Chemical Engineering Journal* 399: 125711.

40 Pham, H.D., Mahale, K., Hoang, T.M.L. et al. (2020). Dual carbon potassium-ion capacitors: biomass-derived graphene-like carbon nanosheet cathodes. *ACS Applied Materials & Interfaces* 12: 48518–48525.

41 Patil, A.M., Lokhande, A.C., Shinde, P.A., and Lokhande, C.D. (2018). Flexible asymmetric solid-state supercapacitors by highly efficient 3D nanostructured α-MnO_2 and h-CuS electrodes. *ACS Applied Materials & Interfaces* 10: 16636–16649.

42 Technologies, M. (2022). Ultracapacitor overview. https://maxwell.com/products/ultracapacitors/ (accessed 26 November 2021).

43 Li, H., Hou, Y., Wang, F. et al. (2017). Flexible all-solid-state supercapacitors with high volumetric capacitances boosted by solution processable MXene and electrochemically exfoliated graphene. *Advanced Energy Materials* 7: 1601847.

44 Yunasko (2022). Supercapacitor. https://yunasko.com/en/ (accessed 26 November 2021).

45 Zhao, F., Xie, D., Huang, W. et al. (2020). Defect-rich honeycomb-like nickel cobalt sulfides on graphene through rapid microwave-induced synthesis for ultrahigh rate supercapacitors. *Journal of Colloid and Interface Science* 580: 160–170.

46 Hong, J., Kim, B.-S., Yang, S. et al. (2019). Chalcogenide solution-mediated activation protocol for scalable and ultrafast synthesis of single-crystalline 1-D copper sulfide for supercapacitors. *Journal of Materials Chemistry A* 7: 2529–2535.

47 Gopalakrishnan, A., Sahatiya, P., and Badhulika, S. (2018). Template-assisted electrospinning of bubbled carbon nanofibers as binder-free electrodes for high-performance supercapacitors. *ChemElectroChem* 5: 531–539.

48 Mahmood, N., Zhang, C., Yin, H., and Hou, Y. (2014). Graphene-based nanocomposites for energy storage and conversion in lithium batteries, supercapacitors and fuel cells. *Journal of Materials Chemistry A* 2: 15–32.

49 Wu, Z.-S., Tan, Y.-Z., Zheng, S. et al. (2017). Bottom-up fabrication of sulfur-doped graphene films derived from sulfur-annulated nanographene for ultrahigh volumetric capacitance micro-supercapacitors. *Journal of the American Chemical Society* 139: 4506–4512.

50 Orangi, J., Hamade, F., Davis, V.A., and Beidaghi, M. (2020). 3D printing of additive-free 2D Ti_3C_2Tx (MXene) ink for fabrication of micro-supercapacitors with ultra-high energy densities. *ACS Nano* 14: 640–650.

51 Zhang, Q., Wang, R., Zou, Y. et al. (2022). Turning industrial waste-flax noil into regenerated cellulose fiber electrodes for eco-friendly supercapacitors. *Industrial Crops and Products* 176: 114377.

52 Samsung (2022). Samsung Begins Chip Production Using 3nm Process Technology With GAA Architecture.

53 Wang, Y., An, N., Wen, L. et al. (2021). Recent progress on the recycling technology of Li-ion batteries. *Journal of Energy Chemistry* 55: 391–419.

54 Niu, B., Chen, Z., and Xu, Z. (2017). Method for recycling tantalum from waste tantalum capacitors by chloride metallurgy. *ACS Sustainable Chemistry & Engineering* 5: 1376–1381.

55 Wang, J., Sun, C., Guo, J., and Xu, Z. (2017). Integrated process for recycling aluminum electrolytic capacitors from waste printed circuit boards: disassembly, heat treatment and magnetic–eddy current–electrostatic separating. *Journal of Cleaner Production* 165: 334–345.

56 Huang, X., Wang, D., Yuan, Z. et al. (2018). A fully biodegradable battery for self-powered transient implants. *Small* 14: 1800994.

Part V

Waste Prevention and Recycling

12

Waste Prevention for Energy Storage Devices Based on Second-Life Use of Lithium-Ion Batteries

Oliver Pohl[1,3], Gavin Collis[2], Peter Mahon[3], and Thomas Rüther[1]

[1] CSIRO, Energy, Research Way, Clayton, VIC, 3168, Australia
[2] CSIRO, Manufacturing, Research Way, Clayton, VIC, 3168, Australia
[3] Swinburne University of Technology, Faculty of Science, Engineering and Technology, Hawthorn, VIC, 3122, Australia

12.1 Introduction

In the normal life cycle of a battery, the amount of energy it can store will decrease with use and storage and eventually, the battery will no longer be able to sustain the performance required for its desired application. This is the battery's end-of-life (EoL) and it would now be considered unusable and defined as waste. There are currently three options for EoL, which are landfill, recycle and reuse. Battery waste contains valuable materials, such as nickel, cobalt, manganese, lithium, and lithium hexafluorophosphate ($LiPF_6$) where some materials, such as cobalt and $LiPF_6$, are toxic and can leach into the environment [1–3]. However, recycling is the ideal outcome for lithium-ion batteries (LIBs) but currently, worldwide the overwhelming majority (>90%) are sent to landfill or stored indefinitely [4–7]. This is mostly due to the battery recycling industry not being set up for LIBs [8, 9]. The third option is reuse, more commonly known as "second-life" and is the topic of this chapter, specifically considering LIBs.

The idea for second-life LIBs has been around since 2003 when it was proposed by Cready et al. to use degraded electric vehicle (EV) batteries for stationary energy storage [10–12]. Currently, an EV battery is considered to be no longer fit-for-purpose when its nominal capacity is reduced to 80% of the initial capacity [13]. This is also known as 80% State of Health (SoH) because a 20% reduction in capacity significantly impacts the vehicle's performance and range. The remaining 80% capacity is still useful for less demanding applications such as grid storage, off-grid solar storage [14], industrial EV (e.g. Forklifts), charging ports for EV [15], or low-cost household batteries [16, 17]. This chapter will discuss the advantages, disadvantages, challenges, and current successes of so-called second-life batteries.

Sustainable Energy Storage in the Scope of Circular Economy: Advanced Materials and Device Design, First Edition. Edited by Carlos Miguel Costa, Renato Gonçalves, and Senentxu Lanceros-Méndez.

As the term suggests, an electric vehicle is a vehicle whose motor is powered by electricity. There are multiple types of electric vehicles, such as battery electric vehicles (BEV), where an electric motor is powered by a battery. Plug-in hybrid electric vehicles (PHEV) use an electric motor powered by a battery that can be charged from the electricity grid or an onboard internal combustion engine. Hybrid electric vehicles (HEV) have an electric motor powered by a battery and an internal combustion engine that work together to achieve greater fuel efficiency than the internal combustion engines alone. BEVs will tend to have the largest batteries, where the average usable battery capacity in 2021 is 60 kWh, with luxury BEVs having capacities of around 100 kWh [18]. PHEVs and HEV have smaller batteries with HEVs having the smallest due to the alternative power source. Most of these batteries are LIBs, however, some are lead-acid or nickel-metal hydride types.

Due to the power requirement, these batteries are made up of individual electrochemical cells, most often referred to as just, "cells." Cells can be combined to gain increased voltage and/or capacity. The voltage range of a galvanic cell is dependent on the specific combination of redox couples that comprise the cell. By connecting the cells in series, the voltage, which is usually less than 5 V per cell, can be summed together. When connecting in parallel the capacity sums. Groups of cells are called battery modules, which can be further combined into battery packs. Figure 12.1 is a visual explanation of cells wired up into modules and packs. In EVs, the number of individual cells assembled into packs can be in the thousands. As the supply of EoL EV batteries will continue to increase, EV batteries are ideal for reuse at the pack, module, or cell level in a variety of applications.

Generally, there are multiple second-life strategies for degraded batteries but there is no consensus on definitions in the literature [19–24]. The strategies in question are called; "reuse," "repurpose," and "second-life." These terms have been used interchangeably and multiple definitions have been proposed. Faessler [19] and RECHARGE (The European Association for Advanced Rechargeable Batteries) [20] define "reuse" as degraded batteries or degraded battery parts used for the same application as their first-life. They define

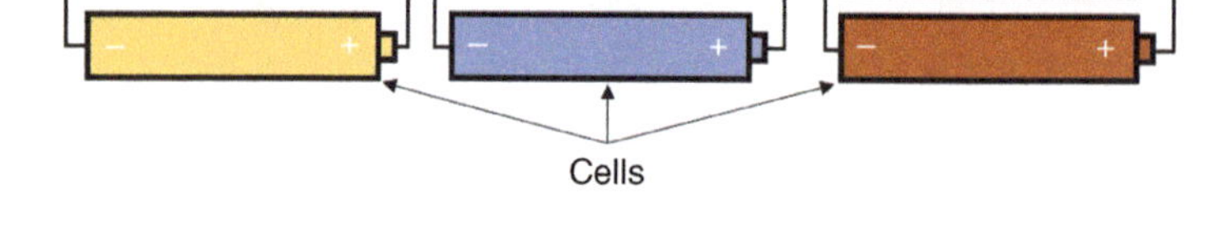

Figure 12.1 Arrangement of cells and modules inside a battery pack. The battery pack in this example contains three series modules, which contain three cells in parallel. If the cells here have an open circuit potential of 3.6 V and a capacity of 2 Ah, then each module will have an OCV of 3.6 V and a capacity of 6 Ah and the battery pack will have an OCV of 10.8 V and a capacity of 6 Ah. Cells may be arranged into series and/or parallel for modules to suit the performance requirements of the pack.

"second-life" as degraded batteries or degraded battery parts used for a different application than the first-life. Reinhardt [23] defines "second-life" or second use as "transforming a battery after its vehicle application for use in off-road applications such as energy storage systems." Casals [24, 25] defines "direct reuse" as taking a degraded battery, performing a few tests such as state of health and starting its second-life with the same shell/cover, cooling systems, and cells. While "repurposing" was defined as dismantling the degraded battery pack into modules or cells to reassemble them into a new battery [24, 25]. These terms are still yet to be fully adopted and are still, unfortunately, used interchangeably. Additionally, it can be argued they do not satisfy all criteria. Using "second-life" to refer to some continually used degraded batteries but not all of them implies that some batteries have not finished their first-life when they have significantly degraded. For Faessler's [19] and RECHARGE's [20] definition of reuse, if a battery is still able to be used for its original intended purpose without any alterations to the battery, it has arguably not finished its first-life. Using the term "repurposed" implies the second-life application being different for first-life and second-life. Additionally, most definitions also assume a repurposed battery was remanufactured from degraded cells, which is not clear from the name. Therefore, in this chapter, "second-life" refers to all batteries being used after degradation has rendered them unfit for use in their initial application. A "reused battery" is a second-life battery that has been taken out of its initial application for a second-life without further processing other than basic electrical checks. A "reassembled battery" is a second-life battery made from components of other degraded batteries and this can occur at the module or cell level and can occur at any point in their life to ensure maximum utilization.

A *reused battery* is the less expensive of the two options because it is less labor-intensive to implement. After battery collection, the process involves an SoH test, transport, and installation into a new role. Therefore, most second-life projects currently underway would be classified as reuse, as most of them are using unmodified EV batteries [19, 26]. The drawbacks of reused batteries are the bulky size and lack of smaller package options. Electric vehicle batteries are typically designed to fit into limited space constraints and therefore are manufactured with irregular shapes, see Figure 12.2. Stacking multiple EV battery packs together may produce a lot of void space, which reduces the total volumetric power density. Although a typical residential battery will be lower than 20 kWh and an EoL EV battery at 70% SoH could be used, the geometry and size of the battery pack would not be ideal for residential use but be better suited for stationary grid energy storage applications.

A *reassembled second-life battery* is more technically challenging and expensive to make than reusing a battery but does offer some key benefits. Making a reassembled battery involves disassembling it down to the cell level and using the degraded but still functioning cells to manufacture a new battery. Within a battery, each cell is unlikely to age at the same

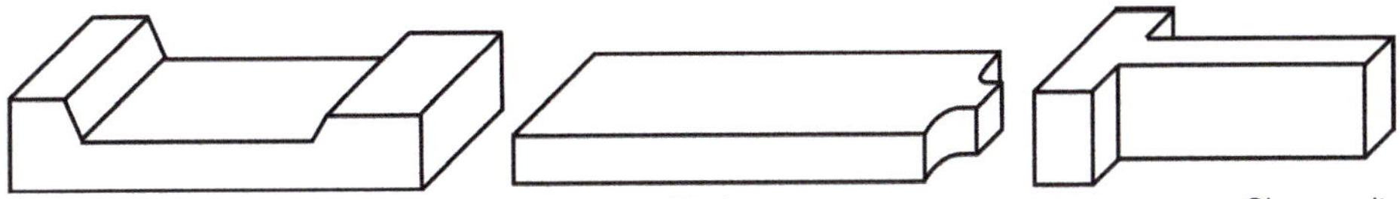

Figure 12.2 Common shapes of the batteries from some popular electric cars. Not to scale.

rate [27] and some cells may have failed early in the battery's lifespan, so the manufacture of a reassembled second-life battery requires sorting the healthy cells from the degraded and failed cells. Assembling batteries with cells that have the same SoH can increase the lifespan of the second-life battery [28]. However, it is the disassembly and determination of SoH where the major costs of the process occur. Electric vehicle batteries are not designed to be disassembled, so human input is required. This labor cost is further increased because safely disassembling and reassembling a battery requires specialized high-voltage training and can take upwards of 10 hours to complete [25, 29]. This makes reassembled batteries significantly more expensive than reused batteries. Some of the expenses can be reduced if disassembly is limited to the module level but this raises the possibility that some good cells will be rejected and some bad cells will be retained.

Due to these differences, reused and reassembled batteries fill different niches within battery EoL management in the circular economy. Reused batteries are fixed to 70–80% of their designed capacity and will retain their original shell. This tradeoff is likely to be acceptable for grid-scale stationary energy storage, where cheaper batteries are preferred and space limitations are negligible, while reassembled batteries will cost more but have greater versatility in capacity and better performance. This makes them better suited to smaller household battery storage units paired with rooftop solar photovoltaics or if the cells are above 80% SoH, a remade EV battery.

It has also been suggested to disassemble the cells of the EoL battery to extract, clean, reactivate and reuse the electrodes in a new battery. Cleaned electrodes have been shown to operate at performances with a 5% reduction from their initial specifications [30, 31]. The cleaning procedure aims to remove the solid–electrolyte interphase (SEI) layer on the surface of the aged electrode, which can be removed by washing with a solvent such as dimethyl carbonate or using a laser [30–34]. Cleaning can only recover capacity lost from SEI growth, not particle exfoliation, cracking, or agglomeration. However, this method is significantly more challenging and there has been minimal reporting in this area since 2017.

12.1.1 Benefits of Second-Life

There is a range of benefits for EoL processing of LIBs that are economic and environmental in nature. While a lot of these are the same benefits as for any circular economy, such as minimizing waste, some benefits are specific to second-life batteries (Figure 12.3). Unlike recycling, second-life manufacture is not reliant on the resale value of the material components because not all of the components are profitable [35]. This could make second-life more attractive than recycling for certain battery chemistries, like lithium iron phosphate (LFP), or chemistries with small amounts of cobalt and nickel. Thus currently, the recycling and second-life industries are not necessarily competitors with the possibility that these two industries could be integrated into one industry with multiple products. However, this depends on the availability and fluctuating stock market prices of raw materials, the viability and supply of second-life batteries compared to new or next-generation batteries and the cost of making an operable second-life battery pack.

As regulations are added to mandate LIBs' EoL management [29, 36–38], battery manufacturers could become large players in these industries. Entering the battery EoL industry could be beneficial for battery and EV manufacturers with many, such as Nissan,

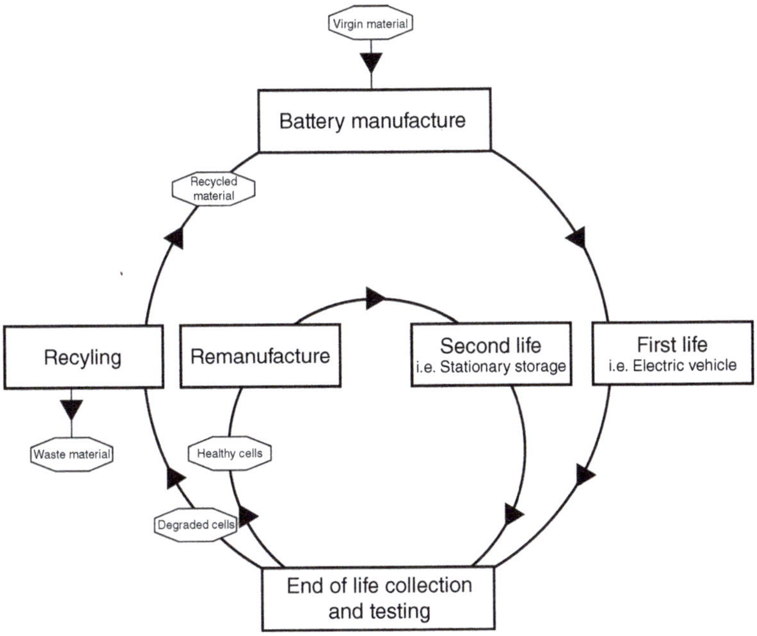

Figure 12.3 Battery life management in a circular economy.

Volkswagen, and Renault, having already entered this domain [39–44]. Most of these second-life projects are in Europe, China, or Japan, with a handful in the United States, which are the regions with the greatest numbers of EVs [45]. These projects are partially funded by the companies and often in partnership with local governments. Some of these plans, like the Johan Cruyff Arena (Amsterdam, Netherlands), are on the megawatt scale used to reduce electricity costs during peak times [39]. Others are energy security plants on remote islands and others are grid storage [46, 47].

Currently, reused second-life batteries are financially viable for large-scale (>1 MWh) [39] and small-scale (<200 kWh) [19, 26] projects but reassembled batteries are still not feasible. Challenges faced include determining the effective life span of the second-life battery, determining the safety of a second-life battery, and determining how to do all the required testing economically. Solving these issues also benefits reused batteries, especially in determining safety. Additionally, there are some legal uncertainties that could prevent the development of second-life and these will be discussed below.

12.1.2 Economic Benefits

According to some modeling studies, the first major economic benefit of reused and second-life batteries is that they should be cheaper than new batteries and this should drastically reduce the cost of building grid storage batteries [26, 29]. Grid storage is predicted to be the primary application for second-life batteries, especially when sourced from EVs. Grid storage batteries purchase electricity to recharge the battery when electricity is cheap and sell it back when electricity is expensive. They can also stabilize the frequency of the grid and hold energy in reserve in the event of a blackout. It is also common to see them combined with renewable energy generators. These batteries serve to smooth out the peaks and dips in electricity prices, making electricity cheaper for the consumer and keeping energy in reserve to power the grid during a sudden reduction in power supply. Coupled with renewable energy sources, grid storage batteries on the megawatt scale have already proven themselves to be profitable, such as the Hornsdale power reserve in South Australia, which returned the construction cost of AUD90 Million (US$72 Million) within 2.5 years [48–50]. Hornsdale was built using new batteries but using second-life batteries to build grid storage could reduce the capital cost, which will encourage more to be built. In addition, second-life batteries could be used as a low-cost alternative for household electricity storage [15] as presently and depending on the region, a household solar array paired with a first-life battery is not financially viable because of the battery cost, with payback periods on the scale of 10 years [51].

The second benefit is one that can be included in the first-life where instead of going to landfill, a degraded EV battery retains some value [11, 23]. This extra value can be factored into the price of EVs to reduce the cost and make them more affordable [26]. This can be achieved in several ways. In the first scenario, an EV owner sells their degraded battery to a second-life manufacturer. The second scenario is the EV manufacturer sells an EV but retains ownership of the battery, requiring it to be returned at its EoL. The third possibility is as a supplement to the recycling industry.

In the first scenario, where the EV owner sells their battery has not been considered widely because of its decentralized nature and a lack of clarity on liability. Additionally, manufacturers of second-life batteries and recyclers would be at a disadvantage because they collect the batteries from individuals without a guarantee of the battery's condition. Furthermore, in this scenario, the battery and EV manufacturers lose out on a potential revenue stream. Overall, due to these reasons, this scenario is the least likely to occur.

In the second scenario, a manufacturer retains ownership of the battery and would be able to reduce the upfront price of the EV and recoup that loss from the resale of the second-life battery. In this case, the consumer would be leasing the battery from the EV manufacture [52] or agreeing to return it for a lower purchase price. Leasing systems vary, a consumer could rent the battery for the duration of its first-life, or purchase a subscription to a battery exchange service where a flat battery is exchanged for a charged one [53]. However, the latter option is unlikely to occur because of fast charging and home charging. Cost-savings in this scenario are dependent on many factors such as remanufacturing cost, degraded battery acquisition costs, remaining capacity of degraded batteries, and expected lifetime. Of course, nothing is stopping EV manufacturers from selling the battery at a profit, retaining ownership of the degraded battery, and profiting from the sale of the second-life battery.

The third way is as a supplement to the recycling industry. The current LIB recycling industry is underdeveloped [8, 9]. Facilities in Germany are being set up [54, 55] but it will take some years before they are capable of processing the 290 000 tonnes of waste batteries generated in Germany in 2018 [56] plus the batteries currently in storage. Second-life could be used to ease the pressure on the recycling industry until it ramps up. If attached to a recycling plant, second-life could bring in more money from the resale of batteries allowing them to invest more into their equipment and processes. Additionally, certain types of LIBs with low-value components, such as LFP, are not profitable to recycle thus making second-life the only financially viable EoL management option for these types of batteries.

12.1.3 Environmental Benefits

The primary environmental benefit of second-life projects is that they can be implemented within the short term for EoL battery management. It should be noted that globally, the overwhelming majority (>90%) of spent LIBs are sent to landfill or stored indefinitely rather than recycled [5, 6]. In a landfill, LIBs are dangerous because they contain toxic and reactive materials that pose a fire risk. Materials within LIBs include lithium, carbon, graphite, aluminum, copper, iron, cobalt, manganese, nickel, phosphorus, carbonate solvents, and lithium hexafluorophosphate. Some of these, such as carbon, graphite, iron, and aluminum are benign. However, others, such as lithium hexafluorophosphate, which is used as an electrolyte salt, degrades in the presence of water into hydrofluoric acid (HF) [3]. Additionally, cobalt ions are toxic to humans [1] and aquatic life [2], while other transition metal ions, like nickel and manganese, can bio-accumulate. Thermal decomposition of a LIB releases toxic gases such as hydrogen cyanide (HCN), HF, nitric oxides (NO_x), and sulfur dioxide (SO_2) [57]. Fires in waste disposal can be started from internal or external battery short circuits, which can occur in general waste streams if they are crushed/shredded, which are common practices [58–62].

Whilst storing degraded batteries in warehouses is preferred over landfill, warehouses are not designed to store batteries indefinitely, which is currently the case in some regions [8, 9]. Such stockpiling occurs because local battery recycling facilities are unable to process the volume of EoL batteries. Recycling facilities may also store EoL batteries before processing [62]. Stockpiling EoL batteries is a fire risk as batteries may have been damaged in transport or during their operating life. Larger stockpiles increase the risk of a fire and the consequences of having one. Second-life projects can take degraded batteries out of indefinite storage, alleviating the issues until the recycling industry is sufficiently developed.

12.2 Challenges

Challenges for second-life batteries can be divided into multiple categories; Chemical, Engineering, Economic, and Legal.

12.2.1 Chemical Challenges

These relate to the SoH of the battery, safety, and expected lifespan. All of these are directly related to the amount and type of degradation that occurs over the battery's lifespan. Degradation occurs within the cell and serves to reduce the total amount of energy that can be stored within the battery. This is known as a battery's state of health and it is a comparison of the capacity that the battery has when new ($Cap_{Initial}$) versus the actual capacity at any time in the life-cycle ($Cap_{Nominal}$) and is typically expressed as a percentage, see Equation (8.1):

$$SoH\% = \frac{Cap_{Nomial}}{Cap_{Inital}} \times 100 \tag{12.1}$$

The SoH is critically dependent on the battery chemistry, which is dictated by the chemical composition of materials used for the cathode, anode, electrolyte, and separator. Processes such as the formation of the SEI layer [63–65], lithium plating [66–68], dendrite formation [69–71], and strain on the electrodes from lithium intercalation [72–75], all contribute to a loss of capacity. These processes are also impacted by environmental and operational conditions, such as temperature, charging/discharging rates, frequency of operation, depth of discharge, and resistances of wires connecting cells. High temperatures will encourage SEI growth and lower battery internal resistance [76, 77] while high currents and reduced temperatures encourage lithium plating [69, 78]. Additionally, environmental and operating conditions are not uniform throughout a battery, especially EV batteries which are quite large. During operation, it is harder to cool the center of the battery, therefore the center is often at a higher temperature. Additionally, for large battery packs made of small cells, it is not uncommon for a few cells to fail early in the battery's first-life due to manufacturing defects. All this can cause cells to degrade at different rates within the battery pack [27]. The non-uniformity across the battery needs to be managed to prevent overcharging and over-discharging of the cell and maintain a voltage balance.

To keep the battery balanced, large batteries will utilize a battery management system (BMS), which controls currents and voltages across the battery. Failure to properly manage a battery with uneven degradation can result in faster degradation of the entire battery [28] and can lead to safety issues such as thermal imbalance and potentially thermal runaway [76, 79]. As the spread of SoH increases, properly balancing a battery becomes harder. Thus, an advantage of reassembled second-life batteries is the ability to match cells with similar SoHs. However, measuring battery SoH comes with nuances that make complete state-of-health measurements difficult.

There are many descriptions within the literature to determine battery SoH that focus on testing methods that are economical and non-destructive [80–84]. The most common method of estimating the SoH is coulomb counting where a battery is discharged at a slow rate, such as a period of 10 hours (this is known as a C/10 discharge). However, this is not viable for industrial-scale second-life production for two reasons. Firstly, it is very time-consuming if the charge/discharge time is longer than a typical workday (8 hours) and might require personnel to be present in case of battery failure. Secondly, instruments that can handle high currents (>10 A) and perform an accurate measurement with multiple channels are expensive. Other methods of determining battery SoH have been proposed, such as differential voltage analysis, electrochemical impedance spectroscopy, and extracting data from the BMS. The various methods mentioned here are discussed in more detail in a later section. For a battery pack, the SoH is a combination of all the cells.

Calculating the lifespan of a second-life battery is complex because battery aging is widely known to follow a distinctive profile, which can be divided into three phases as shown in Figure 12.4 [85, 86]. Phase I is an initial sharp drop in the SoH that quickly

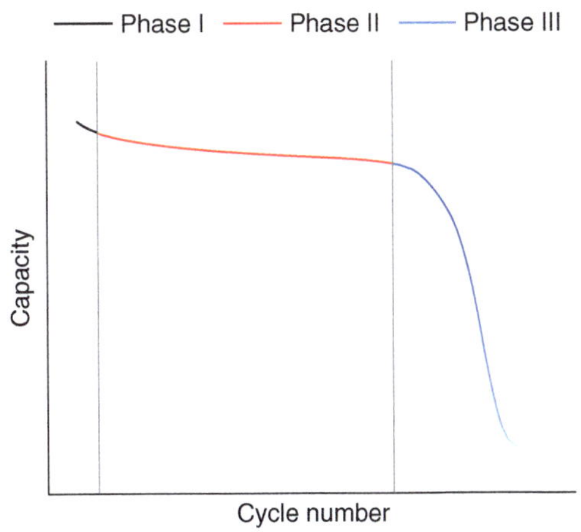

Figure 12.4 Capacity decrease behavior. Not to scale.

plateaus to become Phase II [87], which can be approximated as either linear or two linear sections of different gradients [88]. It is also possible to describe Phases I and II together as a logarithmic process [89]. The final phase, Phase III, is a steep drop off in capacity as the battery begins to fail and this mostly occurs after the battery is below 80% SoH. It is also assumed that the rapid decline in capacity will eventually stabilize but there is a lack of data to confirm or reject this hypothesis. It is Phase III that makes calculating the total battery lifespan difficult due to the unpredictability of when Phase III will occur.

Phase I involves the formation of the SEI [87]. Manufacturers of LIBs have developed preconditioning cycles for the battery to minimize capacity loss during Phase I and promote long cycle life during Phase II [90–92]. In ideal circumstances, it only lasts two to five cycles, so by the time the battery is sold, it will be in Phase II. Phase III is the least studied region because experiments on battery aging are typically focused on the 80–100% SoH region, where Phase III is unlikely to occur and instances when it does occur are usually not published [88, 93–95]. The start of Phase III is the indication that the battery is about to reach the end of its useful life, be it first or second. Within a battery pack, Phase III would begin when a significant number of cells within the pack have started Phase III. At this point, it is still possible to disassemble the pack to extract the cells that are still in Phase II. So, knowing when Phase III starts is critical to second-life performance. If Phase III starts just below 80% SoH, then second-life usage of a battery is extremely limited. The best-case scenario is for Phase III to begin after the battery has reached 50% SoH, which is expected to be at the end of the battery's second-life.

Studies of commercial lithium-ion cells that have analyzed SoH below 80% show that in many cases Phase III is not encountered as the cell degrades from 100% to 50% SoH [89, 96, 97], while in some cases Phase III did occur before 50% SoH was reached [98]. Based on the currently published literature, most modern lithium-ion cells do not experience the Phase III capacity drop during their second-life application, so it is unlikely that the second-life lifespan will be cut short by a sudden loss of capacity. Since Phase III is not expected to happen during the battery's second-life, it can be reasonably assumed that the SoH reduction linearly occurs during Phase II. From here, the life span of the second-life battery depends on the application with times ranging from 3 to 15 years depending on application and estimations [99].

It should be noted that while the current wisdom is that EV batteries are considered degraded at 80% SoH, consumers are likely to continue to use their batteries beyond this point. Reasons for this could be that 80% of SoH can fulfill the needs of the owner, is unable to afford a replacement battery, or is unwilling to change the battery. Therefore, 80% SoH should be considered the upper range of second-life battery cells. Saxena et al. suggest that 55% of U.S. drivers will have their driving needs met by batteries with SoH as low as 30% [100]. If consumers are satisfied with highly degraded batteries in their EVs, then second-life batteries will be less viable. However, manufacturers could provide incentives to encourage a battery replacement at 80% state of health to counter this.

12.2.2 Methods of Investigating Lithium-Ion Battery State of Health

12.2.2.1 Coulomb Counting

The simplest form of determining battery SoH is to measure the charge during the charging and/or discharging of the battery. It is the most accurate method for determining a battery's SoH but only when the charging and discharging currents are small. Ideally, the current should discharge a battery from full charge to flat in 20 hours (0.05 C). When measured at higher currents, this method underestimates the capacity of the battery and is known as the Peukert effect, which was initially observed in the study of lead-acid batteries but also applies to LIBs to a lesser extent [101]. In a LIB, the effect is caused by the internal resistance of the battery [102]. Low currents are necessary to reduce the magnitude of this effect but consequently, make this method impractical for industrially determining a battery SoH as a 20-hour discharge is too slow a process for large quantities of battery cells.

12.2.2.2 Battery Management System Data Extraction

The BMS is a device connected to the battery to manage charging and discharging behavior. The critical functions are to prevent overcharge and over-discharge, maintain a charge balance between cells and estimate the remaining charge [83, 84, 103]. Extracting performance data from the BMS is ideal because it is rapid and contains a large amount of data on which to base a more accurate estimation of capacity. However, there are difficulties when extracting data from the BMS:

1) Not all systems use the same communication protocols, although most use i2C [83].
2) Differences in connector pin layouts that physically prevent communication or require specific adaptors.
3) The BMS may be damaged and therefore the SoH cannot be easily determined [83].

Fortunately, there are methods around these, and BMS data extraction has been performed with some success [14, 83]. Salinas et al. were able to extract the data from 68% of BMSs that were tested, with the remaining 32% having corroded connectors or using non-standard data formats that could not be read. Capacity estimations from the BMS correctly predicted if a cell was useable 75% of the time, with no significant bias toward false positives or negatives. To achieve greater levels of accuracy will require a BMS that is designed to remain accessible after a potential decade of neglect and are better at estimating the SoH of the battery during operation [84]. Additionally, standardization of BMS architecture could make this method more robust, or else this method will only be used by second-life manufacturers who specialize in specific battery designs.

12.2.2.3 Electrochemical Impedance Spectroscopy (EIS)

Impedance is the total opposition to an alternating current (AC) in an electrical circuit [104]. It is a combination of resistance, capacitance, and inductance that can be measured by applying an alternating potential and measuring the current, or vice versa. Additionally, the current will lag or lead the potential, which can also be measured in the form of a phase angle. Impedance is also dependent on the frequency of the alternating potential/current and thus, can be viewed as a spectrum. The technique is sensitive to minor changes in resistance, capacitance, and inductance of a sample where the frequency dependence can be seen in the following equations:

$$\text{Resistor } Z_R = R \tag{12.2}$$

$$\text{Capacitor } Z_C = \frac{-1}{j\omega C} \tag{12.3}$$

$$\text{Inductor } Z_L = j\omega L \tag{12.4}$$

Here, Z is the impedance (Ohms), R is the resistance (Ohms), j is an imaginary number, ω is the frequency (Hz), C is the capacitance (F), and L is the inductance (H). Generally, impedance measurements are plotted as either Nyquist plots or Bode plots, see Figure 12.5. Both plots contain the same information, so typically only one is published. Nyquist plots separate the magnitude of the impedance into real and imaginary parts and a Bode plot

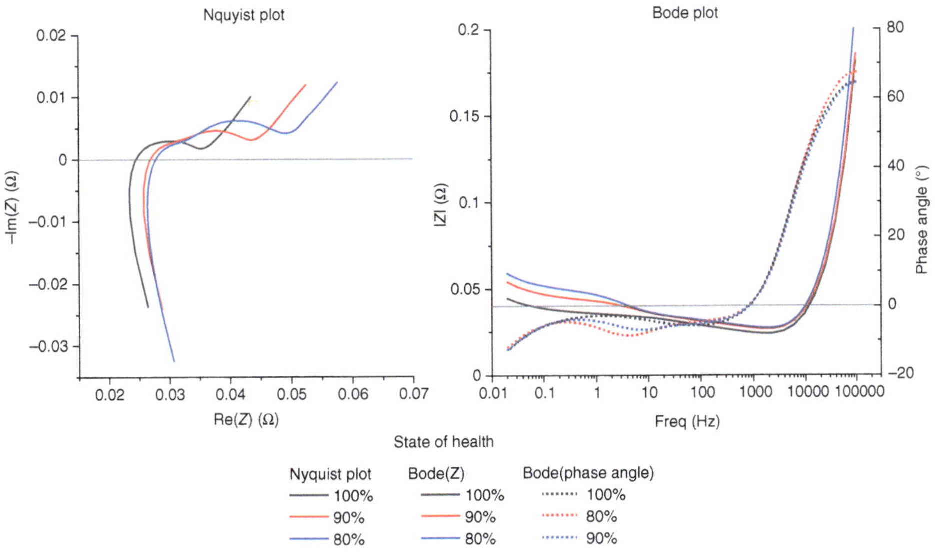

Figure 12.5 Example of electrochemical impedance spectra of a lithium-ion battery at various states of health.

displays the magnitude of the impedance and the phase angle vs. the frequency. However, to fully understand the impedance result, EIS is analyzed by modeling the response of an equivalent circuit, which simulates the resistance, capacitance, and inductance in a sample. For a battery, this can be used to determine internal resistance, double-layer capacitance of the cell electrodes, and diffusion of lithium ions [105, 106].

The advantage that EIS has over Coulomb counting is that it is faster, potentially taking 20 minutes whereas the latter can take up to one day. Additionally, the equivalent circuit can provide insight into the type of degradation, such as excessive SEI growth, limited diffusion rates, and lithium metal plating, which can be used to identify potential safety risks. Studies using electrochemical impedance spectroscopy have been able to determine the SoH with errors below 5% [107–113]. Results are promising but the drawback is that each battery design needs to have its own model of degradation, as minor differences in the cell chemistry can change the results. A model of degradation needs to be built for each chemistry type, which makes this method only applicable for second-life manufacturers where there is a known battery type and this can be difficult to determine without adequate labeling.

12.2.2.4 Incremental Capacity Analysis (ICA)

Incremental capacity is a method that uses the derivative of the charge-voltage plot, mathematically expressed in Eq. (8.5), plotted against the voltage [114]:

$$IC = \frac{\mathrm{d}Q}{\mathrm{dOCV}} \tag{12.5}$$

where Q is the charge capacity (Ah) and OCV is the open circuit potential (V). However, in practical applications, the OCV data are substituted with voltage [81, 114]. Both the charge and voltage can be collected from a BMS. Q–V plots are usually flat but by plotting the derivative, defined peaks can be seen that relate to phase transformations within the sample. The main disadvantage of ICA is that using the derivative increases the amount of noise and data smoothing methods, such as support vector regression, a form of machine learning, have been the focus of cleaning up the data [115, 116].

SoH can be calculated from the incremental capacity peaks which change in intensity and voltage during the aging process. An example of incremental capacity peaks can be seen in Figure 12.6. Studies using ICA have been able to calculate the state of health within a 5% error margin [113, 117, 118]. However, similar to EIS, a model of degradation needs to be built to use this method effectively and the model is specific to each cell architecture.

12.2.3 Engineering Challenges

There are two main bottlenecks for second-life batteries that can be solved via engineering, disassembly cost, and SoH analysis. The primary cost of disassembly is the cost of manual labor because it can take upwards of 10 hours to discharge, then disassemble and reassemble a second-life battery pack [25]. For reassembled second-life batteries to be profitable, greater levels of automation are necessary [119]. The two main difficulties of automated disassembly are the lack of standardization between battery designs and the complexity of some of the tasks [120]. Work on developing software that can adapt to the varied designs

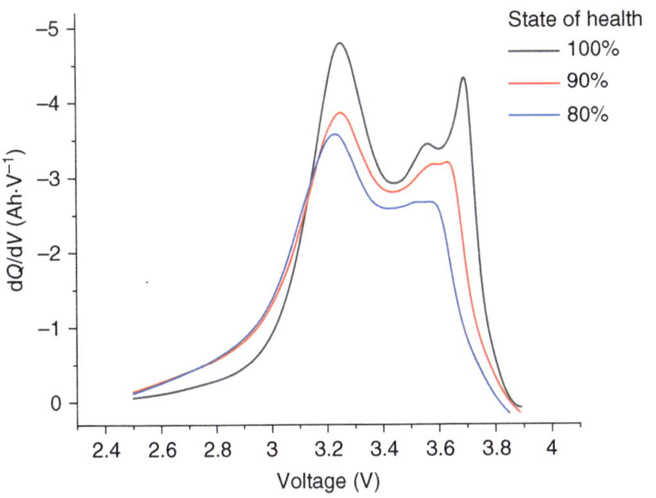

Figure 12.6 Example of the incremental capacity peaks of a commercial lithium-ion battery.

of EV batteries has been performed so that humans and robots can work cooperatively [121]. However, human–robot collaboration comes with added safety risks [120]. By designing the battery to be disassembled fully by a robot, human manual labor costs can be minimized. To fully utilize robots, disassembly needs to be factored into the design of the battery pack to simplify the structure and avoid glues, rivets, and minimize welds. Engineering can also determine the battery SoH through the use of the BMS, which has already been discussed.

12.2.4 Economic Challenges

This challenge is to produce a second-life battery at a cost that the consumer would consider acceptable, which means that second-life batteries need to compete with new batteries. Issues arise for second-life batteries here because battery technology is progressing rapidly [122] and their cost is still projected to decrease [123]. Lifespans for second-life batteries need to be comparable to new batteries and also meet similar safety ratings. All of this needs to be met on a budget and unfortunately, some factors make second-life manufacturing quite expensive.

The main cost of end-of-life processing is the labor cost of disassembling batteries [124], which for reassembled batteries is approximately 50% of the cost. Most EV batteries from 2011 have not been designed with EoL in mind and hence it is not currently possible to automate the EoL processing, which requires dismantling the battery for second-life or recycling, SoH testing, and to fully discharge the battery. All this must be done manually and is a labor-intensive process. Additionally, personnel will need training to safely work on high-voltage products, where EV batteries can be up to 800 V (Porsche Taycan). However, government bodies can legislate manufacturers to implement and comply to design standards for EoL management. This has occurred before in the case of lead-acid batteries and it is now possible to recycle 98% by mass of a lead-acid battery [125, 126]. So, in the future, automation could be possible for the disassembly of LIBs [119, 121, 127]. The labor cost is less of an issue for reused second-life batteries, which do not require disassembly and as a result, can already be profitable when installed into larger systems [39].

The other financial hindrance to second-life is how the SoH for individual cells is to be evaluated accurately. Anything that slows down the cell evaluation process will increase the cost of production. Unlike the lack of standardization, battery cell health is a scientific challenge. While there are numerous publications on measuring SoH, they exclusively test single cells [66, 80, 81, 83, 112]. Packs and modules are more complicated to measure the SoH due to the extra variables with connectors and variations between cells. Coulomb counting methods only measure the total capacity of the cells in the battery pack. Unfortunately, there are minimal studies attempting to measure individual cells in packs [114] and the results are promising for single chemistry types with there being no obvious reasons as to why this cannot be expanded using methods such as EIS or ICA.

The next challenge for second-life batteries is reducing the safety risks. Degraded batteries have a greater chance of being faulty as they have lower thermal runaway temperatures and a higher chance of an internal short circuit [58]. This is potentially catastrophic to the second-life manufacturer but also to the original manufacturer via proxy. A recent example is the faulty batteries in Samsung's Galaxy Note 7 that caused phones to overheat, catch fire and explode in some incidents [128, 129]. Battery manufacturers may even prevent third-party companies from remanufacturing their batteries to protect their reputations. For second-life batteries to be considered by companies, health tests need to detect faulty cells at accuracies comparable to the failure rates of new cells. However, new developments towards safer high-performance electrolytes may present a chance to mitigate the safety risk in the future.

All the challenges discussed above must be addressed at a cost that is commercially viable and estimations of this price vary. In 2012, Neubauer et al. suggested US$38–132 kWh^{-1} [12] and Rohr et al. suggested US$41–76 kWh^{-1} [124]. The large discrepancies are partially caused by the price and performance of new batteries, where energy density is increasing by 3–5% per year [122]. Additionally, the price of LIBs per kWh is decreasing and is expected to be less than US$150 by the late 2020s-early 2030s [130]. This is a greater problem for a 10-year-old reused second-life battery as they will have 70–80% of their initial capacity. Meanwhile, over the same ten years, the capacity of a new battery will increase by 130–160%. For a reassembled battery, this is less of a problem because useable cells can be manufactured into batteries larger than the original battery they have occupied, with the only drawbacks being a more complex BMS and larger pack size due to the lower energy density.

12.2.5 Legal Challenges

The final main challenge for second-life batteries is the legalities involving the risks. If a third-party manufacturer purchases old batteries and resells them, who is responsible if the battery fails? Is there a legal right to sell another company's product if it is considered waste? How is the warranty affected if applicable? Answers depend on yet to be established local and international laws, and how far battery manufactures will go to protect themselves. Unfortunately, this is dependent on the local political climate but regions like Germany and Canada, are already forcing battery EoL management onto the battery manufacturer or EV manufacturer [29, 36–38]. These laws will most likely result in battery/EV manufacturers starting their own second-life battery projects, which circumvents the questions of ownership and liability.

12.2.6 Current Implementations

Second-life batteries have been implemented as small-scale energy storage systems (capacities less than 100 kWh) [19]. These systems are made of a single reused EV battery, occasionally two. Most of these batteries store locally generated electricity or reduce the energy from the grid when prices are high. Overwhelmingly, they are sourced from electric vehicles, primarily BEV, with some hybrid batteries.

Larger examples of reused second-life batteries are the 2.8 MWh battery in the Johan Cruyff Arena (Netherlands) [39], the 2.8 MWh Hamburg battery (Germany), and a 1.9 MWh battery in the EUREF campus (Germany) [43]. These projects were developed by car manufacturers (Nissan, BMW, and Audi, respectively) in cooperation with other companies and/or researchers. The batteries are designed to capture locally generated energy from photovoltaics (Johan Cruyff, EUREF), provide at least one hour of energy to the building (Johan Cruyff, EUREF), provide energy to the grid (Johan Cruyff, EUREF, Hamburg), and act as large-scale laboratories (EUREF, Hamburg). The Johan Cruyff Arena project had trouble sourcing enough EoL Nissan Leaf batteries, so only 40% of the 148 batteries are second-life with the remainder being new batteries that had greater capacity than the second-life ones. In addition, the second-life batteries were charging at a slower rate [21, 39], and to account for this, the control software had to be upgraded. The EUREF battery was designed to test the interaction between electric vehicles and the grid when using electric vehicles as energy storage [43].

Other EoL projects are the zero-emission islands being developed on Koshiki Island, Japan [44, 46] and Porto Santo, Portugal [40, 41], which are being funded by Nissan and Renault, respectively. Both islands have been equipped with a fleet of EVs, EV charging stations, and a 0.8 MWh battery (Koshiki island). The hope is that both islands can eventually be zero-emission islands.

EV manufacturers have been setting up subsidiaries or pairing with energy storage companies to manage battery EoL, such as Nissan and Sumitomo Corp creating the 4R Energy Corporation. In 2018, 4R opened a battery recycling factory in Namie, Japan that is capable of sorting EoL batteries for recycling or second-life [42, 47, 131]. At the factory, EoL batteries are graded to determine what second-life application they could perform. Second-life roles include new EV batteries for the best performing batteries, industrial equipment, solar/grid storage and finally repurposing for the batteries with lower performances [42].

Reassembled batteries have received less attention, with smaller companies leading the field [132–137]. The focus here has been energy storage for households, commercial buildings, and factories. The technologies these companies have been developing are mass SoH testing and advancements in BMSs [136].

12.2.7 Outlook

Second-life is an excellent way of managing EoL LIBs due to the environmental and economic benefits, as well as their ability to supplement the recycling industry. It has the support of EV manufacturers, who have been investing in second-life projects which are already proving financially viable. EV manufacturers are the main drivers behind second-life because EVs are expected to be the dominant source of LIB waste and thus be the primary source for second-life batteries. The large-scale second life projects by EV manufacturers have mainly been using the "direct reuse" strategy for their second-life batteries with the whole battery pack being reused. However, smaller companies have begun manufacturing second-life batteries with a focus on reassembled batteries. Work is still needed to scale up the reassembly of batteries with advances in robotics, BMSs, and cooperation from battery manufacturers being critical. The primary focus should be on the automation of the disassembly process, which needs to be robust and flexible to disassemble multiple battery architectures, while battery manufactures should design EV batteries with EoL management in mind to ease disassembly. Outside of economic reasons, the lack of EoL EV batteries limits the growth of second-life but this will be solved as more EV batteries reach their EoL and more EVs are sold.

The field of second-life has progressed in the last 20 years, while the foundational ideas first proposed by Cready et al. [10], where EV batteries are used for stationary storage, remain intact. Since then, more realistic estimates of the number of EVs have been calculated and the recent developments for second-life revolve around automation, economic assessment, and SoH analysis but still, there is yet to be a clear winner in each of these areas.

References

1 Simonsen, L.O., Harbak, H., and Bennekou, P. (2012). Cobalt metabolism and toxicology – a brief update. *Science of the Total Environment* 432: 210–215. https://doi.org/10.1016/j.scitotenv.2012.06.009.

2 Zeeshan, M., Murugadas, A., Ghaskadbi, S. et al. (2017). Ecotoxicological assessment of cobalt using Hydra model: ROS, oxidative stress, DNA damage, cell cycle arrest, and apoptosis as mechanisms of toxicity. *Environmental Pollution* 224: 54–69. https://doi.org/10.1016/j.envpol.2016.12.042.

3 Campion, C.L., Li, W.T., and Lucht, B.L. (2005). Thermal decomposition of LiPF6-based electrolytes for lithium-ion batteries. *Journal of the Electrochemical Society* 152: A2327–A2334. https://doi.org/10.1149/1.2083267.

4 O'Farrell, K., Veit, R., and Vard, D. (2014). Study into Market Share and Stocks and Flows of Handheld Batteries in Australia, Melbourne.

5 Heelan, J., Gratz, E., Zheng, Z.F. et al. (2016). Current and prospective Li-ion battery recycling and recovery processes. *JOM* 68: 2632–2638. https://doi.org/10.1007/s11837-016-1994-y.

6 Gu, F., Guo, J.F., Yao, X. et al. (2017). An investigation of the current status of recycling spent lithium-ion batteries from consumer electronics in China. *Journal of Cleaner Production* 161: 765–780. https://doi.org/10.1016/j.jclepro.2017.05.181.

7 King, S., Boxall, N.J., and Bhatt, A.I. (2018). *Lithium Battery Recycling in Australia*; Report EP181926, 55. CSIRO.

8 Randell, P. (2016). *Waste Lithium-Ion Battery Projections. Lithium-Ion Forums: Recycling, Transport and Warehousing*, 18. Randell Enviromental Consulting.

9 Lewis, H. (2016). *Lithium-Ion Battery Consultation Report*, 21. Helen Lewis Research.

10 Cready, E., Lippert, J., Pihl, J. et al. (2003). Technical and economic feasibility of applying used EV batteries in stationary applications. In: *A study for the DOE Energy Storage Systems Program, SAND Report, SAND2002-4084* (ed. E. Cready, J. Lippert, J. Pihl, et al.), 1–130. Sandia National Laboratories.

11 Neubauer, J. and Pesaran, A. (2011). The ability of battery second use strategies to impact plug-in electric vehicle prices and serve utility energy storage applications. *Journal of Power Sources* 196: 10351–10358. https://doi.org/10.1016/j.jpowsour.2011.06.053.

12 Neubauer, J., Pesaran, A., Williams, B. et al. (2012). *Techno-Economic Analysis of PEV Battery Second Use: Repurposed-Battery Selling Price and Commercial and Industrial End-User Value*; 0148-7191. Golden, CO: National Renewable Energy Lab.(NREL).

13 Hunt, G. (1996). *USABC Electric Vehicle Battery Test Procedures Manual*. Washington, DC: *United States Department of Energy*.

14 Birkl, C.R., Frost, D.F., Bizeray, A.M. et al. (2014). Modular converter system for low-cost off-grid energy storage using second life Li-ion batteries. In: *Proceedings of the Fourth Ieee Global Humanitarian Technology Conference*, 192–199. IEEE.

15 Tong, S.J., Same, A., Kootstra, M.A., and Park, J.W. (2013). Off-grid photovoltaic vehicle charge using second life lithium batteries: An experimental and numerical investigation. *Applied Energy* 104: 740–750. https://doi.org/10.1016/j.apenergy.2012.11.046.

16 Assuncao, A., Moura, P.S., and de Almeida, A.T. (2016). Technical and economic assessment of the secondary use of repurposed electric vehicle batteries in the residential sector to support solar energy. *Applied Energy* 181: 120–131. https://doi.org/10.1016/j.apenergy.2016.08.056.

17 Melin, H.E. (2020). *The Lithium-Ion Battery Life Cycle Report 2021*, 89. London: Circular Energy Storage.

18 Database EV (2021) Useable battery capacity of full electric vehicles. https://ev-database.org/cheatsheet/useable-battery-capacity-electric-car (accessed on 14 October 2021).

19 Faessler, B. (2021). Stationary, second use battery energy storage systems and their applications: a research review. *Energies* 14: https://doi.org/10.3390/en14082335.

20 RECHARGE (2014). *Re-use and Second Use of Rechargeable Batteries*, vol. 2014, 35. RECHARGE aisbl.

21 Bobba, S., Podias, A., Di Persio, F. et al. (2018). *Sustainability Assessment of Second Life Application of Automotive Batteries (SASLAB)*; JRC112543, 140. Luxembourg: Joint Research Centre & The European Commission's Science and Knowledge Service.

22 Casals, L.C., Garcia, B.A., and Cremades, L.V. (2017). Electric vehicle battery reuse: preparing for a second life. *Journal of Industrial Engineering and Management* 10: 266–285. https://doi.org/10.3926/jiem.2009.

23 Reinhardt, R., Garcia, B.A., Casals, L.C., and Domingo, S.G. Critical evaluation of European Union legislation on the second use of degraded traction batteries. In: *Proceedings of 13th International Conference on the European Energy Market (EEM)*, 1–5. Porto, Portugal: IEEE.

24 Canals Casals, L. and Amante García, B. (2016). Assessing electric vehicles battery second life remanufacture and management. *Journal of Green Engineering* 6: 77–98. https://doi.org/10.13052/jge1904-4720.614.

25 Casals, L.C., Garcia, B.A., and Benitez, M.M.G. (2016; https://doi.org/10.1007/978-3-319-26459-2_10pp). A cost analysis of electric vehicle batteries second life businesses. In: *2014 Project Management and Engineering Research*, 129–141. SPAIN: Alcaniz.

26 Reinhardt, R., Christodoulou, I., Gasso-Domingo, S., and Garcia, B.A. (2019). Towards sustainable business models for electric vehicle battery second use: a critical review. *Journal of Environmental Management* 245: 432–446. https://doi.org/10.1016/j.jenvman.2019.05.095.

27 Gallardo-Lozano, J., Romero-Cadaval, E., Milanes-Montero, M.I., and Guerrero-Martinez, M.A. (2014). Battery equalization active methods. *Journal of Power Sources* 246: 934–949. https://doi.org/10.1016/j.jpowsour.2013.08.026.

28 Song, Z., Yang, X.-G., Yang, N. et al. (2021). A study of cell-to-cell variation of capacity in parallel-connected lithium-ion battery cells. *eTransportation* 7: 100091. https://doi.org/10.1016/j.etran.2020.100091.

29 Zivcak, J., Kadarova, J., Kocisova, M. et al. (2021). Economic analysis of potential secondary use of batteries from electric vehicles. *Applied Sciences* 11: 23. https://doi.org/10.3390/app11093834.

30 Abraham, D.P., Knuth, J.L., Dees, D.W. et al. (2007). Performance degradation of high-power lithium-ion cells - electrochemistry of harvested electrodes. *Journal of Power Sources* 170: 465–475. https://doi.org/10.1016/j.jpowsour.2007.03.071.

31 Liu, W.W., Zhang, H., Liu, L.H. et al. (2016). Remanufacturing cathode from end-of-life of lithium-ion secondary batteries by Nd:YAG laser radiation. *Clean Technologies and Environmental Policy* 18: 231–243. https://doi.org/10.1007/s10098-015-1010-1.

32 Ramoni, M.O., Zhang, Y., Zhang, H.C., and Ghebrab, T. (2017). Laser ablation of electrodes for Li-ion battery remanufacturing. *International Journal of Advanced Manufacturing Technology* 88: 3067–3076. https://doi.org/10.1007/s00170-016-8986-5.

33 Zhang, Y., Zhang, Y., Liu, Z.C. et al. (2017). Laser ablation on lithium-ion battery electrode solid electrolyte interface removal. *Journal of Laser Applications* 29: https://doi.org/10.2351/1.5003335.

34 Broussely, M., Herreyre, S., Biensan, P. et al. (2001). Aging mechanism in Li ion cells and calendar life predictions. *Journal of Power Sources* 97, 98: 13–21. https://doi.org/10.1016/s0378-7753(01)00722-4.

35 Wang, X., Gaustad, G., Babbitt, C.W., and Richa, K. (2014). Economies of scale for future lithium-ion battery recycling infrastructure. *Resources, Conservation and Recycling* 83: 53–62. https://doi.org/10.1016/j.resconrec.2013.11.009.

36 Leclerc, S.H. and Badami, M.G. (2020). Extended producer responsibility for E-waste management: policy drivers and challenges. *Journal of Cleaner Production* 251: 10. https://doi.org/10.1016/j.jclepro.2019.119657.

37 Règlement sur la récupération et la valorisation de produits par les entreprises Loi sur la qualité de l'environnement. Ministère de l'Environnement, de la Lutte contre les changements climatiques, de la Faune et des Parcs. Québec, B., Ed. 2019; Vol. D.597-2011; N.I. 2019-12-01.

38 Gesetz über das Inverkehrbringen, die Rücknahme und die umweltverträgliche Entsorgung von Batterien und AkkumulatorenBatteriegesetz-2. Bundesministerium der Justiz. Federal Ministery for the Enviroment, N.C.A.N.S. 2020; p 18.

39 Warmerdam, J., van der Hoogt, J., and Kotter, R. (2020). *Final Report – Johan Cruijff ArenA Operational Pilot*, 36. Amsterdam: Amsterdam University of Applied Sciences.

40 Dent, S. Renault's 'smart island' runs on wind power and recycled batteries. https://www.engadget.com/2018-02-21-renault-ev-second-life-batteries-smart-island.html (accessed on 28 July 2021).

41 Hampel, C. Renault's Mobilize reveals business plans. https://www.electrive.com/2021/06/17/renaults-mobility-brand-mobilize-presents-business-plans/ (accessed on 28 July 2021).

42 Nissan. Nissan gives EV batteries a second life. https://global.nissanstories.com/en/releases/4r (accessed on 27 July 2021).

43 Still, B. Audi Opens Battery Storage Unit on Berlin EUREF Campus. https://www.audi-mediacenter.com/en/press-releases/audi-opens-battery-storage-unit-on-berlin-euref-campus-11681 (accessed on 21 September 2021).

44 Sumitomo Corporation. Spread and Expansion of Renewable Energy through Power Storage Business –Reused EV Battery Project on Koshiki Islands. https://www.sumitomocorp.com/en/africa/business/case/group/project_infra01 (accessed on 27 July 2021).

45 Zhou, Y., Wang, M., Hao, H. et al. (2015). Plug-in electric vehicle market penetration and incentives: a global review. *Mitigation and Adaptation Strategies for Global Change* 20: 777–795. https://doi.org/10.1007/s11027-014-9611-2.

46 Nissan. Koshikishima: EV 'Islands of the Future'. https://reports.nissan-global.com/EN/?p=18089 (accessed on 27 July 2021).

47 Nissan. Nissan, Sumitomo Corp. and 4R set up plant to recycle electric-car batteries. https://global.nissannews.com/en/releases/release-487297034c80023008bd9722aa069598-180326-01-e? (accessed on 27 July 2021).

48 Parkinson, G. Tesla big battery in South Australia delivers stunning windfall profits. https://reneweconomy.com.au/tesla-big-battery-in-south-australia-delivers-stunning-windfall-profits-77644/ (accessed on 19 August 2021).

49 Fox, E. Tesla Big Battery in South Australia Earns Huge Profits, Pays for Itself in 2.5 Years. https://www.tesmanian.com/blogs/tesmanian-blog/tesla-big-battery-in-south-australia-gets-colossal-profits (accessed on 19 August 2021).

50 Parkinson, G. Big battery in Australia proves profitable as Neoen recovers capital costs in just two years. https://ieefa.org/big-battery-in-australia-proves-profitable-as-neoen-recovers-capital-costs-in-just-two-years/ (accessed on 19 August 2021).

51 Trusted Solar. Are Solar Batteries Worth It In 2021? https://www.trustedsolar.com.au/post/are-solar-batteries-worth-it-in-2021/ (accessed on 9 September 2021).

52 Li, Z. and Ouyang, M.G. (2011). A win-win marginal rent analysis for operator and consumer under battery leasing mode in China electric vehicle market. *Energy Policy* 39: 3222–3237. https://doi.org/10.1016/j.enpol.2011.03.014.

53 Zhang, X.P. and Rao, R. (2016). A benefit analysis of electric vehicle battery swapping and leasing modes in China. *Emerging Markets Finance and Trade* 52: 1414–1426. https://doi.org/10.1080/1540496x.2016.1152798.

54 Vorrath, S. Battery recycling plant starts "shredding" in Germany, using Australian technology. https://reneweconomy.com.au/battery-recycling-plant-starts-shredding-in-germany-using-australian-technology/ (accessed on 10 September 2021).

55 Mining.Com. German battery recycling project awarded $1 million in public funding. https://www.mining.com/german-battery-recycling-project-awarded-1-million-in-public-funding/.

56 Eurostat. Generation of waste by waste category, hazardousness and NACE Rev. 2 activity. https://appsso.eurostat.ec.europa.eu/nui/submitViewTableAction.do (accessed on 27 October 2021).

57 Eshetu, G.G., Bertrand, J.P., Lecocq, A. et al. (2014). Fire behavior of carbonates-based electrolytes used in Li-ion rechargeable batteries with a focus on the role of the LiPF6 and LiFSI salts. *Journal of Power Sources* 269: 804–811. https://doi.org/10.1016/j.jpowsour.2014.07.065.

58 Wu, M.S., Chiang, P.C.J., Lin, J.C., and Jan, Y.S. (2004). Correlation between electrochemical characteristics and thermal stability of advanced lithium-ion batteries in abuse tests - short-circuit tests. *Electrochimica Acta* 49: 1803–1812. https://doi.org/10.1016/j.electacta.2003.12.012.

59 Larsson, F., Bertilsson, S., Furlani, M. et al. (2018). Gas explosions and thermal runaways during external heating abuse of commercial lithium-ion graphite-$LiCoO_2$ cells at different levels of ageing. *Journal of Power Sources* 373: 220–231. https://doi.org/10.1016/j.jpowsour.2017.10.083.

60 Fleischhammer, M., Waldmann, T., Bisle, G. et al. (2015). Interaction of cyclic ageing at high-rate and low temperatures and safety in lithium-ion batteries. *Journal of Power Sources* 274: 432–439. https://doi.org/10.1016/j.jpowsour.2014.08.135.

61 Fattal, A., Kelly, S., Liu, A., and Giurco, D. (2016). *Waste Fires in Australia: Cause for Concern?* 33. Sydney: University of Technology Sydney Institute for Sustainable Futures.

62 U.S.E.P. Agency (2021). *An Analysis of Lithium-ion Battery Fires in Waste Management and Recycling*, vol. 2021, 79. USA: United States Enviromental Protection Agency.

63 Agubra, V. and Fergus, J. (2013). Lithium ion battery anode aging mechanisms. *Materials* 6: 1310–1325. https://doi.org/10.3390/ma6041310.

64 Peled, E. (1979). The electrochemical-behavior of alkali and alkaline-earth metals in non-aqueous battery systems - the solid electrolyte interphase model. *Journal of the Electrochemical Society* 2047-2051: 126. https://doi.org/10.1149/1.2128859.

65 Lu, M., Cheng, H., and Yang, Y. (2008). A comparison of solid electrolyte interphase (SEI) on the artificial graphite anode of the aged and cycled commercial lithium ion cells. *Electrochimica Acta* 53: 3539–3546. https://doi.org/10.1016/j.electacta.2007.09.062.

66 Smart, M.C. and Ratnakumar, B.V. (2011). Effects of electrolyte composition on lithium plating in lithium-ion cells. *Journal of the Electrochemical Society* 158: A379–A389. https://doi.org/10.1149/1.3544439.

67 Legrand, N., Knosp, B., Desprez, P. et al. (2014). Physical characterization of the charging process of a Li-ion battery and prediction of Li plating by electrochemical modelling. *Journal of Power Sources* 245: 208–216. https://doi.org/10.1016/j.jpowsour.2013.06.130.

68 Petzl, M. and Danzer, M.A. (2014). Nondestructive detection, characterization, and quantification of lithium plating in commercial lithium-ion batteries. *Journal of Power Sources* 254: 80–87. https://doi.org/10.1016/j.jpowsour.2013.12.060.

69 Kim, J.S., Yoon, W.Y., Yi, K.Y. et al. (2007). The dissolution and deposition behavior in lithium powder electrode. *Journal of Power Sources* 165: 620–624. https://doi.org/10.1016/j.jpowsour.2006.10.033.

70 Brissot, C., Rosso, M., Chazalviel, J.N. et al. (1998). In situ study of dendritic growth in lithium/PEO-salt/lithium cells. *Electrochimica Acta* 43: 1569–1574. https://doi.org/10.1016/s0013-4686(97)10055-x.

71 Raj, R. and Wolfenstine, J. (2017). Current limit diagrams for dendrite formation in solid-state electrolytes for Li-ion batteries. *Journal of Power Sources* 343: 119–126. https://doi.org/10.1016/j.jpowsour.2017.01.037.

72 Mickelson, L., Castro, H., Switzer, E., and Friesen, C. (2014). Bulk stress evolution during intercalation of lithium in graphite. *Journal of the Electrochemical Society* 161: A2121–A2127. https://doi.org/10.1149/2.0591414jes.

73 Renganathan, S., Sikha, G., Santhanagopalan, S., and White, R.E. (2010). Theoretical analysis of stresses in a lithium ion cell. *Journal of the Electrochemical Society* 157: A155–A163. https://doi.org/10.1149/1.3261809.

74 Sethuraman, V.A., Van Winkle, N., Abraham, D.P. et al. (2012). Real-time stress measurements in lithium-ion battery negative-electrodes. *Journal of Power Sources* 206: 334–342. https://doi.org/10.1016/j.jpowsour.2012.01.036.

75 Zhang, X.C., Sastry, A.M., and Shyy, W. (2008). Intercalation-induced stress and heat generation within single lithium-ion battery cathode particles. *Journal of the Electrochemical Society* 155: A542–A552. https://doi.org/10.1149/1.2926617.

76 Yang, N.X., Zhang, X.W., Shang, B.B., and Li, G.J. (2016). Unbalanced discharging and aging due to temperature differences among the cells in a lithium-ion battery pack with parallel combination. *Journal of Power Sources* 306: 733–741. https://doi.org/10.1016/j.jpowsour.2015.12.079.

77 Amine, K., Liu, J., and Belharouak, I. (2005). High-temperature storage and cycling of C-$LiFePO_4$/graphite Li-ion cells. *Electrochemistry Communications* 7: 669–673. https://doi.org/10.1016/j.elecom.2005.04.018.

78 Seong, I.W., Hong, C.H., Kim, B.K., and Yoon, W.Y. (2008). The effects of current density and amount of discharge on dendrite formation in the lithium powder anode electrode. *Journal of Power Sources* 178: 769–773. https://doi.org/10.1016/j.jpowsour.2007.12.062.

79 Wang, Q.S., Wang, Z.P., Zhang, L. et al. (2021). A novel consistency evaluation method for series-connected battery systems based on real-world operation data. *IEEE Transactions on Transportation Electrification* 7: 437–451. https://doi.org/10.1109/tte.2020.3018143.

80 Galeotti, M., Cina, L., Giammanco, C. et al. (2015). Performance analysis and SOH (state of health) evaluation of lithium polymer batteries through electrochemical impedance spectroscopy. *Energy* 89: 678–686. https://doi.org/10.1016/j.energy.2015.05.148.

81 Jiang, Y., Jiang, J.C., Zhang, C.P. et al. (2018). State of health estimation of second-life $LiFePO_4$ batteries for energy storage applications. *Journal of Cleaner Production* 205: 754–762. https://doi.org/10.1016/j.jclepro.2018.09.149.

82 Li, Y., Abdel-Monema, M., Gopalakrishnan, R. et al. (2018). A quick on-line state of health estimation method for Li-ion battery with incremental capacity curves processed by Gaussian filter (vol 373, p. 40, 2018). *Journal of Power Sources* 393: 230–230. https://doi.org/10.1016/j.jpowsour.2018.05.035.

83 Salinas, F., Krüger, L., Neupert, S., and Kowal, J. (2019). A second life for Li-ion cells rescued from notebook batteries. *Journal of Energy Storage* 24: 100747.

84 Monhof, M.; Beverungen, D.; Klor, B.; Brauer, S. Extending battery management systems for making informed decisions on battery reuse. In *New Horizons in Design Science:*

Broadening the Research Agenda, Donnellan, B., Helfert, M., Kenneally, J., VanderMeer, D., Rothenberger, M., Winter, R., Eds. Springer-Verlag Berlin, 2015; Vol. 9073, pp. 447–454.

85 Spotnitz, R. (2003). Simulation of capacity fade in lithium-ion batteries. *Journal of Power Sources* 113: 72–80. https://doi.org/10.1016/s0378-7753(02)00490-1.

86 Sathre, R., Scown, C.D., Kavvada, O., and Hendrickson, T.P. (2015). Energy and climate effects of second-life use of electric vehicle batteries in California through 2050. *Journal of Power Sources* 288: 82–91. https://doi.org/10.1016/j.jpowsour.2015.04.097.

87 Pinson, M.B. and Bazant, M.Z. (2013). Theory of SEI formation in rechargeable batteries: capacity fade, accelerated aging and lifetime prediction. *Journal of the Electrochemical Society* 160: A243–A250. https://doi.org/10.1149/2.044302jes.

88 Ruiz, V., Kriston, A., Adanouj, I. et al. (2017). Degradation studies on lithium iron phosphate - graphite cells. The effect of dissimilar charging - discharging temperatures. *Electrochimica Acta* 240: 495–505. https://doi.org/10.1016/j.electacta.2017.03.126.

89 Bloom, I., Cole, B.W., Sohn, J.J. et al. (2001). An accelerated calendar and cycle life study of Li-ion cells. *Journal of Power Sources* 101: 238–247. https://doi.org/10.1016/s0378-7753(01)00783-2.

90 An, S.J., Li, J., Du, Z. et al. (2017). Fast formation cycling for lithium ion batteries. *Journal of Power Sources* 342: 846–852. https://doi.org/10.1016/j.jpowsour.2017.01.011.

91 An, S.J., Li, J.L., Daniel, C. et al. (2016). The state of understanding of the lithium-ion-battery graphite solid electrolyte interphase (SEI) and its relationship to formation cycling. *Carbon* 105: 52–76. https://doi.org/10.1016/j.carbon.2016.04.008.

92 Wang, F.M., Wang, H.Y., Yu, M.H. et al. (2011). Differential pulse effects of solid electrolyte interface formation for improving performance on high-power lithium ion battery. *Journal of Power Sources* 196: 10395–10400. https://doi.org/10.1016/j.jpowsour.2011.08.045.

93 Han, X.B., Ouyang, M.G., Lu, L.G., and Li, J.Q. (2014). A comparative study of commercial lithium ion battery cycle life in electric vehicle: capacity loss estimation. *Journal of Power Sources* 268: 658–669. https://doi.org/10.1016/j.jpowsour.2014.06.111.

94 Smith, A.J., Dahn, H.M., Burns, J.C., and Dahn, J.R. (2012). Long-term low-rate cycling of $LiCoO_2$/graphite Li-ion cells at 55 °C. *Journal of the Electrochemical Society* 159: A705–A710. https://doi.org/10.1149/2.056206jes.

95 Liu, D.T., Song, Y.C., Li, L. et al. (2018). On-line life cycle health assessment for lithium-ion battery in electric vehicles. *Journal of Cleaner Production* 199: 1050–1065. https://doi.org/10.1016/j.jclepro.2018.06.182.

96 Swierczynski, M., Stroe, D.I., and Kaer, S.K. (2017). Calendar ageing of $LiFePO_4$/C batteries in the second life applications. In: *2017 19th European Conference on Power Electronics and Applications*, 1–8. New York: IEEE.

97 Genieser, R., Ferrari, S., Loveridge, M. et al. (2018). Lithium ion batteries (NMC/graphite) cycling at 80 °C: different electrolytes and related degradation mechanism. *Journal of Power Sources* 373: 172–183. https://doi.org/10.1016/j.jpowsour.2017.11.014.

98 Martinez-Laserna, E., Sarasketa-Zabala, E., Stroe, D.I. et al. (2016). Evaluation of lithium-ion battery second life performance and degradation. In: *2016 IEEE Energy Conversion Congress and Exposition*, 1–7. New York: IEEE.

99 Strickland, D., Chittock, L., Stone, D.A. et al. (2014). Estimation of transportation battery second life for use in electricity grid systems. *IEEE Transactions on Sustainable Energy* 5: 795–803. https://doi.org/10.1109/tste.2014.2303572.

100 Saxena, S., Le Floch, C., MacDonald, J., and Moura, S. (2015). Quantifying EV battery end-of-life through analysis of travel needs with vehicle powertrain models. *Journal of Power Sources* 282: 265–276. https://doi.org/10.1016/j.jpowsour.2015.01.072.

101 Doerffel, D. and Abu Sharkh, S. (2006). A critical review of using the Peukert equation for determining the remaining capacity of lead-acid and lithium-ion batteries. *Journal of Power Sources* 155: 395–400. https://doi.org/10.1016/j.jpowsour.2005.04.030.

102 Xie, J.L., Ma, J.C., and Chen, J. (2018). Peukert-equation-based state-of-charge estimation for $LiFePO_4$ batteries considering the battery thermal evolution effect. *Energies* 11: https://doi.org/10.3390/en11051112.

103 Rahn, C.D. and Wang, C.-Y. (2013). *Battery Systems Engineering*. Wiley.

104 Bard, A.J.A. (2001). *Electrochemical Methods: Fundamentals and Applications*, 2e. New York: Wiley.

105 Andre, D., Meiler, M., Steiner, K. et al. (2011). Characterization of high-power lithium-ion batteries by electrochemical impedance spectroscopy. II Modelling. *Journal of Power Sources* 196: 5349–5356. https://doi.org/10.1016/j.jpowsour.2010.07.071.

106 Andre, D., Meiler, M., Steiner, K. et al. (2011). Characterization of high-power lithium-ion batteries by electrochemical impedance spectroscopy. I. Experimental investigation. *Journal of Power Sources* 196: 5334–5341. https://doi.org/10.1016/j.jpowsour.2010.12.102.

107 Ecker, M., Gerschler, J.B., Vogel, J. et al. (2012). Development of a lifetime prediction model for lithium-ion batteries based on extended accelerated aging test data. *Journal of Power Sources* 215: 248–257. https://doi.org/10.1016/j.jpowsour.2012.05.012.

108 Stroe, D.I., Swierczynski, M., Stroe, A.I. et al. (2017). Lithium-ion battery power degradation modelling by electrochemical impedance spectroscopy. *IET Renewable Power Generation* 11: 1136–1141. https://doi.org/10.1049/iet-rpg.2016.0958.

109 Jalkanen, K., Karppinen, J., Skogstrom, L. et al. (2015). Cycle aging of commercial NMC/graphite pouch cells at different temperatures. *Applied Energy* 154: 160–172. https://doi.org/10.1016/j.apenergy.2015.04.110.

110 Stroe, D.I., Swierczynski, M., Kaer, S.K., and Teodorescu, R. (2018). Degradation behavior of lithium-ion batteries during calendar ageing-the case of the internal resistance increase. *IEEE Transactions on Industry Applications* 54: 517–525. https://doi.org/10.1109/tia.2017.2756026.

111 Itagaki, M., Honda, K., Hoshi, Y., and Shitanda, I. (2015). In-situ EIS to determine impedance spectra of lithium-ion rechargeable batteries during charge and discharge cycle. *Journal of Electroanalytical Chemistry* 737: 78–84. https://doi.org/10.1016/j.jelechem.2014.06.004.

112 Schmitt, J., Maheshwari, A., Heck, M. et al. (2017). Impedance change and capacity fade of lithium nickel manganese cobalt oxide-based batteries during calendar aging. *Journal of Power Sources* 353: 183–194. https://doi.org/10.1016/j.jpowsour.2017.03.090.

113 Pastor-Fernandez, C., Uddin, K., Chouchelamane, G.H. et al. (2017). A comparison between electrochemical impedance spectroscopy and incremental capacity-differential voltage as Li-ion diagnostic techniques to identify and quantify the effects of degradation modes within battery management systems. *Journal of Power Sources* 360: 301–318. https://doi.org/10.1016/j.jpowsour.2017.03.042.

114 Weng, C.H., Feng, X.N., Sun, J., and Peng, H. (2016). State-of-health monitoring of lithium-ion battery modules and packs via incremental capacity peak tracking. *Applied Energy* 180: 360–368. https://doi.org/10.1016/j.apenergy.2016.07.126.

115 Weng, C.H., Cui, Y.J., Sun, J., and Peng, H. (2013). On-board state of health monitoring of lithium-ion batteries using incremental capacity analysis with support vector regression. *Journal of Power Sources* 235: 36–44. https://doi.org/10.1016/j.jpowsour.2013.02.012.

116 Feng, X.N., Weng, C.H., He, X.M. et al. (2018). Incremental capacity analysis on commercial lithium-ion batteries using support vector regression: a parametric study. *Energies* 11: https://doi.org/10.3390/en11092323.

117 Li, Y., Abdel-Monem, M., Gopalakrishnan, R. et al. (2018). A quick on-line state of health estimation method for Li-ion battery with incremental capacity curves processed by Gaussian filter. *Journal of Power Sources* 373: 40–53. https://doi.org/10.1016/j.jpowsour.2017.10.092.

118 Agudelo, B.O., Zamboni, W., and Monmasson, E. (2021). Application domain extension of incremental capacity-based battery SoH indicators. *Energy* 234: https://doi.org/10.1016/j.energy.2021.121224.

119 Gloser-Chahoud, S., Huster, S., Rosenberg, S. et al. (2021). Industrial disassembling as a key enabler of circular economy solutions for obsolete electric vehicle battery systems. *Resources, Conservation and Recycling* 174: https://doi.org/10.1016/j.resconrec.2021.105735.

120 Arents, J., Abolins, V., Judvaitis, J. et al. (2021). Human-robot collaboration trends and safety aspects: a systematic review. *Journal of Sensor and Actuator Networks* 10: https://doi.org/10.3390/jsan10030048.

121 Wegener, K., Chen, W.H., Dietrich, F. et al. (2015). Robot assisted disassembly for the recycling of electric vehicle batteries. In: *Proceedings of 22nd CIRP Conference on Life Cycle Engineering (LCE)*, 716–721. Sydney, Australia: Univ New S Wales.

122 Zu, C.X. and Li, H. (2011). Thermodynamic analysis on energy densities of batteries. *Energy & Environmental Science* 4: 2614–2624. https://doi.org/10.1039/c0ee00777c.

123 Bloomberg NEF (2019). *Bloomberg NEF Battery price survey 2019.*

124 Rohr, S., Wagner, S., Baumann, M., and Mullier, S. A techno-economic analysis of end of life value chains for lithium-ion batteries from electric vehicles. In: *Proceedings of 12th International Conference on Ecological Vehicles and Renewable Energies (EVER)*, 11–13. Monaco: IEEE.

125 Quirijnen, L. (1999). How to implement efficient local lead-acid battery recycling. *Journal of Power Sources* 78: 267–269. https://doi.org/10.1016/s0378-7753(99)00032-4.

126 Ballantyne, A.D., Hallett, J.P., Riley, D.J. et al. (2018). Lead acid battery recycling for the twenty-first century. *Royal Society Open Science* 5: 12. https://doi.org/10.1098/rsos.171368.

127 Schwarz, T.E., Rubenbauer, W., Rutrecht, B., and Pomberger, R. (2002). Forecasting real disassembly time of industrial batteries based on virtual MTM-UAS data. In: *Proceedings of 25th CIRP Life Cycle Engineering (LCE) Conference*, 927–931. Copenhagen, Denmark: IEEE.

128 BBC. Samsung confirms battery faults as cause of Note 7 fires. https://www.bbc.com/news/business-38714461 (accessed on 6 March 2020).

129 Loveridge, M.J., Remy, G., Kourra, N. et al. (2018). Looking Deeper into the Galaxy (Note 7). *Batteries (Basel)* 4: https://doi.org/10.3390/batteries4010003.

130 Reichmuth, D. and Goldman, J. (2017). Accelerating US leadership in electric vehicles. *Union of concerned Scientists* 1–5.

131 CEFC. 4R Energy and Relectrify collaborate in second-life battery storage products. https://www.cefc.com.au/media/media-release/4r-energy-and-relectrify-collaborate-in-second-life-battery-storage-products/ (accessed on 27 July 2021).

132 Sánchez Molina, P. Second-life batteries turned into storage system with capacity of up to 1 MWh. https://www.pv-magazine.com/2021/05/18/second-life-batteries-turned-into-storage-system-with-capacity-of-up-to-1-mwh/ (accessed on 13 October 2021).

133 BeeBattery. BeeBattery Home. https://beeplanetfactory.com/wp-content/uploads/2019/12/190727-Datasheet-BeeBatteryHome-EN.pdf.

134 Relectrify. 4R Energy and Relectrify collaborate in second-life battery storage products. https://www.relectrify.com/newsblog/4r-energy-and-relectrify-collaborate-in-second-life-battery-storage-products (accessed on 27 July 2021).

135 Vorrath, S. Australian made second-life EV battery system to be tested on New Zealand grid. https://reneweconomy.com.au/australian-made-second-life-ev-battery-system-to-be-tested-on-new-zealand-grid-58371/ (accessed on 27 July 2021).

136 Australian Renewable Energy Agency. New life for electric vehicle batteries. https://arena.gov.au/news/new-life-for-electric-vehicle-batteries/ (accessed on 27 July 2021).

137 Guthrie, S. Melbourne-based company to launch hybrid battery recycling program. https://www.drive.com.au/news/melbourne-based-company-to-launch-hybrid-battery-recycling-program/ (accessed on 23 August 2021).

13

Recycling Procedures for Energy Storage Devices in the Scope of the Electric Vehicle Implementation

Carlos Miguel Costa[1,2,3]*, Yifeng Wang*[4,5]*, Eider Goikolea*[5]*, Qi Zhang*[4,6]*, Hélder Castro*[7]*, Renato Gonçalves*[8]*, and Senentxu Lanceros-Méndez*[4,6]

[1] *Physics Centre of Minho and Porto Universities (CF-UM-UP), University of Minho, 4710-053, Braga, Portugal*
[2] *Laboratory of Physics for Materials and Emergent Technologies, LapMET, University of Minho, 4710-057, Braga, Portugal*
[3] *Institute of Science and Innovation for Bio-Sustainability (IB-S), University of Minho, 4710-057, Braga, Portugal*
[4] *BCMaterials, Basque Center for Materials, Applications and Nanostructures, UPV/EHU Science Park, 48940, Leioa, Spain*
[5] *Organic and Inorganic Chemistry Department, Faculty of Science and Technology, University of the Basque Country UPV/EHU, 48940, Leioa, Spain*
[6] *IKERBASQUE, Basque Foundation for Science, Bilbao, 48009, Spain*
[7] *Instituto de Soldadura e Qualidade, 4415-491, Grijó, Portugal*
[8] *Center of Chemistry, University of Minho, 4710-057, Braga, Portugal*

13.1 Introduction

The concern about energy and environmental issues is a growing and relevant subject in modern society. When it comes to climate change and air quality to reduce CO_2 emissions, decreasing the economic dependence on fossil fuels plays an important role [1].

This dependence is strongly related to the increasing mobility of the population, overall improvement in the quality of life, and industrial and electricity production processes, among others [2]. For an increasing number of applications, after the electricity is generated, it must be stored efficiently in energy storage devices where high energy density batteries stand out, more specifically lithium-ion batteries (LIBs) [3].

Their main characteristics are being cheaper, showing a higher number of charge/discharge cycles, higher energy and power density, and no memory effect, among others, when compared with related batteries, so they play an essential role in the present and future technologies as the most suitable system for effective energy storage systems for a wide range of applications, such as smartphones laptops, electric vehicles (EVs), or even home storage systems [4].

As a consequence of its massive use, a heavy reliance on this technology leads to increasing demand for lithium and other metals/materials needed to manufacture batteries. To reduce this dependence, recycling procedures are necessary to decrease the environmental impact [5, 6].

Sustainable Energy Storage in the Scope of Circular Economy: Advanced Materials and Device Design, First Edition. Edited by Carlos Miguel Costa, Renato Gonçalves, and Senentxu Lanceros-Méndez.

In this chapter, the current state of the art on LIB recycling processing methods is presented, being reported the latest developments and proposed processes.

Recycling methods for each battery component are presented, the main focus being on the cathode materials due to their high value and because it is difficult to obtain effective recycling of their components, making it the most widely studied battery component in the scope of recycling. This issue is extremely relevant in modern society in a transition to a more sustainable economic model based on circular economy paradigms and renewable energy sources, allowing to reduce the necessity of extract more resources and also reduce the landfill disposal of used devices.

One of the main applications of LIBs is as energy storage systems in the scope of electric mobility. Electrified vehicles that use LIBs are strongly increasing in nowadays public and private transportation. In this work, we will present the status of the battery electric vehicle (BEV) fleet, together with the current LIB demand to portray a snapshot of the status. Additionally, an outlook on BEVs is presented to forecast the future demand for LIBs in this sector. Finally, it is presented for the life-cycle assessment for LIBs.

13.2 Lithium-Ion Batteries: Environmental Impact and Sustainability

In the scope of LIBs, environmental impact and sustainability can be linked not only to the lithium and metals mining process but also to the use of sustainable materials and processes in their production. Typically, in the extraction of lithium, the main processes rely on brine and hard rock mining that depend on fossil fuels [7]. Whereas brine is similar to seawater and represents more than 60% of lithium resources, hard rock mining is basically present in spodumene and lepidolite [7, 8]. These processes need not only a large amount of energy (~1.01 GJ of electricity) but also large amounts of water (more than 240001 of water) to extract one ton of lithium [9]. Considering that the main lithium reserves are in brine, the main processes of this extraction are solar evaporation–precipitation process, ion sieves adsorption, membrane technology, electrochemical-battery approach, and solvent extraction [7]. These processes result in serious environmental problems, such as soil contamination, depletion of the water table, reduction of the amount of potable water, and a significant impact on the extraction zone [10, 11]. Furthermore, the brine process affects marine life due to the increase in salinity of seawater which contains toxic chemicals as a result of this process, the focus being in the development of sustainable processes to reduce the ecosystem impact [12].

Regarding hard rock mining, the cost of extracting lithium is higher when compared to brines, but its concentration is higher. Thus, it is necessary a technical-economic analysis prior to the extraction to demonstrate its feasibility. The environmental concerns are also related to the energy used for its extraction and the potential contamination of drinking water [13].

In addition to lithium extraction, another strategy to increase battery sustainability is the use of sustainable materials and/or manufacturing process for the different battery components: electrodes and separators/electrolytes.

Regarding electrode materials, the focus is directed toward the use of environmentally friendly synthesis of active materials, sustainable solvents, and biopolymers as polymer binders, among others.

For separators/electrolytes, the environmental issues are related to the use of natural and biopolymers, sustainable fillers such as ionic liquids and the development of solid electrolytes, among others.

Combining these issues and the increasing implementation of the EV, which will demand more than 2 million tons of batteries per year [14], strong efforts are needed in the process of recycling battery components in order to reduce battery disposal.

13.3 Lithium-Ion Batteries: Recycling Strategies and Processes

13.3.1 Electrode Recycling Approaches

In the waste management hierarchy, recycling ranks in third place after redesign and reuse, which rank first and second places, respectively [15]. Nevertheless, redesign demands great amount of innovation in already consolidated systems, making it more like a guiding scheme instead of a realistic solution for rapid transformations in specific sectors. With regard to reuse, it has already been a part of battery life. For example, LIBs which are retired from electrical cars can be sorted and applied in household appliances or even child toys [16]. When the reused batteries degrade and become invalid for any type of application, they should be recycled, which seems to be the best approach. This chapter discusses the prevalent recycling methods nowadays, in particular for cathode materials which are more profitable compared to anodes and separators. The different approaches are categorized into pyro-, hydro-, mechanical, and direct recycling, depending on the specific mechanism. Figure 13.1 illustrates the categorization of recycling methods.

13.3.1.1 Pyrometallurgical Methods

Pyrometallurgical methods rely on thermal treatments to recycle or recover spent LIBs. Pyrometallurgy includes chemical reactions under high temperature. It is an efficient and

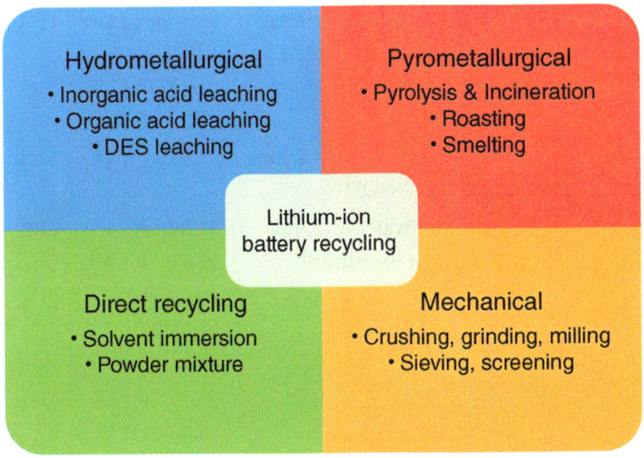

Figure 13.1 General categorization of recycling approaches for lithium-ion batteries where DES is the deep eutectic solvent.

relatively simple process that can handle a large number of batteries at one time. Furthermore, pyro-recycling methods can cope with batteries employing different cathodic materials, such as $LiCoO_2$ (LCO), $LiMnO_2$ (LMO), $LiFePO_4$ (LFP), and $LiNiMnCoO_2$ (NMC), which makes it popular for industrial processes. Thus, Umicore battery recycling and Sony & Sumitomo are the world's leading companies that utilize pyrometallurgical methods [17]. However, the disadvantages, such as the presence of toxic gases and huge energy consumption, cannot be ignored. Popular pyro-recycling methods are summarized in Table 13.1. More details about these methods will be explained in the following sections.

13.3.1.1.1 Thermal Pre-treatment

After being discharged and disassembled, spent LIBs are then pre-treated to liberate cathodic materials from current collectors by thermal treatments.

Incineration is a process that burns the organic compounds in spent batteries, such as carbon and binders, in an oxygen environment to detach active materials from collectors under a specific temperature. In this process, the carbon content works as a reducing reagent, contributing to the carbothermic reduction of the cathodic materials. As a consequence, Co, Mn, and Ni ions are reduced to a lower valence state [26]. Temperature control is critical for the incineration process as it can trigger the burning of organic compounds as well as influence the following extraction of cobalt and lithium [27]. Lombardo et al. [26] reported that 600 °C and 90 minutes were ideal processing conditions for decomposing these organic chemicals in LIBs. As a result, subsequent leaching and

Table 13.1 Summary of pyrometallurgical recycling methods.

Cathode	Additive	Processing conditions	Subsequent operation	Efficiency	Ref.
LCO	HNO_3 (75 °C, 5 h)	250 °C, 60 min, in air	Water leaching (25 °C, S/L 2 : 1)	93% for Li	[18]
LCO	NH_4Cl	350 °C, 20 min, in air	Water leaching (S/L 100 g l^{-1})	99% for Li and Co	[19]
LCO	Graphite, NaOH	520 °C, 180 min, in air	Water leaching	93% for Li	[20]
LCO	Al foil	600 °C, 60 min, in air	Alkaline leaching (NaOH)	93.67% for Li	[21]
LMO	Graphite	800 °C, 45 min, oxygen-free	Water leaching (20–30 min)	91.3% for Li 95.11% for Mn	[22]
NMC	Graphite	500 W (microwave power), 30 min, in nitrogen	Acid leaching (HCl)	97% for Ni, Co, Mn 99% for Li	[23]
LCO	Graphite	1000 °C, 30 min, in nitrogen	Wet magnetic separation	95.75% for Co 98.93% for Li	[24]
$LiCoNiO_2$	Copper slag	1450 °C, 30 min, in Argon	Grinding of alloy and slag	98.83% for Co 98.39% for Ni	[25]

liberation processes were easier due to the removal of hydrophobic carbon and organic compounds. If the temperature goes higher, the Al current collector would begin to melt and cover the surface of the cathodic material which would decrease the cobalt recovery rate [28, 29]. The drawback of incineration is that harmful gases and greenhouse gases are emitted during the process, which requires more investment for gas post-treatment.

With regard to pyrolysis, inert gases or vacuum are used during heating to evaporate electrolytes or decompose organic compounds into chemicals with lower molar weight. In the process of pyrolysis, the gas flow of Ar or N_2 pushes the gas generated to a condenser, where the electrolyte can be condensed and recycled [30]. Subsequently, the gases after condensation are tuned into an alkali solution to remove some harmful components and then collected in gas bottles for further usage [31]. Easy and efficient peeling-off of active materials from Al collectors is achieved after pyrolysis, which also contributes to the recycling of Al collectors as they are still in the metallic form [31]. In order to improve heating efficiency and ensure homogeneous heating, microwave pyrolysis has been investigated [29]. In [32], microwave-assisted pyrolysis was conducted with the addition of biomass from macadamia shells to explore a new possibility of recycling LIBs. It was reported [32] that the process can significantly decrease the decomposition temperature of cathodes to 300 °C, which saves energy. Furthermore, 99.2% of lithium can be recycled under optimum operation conditions [32]. However, Al heats up and melts more easily than other metals, thus, Al current collectors can melt and contaminate cathodic materials if the microwave heating is too powerful [33].

One of the main differences among the pre-treatment methods above mentioned is the participation of oxygen. While the incineration burns organic compounds with oxygen, the pyrolysis is conducted in an oxygen-free environment, whereas the former is exothermic, the latter is endothermic. In order to be greener, the steam from the incineration and the final product of pyrolysis has been used to compensate for the energy consumption, as there is a great amount of energy in them [29]. In addition, a comparison among incineration, dynamic pyrolysis under inert gases and vacuum pyrolysis has been carried out [34]. The results revealed that incineration is the worst choice from the gas emission perspective, as it releases much more harmful gases than the other two. While vacuum pyrolysis has proven to be the most environmentally friendly choice among these three, it consumes more energy. Incineration, on the contrary, requires less processing time and lower processing temperature than the other two.

13.3.1.1.1.1 Smelting Smelting takes advantage of heat to extract target metals with the help of some reductants to decompose the raw materials. Typically, smelting can be categorized into three groups: flash, matte, and reduction [35]. In most spent LIBs, metal oxides, such as LCO, LMO, and NMC, are the "ore" to be dealt with. Consequently, reduction smelting is the most popular method for recycling LIBs. In the industrialized smelting process, spent LIBs are first heated into a furnace at a relatively low temperature, which prevents explosion [36], to vaporize the electrolyte without being pre-treated [37], or passivated [38]. Afterward, the temperature increases to melt the metals as well as burn the organic compounds. The burning of the organics will release heat which will also contribute to the process, thus saving some energy [39]. Metal oxides are reduced and melted into liquid form which is easy to extract. Lithium and some other compounds are often found in slag [40]. Equations (13.1)–(13.3) describe a few examples of this reductive smelting. Al collector and carbon anodes (also act as reducing reagents) react with cathodes, such as LCO and $LiCoNiO_2$ [25]. Li_2O and alumina are transformed into slags that are difficult to recycle [41]. CaO or SiO_2 is often utilized as the flux to produce low melting slags, and it can eliminate contamination in the slag as well [41]:

$$2LiCoO_2 + 2Al \rightarrow Li_2O + 2Co + Al_2O_3 \tag{13.1}$$

$$4LiCoNiO_2 + 4Al \rightarrow 2Li_2O + 4Co + 4Ni + 2Al_2O_3 \tag{13.2}$$

$$2LiCoNiO_2 + 3C \rightarrow Li_2O + 2Co + 2Ni + 3CO(g) \tag{13.3}$$

High efficiency can be achieved by smelting when dealing with many batches of spent LIBs, as a pre-treatment is not mandatory which saves time and energy. However, a subsequent refining process is always needed for smelting to obtain the liquid metal in high purity. The refining process is often hydrometallurgical, and it will be discussed in the following section. Another drawback of smelting is the huge energy consumption, as the furnace has to reach the melting point of the target metals which can be higher than 1000 °C. This limits its applications for some unprofitable batteries, such as LFP. Besides, smelting is also not in favor of LMO batteries, due to the fact that Li and Mn are transformed into slag which is arduous to recover.

As traditional smelting slag systems, such as $CaO–SiO_2–Al_2O_3$ or $FeO–CaO–SiO_2–Al_2O_3$, have a low recycling efficiency of Li and Mn, investigations were focused to pursue a more economical smelting method [10, 25]. Two slag systems were investigated: $FeO–SiO_2–Al_2O_3$ [25] and $MnO–SiO_2–Al_2O_3$ [10]. In $FeO–SiO_2–Al_2O_3$ system, spent LIBs were first cut into pieces, and then copper slag was added as the slag former at the ratio of 1 : 4 (batter:copper slag). The cathodic materials together with the slag former were thermally treated at 1732 K for half an hour, and then naturally cooled down to room temperature. The results revealed that 93.57% copper, 98.39% nickel, and 98.83% cobalt could be recovered when the weight ratio of FeO and SiO_2 in the slag former is between 0.58 : 1 and 1.03 : 1, and Al_2O_3 is between 17.19 and 21.52 wt% of the slag former. In $MnO–SiO_2–Al_2O_3$ system, spent batteries were also first cut into small pieces, and then they were mixed with the slag former (pyrolusite) and slag modifier (CaO and SiO_2). The mixture was heated at 1475 °C for 30 minutes and naturally cooled down to room temperature. When the ratios of pyrolusite, SiO_2, and CaO vs. the slag former are 1.5, 6, and 2 wt%, respectively, the recovering efficiencies of cobalt, copper, and nickel reach 99.79%, 99.3%, and 99.3%, respectively. Subsequent leaching efficiencies of Li and Mn in the slag were reported to be 94.85% and 79.86% [10].

Traditional pyro-recycling methods face the problems of losing lithium in slag, huge energy consumption, and expensive capital investment. A two-staged investigation was done to optimize the smelting process with high recovery efficiency [42, 43], and this method can be fitted to existed furnaces to save money.

In the laboratory stage [42], LCO was mixed with graphite as well as calcium halides ($CaCl_2$ or CaF_2) and then, it was thermally treated in a furnace to 1550 °C at a rate of 2 °C min^{-1}. The temperature was held at 1550 °C for 30 minutes and samples were then cooled at a rate of 2 °C min^{-1} to room temperature. Argon was constantly supplied during the thermal process at a flow rate of 300 ml min^{-1}. Afterward, the powders obtained from the furnace were screened and it was observed that the particles larger than 125 μm were magnetic and contained cobalt. X-ray diffraction analysis (XRD) characterization results revealed that lithium could be reduced to lithium vapor and formed lithium halide as flue dusts containing particles smaller than 125 μm. Cobalt oxide could also be reduced to the pure metal form during the heating process. Almost all the Co, Ni, and Mn elements were reported to be recovered and no slag was formed. In addition, the Al collector can decrease the lithium recovery rate due to the formation of γ-$LiAlO_2$ slag. Consequently, pre-treatment of peeling-off cathodes from Al foils is preferred in this application.

In the pilot stage [43], spent NMC batteries were pre-treated with crushing and screening processes to pieces with different sizes. The obtained batteries were mixed with iron/steel scrap and slag formers as the starting materials, and then thermally treated in an electric arc furnace. During the process, more battery pieces and slag formers (lime and dolomite) were charged into the furnace from the top while some smaller battery pieces were injected through a lane into the furnace. The temperature was set at no less than 1500 °C to melt the metals, and the gas emission was post-combusted and filtered to collect flue dusts. The yields for cobalt, nickel, and manganese were reported to be >97%, >97%, and >85%, respectively. Besides, more than 60% of lithium can be recovered from the flue dusts and the purity of the Li_2CO_3 extracted from the flue dusts can reach 95.8% [43].

13.3.1.1.1.2 Roasting Unlike smelting, pre-treatment is necessary for the roasting process. The spent batteries are first discharged and disassembled, and the cathodic materials are peeled-off from the Al collector. Subsequently, the black mass is roasted in a furnace with some reagents, such as sodium hydrogen sulfate [35], carbon [14, 44], ammonium sulfate [45], and calcium chloride [46].

Roasting reactions for these reagents are shown in Equations (4)–(7):

$$4LiCoO_2 + 12NaHSO_4 \rightarrow 6Na_2SO_4 + 4CoSO_4 + 2Li_2SO_4 + 6H_2O(g) + O_2(g) \quad (13.4)$$

$$4LiCoO_2 + 3C \rightarrow 2Li_2CO_3 + 4Co + CO_2(g) \quad (13.5)$$

$$\frac{2}{3}LiCoO_2 + (NH_4O_4)_2SO_4 \rightarrow \frac{2}{3}CoSO_4 + \frac{1}{3}Li_2SO_4 + \frac{1}{9}N_2(g) + \frac{16}{9}NH_3(g) + \frac{4}{3}H_2O(g)O_4 \quad (13.6)$$

$$2LiAl(SiO_3)_2 + CaCl_2 \rightarrow 2LiCl + 2SiO_2 + CaAlSi_2O_8 \quad (13.7)$$

Unlike carbothermic reduction roasting which results in insoluble metal oxide products, sulfate roasting can produce soluble $LiSO_4$ or $CoSO_4$ with the addition of sodium hydrogen sulfate or ammonium sulfate, which is better for subsequent refining processes. Chlorination [19] and nitration [47] roastings are also available for dealing with spent LIB cathodes. Reducing reagents can break the bonds in cathodic materials and set Li and Co free from the oxygen octahedral microstructures in LCO [14], thus, boosting the kinetics of the roasting and decreasing the temperature needed for roasting, which saves energy [14, 45] compared with smelting process.

The work reported in [48] investigated metal phase changes during a carbothermic reduction roasting via changing the process temperature, time, and the amount of reducing carbon. The raw materials of this roasting are NMC cathodic materials and anode graphite from spent LIBs, graphite working as a reducing reagent. Ar environment is ensured during the roasting. After being heated at 600 °C for three hours, the obtained samples were cooled down to room temperature. The results revealed that CoO, NiO, and MnO were formed with the help of graphite (14 wt%) reduction, and these metal oxides can be easily leached and reused.

Selectivity is always a big concern in the recycling process of LIBs. The work in [49] investigated a selective roasting method and also studied the phase transition, the thermodynamic feasibility, and the subsequent leaching behavior throughout the process. Spent NMC batteries were first discharged and manually dismantled. Electrodes were then immersed in a solution to dissolve the binders, thus, separating cathodes and anodes, and obtaining homogeneous powder samples of each with different mass ratios. The roasting was conducted in a tube furnace with Ar flow, and the subsequent leaching efficiency of lithium increased by 75% as the processing temperature increased from 450 to 650 °C and kept stable with further temperature increase. 82.2% of Li, 97.7% of Ni, and 99.1% of Co can be leached selectively when the electrodes (5 wt% anode) were heated at 650 °C for an hour.

13.3.1.1.2 Hydrometallurgical Methods

Hydrometallurgical extraction could recycle valuable elements from LIB waste (WLIB) through chemical reactions. It consumes less energy and requires less capital investment than pyro-recycling methods, making it a reasonable choice for real industrial applications [50]. Generally, the procedure in hydrometallurgy involves two steps: the dissolution of cathode materials into a solvent and the extraction of dissolved ions via precipitation and filtration or electrodeposition. Figure 13.2 illustrates the most common procedures of a hydrometallurgical approach to recycle WLIBs [51]. Table 13.2 summarizes

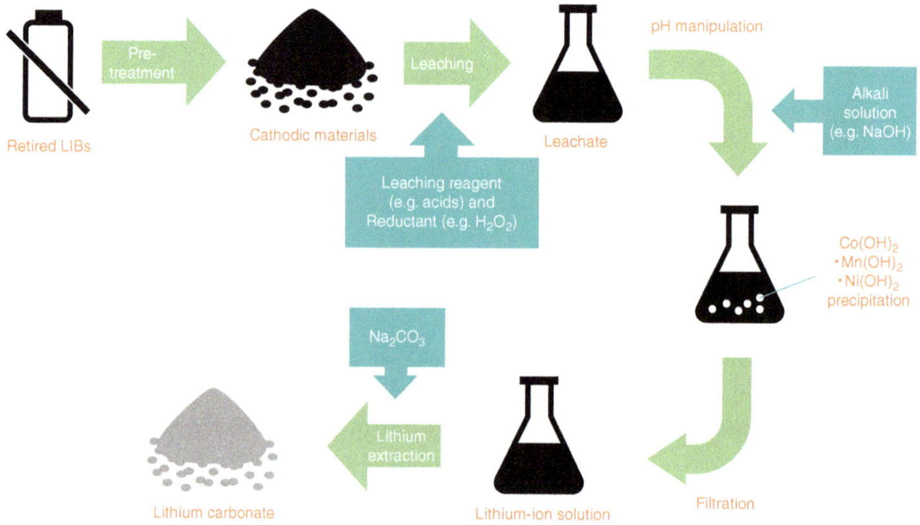

Figure 13.2 Flowchart of a typical hydrometallurgical method.

Table 13.2 Main processing parameters and leaching efficiency of hydrometallurgical methods for Li-ion battery cathodes.

Group	Raw material	Leaching reagent	Reductant	Temperature and time	Leaching efficiency (%)		References
					Li	Co	
Inorganic	LCO	2 M HCl	N/A	60–80 °C, 90 min	100	100	[52]
Inorganic	LCO/ LMO/ NMC	4 M HCl	N/A	80 °C, 60 min	100	100	[53]
Inorganic	LCO	1 M HNO_3	1.7 vol% H_2O_2	75 °C, 60 min	85	85	[54]
Inorganic	LCO	1 M HNO_3	1 vol% H_2O_2	80 °C, 60 min	100	100	[55]
Inorganic	LCO	2 M H_2SO_4	5 vol% H_2O_2	80 °C, 60 min	99	99	[31]
Inorganic	LCO	4 vol% H_2SO_4	1 vol% H_2O_2	40 °C, 60 min	100	97	[56]
Inorganic	Mixed LIBs	2 M H_2SO_4	4 vol% H_2O_2	70 °C, 120 min	98.8	99.6	[57]
Inorganic	LCO	6 vol% H_2SO_4	1 vol% H_2O_2	65 °C, 60 min	95	72	[58]
Inorganic	LCO	4 M H_2SO_4	10 vol% H_2O_2	65 °C, 120 min	96	95	[59]
Inorganic	LCO	2 M H_2SO_4	6 vol% H_2O_2	60 °C, 60 min	97	98	[60]
Inorganic	LCO	2 M H_2SO_4	15 vol% H_2O_2	75 °C, 10 min	100	100	[61]
Inorganic	LCO	2 M H_2SO_4	2 vol% H_2O_2	60 °C, 120 min	87.5	96.3	[62]
Inorganic	LCO	2 M H_2SO_4	5 vol% H_2O_2	75 °C, 30 min	94	93	[63]
Inorganic	LCO	3 M H_2SO_4	3 wt% H_2O_2	70 °C, 300 min	90	90	[64]
Inorganic	NMC	2 M H_2SO_4	10 vol% H_2O_2	70 °C, 90 min	94	91.6	[65]
Inorganic	NMC	3 M H_2SO_4	3 vol% H_2O_2	80 °C, 60 min	97.8	96.5	[66]
Inorganic	NMC	1.25 M H_2SO_4	0.1 M $Na_2S_2O_5$	60 °C, 60 min	>80	>80	[67]
Inorganic	LCO	2 M H_2SO_4	0.11 ascorbic acid	80 °C, 90 min	95.7	93.8	[68]
Inorganic	LCO	2 g×g^{-1} H_2SO_4	50%exc glucose	90 °C, 180 min	98	97	[69]
Inorganic	NMC	0.2 M H_3PO_4	0.4 citric acid	90 °C, 30 min	100	91.63	[70]
Inorganic	LCO	0.7 M H_3PO_4	4 vol% H_2O_2	40 °C, 60 min	99	99	[71]
Organic	LCO	1 M Oxalate	N/A	80 °C, 120 min	98	98	[72]
Organic	LCO	1.5 M DL-malic acid	3 vol% H_2O_2	80 °C, 25 min	98.13	98.86	[73]
Organic	LCO	1.5 M malic acid	0.6 g/g grape seed	80 °C, 180 min	99	92	[74]

(Continued)

Table 13.2 (Continued)

Group	Raw material	Leaching reagent	Reductant	Temperature and time	Leaching efficiency (%)		References
					Li	Co	
Organic	LCO	1.25 M malic acid	0.3 M glucose	80 °C, 180 min	100	99.87	[75]
Organic	LCO	1.25 M ascorbic acid	N/A	70 °C, 20 min	98.5	94.8	[76]
Organic	NMC	1.5 M citric acid	2 vol% H_2O_2	95 °C, 30 min	97	95	[77]
Organic	NMC	0.5 M citric acid	N/A	90 °C, 80 min	91	90	[78]
DES	LCO	Choline chloride: ethylene glycol (1:2)	N/A	220 °C, 24 h	99.3	99.3	[79]
DES	Ni-MH	Choline chloride: Urea (1:2)	N/A	95 °C, 24 h	N/A	53.3	[80]
DES	LCO	polyethylene glycol: thiourea (2:1)	N/A	160 °C, 24 h	N/A	60.2	[81]

various extraction methods used in hydrometallurgy, including inorganic and organic acids as well as deep eutectic solvents (DES). More details of the approaches are presented in the following sections.

13.3.1.1.2.1 Hydrometallurgy Pre-treatment Cathode materials in a battery are closely attached to Al collectors, typically by the use of polyvinylidene fluoride (PVDF) binder, and they must be liberated prior to recovery. Several solvents that could dissolve PVDF have been investigated so that the Al current collector can be reused after the removal of cathode materials. The dissolved cathodic materials are collected via filtration. The most used solvent is *N*-methyl-2-pyrrolidone (NMP) which achieves the best performance when compared with ethanol, *N,N*-dimethylacetamide (DMAC), *N,N*-dimethylformamide (DMF), and *N,N*-dimethyl sulfoxide, with high recoverability [82] and solubility of 200 g/1 kg of PVDF [83, 84]. To assist in the dissolution of cathodic materials in NMP, ultrasonication is often applied to shorten the dissolving time via mechanically diminishing the adhesive force [82, 85]. Moreover, scissoring the electrodes into small pieces can also help to increase the ultrasonic efficiency.

However, the use of NMP has several disadvantages, including toxicity and high cost, which impels to replace it with an environmentally friendlier and more affordable solvent. The work in [86] investigated the use of a DES for the dissolution of PVDF. The price of a DES is much lower than that of NMP, and the process to synthesis a DES is quite simple: just mixing the chemicals together at moderate heating for some time to acquire a clear solution. A DES is commonly composed of a hydrogen bond donor (HBD) and a hydrogen bond acceptor (HBA) in a specific stoichiometric ratio [87]. The most prevalent HBA is choline chloride, while the choices for HBD are quite numerous, frequently used examples are urea, ethylene glycol as well as bio-acids [88]. In the work of [86], choline chloride and glycerol were mixed at a ratio of 23 : 10 to form a DES, and then WLIB was immersed into the DES at 190 °C for 15 minutes. A separation with an efficiency of 99.86 wt% was accomplished, mainly due to an alkali degradation mechanism, without disrupting the crystal structure of NMC, thus, providing a cheaper, greener alternative to NMP.

Further, the work in [89] used a eutectic system of lithium salts to set the cathode materials free from the Al collector. The $LiNO_3$–LiOH system with a molar ratio of $LiNO_3$:LiOH of 2 : 3 can reach a liberation efficiency of 98.3% when the system is heated at 260 °C for 30 minutes. Lithium ions from the salts can compensate for the lithium loss during the lithiation and de-lithiation processes and this system can drive PVDF to decompose at lower temperatures. Although hydrofluoric acid (HF) was observed during the separation process, the LiOH can capture it and suppress its formation in large quantities.

Another approach to detach cathode materials from Al foils is to dissolve Al foils using NaOH solution and to burn the PVDF binder later at around 600 °C [90, 91]. Despite the fact that NaOH is inexpensive, the subsequent thermal treatment to remove carbonaceous components from cathodic materials consumes much energy.

13.3.1.1.2.2 Acid Leaching

Sulfuric Acid Leaching H_2SO_4 was used as the reagent to leach the LCO cathode material and then H_2O_2 was added to reduce the cobalt(III) ions [31]. It was reported that the leaching efficiency improved significantly with the addition of H_2O_2, and an optimal leaching efficiency of 99% for cobalt and lithium can be reached when the operation parameters are optimized. The reactions with and without H_2O_2 addition can be expressed by Equations (13.8) and (13.9), respectively [31]:

$$2LiCoO_2 + 3H_2SO_4 + H_2O_2 \rightarrow Li_2SO_4 + 2CoSO_4 + 4H_2O + O_2 \tag{13.8}$$

$$4LiCoO_2 + 6H_2SO_4 \rightarrow 2Li_2SO_4 + 4CoSO_4 + 6H_2O + O_2 \tag{13.9}$$

The high leaching efficiency was considered to be due to the fact that H_2O_2 reduces Co(III) in LCO to Co(II), which facilitates the leaching procedure. The best leaching conditions and processing parameters were found to be: 2 M H_2SO_4 with 5 vol% H_2O_2, a solid/liquid ratio of 50 g l^{-1}, together with heating at 60–80 °C for 60–90 minutes [51].

The reducing performance of other reductants, including $Na_2S_2O_5$, Na_2SO_3, H_2O_2 as well as iron scrap, has also been investigated [67]. Among them, $Na_2S_2O_5$ stands out as the optimum choice, as it allows to improve the dissolving efficiency of active materials and also prevents the dissolution of Cu, Al, Fe, etc., which makes the subsequent purification relatively easier. It has been also reported [67] that the solubility of cathodic metals largely increases when increasing H_2SO_4 acid concentration, temperature, stirring frequency, and liquid/solid ratio. However, the increase of H_2SO_4 concentration above 1.25 M no longer makes a significant impact on the dissolution of cathodic metals. The reaction can be described as:

$$4LiCoO_2 + 6H_2SO_4 + Na_2S_2O_5 \rightarrow 2Li_2SO_4 + 4CoSO_4 + 5H_2O + 2NaHSO_4 \tag{13.10}$$

Ascorbic acid (as the reductant) is found to be a weak solvent for copper in comparison with the solubility of Li and Co, which makes ascorbic acid a good choice upon selectively recycling metals, as reported in the work of [68]. The raw materials in [68] were WLIB scraps from real industry, containing 23.6% cobalt, 3.6% lithium, and 6.2% copper. It was found that the industrial scrap dissolved more efficiently in H_2SO_4 than the pure LCO. Moreover, the addition of ascorbic acid as reducing agent lead to a higher leaching efficiency for cobalt and lithium and a lower leaching efficiency for copper, demonstrating the selectivity of the leaching approach. Under optimized conditions (2 M H_2SO_4, 0.11 M ascorbic acid, S/L ratio 200 g l^{-1}, 80 °C, 90 minutes), 95.7% of lithium and 93.8% of cobalt can be extracted but only 0.7% of copper was dissolved, making it a satisfactory option for further treatment as a selective recycling method. The basic chemical reaction in this approach is the following:

$$2LiCoO_2 + C_6H_8O_6 + 6H^+ \rightarrow 2Li^+ + 2Co^{2+} + 4H_2O + C_6H_6O_6 \quad (13.11)$$

Another simplified approach is using a concentrated sulfuric acid of 18.4 M to liberate the NMC cathode material from the Al collector and to leach it simultaneously at 80 °C for one hour [92]. Because PVDF is only partially dissolved in the concentrated sulfuric acid and a passivation film on the surface of Al foils forms, the cathodic materials can be easily peeled off from Al foils that can be collected without any consumption. H_2O_2, 4.5 vol%, was then added to the sulfuric acid solution, after the removal of the Al foils, to leach all the NMC when heated to 70 °C for 70 minutes. This approach not only saves time and costs for the separation of Al collectors and leaching active materials, but it also consumes less amount of acid than traditional methods [92]. However, the generation and emission of HF during the decomposition of PVDF and the potential harm brought by the use of concentrated sulfuric acid itself should be taken into consideration for the broader applications of this approach.

Nitric Acid Leaching The use of nitric acid for dissolving cathodic materials is not particularly popular, mainly because the process generates and releases NO_x gases that are harmful to the environment. NO_x gases can be recycled for producing nitric acid through a close loop procedure, however, it requires a large investment and intensive energy to cope with.

Anyhow, the use of nitric acid for leaching cathodic materials was investigated [54, 55, 93]. It was found that leaching with nitric acid is a time-saving process in comparison with the use of sulfuric acid. It was reported that the addition of H_2O_2 as a reductant to nitric acid increases the leaching efficiency of lithium and cobalt by 10% and 45%, respectively, owing to the reduction of Co^{3+} to the soluble Co^{2+} [54, 92]. In the work of [54], the raw cathodic powder obtained from spent LIBs via thermal pre-treatment was leached in nitric acid and H_2O_2 solution. The molar ratio of Li:Co was then adjusted to 1 : 1 by adding $LiNO_3$ solution. Afterward, citric acid was added to the solution to prepare an amorphous gel precursor in a rotary vacuum dryer. The gel was then calcinated at 950 °C for 24 hours to get a well-crystalized LCO. The charge and discharge capacities were 165 and 154 mAh g^{-1}, respectively, and there was only less than 10% loss in capacity after 30 cycles [54]. Under optimum conditions (1 M HNO_3, 75 °C for an hour, addition of 1.7 vol% H_2O_2, and solid/liquid ratio of 20 g l^{-1}), the leaching efficiency for lithium and cobalt could reach as high as 85%. The kinetics were also studied, and it was concluded that the surface chemical reactions are key to the dissolution of LCO and the dissolution rate of Li and Co were found to be inversely proportional to their ionic concentrations [93].

Castillo et al. proposed a simple approach to recycle LMO batteries [94]. The batteries were first opened manually, and then, cathode materials were dissolved in a nitric acid solution. Next, sodium hydroxide was added to raise the pH value in an alkaline environment, therefore, Mn precipitated as a metal hydroxide. The metal hydroxide precipitate was then thermally treated for characterization, and all the lithium was recovered from the alkaline solution. When the concentration of nitric acid reached 2 M, the leaching efficiency of LMO cathode materials was greatly improved. In this case, optimum conditions were: immersing LMO in 2 M nitric acid at 80 °C for two hours. The leaching efficiency of lithium and manganese reached 100% and 95%, respectively. All the leached Mn and Li can be recovered through precipitation and filtration, respectively. The leaching reaction can be represented by:

$$LiMn_2O_4 + 10HNO_3 \rightarrow 2Mn(NO_3)_2 + LiNO_3 + 5H_2O + 5NO_2 + 2O_2 \quad (13.12)$$

Hydrochloric Acid Leaching One of the advantages of using HCl as a leaching agent is that a leaching efficiency above 99% without the addition of any reducing agent can be obtained.

The work in [53] investigated an approach aiming at recovering all the valuable metals from WLIB separately. Cycled cathodic materials, such as LCO, LMO, and NMC, were immersed in a 4 M hydrochloric acid solution at 80 °C for one hour with a solid/liquid ratio of 0.02 g ml^{-1}. Over 99% of lithium, cobalt, manganese, and nickel were dissolved. Mn precipitated first as MnO_2 and $Mn(OH)_2$ with the addition of $KMnO_4$. Ni was subsequently extracted as a red solid complex via the addition of dimethylglyoxime. Finally, sodium hydroxide was added to the solution to raise the pH value to 11 and the cobalt was then extracted as $Co(OH)_2$. The purities of the obtained Li, Mn, Co, and Ni compounds were 96.97, 98.23, 96.94, and 97.43 wt%, respectively [53].

To compare the leaching performance of HCl and H_2SO_4, as well as to study the thermodynamics of Li and Co leaching processes without the addition of any reducing agent, Takacova et al. [52] investigated the kinetics of Li and Co extractions using both acids under the same conditions. WLIBs were first treated by mechanically crushing and sieving to obtain the cathodic materials, and then, HCl and H_2SO_4 solutions with various concentrations were used to leach the cathodic metals. It was found that HCl worked more efficiently than H_2SO_4 and that 100% of lithium and cobalt can be leached by using 2 M HCl at 60–80 °C for 90 minutes. with a liquid:solid ratio of 50 : 1. The cobalt leaching in HCl solution goes into two stages: chemical reaction (activation energy, E_a = 40–44 kJ mol^{-1}) and mixed mechanism (chemical reaction and diffusion; E_a = 20–26 kJ mol^{-1}). The lithium leaching process is controlled by diffusion or with mixed mode (E_a between 2–20 kJ mol^{-1}). The structure of the active mass was analyzed by XRD, and it was found that the lithium defects in the crystal lattice increased the cobalt extraction rate. The above result indicated that lithium was extracted first, and the caged cobalt was then able to react or diffuse.

Barik et al. [95] studied the performance of sodium hypochlorite as a precipitating agent solely for Mn. The shredded WLIB (11.73%Co, 8.48%Mn, 1.95%Li, 0.58%Al, 0.26%Ni, and 0.004%Cu) was immersed in HCl solution to leach the cathodic materials, obtaining the best performance when using a 1.75 M HCl solution heated at 50 °C for two hours (over 99% leaching efficiency of lithium, cobalt, and manganese) with a 20% (w/v) of active material in acid. The sodium hypochlorite was added into the solution to precipitate out Mn as MnO_2, Mn_3O_4, and $Na_{0.55}Mn_2O_4{\cdot}1.5H_2O$. Co was then extracted by adding a sodium carbonate solution into the Mn-free solution. The recovery rate of Mn and Co was 95% and 90%, respectively.

The chemical reaction between LCO and hydrochloric acid can be described as follows:

$$2LiCoO_2 + 8HCl \rightarrow 2LiCl + 2CoCl_2 + 2H_2O + Cl_2 \uparrow \tag{13.13}$$

It is worth noticing that chlorine gas is generated in Eq. (13.13), which is a harmful gas to the human body. Further treatment for chlorine gas has to be taken into consideration if hydrochloric acid is applied for leaching.

Phosphoric Acid Leaching Phosphoric acid can also be used for dissolving WLIB cathodic materials. Because it is a mild acid, reductants are mandatory in order to achieve a satisfactory leaching efficiency, being hydrogen peroxide one of the most widely used [96]. In this case, phosphoric acid acts not only as a leaching reagent but also as a precipitating agent. Anwani et al. first discharged WLIBs by immersing in a Na_2SO_4 solution for 24 hours and then manually dismantled. NMP solution was used subsequently to remove the PVDF, liberating the cathode material from the collector. The obtained cathode material was leached in a flask under various conditions. It was reported that more than 99% of lithium can be leached into a mixed solution of 0.7 M H_3PO_4 and 4 vol% H_2O_2 at 40 °C for one hour with the liquid-to-solid ratio being 20 ml g^{-1}. Besides, 99% of cobalt would precipitate as $Co_3(PO_4)_2$ with a purity of 97.1%, leaving lithium ions in the unreacted phosphoric acid solution, which can be recycled and reused later. The reactions can be explained by Eq. (13.14):

$$Co^{2+} + Li^+ + PO_4^{\ 3-} \rightarrow \frac{1}{3}Co_3\left(PO_4\right)_2 + \frac{1}{3}Li_3PO_4 \tag{13.14}$$

Comparing to other reducing agents, glucose is easier to get and cheaper. The leachability of glucose and phosphoric acid system for recycling WLIB has been studied by Meng et al. [90], who successfully applied glucose as the reducing agent and phosphoric acid as the reagent to leach LCO. The dismantled cathode from WLIB was cut into pieces and immersed in NaOH solution to dissolve the Al collector, and then, the cathodic powder was filtered out and burnt to remove organic compounds. The burnt powder was ground and sieved to get fine particles for the leaching experiment. The sample was immersed in a glucose and phosphoric acid mixture that was heated in a water bath. Approximately 100% of lithium and cobalt could be dissolved in a solution of 1.5 M H_3PO_4 and 0.02 M glucose at 80 °C for two hours with the solid-to-liquid ratio of 2 g l^{-1}. It was reported that glucose was oxidized and decomposed to monocarboxylic acid, and Co(III) was reduced to Co(II), which promoted the leaching efficiency. Oxalic acid was then added to the solution to precipitate out the Co(II).

In another work [70], LMO cathode materials were treated with a mixture of phosphoric acid and citric acid, in which the phosphoric acid acted as a leaching agent and the citric acid performed as a reductant. This phosphoric acid and citric acid leaching system is said to be time-saving and cheaper, as it requires less amount of acid than traditional single-acid methods. Besides, this system has little impact on the ecosystem. It was reported that when cathodic materials were immersed into the mixed solution of 0.2 M H_3PO_4 and 0.4 M citric acid and heated at 90 °C for 0.5 hour with a solid–liquid ratio of 20 g l^{-1}, the leaching efficiencies for Li, Ni, Mn, and Co were 100%, 93.38%, 92.00%, and 91.63%, respectively. The kinetic study illustrated that the surface chemical reaction dominated the leaching process. Besides, huge amounts of citrate and small amounts of dihydrogen phosphate were detected in the leachate. The whole process is reasonably eco-friendly.

Organic Acids Leaching Organic acids are also widely used for recycling WLIBs as leaching reagents and reductants because they are more environmentally friendly [97]. There are many methods for categorizing organic acids, but here, they will be discussed according to the number of carboxylic groups.

Lactic acid is a common monocarboxylic acid in nature and its industrial manufacturing of it is also environmentally friendly, making it a competitive candidate for the green leaching of WLIB [91]. A sol–gel method was conducted to extract the valuable metals from WLIB [91], in which lactic acid was used as a leaching reagent and H_2O_2 as a reductant. First, wasted NMC batteries were discharged and dismantled, and then, alkaline leaching treatment was used to dissolve the Al foils. Next, the filtrate from the alkaline solution was calcinated to remove all the carbonaceous components and the obtained powders were ground and sieved for further leaching. The samples were treated with 1.5 M lactic acid and 0.5 vol% H_2O_2 at 70 °C for 20 minutes, while the solid-to-liquid ratio was 20 g l^{-1}. Consequently, more than 98% of lithium, nickel, manganese, and cobalt could be leached following the reaction:

$$3LiNi_{\frac{1}{3}}Co_{\frac{1}{3}}Mn_{\frac{1}{3}}O_2(s) + 9C_3H_6O_3(aq) + \frac{1}{2}H_2O_2 \rightarrow 3C_3H_5O_3Li(aq) + (C_3H_5O_3)_2Ni(aq)$$
$$+ (C_3H_5O_3)_2Co(aq) + (C_3H_5O_3)_2Mn(aq) + 5H_2O(l) + O_2(g) \quad (13.15)$$

Lactic acid also acted as a chelating agent, and in a sol–gel extraction method, it also helps to catch the metals through chelating [91]. In this case, the leached liquid solution was first characterized by inductively coupled plasma-optical emission spectrometry (ICP-OES) to acquire the concentration of various metal ions, and then, the molar ratio of these ions was adjusted to 3.05 : 1 : 1 : 1 for Li^+, Ni^{2+}, Co^{2+}, Mn^{2+} by the addition of CH_3COOM (M = Li, Ni, Co, Mn). Then, the pH of the solution was adjusted to 7 by adding NaOH, and the resultant solution was heated at 80°C to form a gel that was subsequently dried and calcinated to achieve NMC. This method skips the metal separation steps. The batteries built with this recycled NMC were tested, and they performed well in terms of specific capacity and cycling performance, with 128.3 mA h g^{-1} after 100 cycles at 1C.

Many dicarboxylic acids were studied as the leaching reagents for WLIB, such as malic acid [98], tartaric acid [99], succinic acid [100], maleic acid [101], and iminodiacetic acid [101]. DL-Malic acid is one of the most competitive candidates for dissolving the active materials in WLIB, and it is cheap compared to pure L-malic acid [51, 98]. It can degrade efficiently under aerobic and anaerobic conditions, which makes waste treatment much easier [51, 97]. In the work of [98], DL-malic acid was used to leach the metals from waste NMC batteries under ultrasound assistance and its leaching conditions and kinetics were investigated. The WLIB was first manually dismantled and then immersed in Na_2SO_4 for 24 hours to get it fully discharged. Then, alkaline leaching was conducted to peel off cathodic materials from Al foils. The obtained powder was dried for the subsequent leaching. It was reported that the leaching efficiencies of nickel, cobalt, manganese, and lithium can reach 97.8%, 97.6%, 97.3%, and 98%, respectively. The optimum leaching conditions are 90 W ultrasound power, 1.0 mol l^{-1} DL-malic acid, 5 g l^{-1} pulp density, 80 °C, 4 vol% H_2O_2, and 30 minutes. It was found that the ultrasonic treatment could decrease the leaching time. After leaching, the molar ratio of the metals in the solution was measured via ICP-OES and the ratio was set to 3 : 1 : 1 for Ni:Co:Mn with the addition of CH_3COOM (M = Li, Ni, Co, and Mn). The solution was then mixed with aqueous Na_2CO_3 and heated to 60 °C for 10 hours while the pH value was controlled between 8 and 9. The carbonate precipitation was dried in a vacuum at 120 °C for 12 hours to obtain $Ni_{0.6}Co_{0.2}Mn_{0.2}CO_3$. At last, the obtained product and $LiOH{\cdot}H_2O$ were mixed at a Li/M (M is the sum of Ni, Mn, and Co) molar ratio of 1.05, and a two-step thermal treatment was conducted to get the final $LiNi_{0.6}Co_{0.2}Mn_{0.2}O_2$ product which created a close loop approach of recycling at 450 °C for five hours and then at 800 °C for 15 hours in the air [98].

A mixed system of tartaric acid and ascorbic acid was studied [99] to understand the leaching behavior of Co and Li in this system. WLIB was first discharged and manually dismantled, and then, an NMP solution was used to remove the PVDF binder and release the cathodic materials from the Al foils. The obtained powders were burnt at 700 °C for two hours to eliminate organic compounds, followed by grinding and sieving to increase the surface area of the powders for better leaching. The complete dissolution of the samples was achieved when immersed in a solution containing 0.4 M tartaric acid and 0.02 M ascorbic acid at 80 °C for five hours. The UV-vis results proved that ascorbic acid reduced Co(III) to Co(II). Finally, the cobalt was separated as cobalt oxalate.

In the work of [100], succinic acid was investigated as a leaching reagent for WLIB recycling. This reagent has the advantages of little damage to the experimental equipment and easy degradation. WLIBs were first discharged and dismantled to obtain the cathode materials. These were then uncurled and immersed in an NMP solution to remove the PVDF binder and to them from the Al foil. Subsequently, the cathodic materials were calcinated to remove organic compounds and the obtained powders were immersed in a mixed solution of succinic acid and H_2O_2. It was reported that about 100% of cobalt and over 96% of lithium were leached. The optimum leaching conditions were using a mixed solution containing 1.5 M succinic acid and 4 vol% H_2O_2 at 70 °C for 40 minutes, with the solid to liquid ratio of 15 g l^{-1} [100]. The kinetics investigation revealed that the leaching in the first 20 minutes was dominated by chemical reactions and afterward the leaching was controlled by diffusion.

Another two dicarboxylic acids, iminodiacetic acid and maleic acid, were investigated for their leaching performance to recycle WLIB [101]. The WLIB was first discharged and dismantled. The cathode was then uncurled and cut into small pieces. A solution of NMP was used to remove the PVDF binder to liberate cathodic materials from Al foil. Afterward, the obtained cathodic materials were thermally treated to burn off carbonaceous components. The powders were ground to increase surface area for efficient leaching. As for the leaching experiments, 0.2 g cathodic powder was immersed in both acids, and ascorbic acid was then added as a reducing agent. Consequently, almost all cobalt and lithium were leached at 80 °C for six hours. It is worth noticing that this method is time-consuming compared to the method using succinic acid or tartaric acid.

Citric acid, a tricarboxylic organic acid, is one of the most popular choices both in food production and pharmaceutical industries because of its capability to prolong the food shelf-time and its magnificent chelating effect [97]. It also receives great attention as a reagent for WLIB leaching, for its relatively low price and better leachability compared to other organic acids even without reductants [77, 78, 102]. The leaching performance of citric acid for different WLIB materials was investigated [77]. WLIBs were discharged in a NaCl solution for 48 hours and then manually dismantled. Subsequently, a NaOH solution was used to dissolve Al foils and liberate cathodic materials. The obtained powders were vacuum dried and ground for the leaching experiment. It was reported that the leaching performance of citric acid was better with the addition of 2 vol% of H_2O_2, which results in an increase of more than 20% in the cobalt extraction, compared to the samples without H_2O_2. The highest leaching speed was acquired when cathode materials were immersed in a solution of 1.5 M citric acid and 2 vol% of H_2O_2 and heated to 95 °C. 95% of cobalt, 97% of lithium, and 99% of nickel could be leached under the aforementioned conditions within 30 minutes. In another example of citric acid leaching [78], WLIB was cut and ball-milled into fine powders in order to improve leaching efficiency. It was reported for the milled fine powders that leaching efficiencies by using citric acid without any reductants were 91%, 90.9%, 94.1%, and 88.6% for Li, Co, Ni, and Mn, respectively, providing an easier and more economical option of citric acid leaching. The optimum conditions are 0.5 M of citric acid, 80 g l^{-1} solid-to-liquid ratio, and 80 minutes of heating at 90 °C. H_2O_2 can significantly shorten the leaching time, but cannot increase the leaching efficiency when it is added at the beginning of the leaching [78]. Thus, a stepwise addition of reductants into the citric acid was investigated and the results indicated that the leaching efficiency could be slightly improved because H_2O_2 decomposed quickly at 90 °C and, thus, the stepwise addition ensures a longer reaction time for the reduction.

After leaching, valuable metals were extracted out to create a close-loop recycling procedure. The most commonly used hydrometallurgical extraction method is [51]: first, precipitate Mn^{2+}, Co^{2+}, and Ni^{2+} using NaOH as $Mn(OH)_2$, $Co(OH)_2$, and $Ni(OH)_2$, respectively; then, precipitate Li^+ using Na_2CO_3 as Li_2CO_3.

However, in order to better utilize cathodic materials, a stepwise approach for the extraction of Mn^{2+}, Co^{2+}, and Ni^{2+} was separately investigated. In the work of [103], up to 99% of Mn was first extracted with the help of *N,N,N′,N′-tetra-n*-octyldiglycolamide diluted in 1-butyl-3-methylimidazolium bis(trifluoromethylsulfonyl)imide. Then, the Mn-free solution was mixed with tri-hexyltetradecylphosphonium chloride which enabled the Co to be separated from the solution at an efficiency of over 90%. Subsequently, they used a DES (lidocaine:decanoic acid = 1 : 2) with a high extraction selectivity for Ni with an efficiency of over 90%, leaving the Li^+ in the solution for further recovery. However, this method is not suitable for industrial production, because the reagents used in the recovery operation require a complicated synthesis procedure. Another stepwise method based on sulfuric acid and H_2O_2 leaching system was reported [104]. This method allows to extract different metals separately with high purity. First, cobalt and manganese were extracted out via the addition of PC88A (2-ethylhexyl phosphoric acid-mono-2-ethylhexyl ester) diluted in kerosene, with an efficiency of 98% and 90% for cobalt and manganese, respectively.

Subsequently, Co^{2+} and Mn^{2+} were separated via the addition of $KMnO_4$, which enabled over 90% of Mn^{2+} to precipitate as MnO_2 while only 2.73% of cobalt was lost during the process. The cobalt- and manganese-free solution was treated with Versatic 10 (neodecanoic acid) extracted 99.79% of nickel out while lithium remained in the liquor. However, the contamination from metallic current collectors, copper, and aluminum, affected the purity of the final product, thus, calling for further investigations to remove impurities.

Comparison of Different Acids The work reported in [105] compared the leaching performance of four inorganic acids: H_3PO_4, H_2SO_4, HNO_3, and HCl. The variables were the concentration of the acids, the temperature (from 40 to 80 °C), time (from 5 to 120 minutes), and pulp density (10–50 g l^{-1}). No reductants were used in these trials. Because of its powerful reduction ability, hydrochloric acid can reduce metal ions to a lower valance without the help of reductants, thus increasing the solubility of metals and the leaching efficiencies. A maximum pulp density of 20 g l^{-1} could be obtained under optimum conditions: 3 M HCl, 70 °C for 50 minutes, and the leaching efficiencies could reach as high as 100%, 99.7%, 99.3%, and 99.7% for lithium, nickel, cobalt, and manganese, respectively.

The work reported in [97] compared the leaching performance of organic acids, such as DL-malic acid, citric acid, oxalic acid, and acetic acid, with the addition of H_2O_2 as a reducing agent in all cases. Four parameters were chosen for each variable: 25, 50, 70, and 90 °C for the temperature; 0, 1, 2.5, and 5 vol% for the H_2O_2; 15, 30, 40, and 50 g l^{-1} for the solid to liquid ratio; and 1, 2, 4, and 6M for the acid concentration. Citric acid achieved the best performance among all the investigated organic acids with 92.53% of lithium and 81.50% of cobalt being leached when the sample was immersed in a mixed solution of 2 M citric acid and 1.25 vol% H_2O_2 and heated at 60 °C for two hours with a solid to liquid ratio of 30 g l^{-1}. One mole of citric acid can generate three moles of H^+ ions, thus ensuring the highest leaching efficiencies among these organic acids.

In order to have a comprehensive understanding on the uses of different acids (inorganic and organic) from an industrial perspective, rather than only leaching ability, Anwani et al. [96] introduced more parameters for comparison, such as economical index, amount of CO_2 released during the whole process of recovery, product purity and overall efficiency. Both inorganic acids (H_3PO_4, H_2SO_4, HNO_3, and HCl) and organic acids (oxalic acid, acetic acid, and formic acid) were selected as the candidates. The economic analysis was based on the recycling of 10 g of cathode materials and the costs included the leaching solvent, the reagents for precipitation, and the electricity bills. The CO_2 release data was obtained via GaBi software simulation [96], and the whole recycling process for different acids was programmed in the software; the product purity P was defined as below,

$$P = \left(P_{\mathrm{Co}} * P_{\mathrm{Li}}\right) * 100\% \tag{13.16}$$

where the P_{Co} and P_{Li} represented the purities of Co product and Li product, respectively.

The overall efficiency η was calculated via:

$$\eta = \left(\eta_d * \eta_{\mathrm{Co}} * \eta_{\mathrm{Li}}\right) * 100\% \tag{13.17}$$

where the η_d stood for the dissolution efficiency of cathode materials, η_{Co} and η_{Li} represented the extraction efficiencies for cobalt and lithium, respectively.

The analysis of the data on the product purity *P* and the overall efficiency η reveals how an acid system performs in the leaching of cathodic materials. Sulfuric acid and oxalic acid were the two candidates entering the final stage with an eye-catching performance in those four filters (economical index, CO_2 release during process, product purity, and overall efficiency). It was recommended [96] that oxalic acid was the better choice because the use of sulfuric acid consumed much more NaOH than the use of oxalic acid in the extraction stage. However, if the product purity was the priority, then sulfuric acid would win with a product purity of 90.9% than 75.52% of oxalic acid.

13.3.1.1.3 Deep Eutectic Solvents Leaching

DESs have also been investigated as leaching reagents to dissolve cathode materials. The advantages of the use of DESs as leaching reagents include: unharmful, cost-effective as well as recyclable. In the work of [79], a typical DES prepared by mixing choline chloride and ethylene glycol in a molar ratio of 1 : 2 at room temperature, could effectively dissolve LCO and the leaching efficiency reached 94.1% when heated in a sealed vial at 220 °C for 24 hours. The DES was recovered and re-used to extract cobalt [79]. The recovered DES worked significantly well, achieving almost the same efficiency as the new DES. This indicated that the DES could maintain its leachability after several cycles. However, as the electrode deposition consumes much energy, an alternative method was used to precipitate cobalt by a Na_2CO_3 extraction, and the possible products were reported to be $CoCO_3$, $Co(OH)_2$, and Co_3O_4. The final products were calcinated into Co_3O_4 with a 74% of cobalt recovery.

Another DES, polyethylene glycol (PEG):thiourea in a 2 : 1 molar ratio, was investigated for its leaching ability of LCO [81]. The DES was prepared by mixing the two components at >50 °C until a clear and homogeneous solution was achieved. Then the LCO powder was immersed in this DES in a sealed vial and heated. The temperatures chosen for the leaching experiments were relatively low: 25–80 °C. The solubility of LCO in the PEG:thiourea DES at 80 °C is 2121.14 ppm of cobalt, which is much higher than that of the choline chloride:ethylene glycol DES (1 : 2 ratio) under the same conditions [81].

However, the choline chloride: ethylene glycol DES was found to have a different leaching ability towards different metals at certain temperatures, making it an option for selective leaching and extraction [106]. For instance, the choline chloride:ethylene glycol DES was prepared by mixing the components at 50 °C until a homogeneous clear liquid was achieved and stored at 50 °C before the leaching experiments. Then, WLIB electrode powders were added to the DES in a sealed bottle and heated in an oil bath to get better temperature stability. The time required in these experiments was from 6 to 72 hours while the temperature varied from 90 to 180 °C. When the temperature reached 90 °C, exclusive leaching of copper was achieved, while both copper and manganese dissolved much more when the temperature rose to 160 °C. The extraction yield for the nickel was only 10% compared to the 90% obtained for cobalt at 180 °C and 24 hours. After the leaching, cobalt oxalate was recovered via a solvent extraction and it was implemented into batteries which presented a capacity of 150 $mA\,h\,g^{-1}$ and a capacity retention of 83% between 10 and 100 galvanostatic cycles [106].

DESs can be a more ecological, economical, and recyclable alternative to acids; however, it takes a longer time to react with metal oxides and more energy is consumed if electrode deposition is adopted to precipitate valuable metals and to recover DES. More investigations are needed to better evaluate the leaching performance of DESs in the recycling of batteries.

13.3.1.1.4 Mechanical Methods

Mechanical methods are often utilized to liberate cathodic materials from Al collectors after spent LIBs are discharged and dismantled. They can be generally divided into two steps: milling and grinding to reduce the size and sieving to select valuable metals [5, 39, 50].

PVDF removal has always been an issue when dealing with spent LIBs, and NMP solution is one of the most popular choices to dissolve PVDF. However, due to the toxicity and price of NMP solution, this is often not recommended. The work in [107] took advantage of the ultrasonic Fenton reaction to peel off cathodic materials from the Al collector. In this work, spent batteries were discharged and dismantled in a glove box to avoid contact with oxygen, which could pose danger to the operator. The cathodes were then cut into smaller pieces and immersed in Fenton's reagent, which is hydroperoxide with a ferrous iron solution. The samples were located in an ultrasonic bath for the liberation of cathodic black slurry. This simplified approach can reach a cathode peeling-off efficiency of 97% as Fenton's reagent can decompose the binder, thus, making the peeling-off much easier.

In the work of [108], spent LIBs were shredded in a granulator in a fume hood to prevent potential harm from electrolyte evaporation. The size of the shredded batteries was less than 7.5 mm and they were put on top of an Al foil in a fume hood for a week to get rid of all volatile compounds. Subsequently, the samples were categorized by their sizes: <0.5, 0.5–1, 1–2.5, 2.5–6, and >6 mm, with the help of sieving and a vibrator. These groups were analyzed by X-ray fluorescence, and the results revealed that (i) size sieving cannot selectively recycle the metals; (ii) Cu and Al contamination is inevitable and can decrease the purity of the final product. A more selective method is required, and cryogenic grinding was investigated to improve the selectivity [109]. A nitrogen tank was attached to a ball mill grinding equipment, and it can decrease the operating temperature to 77 K. It was reported that when the temperature was 77 K during the grinding, the peeling-off efficiency was 87.3% which was much higher than 25.0% at room temperature. The reason for this efficiency jump was believed to be the mechanical properties of Al at 77 K, which make the metal strong enough to maintain as large-size chips and cathodic materials can be easily sieved out as powders. More investigation has been done by [110] to acquire knowledge of this cryogenic grinding. Specifically, the crystal transition of PVDF, mechanical properties of metal collectors, and battery shells are included in this work. Differential scanning calorimetry results revealed that at 235 K PVDF started to change its crystal structure to a more brittle form, reducing the binding force between cathodic materials and the Al collector. As for collectors and battery shells, impact test was conducted after these samples were first cooled in a mixture of alcohol and liquid nitrogen for half an hour. The results for collectors and battery shells are totally different, while the former has increased impact strength with decreasing temperature from 300 to 75 K, the latter suffered a loss of impact strength by 94.4% when compared with that at room temperature [110]. With the different changes in impact strength at low temperatures, selective mechanical recycling is easier to achieve.

Density, electrical conductivity as well as magnetic properties are the key factors of mechanical recycling approaches [111], which can help to separate different parts of spent batteries. The advantages of mechanical methods are the size reduction of spent LIBs and the dividing and concentrating of target materials [111]. Nevertheless, safety concern is always an issue for mechanical methods when removing electrolyte and crushing batteries [112]. Furthermore, the purity of the products and the selectivity of these methods still need to be improved.

13.3.1.1.5 Direct Recycling

Direct recycling is an efficient and cheap approach as it directly heals the spent cathode without conducting chemical reactions [113-115]. The healing process is often achieved by the addition of lithium into spent LIBs because these batteries age when lithium is lost during the operation. Lithium sources for the compensation can be $LiCO_3$ powders [115] or a saturated lithium solution [114]. Nevertheless, the liberation of cathodic materials from the Al collector is necessary and it takes a lot of time and effort, making it not so "direct" [116].

An industrialized direct recycling approach was presented in the work of [114]. Spent LIBs were discharged in sodium hydrogen carbonate brine solution, and a liquid CO_2 system was utilized to remove the electrolyte. The batteries were then shredded, and the collected cathodes were immersed in a lithium-saturated solution for healing. A floatation process was then conducted to remove impurities like carbon. The cathodes in the solution were thermally treated, and unwanted trace metals were dissolved by the addition of nitric acid. Finally, the cathodes were washed with deionized water and dried in vacuum.

The work of [115] presented another direct recycling method with alumina coated on the healed cathode. The spent batteries were discharged, dismantled first, and then the LCO powders were separated from the collector. The LCO powders were mixed with $LiOH*H_2O$ or Li_2CO_3 which were lithium supplements. The mixture was thermally treated for cathode healing at 800 °C for eight hours in a tube furnace. The healed cathode was then mixed with alumina powders, while the molar ratio of Al:Co was 0.01 : 1. Finally, the obtained powders were sintered at 800 °C for eight hours in a tube furnace. The results revealed that with the Li_2CO_3 supplement, LCO had a layered structure after healing. The coating of alumina can help to improve the performance of healed LCO to the standard of pristine LCO [115]. The reason for the improvement is that the absence of a phase transition from a hexagonal to a monoclinic phase causes the decrease of *c*-axis strain in the LCO microstructure [117].

Another direct recycling process presented in the work of [108] was based on LMO batteries. Spent batteries were discharged and dismantled. The separation of cathodic materials and collectors was achieved by immersing in NMP solvent at 50 °C for six hours. Centrifugation was then conducted to obtain the cathodic powders, and the obtained powders were soaked in LiOH solution. LiOH solution with cathode powders was heated at 180 °C for 12 hours and naturally cooled down to room temperature. These regenerated powders were rinsed and dried in a vacuum at 80°C overnight. Coin cell batteries were fabricated from this powder for electrochemical characterization. It was reported that with regard to specific capacity, cycling stability as well as rate performance, the relithiated cathode showed a similar performance as pristine materials. It was also stated that with the analysis of the EverBatt model of Argonne National Laboratory, this one-step recycling method is more economical as well as ecological [108].

The cathodes of different spent batteries can have totally different extents of damage, causing extra problems when healing them with the sample process. With batteries that have various performances in the same battery pack, inhomogeneous heat release can cause potential dangers. Besides, lithium supplement is mandatory for direct recycling process which maybe run out in the future. More direct recycling approaches are listed in Table 13.3.

13.3.2 Separators/electrolytes

In conventional LIBs, the component placed between the electrodes is composed of a separator (porous polymer membrane) soaked in an electrolyte solution in which electrolyte recycling is a critical concern considering that it is composed of an organic solvent and lithium salts which are toxic and harmful [124].

An electrolyte recycling method based on supercritical CO_2 extraction, molecular sieve, and resin purification has been reported with the aim of maintaining the ionic conductivity value [125]. This process was carried out in a glovebox and after purification, the electrolyte was again tested in batteries with adequate electrochemical performance [125].

Another method of recovering the organic solvent present in the electrolyte solution was reported by Mu et al. based on an extraction method of transcritical conditions CO_2 at high pressure [126]. Another demonstrated method for electrolyte recovery was the use of a suitable solvent with a boiling temperature at low pressures for the lithium salt decomposition temperature (~80 °C) [127].

Considering that the separator is a porous polymer membrane, its recovering is through the pneumatic separation method with flow velocities between 3 and 7.8 $m\,s^{-1}$, with a recovery efficiency rate of about 98% [128].

More sustainable methods for their recovery are still to be explored.

13.4 Status of the Battery Electric Vehicle Fleet

EV sales are surging due to a combination of policy support, improvements in battery technology, increased charging infrastructure, and new compelling models from automakers. Electrification is also spreading to new segments of road transport, setting the stage for large changes ahead. This phenomenon can be visualized in Figure 13.3 which presents the global passenger BEV by market, provided by BloombergNEF [129].

It is noticeable that in 2021 the global passenger BEV sales increased by 103% when compared with 2020. It is observed that an exponential increase suggests continuation over the next years. China still dominates the global EV market, but sales are rising quickly elsewhere too.

13.4.1 Battery Demand

LIB demand for automobility was 340 gigawatt-hours (GWh) in 2021, which represents more than twice the level recorded in 2020 (Figure 13.4a). This increase is mainly driven by the increase in electric passenger cars. Battery demand for other transport modes increased by 65% when compared to 2020.

Table 13.3 Main process parameters, conditions, and characteristics of direct recycling approaches.

References	Pre-treatment	Operation condition	Results	Significance
[114]	1) Batteries discharged via sodium bicarbonate brine solution 2) Electrolyte removed by liquid CO_2 system 3) Batteries shredded into smaller pieces	1) Black mass immersed and then heated in a saturated Lithium solution 2) Cathodic materials separated via floatation 3) Cathodic materials were heated under air with a ramp rate of 5 °C min^{-1} for 9 h and then cooled down to room temperature 4) Cathodic materials were rinsed with 0.25 M nitric acid and then deionized water 5) Cathodic materials dried at 120 °C in vacuum and then heated at 500 °C under nitrogen	The full cell made of the regenerated cathode demonstrated useful capacity.	It can minimize environmental and safety hazards and process costs.
[115]	1) Discharging and disassembling 2) Cathode soaked in *N,N*-dimethylacetamide (DMAC) for 24 h to dissolve binders 3) Black mass heated at 120 °C for 12 h to remove DMAC 4) Black mass ground into fine powder	1) Cathode materials were mixed with $LiOH*H_2O$ powder and then ball milled for 3 h at 250 rpm 2) Mixture calcinated at 800 °C for 8 h (heating rate 5 °C min^{-1}) 3) Al_2O_3 powder was added to the mixture and then ball milled for 3 h at 250 rpm 4) Al_2O_3-coated cathode heated at 800 °C for 5 h in the air then naturally cooled	Discharge specific capacity for the Al_2O_3-coated regenerated LCO is 136.8 mAh g^{-1}, and it remains at 90.1% after 100 cycles.	Great formation of a layered structure in the LCO. Al_2O_3 coating enhanced the electrochemical properties of the LCO.
[118]	1) Discharging and disassembling 2) Batteries immersed in NMP at 50 °C for 6 h to dissolve PVDF 3) Centrifuging to get cathodic materials and then dried at 80 °C overnight under vacuum	1) The powder was soaked in 0.1 M LiOH solution and then heated in an autoclave at 180 °C for 12 h 2) Powder washed with deionized water and dried at 80 °C overnight	Discharge capacity of 111 mAh g^{-1} and 98 mAh g^{-1} after 100 cycles at 0.5C rate.	A simple one-step hydrothermal direct recycling method. Regenerated LMO with high capacity and great cycling stability as well as high-rate performance

(Continued)

Table 13.3 (Continued)

References	Pre-treatment	Operation condition	Results	Significance
[119]	1) New batteries cycled, discharged, and disassembled 2) Cathode strip immersed in NMP to remove PVDF for 30 min and then 20 min sonication 3) Black mass centrifuged, separated, and dried	Method 1: Black mass immersed in LiOH and then heated in a tube furnace. Powder washed, dried, and then annealed. Method 2: Measure composition via ICP-OES. Powder mixed with Li_2CO_3 and then sintered.	91.2% capacity retention after 100 cycles at 1C 141.9 $mAh\,g^{-1}$ at 2 C 130.3 $mAh\,g^{-1}$ at 5 C	A simple and efficient non-destructive recycling approach. It can fully recover the specific capacity and cycling stability of LCO (method 1). Different cathodes can be treated together.
[120]	1) Commercial and homemade batteries were used as raw materials	Hydrothermal treatment combined with a short annealing step in controlled atmospheres (hydrothermal treatment at 220 °C for 4 h, post-annealing at 850 °C in oxygen for 4 h) Regenerated materials were made into new coin cells to characterize their properties.	NMC 111: discharge capacity 158.4 $mAh\,g^{-1}$ at 1C and 122.6 $mAh\,g^{-1}$ after 100 cycles NMC 532: discharge capacity 128.3 $mAh\,g^{-1}$ after 100 cycles	A more efficient and non-destructive process to recycle NMC cathodes. The electrochemical properties of the regenerated NMC cathode can reach the virgin commercial materials' level.
[121]	1) Discharging and dismantling 2) Cathode plates were cut into pieces and immersed in organic solvents to remove PVDF	1) Spent cathode heated in a tube furnace under N_2 2) Solid phase sintered for 8 h with the doping of new LPF powder	Specific capacity: 144 $mAh\,g^{-1}$ at 0.1 C and 135 $mAh\,g^{-1}$ after 100 cycles	The regenerated LPF cathode materials can meet the basic requirement for reuse. 800 °C sintering behaved worse than 600 and 700 °C.

[122]	1) Discharging and dismantling 2) Immersed in NaOH solution and then washed and centrifuged 3) Dried and ball milled and sieved	Solid phase heat treatment at 650 °C for an hour under Ar/H_2 Li_2CO_3 as lithium supplement	First-discharge capacity of 147.3 $mAh\,g^{-1}$ and of 140.4 $mAh\,g^{-1}$ after 100 cycles at 0.2 C rate, while capacity retention is 95.32%	A more ecological approach of direct recycling LPF without any second pollution. 650°C appeared to be the optimum sintering temperature for LPF.
[123]	1) Batteries are cycled, discharged, and dismantled. 2) Scraping the cathode and grinding	Electrochemical lithiation: cycling spent cathode and pure lithium in a coin cell. Chemical lithiation: spent cathode powder was soaked in 1 M lithium iodide solution for 20 h.	The re-lithiated samples regained a capacity of 150–155 $mAh\,g^{-1}$.	The main failure mechanism of LIB is the loss of Li during cycles. Regenerated cathode showed 50% less cumulative energy demand than the synthesis of virgin materials.

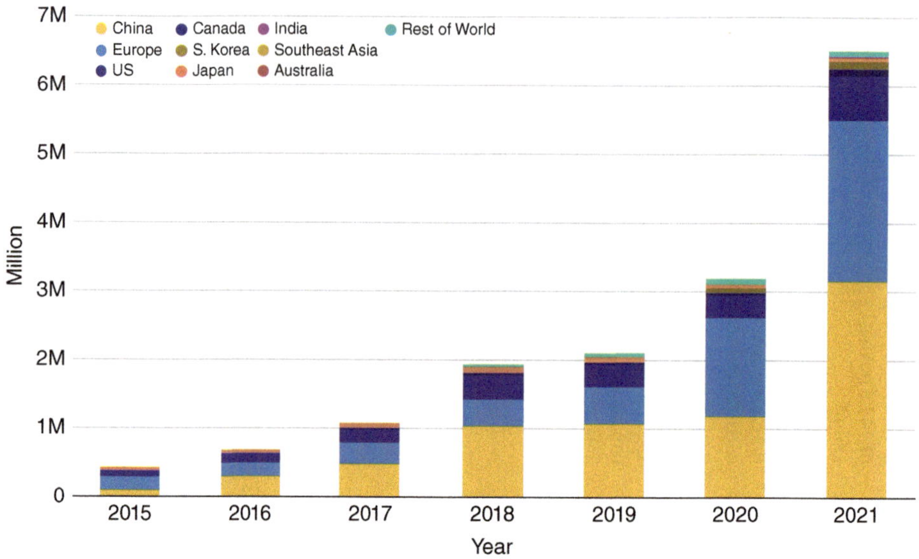

Figure 13.3 Global passenger BEV sales by market. *Source:* Data from Ref. [129].

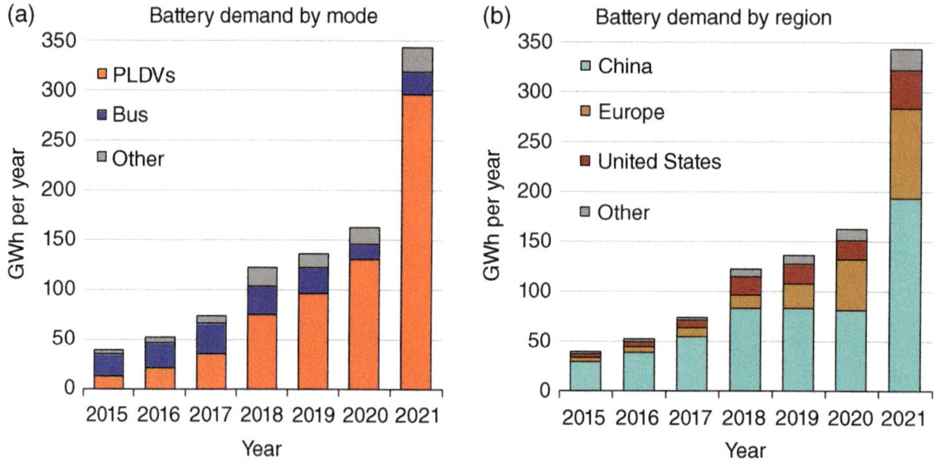

Figure 13.4 Battery demand by (a) vehicle type and (b) region (PLDVs = passenger light-duty vehicles). *Source:* Adapted from Ref. [130].

China represents currently the largest share of automotive battery demand, with almost 200 GWh of battery demand in 2021, an increase of 140% with respect to 2020 (Figure 13.4b). Worth also noting that in the United States the demand more than doubled in 2021, albeit from a lower base. Europe's demand growth was slightly lower than last year, yet still increasing by more than 70%.

13.4.2 Battery Electric Vehicle Outlook

Several studies [129, 130] suggest that passenger BEV sales are set to continue rising sharply in the years ahead as policy pressure continues to increase, more models are introduced in the market, and consumer interest takes off. The rising cost of batteries seems to not derail near-term BEV adoption. Some of the factors that are driving high battery raw material costs – war, inflation, and trade friction – are also pushing the price of gasoline and diesel to record highs, which is driving more consumer interest in EVs. Further, internal combustion engine vehicles (ICEVs) are also becoming more expensive to produce [129].

BloombergNEF [129] in its EV outlook report indicated that by 2025, plug-in vehicles will represent 23% of new passenger vehicle sales globally, up from just under 10% in 2021, while 75% of those will be full BEV (Figure 13.5). The report suggests that the BEV share of sales in some markets will be much higher, with EVs reaching 39% of sales in 2025 in China and in Europe. Also, the report forecasts that some of the major European car markets will go even faster, with Germany, the UK, and France reaching all between 40% and 50% in 2025.

13.4.2.1 Sustainability

EV battery supply chains consist of multiple complex stages. Starting from the extraction of the necessary mineral ores, refining to form sufficient purity chemicals, and advanced materials synthesis to form cathode and anode materials. Similar complex supply chains characterize other battery components such as electrolytes and separators. Cells are then fabricated and housed in modules within a battery pack which is integrated into the EV. Figure 13.6 identifies the different stages present in the LIB life cycle.

LIB life-cycle environmental impact assessment can be analyzed in several categories. Table 13.4 summarizes the environmental impact categories considered, together with the

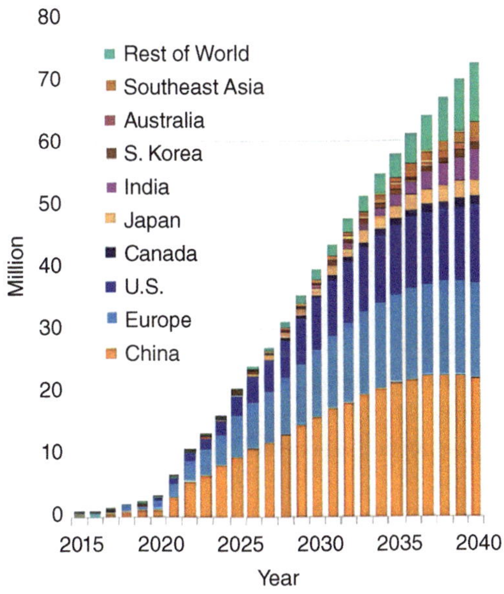

Figure 13.5 Global outlook of passenger BEV sales by market. *Source:* Data from Ref. [129].

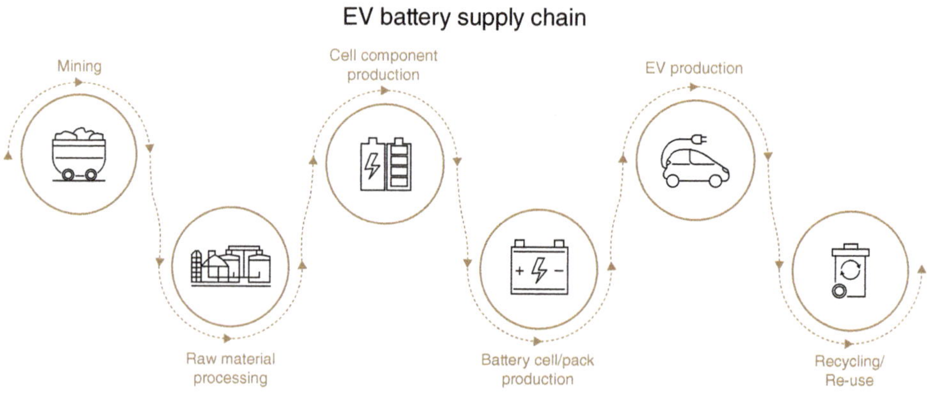

Figure 13.6 Stages involved in lithium-ion batteries for BEVs. *Source:* Adapted from Ref. [130].

Table 13.4 Impact categories and characterization factors that can be used for the Life cycle impact assessment.

Impact category	Characterizations factor
Climate change (kg CO_2-eq)	Global warming potential (GWP)
Terrestrial acidification (kg SO_2-eq)	Terrestrial acidification potential (TAP)
Fine particulate matter formation (kg $PM_{2.5}$-eq)	Particulate matter formation potential (PMFP)
Mineral resource scarcity (kg Cu-eq)	Surplus ore potential (SOP)
Fossil resource scarcity (kg Oil-eq)	Fossil fuel potential (FFP)

Source: Data from Ref. [131].

characterization factors and the units adopted, to quantify the contribution of each category. In this work, the focus will be placed on the global warming potential (GWP).

13.4.2.2 Production Stage

Pipitone et al. [132] studied the environmental impact related to the production phase, the processes related to materials production (e.g. steel production) and to mining (e.g. iron extraction) using the GREET 2020.Net suite [133] developed in the United States and adapted to replace the built-in electricity mix of the United States with the electricity mix available in Table 13.5.

According to the GREET model used, the mass of each vehicle is divided into three categories [133]: components, fluids, and batteries,

In the following, the focus will be set on the emissions related to the production of the battery. BEVs and hybrid electric vehicles (HEVs) are mainly equipped with LIBs of various types, depending on the different compositions of cathode materials.

13.4.2.3 Use Stage

As the BEV employs electric energy previously produced by means of different primary energy sources, for example, natural gas, coal, hydro, wind, solar energy, etc., evaluating

Table 13.5 Composition of the electricity mixes.

Type	EU-28 (2019) (%)	USA (2020) (%)	China (2018) (%)	Germany (2019) (%)	Norway (2019) (%)	Poland (2019) (%)
Coal	15.4	10.2	66.4	30.0	0.121	73.72
Oil	1.64	34.5	0.2	0.8	0.01	1.09
Natural gas	21.9	34.0	3.3	15.3	1.7	9.2
Nuclear	25.3	8.2	4.1	12.1	0	0
Hydro	10.9	2.5	17.1	4.24	93.4	1.6
Wind	13.3	3.0	5.0	20.4	4.1	9.2
Solar photovoltaic	4.07	1.3	2.4	7.7	0.01	0.4
Biofuels	5.27	4.6	1.26	7.2	0.03	4.3
Waste	1.6	—	0.2	2.03	0.3	0.4
Geothermal	0.2	0.2	0.002	—	—	—
Solar thermal	0.1	—	0.004	—	—	—
Tide	0.015	—	0.0002	—	—	—
Others	0.14	1.5		0.17	0.24	0.05

Source: Data from Ref. [134].

the environmental impact of BEVs is highly determined by the energy mix provided by the supplying region. Table 13.5 presented the composition of the electricity mixes by some selected regions.

Regarding the environmental payback of the BEV when compared to ICEVs, Costa et al. [135] presented the results comparing different European countries (Figure 13.7). With exception of Estonia and Poland, the BEV brings environmental benefits in all countries in less than 60 000 km of travel distance. The bad results of Estonia and Poland (more than 100 000 km of travel distance) are explained by the high share of coal in their energetic mixes. On the opposite side, the northern countries (Sweden, Norway, and Iceland) show the fastest environmental payback, in less than 30 000 km of travel distance.

These findings can be explained by the large variety of energy sources and the different energetic mixed presented in each country that reflect the environmental impact of the use of a BEV. Ideally, an energetic mix with a large focus on renewable energies will lead to a larger reduction in BEV emissions. Some countries, such as Poland or the Czech Republic, show a high dependency on fossil fuels, particularly coal, which makes electricity production a high emitter of greenhouse gases. On the other hand, countries such as Norway rely almost exclusively on renewable energies, leading to a lower environmental footprint.

13.4.2.4 End of Life and Analysis

LIB disposal and recycling process were modeled using the impact indicators shown in [135], and assuming the pyrometallurgical method as the recycling process [5]. It was considered that the pyrometallurgical process, however, does not allow the recovery of

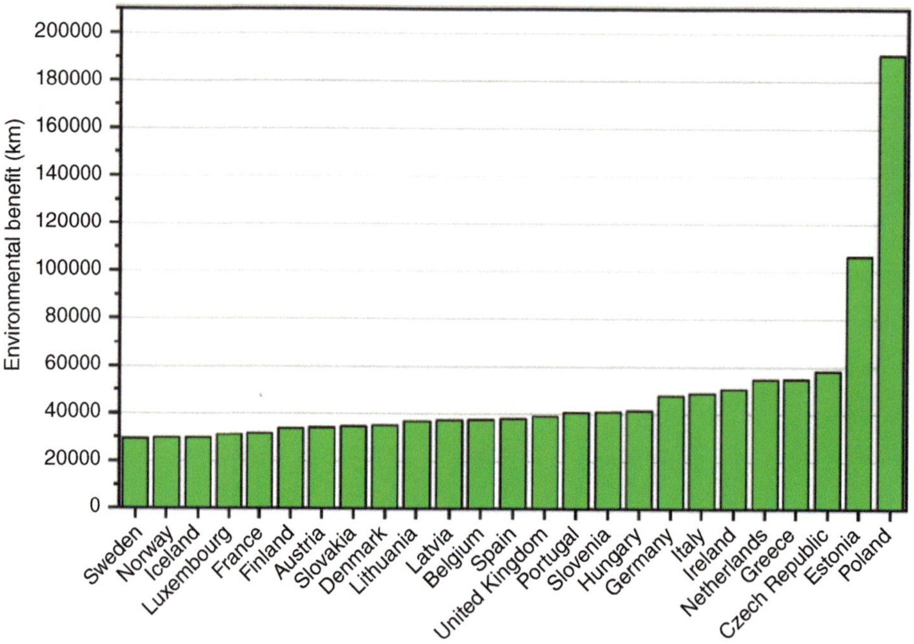

Figure 13.7 Environmental payback of the BEV when compared with the ICEV for each country, in distance traveled. *Source:* Ref. [135]/With permission of Elsevier.

Table 13.6 GWP impact indicators related to the disposal of the batteries of both HEV and BEV reference models.

Indicator	HEV	BEV
GWP (kg CO_2 eq)	28.46	396.0
Energy consumption (MJ)	488.2	6794

Source: Data from Ref. [132].

materials such as graphite, plastic materials, aluminum, lithium, and manganese. The resulting GWP impact indicators related to the disposal of the batteries of both HEV and BEV reference models are shown in Table 13.6.

Figure 13.8 shows the GWP generated by the three reference vehicle models (ICEV, HEV, and BEV) during each phase of their life obtained in the sustainability section.

The GWP analysis shows that, at the end of its life, traditional vehicles with gasoline-fueled internal combustion engines are responsible for an average specific CO_2 equivalent emission of 28 063 kg CO_2 eq. The HEV is characterized by a lower impact (14.0%) with respect to the ICEV, with an overall average specific CO_2 eq emission of 24 097 kg CO_2 eq. Finally, the BEV revealed a CO_2 equivalent emission of 16 434 kg CO_2 eq, which represents about 41.4% and 31.8% reduction when compared to the traditional ICEV and HEV, respectively. Figure 13.8 also shows that the production of the traditional vehicle is the less

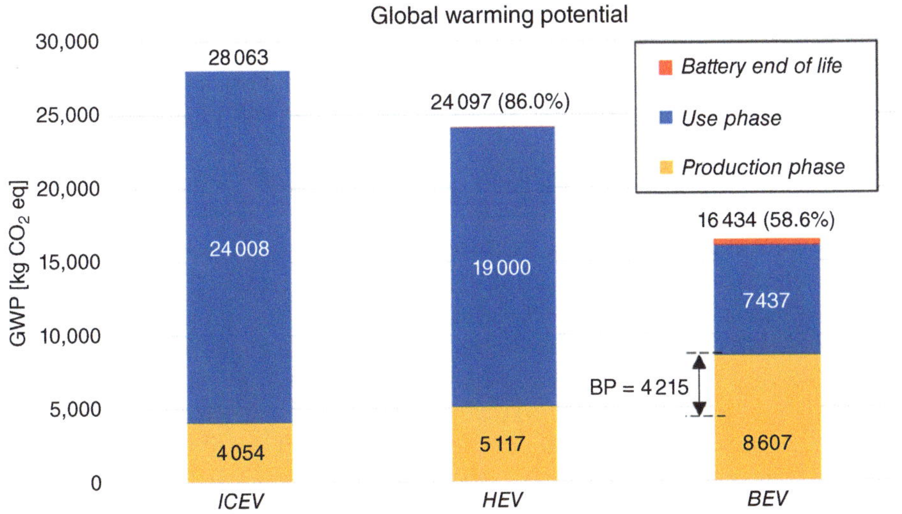

Figure 13.8 Global warming potential of the entire lifecycle of ICEV, HEV, and BEV (BP = battery production). *Source:* Ref. [132]/MDPI/CC BY 4.0.

impacting among the three production phases, where BEV and HEV rely on LIB production. In the case of the BEV, the LIB production implies an additional 4215 kg CO_2 eq.

The use phase represents the major contribution of greenhouse gas emissions on the ICEV with 85.5% of the total contribution, while the HEV shows a contribution of 78.8%. In this phase, the BEV shows a clear reduction in greenhouse gas emissions with only 45.2% of the total BEV emissions. The BEV use phase emissions are highly dependent on the electricity mix, as shown in Table 13.5 and related studies [135], which can compensate for the added greenhouse emissions from LIB production. Finally, battery end-of-life greenhouse emissions only represent 2.3% of the BEV total emissions.

13.5 Conclusions and Outlook

In modern society, efficient LIBs are essential for technological and economic evolution, being present in portable devices and EVs.

With the growth of the EV market, there is an increasing battery demand, settling pressure on different materials for the battery components. Considering environmental issues, it is also necessary to reduce electronic waste in landfills. To solve this problem, efficient and sustainable recycling processes are essential. This chapter presented the commonly used recycling methods for LIB electrodes and separators/electrolytes. Considering the cost of electrodes, most processes are applied to electrodes, primarily to recover their metal elements. Basically, the most used recycling processes in electrodes are pyro- and hydrometallurgical processes with high metal recovery rates (above 90%), despite some environmental issues still being to be addressed. There are only a few recycling processes applied to the separator/electrolyte and more technological efforts are needed to address this issue in the near future. Combining environmental and energy issues, more environmentally friendly materials and manufacturing processes, together with advanced recycling strategies are still needed to comply with the demands of circular economy models, in particular considering the strong increase in battery use related to the implementation of electric mobility.

Acknowledgments

The authors thank the Fundação para a Ciência e Tecnologia (FCT) and COMPETE 2020 for financial support under the framework of Strategic Funding grants UID/FIS/04650/2020, UID/EEA/04436/2020, and UID/QUI/00686/2020 and under projects MIT-EXPL/TDI/0033/2021, POCI-01-0247-FEDER-046985, 2022.03931.PTDC and PTDC/FIS-MAC/28157/2017. The authors also thank the FCT for the investigator contracts CEECIND/00833/2017 (RG) and 2020.04028.CEECIND (C.M.C.). Financial support from the BIDEKO Project, funded by MCIN/AEI, NextGenerationEU, PRTR, and the Basque Government Industry Department under the ELKARTEK program are acknowledged.

References

1 Martins, F., Felgueiras, C., Smitkova, M., and Caetano, N. (2019). Analysis of fossil fuel energy consumption and environmental impacts in European countries. *Energies* 12: 964.

2 Johnsson, F., Kjärstad, J., and Rootzén, J. (2019). The threat to climate change mitigation posed by the abundance of fossil fuels. *Climate Policy* 19: 258–274.

3 Chen, Y., Kang, Y., Zhao, Y. et al. (2021). A review of lithium-ion battery safety concerns: the issues, strategies, and testing standards. *Journal of Energy Chemistry* 59: 83–99.

4 Nitta, N., Wu, F., Lee, J.T., and Yushin, G. (2015). Li-ion battery materials: present and future. *Materials Today* 18: 252–264.

5 Costa, C.M., Barbosa, J.C., Gonçalves, R. et al. (2021). Recycling and environmental issues of lithium-ion batteries: advances, challenges and opportunities. *Energy Storage Materials* 37: 433–465.

6 Shahjalal, M., Roy, P.K., Shams, T. et al. (2022). A review on second-life of Li-ion batteries: prospects, challenges, and issues. *Energy* 241: 122881.

7 Zhang, Y., Sun, W., Xu, R. et al. (2021). Lithium extraction from water lithium resources through green electrochemical-battery approaches: a comprehensive review. *Journal of Cleaner Production* 285: 124905.

8 Liu, G., Zhao, Z., and Ghahreman, A. (2019). Novel approaches for lithium extraction from salt-lake brines: a review. *Hydrometallurgy* 187: 81–100.

9 Talens Peiró, L., Villalba Méndez, G., and Ayres, R.U. (2013). Lithium: sources, production, uses, and recovery outlook. *JOM* 65: 986–996.

10 Guoxing, R., Songwen, X., Meiqiu, X. et al. (2016). Recovery of valuable metals from spent lithium-ion batteries by smelting reduction process based on $MnO–SiO_2–Al_2O_3$ slag system. In: *Advances in Molten Slags, Fluxes, and Salts* (ed. R.G. Reddy, P. Chaubal, P.C. Pistorius and U. Pal), 211–218.

11 Tahil, W. (2007). *The Trouble With Lithium*. Martainville, France: Meridian International Research.

12 Jones, E., Qadir, M., van Vliet, M. T. H., Smakhtin, V., Kang, S.-m. The state of desalination and brine production: a global outlook. *Science of the Total Environment* 657,1343–56, 2019.

13 Liu, H. and Azimi, G. (2021). Process analysis and study of factors affecting the lithium carbonate crystallization from sulfate media during lithium extraction. *Hydrometallurgy* 199: 105532.

14 Mao, J.K., Li, J., and Xu, Z. (2018). Coupling reactions and collapsing model in the roasting process of recycling metals from $LiCoO_2$ batteries. *Journal of Cleaner Production* 205: 923–929.

15 Hansen, W., Christopher, M., and Verbuecheln, M. (2002). EU waste policy and challenges for regional and local authorities. *Ecological Institute for International and European Environmental Policy: Berlin, Germany*.

16 Pagliaro, M. and Meneguzzo, F. (2019). Lithium battery reusing and recycling: a circular economy insight. *Heliyon* 5: e01866.

17 Sonoc, A., Jeswiet, J., and Soo, V.K. (2015). Opportunities to improve recycling of automotive lithium ion batteries. *Procedia CIRP* 29: 752–757.

18 Chen, Y., Shi, P., Chang, D. et al. (2021). Selective extraction of valuable metals from spent EV power batteries using sulfation roasting and two stage leaching process. *Separation and Purification Technology* 258: 118078.

19 Fan, E., Li, L., Lin, J. et al. (2019). Low-temperature molten-salt-assisted recovery of valuable metals from spent lithium-ion batteries. *ACS Sustainable Chemistry & Engineering* 7: 16144–16150.

20 Wang, W., Han, Y., Zhang, T. et al. (2019). Alkali metal salt catalyzed carbothermic reduction for sustainable recovery of $LiCoO_2$: accurately controlled reduction and efficient water leaching. *ACS Sustainable Chemistry & Engineering* 7: 16729–16737.

21 Wang, W., Zhang, Y., Liu, X., and Xu, S. (2019). A simplified process for recovery of Li and Co from spent $LiCoO_2$ cathode using Al foil as the in situ reductant. *ACS Sustainable Chemistry & Engineering* 7: 12222–12230.

22 Xiao, J., Li, J., and Xu, Z. (2017). Recycling metals from lithium ion battery by mechanical separation and vacuum metallurgy. *Journal of Hazardous Materials* 338: 124–131.

23 Fu, Y., He, Y., Yang, Y. et al. (2020). Microwave reduction enhanced leaching of valuable metals from spent lithium-ion batteries. *Journal of Alloys and Compounds* 832: 154920.

24 Li, J., Wang, G., and Xu, Z. (2016). Environmentally-friendly oxygen-free roasting/wet magnetic separation technology for in situ recycling cobalt, lithium carbonate and graphite from spent $LiCoO_2$/graphite lithium batteries. *Journal of Hazardous Materials* 302: 97–104.

25 Ren, G.-X., Xiao, S.-W., Xie, M.-Q. et al. (2017). Recovery of valuable metals from spent lithium ion batteries by smelting reduction process based on $FeO–SiO_2–Al_2O_3$ slag system. *Transactions of Nonferrous Metals Society of China* 27: 450–456.

26 Lombardo, G., Ebin, B., Mark, M.R. et al. (2020). Incineration of EV Lithium-ion batteries as a pretreatment for recycling – determination of the potential formation of hazardous by-products and effects on metal compounds. *Journal of Hazardous Materials* 393: 122372.

27 Ekberg, C. and Petranikova, M. (2015). Chapter 7 - Lithium batteries recycling. In: *Lithium Process Chemistry* (ed. A. Chagnes, J. Światowska, A. Chagnes and J. Światowska), 233–267. Amsterdam: Elsevier.

28 Bridgwater, V.A. (1980). Waste incineration and pyrolysis. *Resource Recovery and Conservation* 5: 99–115.

29 Makuza, B., Tian, Q., Guo, X. et al. (2021). Pyrometallurgical options for recycling spent lithium-ion batteries: a comprehensive review. *Journal of Power Sources* 491: 229622.

30 Lu, M., Zhang, H., Wang, B. et al. (2013). The re-synthesis of $LiCoO_2$ from spent lithium ion batteries separated by vacuum-assisted heat-treating method. *International Journal of Electrochemical Science* 8: 8201–8209.

31 Sun, L. and Qiu, K. (2011). Vacuum pyrolysis and hydrometallurgical process for the recovery of valuable metals from spent lithium-ion batteries. *Journal of Hazardous Materials* 194: 378–384.

32 Zhao, Y., Liu, B., Zhang, L., and Guo, S. (2020). Microwave pyrolysis of macadamia shells for efficiently recycling lithium from spent lithium-ion batteries. *Journal of Hazardous Materials* 396: 122740.

33 Diaz, F., Wang, Y., Moorthy, T., and Friedrich, B. (2018). Degradation mechanism of nickel-cobalt-aluminum (NCA) cathode material from spent lithium-ion batteries in microwave-assisted pyrolysis. *Metals* 8 (8): 565.

34 Lombardo, G., Ebin, B., Steenari, B.-M. et al. (2021). Comparison of the effects of incineration, vacuum pyrolysis and dynamic pyrolysis on the composition of NMC-lithium battery cathode-material production scraps and separation of the current collector. *Resources, Conservation and Recycling* 164: 105142.

35 Zhao, S., He, W., and Li, G. (2019). Recycling technology and principle of spent lithium-ion battery. In: *Recycling of Spent Lithium-Ion Batteries*, 1–26. Cham: Springer.

36 Di, C., Yongming, C., Yan, X. et al. (2020). Selective recovery of lithium from ternary spent lithium-ion batteries using sulfate roasting-water leaching process. In: *Energy Technology 2020: Recycling, Carbon Dioxide Management, and Other Technologies* (ed. X. Chen, Y. Zhong, L. Zhang, et al.), 387–395. Cham: Springer.

37 Chen, M., Ma, X., Chen, B. et al. (2019). Recycling end-of-life electric vehicle lithium-ion batteries. *Joule* 3: 2622–2646.

38 Harper, G., Sommerville, R., Kendrick, E. et al. (2019). Recycling lithium-ion batteries from electric vehicles. *Nature* 575: 75–86.

39 Yun, L., Linh, D., Shui, L. et al. (2018). Metallurgical and mechanical methods for recycling of lithium-ion battery pack for electric vehicles. *Resources, Conservation and Recycling* 136: 198–208.

40 Khalafalla, S.E., Evans, J.W., Koo, C.H. et al. (1979). Pyrometallurgical processes. In: *Rate Processes of Extractive Metallurgy* (ed. H.Y. Sohn, M.E. Wadsworth, H.Y. Sohn and M.E. Wadsworth), 245–428. Boston, MA: Springer.

41 Yin, H. and Xing, P. (2019). Pyrometallurgical routes for the recycling of spent lithium-ion batteries. In: *Recycling of Spent Lithium-Ion Batteries: Processing Methods and Environmental Impacts* (ed. L. An and L. An), 57–83. Cham: Springer.

42 Hu, X., Mousa, E., Tian, Y., and Ye, G. (2021). Recovery of Co, Ni, Mn, and Li from Li-ion batteries by smelting reduction - Part I: a laboratory-scale study. *Journal of Power Sources* 483: 228936.

43 Hu, X., Mousa, E., and Ye, G. (2021). Recovery of Co, Ni, Mn, and Li from Li-ion batteries by smelting reduction - Part II: a pilot-scale demonstration. *Journal of Power Sources* 483: 229089.

44 Haldar, S.K. (2018). Chapter 13 - Mineral processing. In: *Mineral Exploration*, 2e (ed. S.K. Haldar and S.K. Haldar), 259–290.

45 Tang, Y., Zhang, B., Xie, H. et al. (2020). Recovery and regeneration of lithium cobalt oxide from spent lithium-ion batteries through a low-temperature ammonium sulfate roasting approach. *Journal of Power Sources* 474: 228596.

46 Dang, H., Li, N., Chang, Z. et al. (2020). Lithium leaching via calcium chloride roasting from simulated pyrometallurgical slag of spent lithium ion battery. *Separation and Purification Technology* 233: 116025.

47 Peng, C., Liu, F., Wang, Z. et al. (2019). Selective extraction of lithium (Li) and preparation of battery grade lithium carbonate (Li_2CO_3) from spent Li-ion batteries in nitrate system. *Journal of Power Sources* 415: 179–188.

48 Zhang, Y., Wang, W., Fang, Q., and Xu, S. (2020). Improved recovery of valuable metals from spent lithium-ion batteries by efficient reduction roasting and facile acid leaching. *Waste Management* 102: 847–855.

49 Ma, Y., Tang, J., Wanaldi, R. et al. (2021). A promising selective recovery process of valuable metals from spent lithium ion batteries via reduction roasting and ammonia leaching. *Journal of Hazardous Materials* 402: 123491.

50 Lv, W., Wang, Z., Cao, H. et al. (2018). A critical review and analysis on the recycling of spent lithium-ion batteries. *ACS Sustainable Chemistry & Engineering* 6: 1504–1521.

51 Jung, J.C.-Y., Sui, P.-C., and Zhang, J. (2021). A review of recycling spent lithium-ion battery cathode materials using hydrometallurgical treatments. *Journal of Energy Storage* 35: 102217.

52 Takacova, Z., Havlik, T., Kukurugya, F., and Orac, D. (2016). Cobalt and lithium recovery from active mass of spent Li-ion batteries: theoretical and experimental approach. *Hydrometallurgy* 163: 9–17.

53 Wang, R.-C., Lin, Y.-C., and Wu, S.-H. (2009). A novel recovery process of metal values from the cathode active materials of the lithium-ion secondary batteries. *Hydrometallurgy* 99: 194–201.

54 Lee, C.K. and Rhee, K.-I. (2002). Preparation of $LiCoO_2$ from spent lithium-ion batteries. *Journal of Power Sources* 109: 17–21.

55 Li, L., Chen, R., Sun, F. et al. (2011). Preparation of $LiCoO_2$ films from spent lithium-ion batteries by a combined recycling process. *Hydrometallurgy* 108: 220–225.

56 Ferreira, D.A., Prados, L.M.Z., Majuste, D., and Mansur, M.B. (2009). Hydrometallurgical separation of aluminium, cobalt, copper and lithium from spent Li-ion batteries. *Journal of Power Sources* 187: 238–246.

57 Nayl, A.A., Elkhashab, R.A., Badawy, S.M., and El-Khateeb, M.A. (2017). Acid leaching of mixed spent Li-ion batteries. *Arabian Journal of Chemistry* 10: S3632–S3639.

58 Dorella, G. and Mansur, M.B. (2007). A study of the separation of cobalt from spent Li-ion battery residues. *Journal of Power Sources* 170: 210–215.

59 Chen, L., Tang, X., Zhang, Y. et al. (2011). Process for the recovery of cobalt oxalate from spent lithium-ion batteries. *Hydrometallurgy* 108: 80–86.

60 Kang, J., Senanayake, G., Sohn, J., and Shin, S.M. (2010). Recovery of cobalt sulfate from spent lithium ion batteries by reductive leaching and solvent extraction with Cyanex 272. *Hydrometallurgy* 100: 168–171.

61 Shin, S.M., Kim, N.H., Sohn, J.S. et al. (2005). Development of a metal recovery process from Li-ion battery wastes. *Hydrometallurgy* 79: 172–181.

62 Zhu, S.G., He, W.Z., Li, G.M. et al. (2012). Recovery of Co and Li from spent lithium-ion batteries by combination method of acid leaching and chemical precipitation. *Transactions of Nonferrous Metals Society of China (English Edition)* 22: 2274–2281.

63 Swain, B., Jeong, J., Lee, J., and c., Lee, G. H., Sohn, J. S. (2007). Hydrometallurgical process for recovery of cobalt from waste cathodic active material generated during manufacturing of lithium ion batteries. *Journal of Power Sources* 167: 536–544.

64 Nan, J., Han, D., Yang, M. et al. (2006). Recovery of metal values from a mixture of spent lithium-ion batteries and nickel-metal hydride batteries. *Hydrometallurgy* 84: 75–80.

65 Chen, W.S. and Ho, H.J. (2018). Recovery of valuable metals from lithium-ion batteries NMC cathode waste materials by hydrometallurgical methods. *Metals* 8 (5): 321.

66 Yang, X., Dong, P., Hao, T. et al. (2020). Combined method of leaching and Co-precipitation for recycling spent $LiNi_{0.6}Co_{0.2}Mn_{0.2}O_2$ cathode materials: process optimization and performance aspects. *JOM* 72: 3843–3852.

67 Vieceli, N., Nogueira, C.A., Guimarães, C. et al. (2018). Hydrometallurgical recycling of lithium-ion batteries by reductive leaching with sodium metabisulphite. *Waste Management* 71: 350–361.

68 Peng, C., Hamuyuni, J., Wilson, B.P., and Lundström, M. (2018). Selective reductive leaching of cobalt and lithium from industrially crushed waste Li-ion batteries in sulfuric acid system. *Waste Management* 76: 582–590.

69 Granata, G., Moscardini, E., Pagnanelli, F. et al. (2012). Product recovery from Li-ion battery wastes coming from an industrial pre-treatment plant: Lab scale tests and process simulations. *Journal of Power Sources* 206: 393–401.

70 Zhuang, L., Sun, C., Zhou, T. et al. (2019). Recovery of valuable metals from $LiNi_{0.5}Co_{0.2}Mn_{0.3}O_2$ cathode materials of spent Li-ion batteries using mild mixed acid as leachant. *Waste Management* 85: 175–185.

71 Chen, X., Ma, H., Luo, C., and Zhou, T. (2017). Recovery of valuable metals from waste cathode materials of spent lithium-ion batteries using mild phosphoric acid. *Journal of Hazardous Materials* 326: 77–86.

72 Sun, L. and Qiu, K. (2012). Organic oxalate as leachant and precipitant for the recovery of valuable metals from spent lithium-ion batteries. *Waste Management* 32: 1575–1582.

73 Zhou, S., Zhang, Y., Meng, Q. et al. (2021). Recycling of $LiCoO_2$ cathode material from spent lithium ion batteries by ultrasonic enhanced leaching and one-step regeneration. *Journal of Environmental Management* 277: 111426.

74 Zhang, Y., Meng, Q., Dong, P. et al. (2018). Use of grape seed as reductant for leaching of cobalt from spent lithium-ion batteries. *Journal of Industrial and Engineering Chemistry* 66: 86–93.

75 Meng, Q., Zhang, Y., and Dong, P. (2018). A combined process for cobalt recovering and cathode material regeneration from spent $LiCoO_2$ batteries: process optimization and kinetics aspects. *Waste Management* 71: 372–380.

76 Li, L., Lu, J., Ren, Y. et al. (2012). Ascorbic-acid-assisted recovery of cobalt and lithium from spent Li-ion batteries. *Journal of Power Sources* 218: 21–27.

77 Musariri, B., Akdogan, G., Dorfling, C., and Bradshaw, S. (2019). Evaluating organic acids as alternative leaching reagents for metal recovery from lithium ion batteries. *Minerals Engineering* 137: 108–117.

78 Meng, F., Liu, Q., Kim, R. et al. (2020). Selective recovery of valuable metals from industrial waste lithium-ion batteries using citric acid under reductive conditions: leaching optimization and kinetic analysis. *Hydrometallurgy* 191: 105160.

79 Tran, M.K., Rodrigues, M.T.F., Kato, K. et al. (2019). Deep eutectic solvents for cathode recycling of Li-ion batteries. *Nature Energy* 4: 339–345.

80 Landa-Castro, M., Aldana-González, J., Montes de Oca-Yemha, M.G. et al. (2020). Ni–Co alloy electrodeposition from the cathode powder of Ni-MH spent batteries leached with a deep eutectic solvent (reline). *Journal of Alloys and Compounds* 830: 1–9.

81 Chen, Y., Lu, Y., Liu, Z. et al. (2020). Efficient dissolution of lithium-ion batteries cathode $LiCoO_2$by polyethylene glycol-based deep eutectic solvents at mild temperature. *ACS Sustainable Chemistry & Engineering* 8: 11713–11720.

82 He, Y., Yuan, X., Zhang, G. et al. (2021). A critical review of current technologies for the liberation of electrode materials from foils in the recycling process of spent lithium-ion batteries. *Science of The Total Environment* 766: 142382.

83 Contestabile, M., Panero, S., and Scrosati, B. (2001). A laboratory-scale lithium-ion battery recycling process. *Journal of Power Sources* 92: 65–69.

84 Kim, S., Bang, J., Yoo, J. et al. (2021). A comprehensive review on the pretreatment process in lithium-ion battery recycling. *Journal of Cleaner Production* 294: 126329.

85 He, L.-P., Sun, S.-Y., Mu, Y.-Y. et al. (2017). Recovery of lithium, nickel, cobalt, and manganese from spent lithium-ion batteries using l-tartaric acid as a leachant. *ACS Sustainable Chemistry & Engineering* 5: 714–721.

86 Wang, M., Tan, Q., Liu, L., and Li, J. (2019). A low-toxicity and high-efficiency deep eutectic solvent for the separation of aluminum foil and cathode materials from spent lithium-ion batteries. *Journal of Hazardous Materials* 380: 120846.

87 Zhang, Q., De Oliveira Vigier, K., Royer, S., and Jérôme, F. (2012). Deep eutectic solvents: syntheses, properties and applications. *Chemical Society Reviews* 41: 7108–7146.

88 Smith, E.L., Abbott, A.P., and Ryder, K.S. (2014). Deep eutectic solvents (DESs) and their applications. *Chemical Reviews* 114: 11060–11082.

89 Ji, Y., Jafvert, C.T., and Zhao, F. (2021). Recovery of cathode materials from spent lithium-ion batteries using eutectic system of lithium compounds. *Resources, Conservation and Recycling* 170: 105551.

90 Meng, Q., Zhang, Y., and Dong, P. (2017). Use of glucose as reductant to recover Co from spent lithium ions batteries. *Waste Management* 64: 214–218.

91 Li, L., Fan, E., Guan, Y. et al. (2017). Sustainable recovery of cathode materials from spent lithium-ion batteries using lactic acid leaching system. *ACS Sustainable Chemistry & Engineering* 5: 5224–5233.

92 Fan, X., Song, C., Lu, X. et al. (2021). Separation and recovery of valuable metals from spent lithium-ion batteries via concentrated sulfuric acid leaching and regeneration of $LiNi_{1/3}Co_{1/3}Mn_{1/3}O_2$. *Journal of Alloys and Compounds* 863: 158775.

93 Lee, C.K. and Rhee, K.-I. (2003). Reductive leaching of cathodic active materials from lithium ion battery wastes. *Hydrometallurgy* 68: 5–10.

94 Castillo, S., Ansart, F., Laberty-Robert, C., and Portal, J. (2002). Advances in the recovering of spent lithium battery compounds. *Journal of Power Sources* 112: 247–254.

95 Barik, S.P., Prabaharan, G., and Kumar, L. (2017). Leaching and separation of Co and Mn from electrode materials of spent lithium-ion batteries using hydrochloric acid: laboratory and pilot scale study. *Journal of Cleaner Production* 147: 37–43.

96 Anwani, S., Methekar, R., and Ramadesigan, V. (2020). Resynthesizing of lithium cobalt oxide from spent lithium-ion batteries using an environmentally benign and economically viable recycling process. *Hydrometallurgy* 197: 105430.

97 Golmohammadzadeh, R., Rashchi, F., and Vahidi, E. (2017). Recovery of lithium and cobalt from spent lithium-ion batteries using organic acids: process optimization and kinetic aspects. *Waste Management* 64: 244–254.

98 Ning, P., Meng, Q., Dong, P. et al. (2020). Recycling of cathode material from spent lithium ion batteries using an ultrasound-assisted DL-malic acid leaching system. *Waste Management* 103: 52–60.

99 Nayaka, G.P., Pai, K.V., Santhosh, G., and Manjanna, J. (2016). Dissolution of cathode active material of spent Li-ion batteries using tartaric acid and ascorbic acid mixture to recover Co. *Hydrometallurgy* 161: 54–57.

100 Li, L., Qu, W., Zhang, X. et al. (2015). Succinic acid-based leaching system: a sustainable process for recovery of valuable metals from spent Li-ion batteries. *Journal of Power Sources* 282: 544–551.

101 Nayaka, G.P., Pai, K.V., Manjanna, J., and Keny, S.J. (2016). Use of mild organic acid reagents to recover the Co and Li from spent Li-ion batteries. *Waste Management* 51: 234–238.

102 Zheng, Q., Watanabe, M., Iwatate, Y. et al. (2020). Hydrothermal leaching of ternary and binary lithium-ion battery cathode materials with citric acid and the kinetic study. *The Journal of Supercritical Fluids* 165: 104990.

103 Zante, G., Braun, A., Masmoudi, A. et al. (2020). Solvent extraction fractionation of manganese, cobalt, nickel and lithium using ionic liquids and deep eutectic solvents. *Minerals Engineering* 156: 106512.

104 Yang, Y., Lei, S., Song, S. et al. (2020). Stepwise recycling of valuable metals from Ni-rich cathode material of spent lithium-ion batteries. *Waste Management* 102: 131–138.

105 Liu, T., Chen, J., Shen, X., and Li, H. (2021). Regulating and regenerating the valuable metals from the cathode materials in lithium-ion batteries by nickel-cobalt-manganese co-extraction. *Separation and Purification Technology* 259: 118088.

106 Schiavi, P.G., Altimari, P., Branchi, M. et al. (2021). Selective recovery of cobalt from mixed lithium ion battery wastes using deep eutectic solvent. *Chemical Engineering Journal* 417: 129249.

107 Chen, X., Li, S., Wang, Y. et al. (2021). Recycling of $LiFePO_4$ cathode materials from spent lithium-ion batteries through ultrasound-assisted Fenton reaction and lithium compensation. *Waste Management* 136: 67–75.

108 Wang, X., Gaustad, G., and Babbitt, C.W. (2016). Targeting high value metals in lithium-ion battery recycling via shredding and size-based separation. *Waste Management* 51: 204–213.

109 Wang, H., Liu, J., Bai, X. et al. (2019). Separation of the cathode materials from the Al foil in spent lithium-ion batteries by cryogenic grinding. *Waste Management* 91: 89–98.

110 Liu, J., Bai, X., Hao, J. et al. (2021). Efficient liberation of electrode materials in spent lithium-ion batteries using a cryogenic ball mill. *Journal of Environmental Chemical Engineering* 9: 106017.

111 Ordoñez, J., Gago, E.J., and Girard, A. (2016). Processes and technologies for the recycling and recovery of spent lithium-ion batteries. *Renewable and Sustainable Energy Reviews* 60: 195–205.

112 Mayyas, A., Steward, D., and Mann, M. (2019). The case for recycling: overview and challenges in the material supply chain for automotive Li-ion batteries. *Sustainable Materials and Technologies* 19: e00087.

113 Pinegar, H. and Smith, Y.R. (2019). Recycling of end-of-life lithium ion batteries, part i: commercial processes. *Journal of Sustainable Metallurgy* 5: 402–416.

114 Sloop, S., Crandon, L., Allen, M. et al. (2020). A direct recycling case study from a lithium-ion battery recall. *Sustainable Materials and Technologies* 25: e00152.

115 Gao, Y., Li, Y., Li, J. et al. (2020). Direct recovery of $LiCoO_2$ from the recycled lithium-ion batteries via structure restoration. *Journal of Alloys and Compounds* 845: 156234.

116 Larouche, F., Tedjar, F., Amouzegar, K. et al. (2020). Progress and status of hydrometallurgical and direct recycling of Li-ion batteries and beyond. *Materials* 13: 801.

117 Cho, J., Kim, Y.J., and Park, B. (2000). Novel $LiCoO_2$ cathode material with Al_2O_3 coating for a Li ion cell. *Chemistry of Materials* 12: 3788–3791.

118 Gao, H., Yan, Q., Xu, P. et al. (2020). Efficient direct recycling of degraded $LiMn_2O_4$ cathodes by one-step hydrothermal relithiation. *ACS Applied Materials & Interfaces* 12: 51546–51554.

119 Shi, Y., Chen, G., and Chen, Z. (2018). Effective regeneration of $LiCoO_2$ from spent lithium-ion batteries: a direct approach towards high-performance active particles. *Green Chemistry* 20: 851–862.

120 Shi, Y., Chen, G., Liu, F. et al. (2018). Resolving the compositional and structural defects of degraded $LiNixCoyMnzO_2$ particles to directly regenerate high-performance lithium-ion battery cathodes. *ACS Energy Letters* 3: 1683–1692.

121 Song, X., Hu, T., Liang, C. et al. (2017). Direct regeneration of cathode materials from spent lithium iron phosphate batteries using a solid phase sintering method. *RSC Advances* 7: 4783–4790.

122 Li, X., Zhang, J., Song, D. et al. (2017). Direct regeneration of recycled cathode material mixture from scrapped $LiFePO_4$ batteries. *Journal of Power Sources* 345: 78–84.

123 Ganter, M.J., Landi, B.J., Babbitt, C.W. et al. (2014). Cathode refunctionalization as a lithium ion battery recycling alternative. *Journal of Power Sources* 256: 274–280.

124 Lebedeva, N.P. and Boon-Brett, L. (2016). Considerations on the chemical toxicity of contemporary Li-ion battery electrolytes and their components. *Journal of the Electrochemical Society* 163: A821–A830.

125 Liu, Y., Mu, D., Li, R. et al. (2017). Purification and characterization of reclaimed electrolytes from spent lithium-ion batteries. *The Journal of Physical Chemistry C* 121: 4181–4187.

126 Mu, D., Liu, Y., Li, R. et al. (2017). Transcritical CO_2 extraction of electrolytes for lithium-ion batteries: optimization of the recycling process and quality–quantity variation. *New Journal of Chemistry* 41: 7177–7185.

127 Lain, M.J. (2001). Recycling of lithium ion cells and batteries. *Journal of Power Sources* 97–98: 736–738.

128 Zhong, X., Liu, W., Han, J. et al. (2020). Pneumatic separation for crushed spent lithium-ion batteries. *Waste Management* 118: 331–340.

129 BloombergNEF (2022). https://about.bnef.com/electric-vehicle-outlook/ (accessed 15 May 2022).

130 IEA (2022). *Global EV Outlook*. IEA.

131 Huijbregts, M.A.J., Steinmann, Z.J.N., Elshout, P.M.F. et al. (2017). ReCiPe2016: a harmonised life cycle impact assessment method at midpoint and endpoint level. *The International Journal of Life Cycle Assessment* 22: 138–147.

132 Pipitone, E., Caltabellotta, S., and Occhipinti, L. (2021). A life cycle environmental impact comparison between traditional, hybrid, and electric vehicles in the European context. *Sustainability* 13: 10992.

133 De Kleine, R., Keoleian, G., Miller, S. et al. (2014). Impact of updated material production data in the GREET life cycle model. *Journal of Industrial Ecology* 18.

134 IEA (2022). Data and Statistics. https://www.iea.org/data-and-statistics (accessed 30 May 2022).

135 Costa, C.M., Barbosa, J.C., Castro, H. et al. (2021). Electric vehicles: to what extent are environmentally friendly and cost effective? – comparative study by european countries. *Renewable and Sustainable Energy Reviews* 151: 111548.

14

Summary and Outlook

Renato Gonçalves[1], *Arkaitz Fidalgo-Marijuan*[2], *Carlos Miguel Costa*[3,4,5], *and Senentxu Lanceros-Méndez*[2,6]

[1] *Center of Chemistry, University of Minho, 4710-057, Braga, Portugal*
[2] *BCMaterials, Basque Center for Materials, Applications and Nanostructures, UPV/EHU Science Park, 48940, Leioa, Spain*
[3] *Physics Centre of Minho and Porto Universities (CF-UM-UP), University of Minho, Campus de Gualtar, Braga, 4710-057, Portugal*
[4] *Laboratory of Physics for Materials and Emergent Technologies, LapMET, University of Minho, Campus de Gualtar, Braga, 4710-057, Portugal*
[5] *Institute of Science and Innovation for Bio-Sustainability (IB-S), University of Minho, 4710-057, Braga, Portugal*
[6] *IKERBASQUE, Basque Foundation for Science, 48009, Bilbao, Spain*

Environmental issues and energy transition are strongly relevant issues nowadays, taking into account the technological development, the increasing mobility of people, and the improvement of the quality of life.

To reduce dependence on fossil fuels, efficient energy storage systems are required, as once electricity is generated by renewable energy sources, it must be either immediately used or stored efficiently to be later used in specific applications. Thus, it is critical and urgent the development of sustainable energy storage devices, such as batteries, supercapacitors, and fuel cells, based on new-generation materials, processes, and device designs.

Sustainable development is one of the greatest challenges our society is facing. According to the UN World Commission on Environment and Development, "sustainable development is development that meets the needs of the present without compromising the ability of future generations to meet their own needs." A circular economy is often associated with a concept based on one simple rule: the more material is kept in circulation, the better. However, the simple circulation of materials does not necessarily imply a higher degree of sustainability, and in this sense, it has been suggested that the activities belonging to circular economy criteria should be assessed in the light of economic, environmental, and social sustainability [1]. These dimensions should be taken into account in addition to the circulation criteria to assess sustainability (Figure 14.1).

Economic sustainability refers to practices that support long-term economic growth without negatively impacting other dimensions, such as environmental or social. Life cycle costing is one of the possible tools for the assessment of this dimension.

Sustainable Energy Storage in the Scope of Circular Economy: Advanced Materials and Device Design, First Edition. Edited by Carlos Miguel Costa, Renato Gonçalves, and Senentxu Lanceros-Méndez.

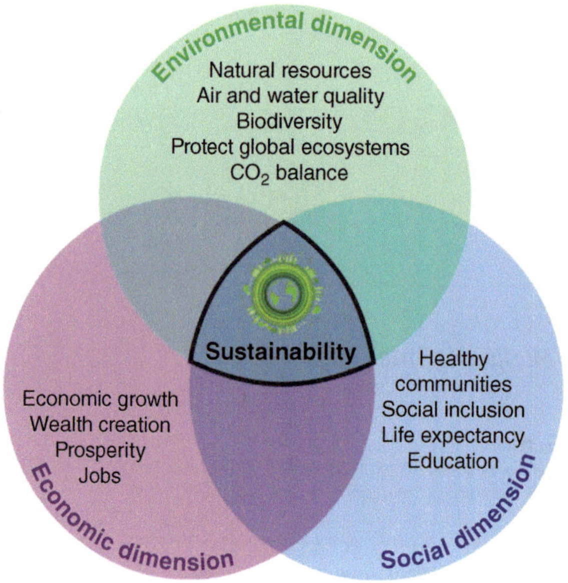

Figure 14.1 Illustration of the dimensions of sustainability. *Source:* Adapted from Ref. [2].

Environmental sustainability is the responsibility to preserve natural resources and protect global ecosystems to support health and well-being, now and in the future. Life cycle assessment is one commonly used tool for evaluating environmental impact. The less damage is caused to the environment, the more positively this criterion is developed.

Circularity provides a strategic and effective way to identify opportunities for both cost savings and value forming. In several cases, switching to renewable energy as an input is the first step. As technologies advance, the adoption of zero-emission energy is becoming more widespread. This shift to renewable energy is a key driver of the circular economy, and a growing number of organizations are seeking to minimize prices, increase sustainability and create long-term strategies. Several parameters are driving renewable energy adoption, and the economic dimension is also becoming less of a barrier at certain points in the power generation sector, but also for end users.

The International Renewable Energy Agency (IRENA) has estimated that renewable energy will represent the main path to carbon neutrality by 2060 with a significant reduction of waste, together with a commitment to recycling [3].

The IRENA study suggests that several technical obstacles need to be overcome before closed circularity for solar panels is feasible, but raw materials are already being recycled at 65–70% in weight [4].

In the case of battery technologies, the expanded use of battery technology is at the forefront of renewable energy storage and transportation decarbonisation strategies. But battery production itself has a significant carbon footprint. The CO_2 footprint of developing a fully electric vehicle, for example, is larger than that of an internal combustion engine vehicle [5]. However, the lower direct and indirect emissions throughout the life of the electric vehicle result in a lower overall CO_2 footprint throughout its lifetime. As with a number of other products and technologies, circularity in the battery chain should be considered in the context of the end market and evaluated against cross-industry benefits for transportation and power generation.

Using resources efficiently, prioritizing renewable inputs, maximizing the effective life of a product, and recovering and reusing what was previously considered waste are the critical, urgent, and necessary ways to implement next-generation technologies. Sticking to circular thinking and sustainability by design principles will lead to powerful new insights and levers for efficiency and costing. The increasing use of renewable energy and energy storage systems will bring new perspectives for environmental friendlier approaches [6] and the necessary steps for a sustainable future.

Acknowledgments

The authors thank the Fundação para a Ciência e Tecnologia (FCT) and COMPETE 2020 for financial support under the framework of Strategic Funding grants UIDB/04650/2020, UID/FIS/04650/2020, UID/EEA/04436/2020, and UID/QUI/00686/2020 and under projects MIT-EXPL/TDI/0033/2021, POCI-01-0247-FEDER-046985, 2022.03931.PTDC and PTDC/FIS-MAC/28157/2017. The authors also thank the FCT for the investigator contracts CEECIND/00833/2017 (RG) and 2020.04028.CEECIND (C.M.C.). Financial support from the BIDEKO Project, funded by MCIN/AEI, NextGenerationEU, PRTR, and Basque Government Industry Department under the ELKARTEK program are acknowledged.

References

1 Blum, N.U., Haupt, M., and Bening, C.R. (2020). Why "Circular" doesn't always mean "Sustainable". *Resources, Conservation and Recycling* 162: 105042.
2 Leat, P., Revoredo-Giha, C., and Lamprinopoulou, C. (2011). Scotland's food and drink policy discussion: sustainability issues in the food supply chain. *Sustainability* 3: 605–631.
3 Tang, Z., Yang, Y., and Blaabjerg, F. (2021). Power electronics: the enabling technology for renewable energy integration. *CSEE Journal of Power and Energy Systems* 8: 39–52.
4 The International Renewable Energy Agency (IRENA) and the International Energy Agency Photovoltaic Power Systems Programme, End-of-Life Management: Solar Photovoltaic Panels. https://www.irena.org/-/media/Files/IRENA/Agency/Publication/2016/IRENA_IEAPVPS_End-of-Life_Solar_PV_Panels_2016.pdf.
5 Buberger, J., Kersten, A., Kuder, M. et al. (2022). Total CO_2-equivalent life-cycle emissions from commercially available passenger cars. *Renewable and Sustainable Energy Reviews* 159: 112158.
6 Sustainability in the energy sector. www.pwc.de/en/sustainability/sustainability-in-the-energy-sector.html.

Index

Sustainable Energy Storage in the Scope of Circular Economy: Advanced Materials and Device Design, First Edition. Edited by Carlos Miguel Costa, Renato Gonçalves, and Senentxu Lanceros-Méndez.

Printed in the USA/Agawam, MA
September 16, 2026

910661.187